Rifted Ocean-Continent Boundaries

NATO ASI Series

Advanced Science Institutes Series

A Series presenting the results of activities sponsored by the NATO Science Committee, which aims at the dissemination of advanced scientific and technological knowledge, with a view to strengthening links between scientific communities.

The Series is published by an international board of publishers in conjunction with the NATO Scientific Affairs Division

A	**Life Sciences**	Plenum Publishing Corporation
B	**Physics**	London and New York
C	**Mathematical and Physical Sciences**	Kluwer Academic Publishers
D	**Behavioural and Social Sciences**	Dordrecht, Boston and London
E	**Applied Sciences**	
F	**Computer and Systems Sciences**	Springer-Verlag
G	**Ecological Sciences**	Berlin, Heidelberg, New York, London,
H	**Cell Biology**	Paris and Tokyo
I	**Global Environmental Change**	

PARTNERSHIP SUB-SERIES

1.	**Disarmament Technologies**	Kluwer Academic Publishers
2.	**Environment**	Springer-Verlag / Kluwer Academic Publishers
3.	**High Technology**	Kluwer Academic Publishers
4.	**Science and Technology Policy**	Kluwer Academic Publishers
5.	**Computer Networking**	Kluwer Academic Publishers

The Partnership Sub-Series incorporates activities undertaken in collaboration with NATO's Cooperation Partners, the countries of the CIS and Central and Eastern Europe, in Priority Areas of concern to those countries.

NATO-PCO-DATA BASE

The electronic index to the NATO ASI Series provides full bibliographical references (with keywords and/or abstracts) to more than 50000 contributions from international scientists published in all sections of the NATO ASI Series.
Access to the NATO-PCO-DATA BASE is possible in two ways:

– via online FILE 128 (NATO-PCO-DATA BASE) hosted by ESRIN,
Via Galileo Galilei, I-00044 Frascati, Italy.

– via CD-ROM "NATO-PCO-DATA BASE" with user-friendly retrieval software in English, French and German (© WTV GmbH and DATAWARE Technologies Inc. 1989).

The CD-ROM can be ordered through any member of the Board of Publishers or through NATO-PCO, Overijse, Belgium.

Rifted Ocean-Continent Boundaries

edited by

E. Banda

Institute of Earth Sciences (J. Almera),
Consejo Superior de Investigaciones Científicas (CSIC),
Barcelona, Spain

M. Torné

Institute of Earth Sciences (J. Almera),
Consejo Superior de Investigaciones Científicas (CSIC),
Barcelona, Spain

and

M. Talwani

Department of Geology and Geophysics,
Rice University,
Houston, Texas, U.S.A.

Springer-Science+Business Media, B.V.

Proceedings of the NATO Advanced Research Workshop on
Rifted Ocean-Continent Boundaries
Mallorca, Spain
11–14 May 1994

A C.I.P. Catalogue record for this book is available from the Library of Congress

ISBN 978-94-010-4024-2   ISBN 978-94-011-0043-4 (eBook)
DOI 10.1007/978-94-011-0043-4

Printed on acid-free paper

Sponsored by:

NATO/ARW Advanced Research Program.
Consejo Superior de Investigaciones Científicas, CSIC (Spain).
European Commission DG-XII.
Dirección General de Investigación Científica y Técnica, DGICYT. Ministerio de
Educación y Ciencia (Spain).

TABLE OF CONTENTS

PREFACE

A major aspect of the Project IV-2 of the "International Lithosphere Program" is the study of the ocean-continent boundary since this is the site where the most basic plate tectonics phenomena occur. In particular the rifted ocean-continent boundaries carry the signature of the manner in which continents split during continental rifting and seafloor spreading. One of the objectives of the working group is to study the nature and structure of the lithosphere under passive continental margins since it holds many unknowns. These include the nature of faulting at the time of break-up, the phenomena of crustal underplating and the significance of seaward dipping reflectors, the geometry and properties of the ocean-continent transitional crust and the rates and causes of subsidence and magmatism. The differences between passive margins created by rifting and those created by shearing, as well as differences and similarities at conjugate margins are key points to a better understanding of the geodynamics of passive margins.

These questions were the theme of the NATO Advanced Research Workshop on "Rifted Ocean-Continent Boundaries", held at Mallorca (Spain), May 11-14, 1994. The workshop was conceived, as part of the activities of Project IV-2 Working Group, to bring together and synthesise various studies of the ocean-continent lithosphere boundary at various continental margins around the world. It was organised by E. Banda, H. Hirata and M. Talwani and benefited from the sponsorship of NATO/ARW Program, CSIC (Consejo Superior de Investigaciones Científicas) and DGICYT (Dirección General de Investigación Científica y Técnica) of Spain, and the European Commission, DG-XII.

A total of 36 scientists from different countries, including leading specialists in this field, attended the meeting. The conference covered different topics mainly related to passive margins, although some aspects of transform and active margins were also discussed. Among the topics were: current knowledge of the structure and evolution of ocean-continent boundaries; case histories of the Atlantic and Pacific oceans margins; modelling of margin formation and evolution -with special attention to the role of the Lithosphere/asthenosphere interaction to the mechanics of rift propagation and to melt generation, magmatism and subsidence.

There were also a number of talks and posters dealing on new advances in processing and interpretation of geophysical data, as well as with ODP results. Proposal of new targets of drilling and instrumentation were described.

This book is a joint effort of the scientists who were invited to participate in the workshop. Its aim is to gather the various lectures given there and to synthesise the major topics that were discussed throughout the workshop.

After an introduction to volcanic margins concepts, the first articles reports the results of numerical models of the mechanics of rift propagation, melt generation and sources of extensional stresses that may cause break-up. A session of the workshop dedicated to the current knowledge of structure and evolution of various margins of the Atlantic is then presented. After a brief incursion to the Mediterranean, the succeeding articles report on transform and active margins with the examples of the Cote d'Ivoire-Ghana transform margin and the Japanese Sea.

We would like to thank the authors who made this study possible. We are grateful to the scientists that reviewed the papers for their efforts and the time spent on improving the original manuscripts. It is our hope that the book will contribute to research related to rifting.

Reviewers

J. Ansorge (Switzerland)	A.W.B. Jacob (Ireland)	W.R. Roest (Canada)
P. Barker (UK)	C. Keen (Canada)	R. Sabadini (Italy)
R. Buck (USA)	J.M. Lorenzo (USA)	D. Sawyer (USA)
J. Collier (UK)	A. Maldonado (Spain)	J-C. Sibuet (France)
M.C. Comas (Spain)	Y. Mart (Israel)	N. Sleep (USA)
J.J. Dañobeitia (Spain)	J.H. McBride (USA)	R. Stephenson (Canada)
O. Eldholm (Norway)	M. McNutt (USA)	M. Talwani (USA)
M. Fernàndez (Spain)	A. Perez-Estaún (Spain)	H. Thybo (Denmark)
R. Govers (USA)	W. Rabbel (Germany)	J. Woodside (UK)
R.W. Hobbs (UK)	T. Reston (UK)	H. Zeyen (Germany)
W.S. Holbrook (USA)		

LIST OF PARTICIPANTS

Prof. Enric BANDA
Institute of Earth Sciences, CSIC
Martí i Franquès s/n
08028-Barcelona (SPAIN)
Phone/Fax: 34-3-4900552/4110012
email: Enric.Banda@u.ija.csic.es

Dr. Boris BARANOV
Russian Academy of Sciences
P.P. Shirshov Institute of Oceanology
23, Krasikova
Moscow, 117218 (RUSSIA)
Phone/Fax: 7-095-124-7942/5983

Dr. Gilbert BOILLOT
Observatoire Océanologique
Laboratoire de Geodynamique Sous-Marine
B.P. 48-06230
Villefranche-sur-Mer (FRANCE)
Phone/Fax: 33-93-763740/763766

Prof. Martin H.P. BOTT
Department of Geology
University of Durham
South Road
Durham DH1 3 LE (ENGLAND)
Phone/Fax: 44-913-742511/743741
email: M.H.P.Bott@durham.ac.uk

Dr. Ramón CARBONELL
Institute of Earth Sciences, CSIC
Martí i Franquès s/n
08028-Barcelona (SPAIN)
Phone/Fax: 34-3-4900552/4110012
email: rcarbonell@u.ija.csic.es

Prof. Sierd CLOETINGH
Faculty of Earth Sciences
Vrije Universiteit
1081 HV Amsterdam (THE NETHERLANDS)
Phone/Fax: 31-20-5484741/6462457

Dr. Juan Jose DAÑOBEITIA
Institute of Earth Sciences, CSIC
Martí i Franquès s/n
08028-Barcelona (SPAIN)
Phone/Fax: 34-3-4900552/4110012
email: jjdc@u.ija.csic.es

Dr. Olav ELDHOLM
University of Oslo
Department of Geology
P.O.Box 1047 Blindern
N-0316 Oslo (NORWAY)
Phone/Fax: 47-22-856676/854215
email: olav.eldholm@geologi.uio.no

Dr. Mustafa ERGÜN
Department of Geophysics
Faculty of Engineering
University of Dokuz Eylül
35100 Bornova-Izmir (TURKEY)
Phone/Fax: 90-232-4631659/4636368

Dr. Manel FERNÁNDEZ
Institute of Earth Sciences, CSIC
Martí i Franquès s/n
08028-Barcelona (SPAIN)
Phone/fax: 34-3-4900552/4110012
email: mfernandez@u.ija.csic.es

Dr. Josep GALLART
Institute of Earth Sciences, CSIC
Martí i Franquès s/n
08028-Barcelona (SPAIN)
Phone/Fax: 34-3-4900552/4110012
email: jgallart@u.ija.csic.es

Prof. Karl HINZ
Federal Institute for Geosciences and
Natural Resources
P.O. BOX 510153
D-30655 Hannover (GERMANY)
Phone/Fax   49-511-6433247/6432304

Prof. Naoshi HIRATA
Earthquake Research Institute
The University of Tokyo
1-1-1 Yayoi Bunkyo-ku
Tokyo, 113 (JAPAN)
Phone/Fax: 11-81-338-183697/161159
email: hirata@eri.u.tokyo.ac.jp

Dr. W. Steven HOLBROOK
Woods Hole Oceanographic Institution
Woods Hole
MA 02543 (USA)
Phone/Fax: 1-508-548140/4572187

Prof. Roland von HUENE
GEOMAR
Wischhofstrasse 1-3
D24148 Kiel (GERMANY)
Phone/Fax: 49-431-7202-272/7202-293

Ingrid JONHSRUD
Norsk Hydro
P.O.Box 200
N-1321 Stabekk (NORWAY)
Phone/Fax   47-22-738100/738861

Dr. Charlotte R. KEEN
Geological Survey of Canada
Atlantic Geoscience Center
Bedford Institute of Oceanography
P.O.Box 1006
Dartmouth NS B2Y 4A2 (CANADA)
Phone/Fax: 1-902-426-3413/426-6152

xii

email: ckeen@agcrr.bio.ns.ca

Dr. Hajumu KINOSHITA
Earthquake Research Institute
The University of Tokyo
1-1-1 Yayoi Bunkyo-ku
Tokyo, 113 (JAPAN)
Phone/Fax: 81-338-12211/161159

Dr. Hans Christian LARSEN
Danish Lithosphere Center
Oster Voldgade 10
1350 Kobenhavn K (DENMARK)
Phone/Fax: 45-33-118866/110878
email: larsenhc@dlc.ggu.min.dk

Dr. Jean MASCLE
Observatoire Océanologique
Laboratoire de Geodynamique Sous-Marine
B.P. 48-06230
Villefranche-sur-Mer (FRANCE)
Phone/Fax: 33-93-763745/763766

Luis MATIAS
Centro de Geofísica da Universidade de Lisboa
Rua da escola Politecnica, 58
1200 Lisboa (PORTUGAL)
Phone/Fax: 351-1-3961521/3953327
email: fceocgul@ptearn.bitnet

Prof. Luis MENDES-VICTOR
Centro de Geofísica da Universidade de Lisboa
Rua da escola Politecnica, 58
1200 Lisboa (PORTUGAL)
Phone/Fax: 351-1-3961521/3953327
email: fceocgul@ptearn.bitnet

Prof. Andrés PEREZ-ESTAUN
Institute of Earth Sciences, CSIC
Martí i Franquès s/n
08028-Barcelona (SPAIN)
Phone/Fax: 34-3-4900552/4110012
email: andres@u.ija.csic.es

Dr. Luis de Menezes PINHEIRO
Departamento de Geociencias
Universidade de Aveiro
2800 Aveiro (PORTUGAL)
Phone/Fax: 351-34-25085/381260
email: lmp@ua.pt

Dr. Tim J. RESTON
GEOMAR
Wischhofstrasse 1-3
D24148 Kiel (GERMANY)
Phone/Fax: 49-431-7202279/7202293
email: treston@geomar.de

Dr. Dale S. SAWYER
Department of Geology and Geophysics
Rice University

P.O.Box 1892
Houston TX 77251 (USA)
Phone/Fax: 1-713-285-5106/285-5214
email: dale@rice.edu

Prof. Giancarlo SERRI
Dipartamento di Scienze della Terra
Universita Degli Studi di Pisa
Via S. Maria 53
56100 Pisa (ITALY)
Phone/Fax: 39-50-568208/500932
email: serri@dst.unipi.it

Dr. Jean Claude SIBUET
IFREMER
Centre de Brest
B.P. 70-29280 Plouzane (FRANCE)
Phone/Fax: 33-98-224233/224549
email: jcsibuet@ifremer.fr

Dr. Shiri P. SRIVASTAVA
Geological Survey of Canada
Atlantic Geoscience Center
Bedford Institute of Oceanography
P.O.Box 1006
Dartmouth NS B2Y 4A2 (CANADA)
Phone/Fax: 1-902-426-3148/426-6152
email: srivasta@agcrr.bio.ns.ca

Dr. Paul S. STOFFA
The University of Texas at Austin
8701 North MoPac Expressway
Austin, Texas 78759-8397 (USA)
Phone/Fax: 1-512-471-0464/471-8844
email: pauls@tau-p.ig.utexas.edu

Prof. Manik TALWANI
Department of Geology and Geophysics
Rice University
P.O.BOX 1892
Houston, TX 77251 (USA)
Phone/Fax: 1-713-363-7917/7924
email: manik@gtri14.gtri.harc.edu

Dr. Montserrat TORNE
Institute of Earth Sciences, CSIC
Martí i Franquès s/n
08028-Barcelona (SPAIN)
Phone/Fax: 34-3-4900552/4110012
email: mtorne@u.ija.csic.es

Prof. Ramón VEGAS
Departamento de Geodinámica
Facultad de Ciencias Geológicas
Universidad Complutense
28040-Madrid (SPAIN)
Phone/Fax: 34-1-3944859/3944845

Prof. Anthony B. WATTS
Department of Earth Sciences
University of Oxford

Parks Road
Oxford OX1 3PR (UK)
Phone/Fax: 44-865-272032/272032
email: tony@earth-sciences.ox.ac.uk

Prof. Robert S. WHITE
Bullard Laboratories
Department of Earth Sciences
University of Cambridge
Madingley Rise
Madingley Road
Cambridge CB3 ꝈEZ (UK)
Phone/Fax: 44-223-337191/60779
email: rwhite@bullard-convex.earth-sciences.cambridge.ac.uk

Robert B. WHITMARSH
Institute of Oceanographic Sciences
Deacon Laboratory
Brook Road, Wormley
Godalming Surrey GU8 5UB (UK)
Phone/Fax: 44-428-684141/683066
email: rbw@unixa-nerc-wormley.ac.uk

LIST OF CONTRIBUTORS

Dr. Juan ACOSTA
Instituto Español de Oceanografía
C. De Maria, 8
28002-Madrid (SPAIN)

Prof. Enric BANDA
Institute of Earth Sciences, CSIC
Martí i Franquès s/n
08028-Barcelona (SPAIN)
Phone/Fax: 34-3-4900552/4110012
email: Enric.Banda@u.ija.csic.es

Dr. C. BASILE
Institut Dolomieu
Université J. Fourier
38031-Grenoble (FRANCE)

Dr. Marie O. BESLIER
Observatoire Océanologique
Laboratoire de Geodynamique Sous-Marine
B.P. 48
06230- Villefranche-sur-Mer (FRANCE)
Phone/Fax: 33-93-763740/763766

Dr. Gilbert BOILLOT
Observatoire Oceanologique
Laboratoire de Geodynamique Sous-Marine
B.P. 48
06230-Villefranche-sur-Mer (FRANCE)
Phone/Fax: 33-93-763740/763766

Prof. Martin H.P. BOTT
Department of Geological Sciences
University of Durham
South Road
Durham DH1 3 LE (ENGLAND)
Phone/Fax: 44-913-742511/743741
email: M.H.P.Bott@durham.ac.uk

Dr. Ross R. BOUTILIER
Geological Survey of Canada
Atlantic Geoscience Center
Bedford Institute of Oceanography
P.O.Box 1006
Dartmouth NS E2Y 4A2 (CANADA)
Phone/Fax: 1-902-426-9999/426-6152
email: ross@agcrr.bio.ns.ca

Jonathan W. BOWN
Bullard Laboratories
University of Cambridge
Madingley Road
Cambridge CB3 0EZ (UK)
Phone: 44-223-337191/60779

Joana CARBONELL
Institute of Earth Sciences, CSIC
Martí i Franquès s/n
08028-Barcelona (SPAIN)
Phone/Fax: 34-3-4900552/4110012

Dr. Ramón CARBONELL
Institute of Earth Sciences, CSIC
Martí i Franquès s/n
08028-Barcelona (SPAIN)
Phone/Fax: 34-3-4900552/4110012
email: rcarbonell@u.ija.csic.es

Dr. William P. CLEMENT
Department of Geology and Geophysics
University of Wyoming
P.O.Box 3006
Laramie 82701, Wyoming (USA)
Phone/Fax: 1-307-7665280/7422649

Dr. Juan Jose DAÑOBEITIA
Institute of Earth Sciences, CSIC
Martí i Franquès s/n
08028-Barcelona (SPAIN)
Phone/Fax: 34-3-4900552/4110012
email: jjdc@u.ija.csic.es

Dr. Olav ELDHOLM
University of Oslo
Department of Geology
P.O.Box 1047 Blindern
N-0316 Oslo (NORWAY)
Phone/Fax: 47-22-856676/854215
email: olav.eldholm@geologi.uio.no

Dr. Mustafa ERGÜN
Department of Geophysics
Faculty of Engineering
University of Dokuz Eylül
35100 Bornova-Izmir (TURKEY)
Phone: 90-232-4631659/4636368

Dr. J. EWING
Woods Hole Oceanographic Institution
Woods Hole
MA 02543 (USA)
Phone/Fax: 1-508-5481400/4572187

Dr. Manel FERNANDEZ
Institute of Earth Sciences, CSIC
Martí i Franquès s/n
08028-Barcelona (SPAIN)
Phone/Fax: 34-3-4900552/4110012
email: mfernandez@u.ija.csic.es

Dr. Josep GALLART
Institute of Earth Sciences, CSIC
Martí i Franquès s/n
08028-Barcelona (SPAIN)
Phone/Fax: 34-3-4900552/4110012
email: jgallart@u.ija.csic.es

Dr. J. GIRARDEAU
Laboratoire de Pétrologie Structurale

Faculté des Sciences et des Techniques
2, rue de la Houssinière
44072 Nantes Cedex (FRANCE)

Dr. Tadeusz P. GLADCZENKO
Department of Geology
University of Oslo
P.B. 1047-Blindern
N-0316 Oslo (NORWAY)
Phone/Fax: 47-28-56678/54215

Dr. L. GLOVER III
Virginia Polytechnic Institute
Dept. of Geological Sciences
Balcksburg, VA 24061
Phone: 1-703-9616213

Prof. Naoshi HIRATA
Earthquake Research Institute
The University of Tokyo
1-1-1 Yayoi Bunkyo-ku
Tokyo, 113 (JAPAN)
Phone/Fax: 11-81-338-183697/161159
email: hirata@eri.u.tokyo.ac.jp

Dr. Richard W. HOBBS
BIRPS
Bullard Laboratories
Mandingley Road
Cambridge CB3 0EZ (ENGLAND)
Phone/Fax   44-22-360376/360779

Dr. W. Steven HOLBROOK
Woods Hole Oceanographic Institution
Woods Hole
MA 02543 (USA)
Phone/Fax: 1-508-5481400/4572187

Dr. Susan J. HORSEFIELD
ESSO Exploration and Production UK Ltd
Leatherhead, KT22 8UY (UK)

Dr. Charlotte R. KEEN
Geological Survey of Canada
Atlantic Geoscience Center
Bedford Institute of Oceanography
P.O.Box 1006
Dartmouth NS B2Y 4A2 (CANADA)
Phone/Fax: 1-902-426-3413/426-6152
email: ckeen@agcrr.bio.ns.ca

Dr. Hajimu KINOSHITA
Earthquake Research Institute
The University of Tokyo
1-1-1 Yayoi Bunkyo-ku
Tokyo, 113 (JAPAN)
Phone/Fax: 81-338-12211/161159

C. M. KRAWCZYK
GEOMAR
Wischhofstrasse 1-3

D24148 Kiel (GERMANY)
Phone/Fax: 49-431-7202279/7202293

Dr. Eiji KURASHIMO
Department of Earth Sciences
Chiba University
Chiba 263 (JAPAN)

Dr. Pierre LEON
IFREMER
Centre de Toulon
B.P. 330
83507-La Seyne/Mer Cedex (FRANCE)

Dr. Véronique LOUVEL
EOPGS
5 rue R. Descartes
67084 Strasbourg Cedex (FRANCE)

C. MARR
Department of Earth Sciences
University of Oxford
Parks Road
Oxford OX1 3PR (UK)
Phone/Fax: 44-865-272032/272032

Dr. Jean MASCLE
Observatoire Océanologique
Laboratoire de Géodynamique Sous-Marine
B.P. 48-06230
Villefranche-sur-Mer (FRANCE)
Phone/Fax: 33-93-763745/763766

Dr. John H. McBRIDE
BIRPS
Bullard Laboratories
Mandingley Road
Cambridge CB3 0EZ (ENGLAND)
Phone/Fax   44-22-360376/360779

Dr. Timothy A. MINSHULL
Department of Earth Sciences
Bullard Laboratories
Mandingley Road
Cambridge CB3 0EZ (ENGLAND)
Phone/Fax   44-22-333400/360779
email: minshull@bull.esc.cam.ac.uk

Dr. Erdeniz OZEL
Department of Geophysics
Faculty of Engineering
University of Dokuz Eylül
35100 Bornova-Izmir (TURKEY)
Phone: 90-232-4631659/4636368

Dr. Luis de Menezes PINHEIRO
Departamento de Geociencias
Universidade de Aveiro
2800 Aveiro (PORTUGAL)
Phone/Fax: 351-34-25085/381260
email: lmp@ua.pt

Dr. Sverre PLANKE
Department of Geology
University of Oslo
P.B. 1047-Blindern
N-0316 Oslo (NORWAY)
Phone/Fax: 47-28-56678/54215
email: planke@granitt.uio.no

Dr. Bernard PONTOISE
Observatoire Océanologique
Laboratoire de Géodynamique Sous-Marine
B.P. 48-06230
Villefranche-sur-Mer (FRANCE)
Phone/Fax: 33-93-763745/763766

Dr. Maurice RECQ
B.P. 809
29285 Brest Cedex (FRANCE)

Dr. Tim J. RESTON
GEOMAR
Wischhofstrasse 1-3
D24148 Kiel (GERMANY)
Phone/Fax: 49-431-7202279/7202293
email: treston@geomar.de

Dr. Walter ROEST
Geological Survey of Canada
1 Observatory Crescent
Ottawa, ON K1A 0Y3 (CANADA)
Phone/Fax: 1-613-9921546/9528987
email: roest@gsc.emr.ca

Dr. F. SAGE
Observatoire Océanologique
Laboratoire de Géodynamique Sous-Marine
B.P. 48-06230
Villefranche-sur-Mer (FRANCE)
Phone/Fax: 33-93-763745/763766

Dr. Coskun SARI
Department of Geophysics
Faculty of Engineering
University of Dokuz Eylül
35100 Bornova-Izmir (TURKEY)
Phone/Fax: 90-232-4631659/4636368

Dr. Dale S. SAWYER
Department of Geology and Geophysics
Rice University
P.O.Box 1892
Houston TX 77251 (USA)
Phone/Fax: 1-713-285-5106/285-5214
email: dale@rice.edu

Prof. Robert E. SHERIDAN
Dept. of Geological Sicences
Rutgers University
Busch Campus, Piscataway
NJ 08855 (USA)

Phone: 1-201-9322044

Dr. Jean Claude SIBUET
IFREMER
Centre de Brest
B.P. 70-29280 Plouzane (FRANCE)
Phone/Fax: 33-98-224233/224549
email: jcsibuet@ifremer.fr

Dr. Bertrand SICHLER
IFREMER
Centre de Brest
B.P. 70-29280 Plouzane (FRANCE)
Phone/Fax: 33-98-224233/224549

Dr. Jakob SKOGSEID
Department of Geology
University of Oslo
P.B. 1047-Blindern
N-0316 Oslo (NORWAY)

Dr. Scott B. SMITHSON
Department of Geology and Geophysics
University of Wyoming
P.O.Box 3006
Laramie 82701, Wyoming (USA)
Phone/Fax   1-307-7665280/7422649

Dr. Marvin A. SPEECE
Geophysical Engineering Department
Montana Tech., W. Park Street
Butte, 5971 Montana (USA)
Phone/Fax   1-406-4964188

Dr. Shiri P. SRIVASTAVA
Geological Survey of Canada
Atlantic Geoscience Center
Bedford Institute of Oceanography
P.O.Box 1006
Dartmouth NS B2Y 4A2 (CANADA)
Phone/Fax   1-902-426-3148/426-6152
email: srivasta@agcrr.bio.ns.ca

Prof. Manik TALWANI
Department of Geology and Geophysics
Rice University
P.O.BOX 1892
Houston, Texas 77251 (USA)
Phone/Fax: 1-713-363-7917/7924
email: manik@gtri 4.gtri.harc.edu

Dr. Montserrat TORNE
Institute of Earth Sciences, CSIC
Martí i Franquès s/n
08028-Barcelona (SPAIN)
Phone/Fax: 34-3-4900552/4110012
email: mtorne@u.ija.csic.es

Dr. Elazar UCHUPI
Woods Hole Oceanographic Inst.
MA 02543 (USA)

Phone/Fax: 1-508-4572000/4572187

Prof. Ramón VEGAS
Departamento de Geodinámica
Facultad de Ciencias Geológicas
Universidad Complutense
28040-Madrid (SPAIN)
Phone/Fax: 34-1-3944859/3944845

Prof. Anthony B. WATTS
Department of Earth Sciences
University of Oxford
Parks Road
Oxford OX1 3PR (UK)
Phone/Fax: 44-865-272032/272032
email: tony@earth-sciences.ox.ac.uk

Prof. Robert S. WHITE
Bullard Laboratories
University of Cambridge
Madingley Road
Cambridge CB3 OEZ (UK)
Phone/Fax: 44-223-337191/60779
email: rwhite@bullard-convex.earth-sciences.cambridge.ac.uk

Robert B. WHITMARSH
Institute of Oceanographic Sciences
Deacon Laboratory
Brook Road, Wormley
Godalming Surrey GU8 5UB (UK)
Phone/Fax: 44-428-684141/683066
email: rbw@unixa-nerc-wormley.ac.uk

VOLCANIC MARGIN CONCEPTS

OLAV ELDHOLM, JAKOB SKOGSEID, SVERRE PLANKE &
TADEUSZ P. GLADCZENKO
Department of Geology, University of Oslo
Pb. 1047 Blindern, N-0316 Oslo
Norway

ABSTRACT. Volcanic margins are part of a tectono-magmatic system in which the margin formation depends on lithospheric and asthenospheric properties before, during and after continental breakup. Whether a volcanic margin develops or not, i.e. the relative magnitude of magmatism during breakup, depends on the temperature and fluid content of the asthenosphere along the incipient plate boundary and the dynamic history of the lithosphere during the syn-rift phase. An adequate description of the system requires analysis of the entire rift. However, the literature commonly shows an emphasis on selected features rather than on a system approach which may introduce bias during interpretation and modeling. For example: 1) seaward dipping reflectors, commonly considered indicative of volcanic margins, are only one element in an igneous succession emplaced during rifting and early sea floor spreading. 2) Excessive melt volumes may be generated by relatively small asthenospheric temperature anomalies ($50\text{-}100°C$). 3) The entire lower crustal high-velocity bodies are not "underplated". 4) Wide-angle seismic experiments appear to overestimate the volumes of the extrusive complexes. 5) The tectono-magmatic dimensions of the NE North Atlantic Late Cretaceous-Paleocene rift challenge the concept of a "non-extensional" rifted margin. Furthermore, asymmetric tectono-magmatic settings need to be considered. 6) Up- or down-scaled "normal" continental and oceanic crustal velocity functions do not apply to the continent-ocean transition. 7) Although a hotspot may result in volcanic margin formation, it is not a necessary condition.

Introduction

Many of the world's passive continental margins bear evidence of excessive, transient magmatic activity during the final breakup of the continents and during the initial period of sea floor spreading. In the past decade these margins have been classified as volcanic margins (Figs. 1-2). Together with continental flood basalts, oceanic plateaus and ocean basin flood basalts the volcanic margins constitute the main categories of transient large igneous provinces (LIP) characterized by voluminous emplacements of predominantly mafic rocks which do not originate at normal seafloor spreading centers. Volumes and emplacement rates for the largest LIPs show that they contribute importantly to the global crustal production budget (Coffin & Eldholm, 1994). Episodic, excessive melting is commonly attributed to hotspot activity (e.g., White & McKenzie, 1989; Duncan & Richards, 1991; Larson, 1991).

Massive extrusive complexes along rifted margins were initially recognized along the Vøring and Lofoten margins off Norway, first by the exceptionally smooth acoustic basement surface near the continent-ocean boundary (Talwani & Eldholm, 1972, 1977), later by the existence of wedges of seaward dipping, intrabasement reflectors below acoustic basement (Hinz & Weber, 1976; Hinz & Schlüter, 1978; Talwani 1978; Eldholm et al., 1979). Subsequent drilling on the North Atlantic Hatton Bank, Vøring and SE Greenland margins (Roberts, Schnitker et al., 1984; Eldholm, Thiede, Taylor et al., 1987; Leg 152 Shipboard Party, 1994) have reported that the dipping wedges consist of numerous subaerially emplaced basalt flows and thin interbedded sediments. The prominent North Atlantic LIP includes onshore flood basalts and volcanic

1

E. Banda et al. (eds.), Rifted Ocean-Continent Boundaries, 1-16.
© *1995 Kluwer Academic Publishers. Printed in the Netherlands.*

	Physiography	Age	Structural framework	Overburden	Pre-rift geology	Magmatism
end-member	Narrow (steep slope)	Young	Rift	Starved	Craton	Volcanic
	Marginal plateau	Intermediate	Shear-rift			
end-member	Wide (gentle slope)	Mature	Shear	Sediment basin	Major basin	Non-volcanic

Fig. 1. Various classifications of passive continental margins.

margins, though the volcanic margins constitute its major part. In addition, deep crustal intrusive companions to flood basalt volcanism represent volumetrically significant contributions to the crust.

As the seismic images improved, seaward dipping reflectors were recognized also outside the North Atlantic (e.g., Hinz, 1981), and the number of volcanic margins has steadily increased (Fig. 2). In addition to the North Atlantic, particularly extensive and voluminous seaward dipping units exist along the U.S. East Coast (e.g., Talwani et al., this volume) and in the South Atlantic south of Walvis Ridge-Rio Grande Rise (Austin & Uchupi, 1982; Gladczenko, 1994; Hinz, pers. comm.). In fact, one may ask whether volcanic margins are the normal rather than the exceptional case.

The early multichannel seismic (MCS) lines revealed many spectacular features along volcanic margins. This led to many concepts and models for margin formation. Some of these initial ideas have indeed proved to be useful. Nonetheless, research on some margins has advanced rapidly, revealing complex, but integrated and consistent, magmatic and tectonic relationships challenging

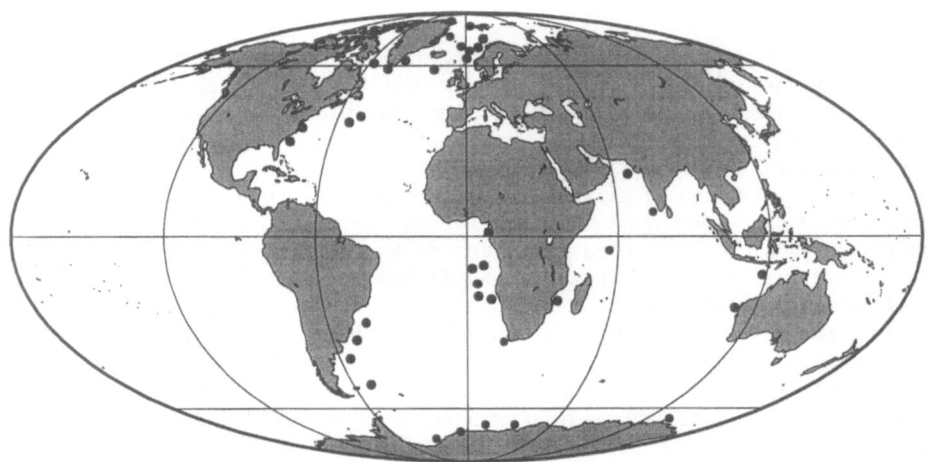

Fig. 2. Global distribution of volcanic margins based on wedges of intrabasement seaward dipping reflectors reported in the literature (black dots). Data sources in Coffin & Eldholm (1994).

some other concepts which commonly seem to have acquired a momentum of their own (Fig. 3). This, in our opinion, may inhibit margin research and we therefore discuss our contention here. A main premise is that we consider volcanic margins as part of a tectono-magmatic system in which margin formation depends on lithospheric and asthenospheric properties before, during and after continental breakup. Hence, it is not sufficient to restrict the analysis to selected features, no matter how spectacular they are.

Seaward dipping reflectors	=	Volcanic margin
Volcanic margin	=	Non-extension
	=	Symmetric
	=	Plume (or plumetail) dependent
	=	High asthenospheric melt temperature
Volcanic margin crust	=	Expanded oceanic crust
Lower high-velocity crust	=	Underplated

Fig. 3. Common inferences and simplifications which may lead to incomplete models for volcanic margin formation and evolution.

Igneous breakup features

Wedges of seaward dipping reflectors are commonly taken as a criterion for identifying a margin as volcanic. However, we recognize a series of other geological features that are genetically related to the breakup magmatism and volcanic margin formation. As such features are most thoroughly studied along North Atlantic margins, most examples are from this region where breakup occurred during chron 24r near the Paleocene-Eocene boundary, ~55 Ma.

Among the igneous breakup features are (cfr., Figs. 2-3 in Skogseid & Eldholm (this volume)): 1) onshore continental flood basalts. 2) Voluminous extrusive basaltic complexes along the continent-ocean transition appearing as intrabasements reflectors in the seismic record. They include the well-known sequences of seaward dipping reflectors (Fig. 4). In seismic records these reflectors vary greatly in character, continuity, dip and amplitude and there are also large variations in reflection pattern and thickness. 3) Sills and low-angle dikes within pre-opening sediments. 4) Volcanic vents and North Atlantic regional tephra horizons in stratigraphic positions coeval with the breakup volcanism. 5) Thicker than normal oceanic crust adjacent to the continent-ocean boundary; 6) Intrusive companions to the extrusive complexes. 7) A lower crustal body (LCB) characterized by relatively high seismic velocities, 7.1-7.7 km/s, on either side of the continent-ocean boundary.

Although the basaltic lavas may locally extend for large distances landward of the continent-ocean boundary, there appears to be a spatial correlation between the LCB and the most voluminous extrusive complexes (Eldholm & Grue, 1994).

Ocean Drilling Program (ODP) Site 642 on the Vøring margin drilled about 800 m through the entire inner part of a major seaward dipping wedge. The site, which yielded 40% recovery of basalt flows and interbedded sediments, was also logged by a series of tools, and a Vertical Seismic Profiling (VSP) experiment was conducted in the hole (Eldholm et al., 1987; Planke, 1994). Thus, it is a reference site for studying compositions and properties of basaltic extrusive

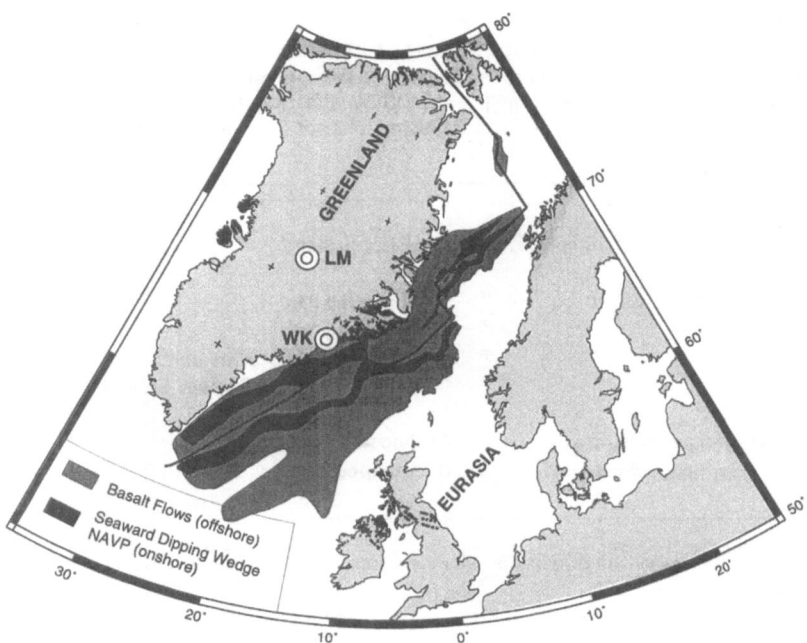

Fig. 4. The North Atlantic LIP in a magnetic anomaly 23 reconstruction (Eldholm & Grue, 1994). WK and LM refer to hotspot locations of White & McKenzie (1989) and Lawver & Müller (1994), respectively.

complexes on volcanic margins.

Although Site 642 clearly drilled into the seaward dipping seismic sequence, typical seaward dipping reflectors are only recognized in the main wedge seaward of the site. Except for some reflectors near the base of the basalt flows, there are no distinct, continuous reflectors at the site. This observation shows that thick flood basalt may exist on some margins without being seismically characterized by distinct continuous intrabasement reflectors.

At Site 642 Planke (1994) determined flow thicknesses from 0.6 to 18.5 m while most sediment layers are less than 1 m, and made a detailed analysis of the physical properties both at the flow-scale (Fig. 5) and at the composite-lava-sequence-scale. The resulting acoustic impedance log was used to model the dipping sequence in the hole proper as well as in the better developed, thicker wedge farther west by Planke & Eldholm (1994). They report that most individual, and composite, lava units do not have the dimensions or the physical property distribution required to produce reflectors imaged by the 40-50 m resolution of standard MCS surveys. The probable candidates for the dipping seismic reflectors are extensive, thick individual flows, which are most likely to extend over large areas, and interference effects. The modeling also suggests that great care is needed when interpreting lava stratigraphy from seismic profiles. For example, seismic interference and thin-layer effects may lead to apparent truncations and pinch-outs without geological significance. It is also important to recognize that all present seismic profiles so far published from extrusive terrain on volcanic margins are two-dimensional, while the combination of point (volcanos) and line sources (fissures) for the lava clearly lead to three-dimensional features.

Therefore, we contend that the basaltic lavas on volcanic margins may extend far beyond the seismically prominent seaward dipping reflectors. In fact, seaward dipping reflectors are only one

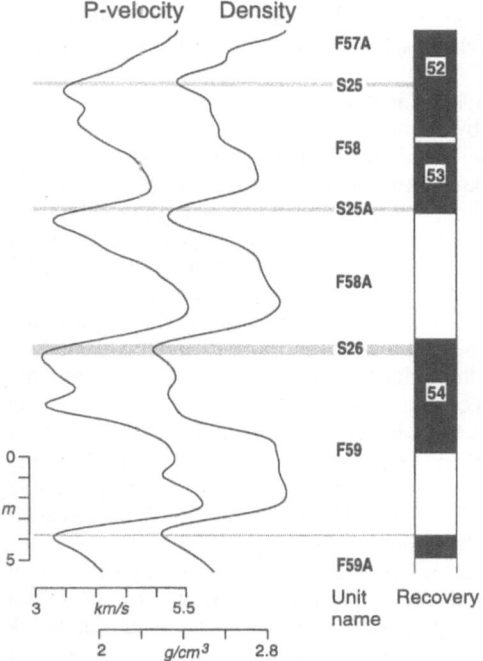

P-velocity Density

F57A
S25   52
F58
             53
S25A
F58A
S26
             54
F59

0
m
5
F59A

3   km/s   5.5   Unit   Recovery
2   g/cm³ 2.8   name

Fig. 5. Typical thicknesses of flows and interbedded sediments and the corresponding density and velocity log response in a wedge of seaward dipping reflectors drilled at ODP Site 642 on the Vøring margin off Norway. Based on Planke (1994). Black columns show core recovery and core number. F and S refer to flow and sediment unit numbers, respectively (Eldholm, Thiede, Taylor et al., 1987).

element in an igneous succession emplaced during breakup and early sea floor spreading. It is conceivable that rifted margin segments may be predominantly volcanic without these features being apparent in seismic records. Thus, seaward dipping reflectors are a sufficient, but not necessary condition for volcanic margin classification. Furthermore, we interpret the great variety in seismic style of the intrabasement volcanic margin reflectors to reflect: 1) volume and rate of magma production, 2) constructional environment, 3) syn- and post-constructional deformation, and 4) subsidence

Tectono-magmatic dimensions

To understand the nature and formation of volcanic margins it is necessary to consider both their magmatic and tectonic dimensions. For LIPs globally, the lack of regional high-quality seismic data and in some cases the choice of emplacement models cause large uncertainties in crustal production volumes (Coffin & Eldholm, 1994). Nonetheless, from a combination of regional MCS and wide-angle velocity profiles along conjugate margin segments it is possible to make first-order estimates of the areal extent and volume of the melts emplaced during the transient breakup event if one makes certain assumptions concerning the breakup process. This includes both the extrusive and intrusive crustal components. Despite the uncertainty in assumptions and the exact seismic velocity field, Eldholm & Grue (1994) inferred from studies of the North Atlantic margins that the order of magnitude of the crustal volumes is maintained for reasonable changes in the choice of geological assumptions. For the North Atlantic LIP they calculated an areal extent of 1.3×10^6 km² and a volume of flood basalts of 1.8×10^6 km³. The total crustal volume is about 6.6×10^6 km³. These values are considered minimum estimates and the onshore flood basalts contribute a small fraction of the extrusive volume. The flood basalt volume is similar to that of the Deccan-Seychelles LIP (Todal, 1994).

No other comparable studies exist elsewhere, but recent work on the U.S. East Coast (Holbrook et al., 1994a,b; Talwani et al., this volume), Argentine and Uruguayan (Hinz, pers.comm.) and Namibian (Gladczenko, 1994) margins support the contention that the dominant component of LIPs at rifted margins is to be found offshore. Thus, rifted margins contribute significantly to the world's inventory of transient large igneous provinces.

Because the basalts emplaced during breakup flow onto the adjacent continental crust, which also may be intruded by sills and low-angle dikes, seismic resolution below the lavas is generally poor. Consequently, the extrusives may conceal evidence for structural deformation during a syn-rift phase leading to breakup. The lack of clearly identified extensional structures has been taken as evidence for a very rapid breakup of the continental lithosphere without a significant phase of continental thinning and deformation preceding breakup (Mutter et al., 1984; Larsen, 1990; Hopper et al., 1992).

In an accompanying chapter (Skogseid & Eldholm, this volume) we point out that the North Atlantic volcanic margin formation was preceded by a syn-rift phase lasting for about 18-20 m.y. prior to breakup (Skogseid et al., 1992 a,b). The syn-rift phase affected a 300 km wide region and separated Eurasia and Greenland by about 140 km (Skogseid, 1994). Support for this model comes from subsidence modeling, extensional faulting and crustal thickness variations (Skogseid & Planke, in press). Moreover, the conjugate Hatton Bank-SE Greenland margins, characterized by almost complete volcanic overprint, have been considered as evidence against this model (Larsen & Marcussen, 1992). New field data from East Greenland (A. Andresen, pers. comm.) and seismic data from the Rockall Plateau (Makris et al., 1991; Keser Neish, 1993) suggest Late Cretaceous-Early Tertiary extension, and we believe the lavas on the SE Greenland margin, drilled during ODP Leg 152, blanket a terrain faulted prior to chron 24r time. We also note that Biswas & Thomas (1992) have proposed rifting preceding emplacement of the Deccan basalts and opening between India and the Seychelles, and the onset of sea floor spreading off the U.S. East Coast follows a Late Triassic-Early Jurassic rifting (e.g., Klitgord et al., 1988). Dynamic modeling of lithospheric extension suggests an extensional period of about 25 m.y. preceding breakup of the Parana-Etendeka region in the South Atlantic (Harry & Sawyer, 1992), a syn-rift development which is supported by seismic data off Namibia (Light et al., 1993).

Thus, a pre-break up phase of continental extension and thinning is compatible with observations from several margins. Together with the ample manifestations of extensional faulting along both continental and oceanic rifts, we propose to consider the tectonic framework of volcanic margins as reflecting a normal rift development rather than "a priori" assumption of an exceptional, non-extensional development. Consequently, there is similarity in structural style and deformation among volcanic margins, non-volcanic margins and continental rifts, the principal differences between these rift types being late rift-early drift intrusive activity and massive extrusive overprint along the volcanic margins.

Margin asymmetry and rifting style

The available data reveal magmatic and tectonic asymmetry on many conjugate volcanic margins; however, except for adjacent continental flood basalt provinces, the implications have been largely ignored in the volcanic margin literature. Admittedly, part of the asymmetry may be apparent due to variable data coverage and data quality along and across conjugate margin segments.

A magmatic asymmetry, expressed by the extrusive distribution and volumes, is commonly observed both along and across the plate boundary. In the North Atlantic for example (Eldholm & Grue, 1994), the area and volume of basaltic lavas on the continental crust are greatest south of Iceland and narrows to the north (Fig. 4). This distribution may suggest that the along-plate-boundary variation is governed by stepwise migration, or propagation, of the plate boundary and the locus of the central melt anomaly (hotspot) with respect to the plate boundary. The extrusive across-plate-boundary asymmetry is expressed by the distribution of flood basalts and wedges of seaward dipping reflectors. In particular, the seaward dipping wedges commonly appear to be best developed on one side of the ocean. A typical case-in-point is the U.S. East Coast margin where wedges, more than 20 km thick, have been reported (Holbrook et al., 1994a) while comparable conjugate features have not been recorded by the available seismic data. We believe that the present distribution of basaltic lavas reflects the combined effects of the magnitude of melt production, vulnerability of the continental crust to melt penetration, multiple transient feeders (Saemundson, 1974), lateral melt migration, constructional environment and erosion.

Reconstructions to the time of crustal breakup, and to the time of onset of syn-rift extension, as well as construction of conjugate transects, provides the framework to evaluate tectonic asymmetry along and across conjugate margins (Skogseid et al., 1992b; Skogseid & Planke, in press). Margin tectonic style is governed by the pre-rift structural configuration and the structural mode of the syn-rift phase. The latter is reflected by fault distribution, detachment geometries and conjugate transfer systems (Lister et al., 1991). In particular, crustal breakup away from the syn-rift axis (Keen, 1987) will amplify the tectonic asymmetry of conjugate margin segments. The U.S. East Coast and its conjugate African margin, where an asymmetric extension model and breakup best satisfy observations, is a case in point (Benson & Doyle 1988; Klitgord et al., 1988).

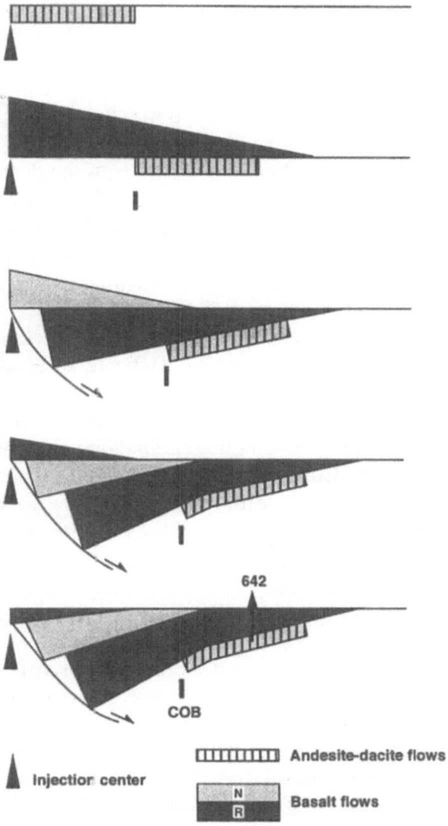

One may query whether a purely symmetric system is required during the period of the initial formation of oceanic crust, i.e during the formation of the most prominent seaward dipping wedges. An important observation is the very abrupt seaward termination of many large dipping wedges. In fact, the termination may be interpreted as a steep, slightly concave fault (Schuepbach & Vail, 1989; Bally 1983). Gibson & Gibbs (1987) have proposed that syn- constructional listric faulting during fissure eruptions will produce the observed lava stratigraphy in Iceland. On the other hand, according to the emplacement model of Mutter et al. (1982) the thick wedges are explained by subsidence of the lava pile due to loading and thermal contraction, hence the model requires a symmetric feature on the conjugate margin. The abrupt wedge termination, the lack of comparable wedges on some conjugate margins as well as the large wedge thickness resulting in very rapid subsidence rates, may be explained by a steep listric, syn-constructional fault (Fig. 6) (Eldholm et al., 1989). This suggests that simple-shear extension should be considered on some volcanic margins.

Fig. 6. Sketch showing the development of the seaward dipping wedge at the Vøring margin drilled by ODP Site 642 based on the Mutter et al. (1982) emplacement model with the addition of a syn-constructional listric fault. The andesite-dacite flows recovered below the wedge were emplaced during the late syn-rift phase. R and N indicate reversed and normal magnetic polarities of the basalts. Not to scale.

8

Crustal structure, seismic velocities and continent-ocean boundary

The expanded crustal thickness of the initial oceanic crust, the extensive extrusive cover, and the LCB question whether seismic velocity could be applied to distinguishing crustal type and composition. In particular, is it valid to make comparisons with linearly up- and down-scaled models of "normal" oceanic and continental crust, or is there a volcanic margin or LIP-type crust? Compared with normal oceanic crust, which has a different velocity structure, the extrusive crustal component on volcanic margins erupt at shallower depths, often subaerially.

Regional consistency of the igneous velocity structure seaward of the continent-ocean boundary, and similarity with the Icelandic upper crust and with other LIPs suggest a relationship between crustal emplacement mode and composition. For example, the three-layer North Atlantic model of Eldholm & Grue (1994), whose main features correspond to those on other volcanic margins,consists of three crustal units (Fig. 7): 1) upper extrusive crust of up to 6 km thick flood basalt and interbedded sediments. The velocity increases rapidly from ~3.7 to >5.0 km/s in the uppermost kilometer with a gentler velocity gradient below, reaching 6.0-6.5 km/s in the deepest

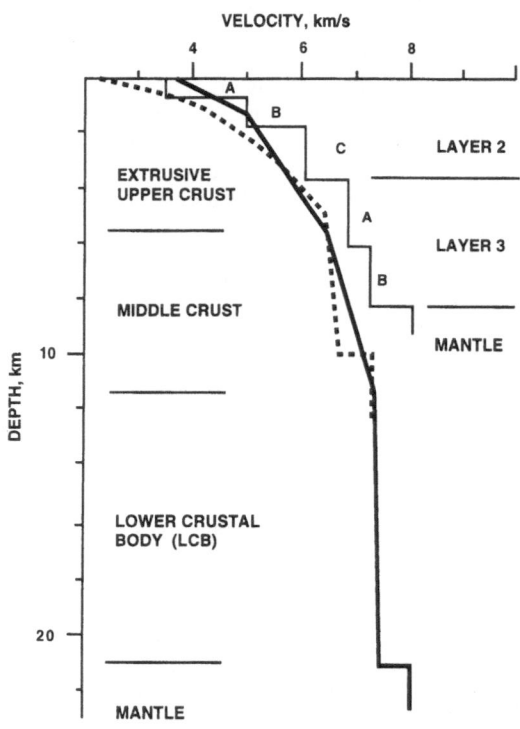

seaward dipping wedges. The high velocities at large depths probably indicate an increasing proportion of dikes with depth. 2) Middle crust with a velocity of 6.5-6.7 km/s at the top and a gentle velocity gradient, resembling a thickened oceanic layer 3A (Ewing & Houtz, 1979). It could consist of dikes near the transition with the extrusive upper layer and gabbro below (Zehnder et al., 1990). 3) Lower crust with high, 7+ km/s, seismic velocities and a very gentle gradient. The LCB velocity is not characteristic of normal oceanic or continental crust (Meissner, 1986), but is reported from oceanic plateau LIPs (Coffin & Eldholm, 1992).

Recent studies show that surface velocity measurements of the upper extrusive crust are not entirely representative. Near ODP Site 642, the velocities determined from wide-angle experiments are 10-20% higher than those from VSP and log data suggesting anisotropic properties of the seaward dipping wedge (Planke & Eldholm, 1994). This observation is corroborated by finite-difference seismic modeling of wave propagation in a model compatible with the drilled basalt sequences which reveals 15-20% transverse isotropy (Planke & Eldholm, 1993), and with the up to 15% anisotropy determined from VSP experiments in Icelandic basalt flows (Planke & Flovenz, 1994).

Fig. 7. Simplified velocity-depth function for North Atlantic "volcanic margin" igneous oceanic crust (thick line), i.e. crustal province 4 in Figure 8, compared with the Ewing & Houtz (1979) oceanic crust model (thin line) and Icelandic crust of Flovenz & Gunnarsson (1991) (stipled line). From Eldholm & Grue (1994).

NATURE OF LCB

Evaluation of LCB dimensions and properties must take into account several observations: 1) a relatively large uncertainty in geometry and velocity. This is principally due to differences in data quality and sampling density and in acquisition and interpretation techniques. Although sophisticated ESP and OBS experiments (e.g., Hinz et al., 1987; Fowler et al., 1989; Olafsson et al., 1992; Mjelde et al., 1993; Holbrook et al., 1994a) have resulted in great improvements in the accuracy of velocities and images, available LCB velocities vary within the 7.1-7.7 km/s range. 2) The apparent lateral continuity from beneath highly extended continental crust and well into oceanic crust (e.g., Fowler et al., 1989; Eldholm & Grue, 1994; Holbrook et al., 1994a). 3) The spatial relationship of the extrusive cover and the LCB. 4) The spatial relationship of the LCB and thickened oceanic crust. 5) Consistently higher LCB velocities than the 7.1-7.2 km/s velocities for ponded hot mantle decompressional basaltic melts due to increased MgO content calculated by White & McKenzie (1989).

These observations suggest that the LCB was most likely emplaced by increased decompressional partial melting during breakup. Possibly, the range of LCB velocities relates to a varying degree of melt fractionation. Compositionally, we suggest the top of the LCB represents a transition from evolved MORB-derived gabbro to olivine picrite. Nonetheless, we also have to consider whether the 7+ km/s wide-angle seismic arrivals originate from a secondary interface. For example, the regionally similar thickness, 11-12 km, of the igneous crust above the LCB in the North Atlantic may imply a metamorphic facies boundary, most likely a transition from gabbros to garnet granulites. However, a metamorphic nature of the LCB requires substantial amounts of fluids shortly after emplacement, for which it is difficult to envisage a viable source (Gladczenko, 1994).

LCBs are commonly described as "underplated" bodies (LASE Study Group, 1986; White et al., 1987). The process of magmatic underplating, which refers to accumulation of mantle derived material below the original (continental) crust, depends on the density contrasts between the old crust and the new melt (Furlong & Fountain, 1986; Herzberg et al., 1983). A density filter of this kind does not exist during the formation of new oceanic crust. For this reason, the term "underplated" is not recommended for characterizing the entire LCB. The existence of the thin continental crust may also explain why the top of the LCB on the Vøring margin produces an intracrustal reflector landward of the continent-ocean boundary (Skogseid & Eldholm, this volume), whereas no interface is imaged within the thick initial oceanic crust where an equally sharp lithological boundary is not expected.

The LCB causes a significant, quantifiable reduction in the subsidence of the outer margin. The magnitude of this effect depends on the dimensions of the body. Thus, the LCB must be taken into account when calculating relative vertical motion and subsidence-derived estimates of lithospheric extension (Skogseid, 1994).

CONTINENT-OCEAN TRANSITION

Since seismic wide-angle profiles image the entire crust, the velocity field along continental margin transects provides a primary data source for studying the transition between the oceanic and the continental crust. Volcanic margin transects show a division into crustal provinces displaying different velocity structure and a varying degree of structural deformation as well as various igneous volumes and emplacement modes associated with breakup. This division is schematically illustrated in Figure 8, which outlines five main provinces: 1) "normal" continental crust; 2) continental crust thinning seaward with a moderate amount of intrusions; 3) strongly extended and intruded continental crust underlain by a LCB; 4) "volcanic margin" crust also with a LCB; and 5) "normal" oceanic crust (White at al., 1992). Note that there are no sharp lateral boundaries between the crustal provinces. Upper crustal plateau basalts may exist in provinces

10

1-4, although the main extrusive volumes appear in provinces 3-4.

The likelihood of lateral structural and compositional changes both along individual, 60-120 km long, wide-angle seismic profiles and along margin transects, and the anisotropic behaviour of the extrusive sequences, argue against using the margin velocity field exclusively to classify crustal type. Furthermore, the unique magmatic emplacement modes during crustal breakup argue against the application of the "accordion principle", i.e. linearly up- or down-scaled province 1 and 5 models, to determine crustal type in the intermediate region (provinces 2-4, Fig. 9).

Province 4 has previously been described as expanded oceanic crust (Zehnder et al., 1990; Mutter & Mutter, 1993), and as Iceland-type oceanic crust (e.g., Eldholm et al., 1989; Hinz et al., 1993). Such comparisons with the standard crustal models have obvious genetic implications, however. In view of the above arguments we do not believe that oceanic layers 2 and 3 have true equivalents in province 4. On the other hand, there is some similarity of the upper and middle crust in province 4 with Icelandic crust (Mutter et al., 1984) (Fig. 8). Since Iceland is underlain by a hot mantle, Moho is not clearly delineated and the lower crust is not directly comparable with province 4. We denote province 4 volcanic margin oceanic crust and suggest that the thickness of its three main crustal units reflects the structural setting and magnitude of melt production during breakup.

There is no overall agreement on a distinct continent-ocean boundary on rifted margins; however, the discussion is partly of a semantic nature. The existence of sharp boundaries on sheared passive margins is, however, well documented by seismic data (e.g., Myhre et al., 1993). A similar definition is viable also on volcanic margins if it refers to the compositional change in the uppermost crystalline crust below the extrusives. In Figure 8 the continent-ocean boundary will mark the change from the intruded, thinned continental crust of province 3 to the crust in province 4 composed entirely of igneous rock emplaced after breakup. On some margins, the province 3-4 boundary corresponds to changes in the intrabasement seismic character below the seaward dipping wedge. The outer part has no seismic base, whereas the inner part lies on a base reflector with deeper seismic stratification (Skogseid & Eldholm, 1989). If breakup took place during a period of reversals in the earth's magnetic field, the magnetic anomaly pattern would also provide important constraints on the location of the continent-ocean boundary.

If we consider the entire crust, the concept "continent-ocean transition" is more applicable. The transitional region will encompass the margin between the continent-ocean and province 2-3 boundaries. Thus, the continent-ocean transition is characterized by extensive extrusive cover, pervasive intrusions, a LCB and considerable syn-rift extension (Fig. 8).

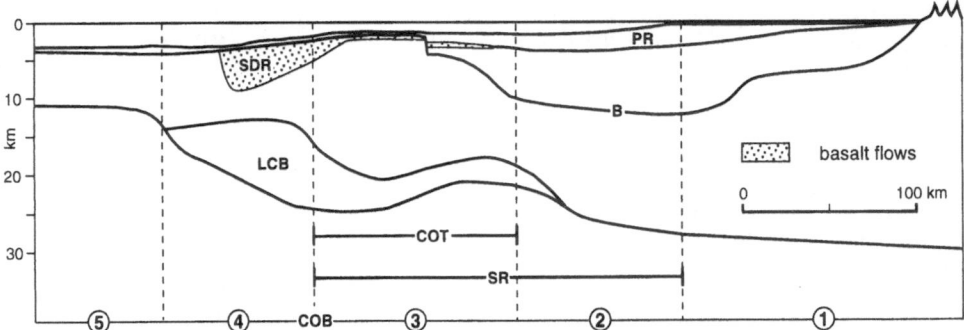

Fig. 8. Schematic volcanic margin transect with crustal provinces described in the text. Based on data from the Vøring and Møre margins off Norway. PR, post-opening sediments; B, crystalline basement; SDR, seaward dipping reflectors; LCB, lower crustal high velocity body; COB, continent-ocean boundary; COT, continent-ocean transition; SR, syn-rift zone.

Volcanic margins and hotspots

The size of dipping reflector sequences varies considerably between margins. If we make the qualitative assumption that their size roughly corresponds to extrusive and total crustal volumes produced during breakup, the inference is that volcanic margins are expressions of asthenospheric melt anomalies of different magnitudes. A similar conclusion is inferred for transient LIPs in general (Coffin & Eldholm, 1994).

There is ample evidence of a relationship between emplacement of LIPs and volcanic margins and mantle plumes expressed as hotspots (White & McKenzie, 1989; Duncan & Richards, 1991). The transient breakup event is associated with impingement of the plumehead on the lithosphere resulting in rapid melt migration to the surface and lithospheric breakup. Subsequently, the excess melting decreases and persistent volcanism is limited to a plumetail, commonly expressed as a submarine ridge or seamount chain. Typical examples are the Iceland hotspot versus North Atlantic margins and the Tristan hotspot versus South Atlantic margins.

Nonetheless, the plume model poses some challenging issues. Among these are: 1) some conjugate volcanic margins exceed reasonable theoretical plumehead diameters. After lithospheric impingement, plume models of Courtney & White (1986) and Griffiths & Campbell (1990), on the basis of different mantle circulation models, yield diameters of about 2000 km for the collapsed plumehead. The rifted margins in the North Atlantic and South Atlantic south of the Tristan plumetail, extend for more than 2700 and 2500 km, respectively. 2) Excessive melt volumes may be generated by relatively smaller asthenospheric temperature anomalies (50-100 °C) (Pedersen & Skogseid, 1989; Skogseid et al., 1992a,b) than those predicted by White & McKenzie (1989). 3) Most observed LCB velocities, 7.1-7.7 on North Atlantic margins (Eldholm & Grue, 1994) and a 7.4-7.5 average value on the U.S. East Coast margin (Holbrook et al., 1994a), require an unrealistically high temperature anomaly if the velocity is related to high-MgO plume derived melt. For example, a 7.2 km/s velocity yields a temperature anomaly of about 200 °C (White & McKenzie, 1989). 4) There are no obvious changes in LCB velocity and thickness with distance from the hotspot, as one would predict from some plume models (Holbrook & Kelemen, 1993). 5) Some margins have no obvious connection to a hotspot. Examples are the U.S. East Coast margin (Holbrook & Kelemen, 1993; Talwani et al., this volume) and the West-Australia margins (Mutter et al., 1984; Colwell et al., 1994). 6) Recent seismic surveys reveal ridgeward dipping reflectors within unequivocal, but thickened oceanic crust away from continental margins and hotspots (Hinz et al., 1993; Hinz pers. comm.).

If the plumes are sourced by thermal instabilities near the core-mantle boundary, the absence of a plumetail may be ascribed to a transient event creating a mantle "blob" (Griffiths & Campbell, 1991) rather than a persistent plume. Thus, a plumetail is not a necessary condition for a deep mantle sourced volcanic margin.

Regardless, the features presented above leave us in doubt as to whether a hotspot is required for volcanic margin formation. To accommodate the models and observational data we suggest that a hotspot is a sufficient, but not necessary condition. This concept explains the variety in magnitude and distribution of igneous volumes on volcanic margins. Accordingly, the asthenospheric melt potential is determined by the thermal and fluid fields in the asthenosphere and the dynamics of the lithosphere, i.e. magnitude and duration of rifting. For example, a combination of relatively small regional temperature and/or fluid content anomalies in the asthenosphere, combined with lithospheric extension, may set the stage for increased melt production during breakup. This scenario is not plume-driven and only requires circulation in the upper mantle (Anderson et al., 1992). Admittedly, a hotspot will provide a primary source for increased melting, particularly if it impinges on lithosphere under extension or thinned by previous rift episodes (Thompson & Gibson, 1991). If a plume impinges on a region which already has increased temperature or fluid content, the effects on the lithosphere may be recorded over a region larger than that affected by the plumehead.

Conclusions

As more high-quality seismic data have become available excessive igneous activity during lithospheric breakup is documented on a number of rifted continental margins; however, the melt dimensions appear to vary between margins. Previously, such volcanic margins have mainly been identified by the occurrence of seaward dipping reflectors in the seismic record, whereas both seismic and drilling data reveal that they are only one element in an igneous succession emplaced during breakup and early sea floor spreading.

An evaluation of the seismic data presently available, suggests that three-dimensional seismic surveying is required to obtain a better understanding of the stratigraphy and constructional history of the voluminous extrusive units. Furthermore, standard procedures for acquisition and interpretation of wide-angle seismic profiles will facilitate comparison of results from different investigators.

We consider the volcanic margin to be part of a tectono-magmatic system whose formation depends on lithospheric and asthenospheric properties before, during and after continental breakup. Therefore, an adequate description of the system requires studies of the entire crust along conjugate margins, or the entire rift, to establish the prerift lithospheric configuration, the syn-rift history and the spatial and temporal tectono-magmatic events during breakup. Such studies should involve evaluation of asymmetric tectono-magmatic settings.

The degree of magmatism during breakup, or whether a volcanic margin develops or not, depends on the melt potential of the asthenosphere along the incipient plate boundary and on the dynamic history of the lithosphere preceding breakup. We infer that breakup is preceded by a syn-rift phase of lithospheric stretching and thinning. Hence, we suggest a similar structural style and deformation of volcanic margins, non-volcanic margins and continental rifts. The different manifestation of these rifts is ascribed to the massive extrusive overprint on volcanic margins.

Despite the obvious spatial and temporal relationships between hotspots and volcanic margins it appears that the variety in magnitude and distribution of igneous volumes is satisfied by a concept in which a hotspot is a sufficient, but not a necessary condition for volcanic margin formation.

Acknowledgements. This work was supported by the Research Council of Norway as part of the IBS (Integrated Basin Studies) project under the JOULE II research programme funded by the Commission of European Communities (contract No. JOU2-CT 92-0110), and in part by the ProPetro program. IBS contribution No. 21.

REFERENCES

Anderson D.L., Zhang Y. & Taniomoto T., 1992. Plume heads, continental lithosphere, flood basalts and tomography. J. Geol. Soc. London Spec. Publ., 68, 99-124.

Austin J.A. & Uchupi E., 1982. Continental-oceanic transition off southwest Africa. Am. Assoc. Petrol. Geol. Bull., 66, 1328-1347.

Bally A.W., 1983. Seismic expression of structural styles. Am. Assoc. Petrol. Geol., Stud. Geol., 15(2).

Benson R.N. & Doyle R.G., 1988. Early Mesozoic rift basins and the development of the United States middle Atlantic continental margin. In Manspeizer W. (ed.), Triassic-Jurassic rifting continental breakup and the origin of the Atlantic passive margins. Elsevier Sci. Publs., Amsterdam, 99-127.

Biswas S.K. & Thomas J., 1992. The Deccan Traps and Indian Ocean volcanism. In Plummer P.S. (ed.), First Indian Ocean Petroleum Seminar, Dept. Tech. Coop. Development, United Nations, Seychelles, 187-209.

Coffin M.F. & Eldholm O., 1994. Large Igneous Provinces: Crustal structure, dimensions, and external consequences. Rev. Geophys., 32, 1-36.

Coffin M.F. & O. Eldholm, 1992. Volcanism and Continental Break-up: A Global Compilation of Large Igneous Provinces. J. Geol. Soc. London Spec. Publ., 68, 21-34.

Colwell J.B., Symonds P.A. & Crawford A.J., 1994. The nature of the Wallaby (Cuvier) Plateau and other igneous provinces of the west Australia margin. AGSO J. Austr. Geol. Geophys., 15, 137-156.

Courtney R.C. & White R.S, 1986. Anomalous heat flow and geoid across the Cape Verde Rise: evidence for dynamic support from a thermal plume in the mantle. Geophys. J., 87, 815-867.

Duncan R.A. & Richards M.A. 1991. Hotspots, mantle plumes, flood basalts, and true polar wander. Rev. Geophys., 29, 31-50.

Eldholm O. & Grue K., 1994. North Atlantic volcanic margins: Dimensions and Production Rates. J. Geophys. Res., 99, 2955-2968.

Eldholm O., Thiede J. & Taylor E., 1989. Evolution of the Vøring Volcanic Margin. In Eldholm O., Thiede, J. et al., Proc. ODP, Sci. Results, 104: College Station (Ocean Drilling Program), 1033-1065.

Eldholm O., Sundvor E. & Myhre A.M., 1979. Continental margin off Lofoten-Vesterålen, Northern Norway. Mar. Geophys. Res., 4, 3-35.

Eldholm O., Thiede J., Taylor E. et al., 1987. Proc., ODP, Init. Repts. 104: College Station, TX (Ocean Drilling Program).

Ewing J. & Houtz R., 1979. Acoustic stratigraphy and structure of the oceanic crust. Maurice Ewing Ser., Am. Geophys. Union, 2, 1-14.

Flovenz O.G. & Gunnarsson K., 1991. Seismic crustal structure in Iceland and surrounding area. Tectonophysics, 189, 1-17.

Fowler S.R., White R.S, Westbrook G.K. & Spence G.D., 1989. The Hatton Bank continental margin, II. Deep structure from two-ship expanded spread seismic profiles. Geophys. J., 96, 295-309.

Furlong K.P. & Fountain D.M., 1986. Continental crustal underplating: thermal considerations and seismic petrological consequences, J. Geophys. Res., 91, 8285-8294.

Gibson I.L. & Gibbs A.D., 1987. Accretionary volcanic processes and the crustal structure of Iceland. Tectonophysics, 133, 57-64.

Gladczenko T.P., 1994. Crustal structure and composition of selected transient large igneous provinces. Cand.scient. thesis Univ. Oslo, 220 pp.

Griffiths R.W. & Campbell I.H., 1991. Interaction of mantle plume heads with the Earth's surface and onset of small-scale convection. J. Geophys. Res., 96, 18295-18310.

Griffiths R.W. & Campbell I.H., 1990. Stirring and structure in mantle starting plumes. Earth Planet. Sci. Lett., 99, 66-78.

Harry D.L. & Sawyer D.S., 1992. Basaltic volcanism, mantle plumes, and the mechanics of rifting: the Paraná flood basalt province of South America., Geology, 20, 207-210.

Herzberg C.T., Fyfe W.S. & Carr M.J., 1983. Density constraints on the formation of the continental Moho and crust. Contrib. Mineral. Petrol., 84, 1-5.

Hinz K., 1981. A hypothesis on terrestrial catastrophes: Wedges of very thick oceanward dipping layers beneath passive margins - their origin and paleoenvironment significance. Geol. Jahrb., E22, 345-363.

Hinz K. & Schlüter H-U., 1978. Der Nordatlantik - ergebnisse geophysikalischer untersuchungen der Bundesanstalt für Geowissenschaften und Rohstoffe an Nordatlantischen kontinentalrändern. Erdöl, Erdgas Z., 94, 271-280.

Hinz K. & Weber J., 1976. Zum geologischen Aufbau des Norwegischen Kontinentalrandes und der Barents-See nach reflexionsseismischen Messungen, Erdöl & Kohle, Erdgas, Petrochem., 3-29.

Hinz K., Eldholm O., Block M. & Skogseid J., 1993. Evolution of N Atlantic volcanic continental margins. In: Parker, J.R. (ed.) Petroleum geology of Northwest Europe: Proc. 4th Conf. Geol. Soc. London, 901-913.

Hinz, K., Mutter J.C., Zehnder C.M. & NGT Study Group, 1987. Symmetric conjugation of continent-ocean boundary structures along the Norwegian and East Greenland margins, Mar. Petrol. Geol., 3, 166-187.

Holbrook W.S. & Kelemen P.B, 1993. Large igneous province on the US Atlantic margin and implications for magmatism during continental breakup. Nature, 364, 433-436.

Holbrook W.S., Reiter E.C., Purdy G.M., Sawyer D., Stoffa P.L., Austin J.A., Oh J. & Makris J., 1994a. Deep structure of the U.S. Atlantic continental margin, offshore South Carolina, from coincident ocean bottom and multichannel data. J. Geophys. Res., 99, 9155-9178.

Holbrook W.S., Purdy G.M., Sheridan R.E., Glover L., Talwani M., Ewing J. & Hutchinson D., 1994b. Seismic structure of the U.S. Mid-Atlantic continental margin. J. Geophys. Res., 99, 17871-17891.

Hopper J.R., Mutter J.C., Larson R.L., Mutter C.Z. & NW Australia Study Group, 1992. Magmatism and rift margin evolution: evidence from northwest Australia. Geology, 20, 853-857.

Keen, C.E., 1987. Some important consequences of lithospheric extension, Geol. Soc. Spec. Publ., 28, 67-73.

Keser Neish, J.C., Seismic structure of the Hatton-Rockall area: an integrated seismic/modelling study from composite datasets. In Parker J.R. (ed.), Petroleum Geology of Northwest Europe: Proc. 4th Conf., Geol. Soc. London, 1047-1056.

Klitgord K.D., Hutchinson D.R. & Schouten H., 1988. U.S. Atlantic continental margin: structural and tectonic framework. In The Geology of North America, vol. I-2, The Atlantic continental margin: U.S. Geological Society of America, Boulder, Colo., 19-55.

Larsen H.C., 1990. The East Greenland Shelf. In The Geology of North America, vol. L, The Arctic Ocean Region. Geological Society of America, Boulder, Colo., 185-210.

Larsen H.C. & Marcussen C., 1992. Sill-intrusion, flood basalt emplacement and deep crustal structure of the Scoresby Sund region, East Greenland. J. Geol. Soc. London Spec. Publ., 68, 365-386.

Larson R.L., 1991. Geological consequences of superplumes. Geology, 19, 963-966.

LASE Study Group, 1986. Deep structure of the U.S. East Coast passive margin from large aperture seismic experiments (LASE), Mar. Petrol. Geol., 3, 234-242.

Lawver L.A. & Müller R.D., 1994. Iceland hotspot track. Geology, 22, 311-314.

Leg 152 Shipboard Party, 1994. Drilling unearths "fire and ice" at southern Greenland margin. Eos, 75, 401-406.

Light M.P.R., Maslanyj M.P., Greenwood R.J. & Banks N.L., 1993. Seismic sequence stratigraphy and tectonics offshore Namibia. Geol. Soc. Spec. Publ., 71, 163-191.

Lister G.S., M.A. Etheridge & P.A. Symonds, 1991. Detachment models for the formation of passive continental margins, Tectonics, 10, 1038-1064.

Makris J., Ginzburg A., Shannon P.M., Jacob A.W.B., Bean C.J. & Vogt U., 1991. A new look at the Rockall region, offshore Ireland. Mar. Petrol. Geol., 8, 410-416.

Meissner, R., 1986. The continental crust. A geophysical approach, Academic Press.

Mjelde R., Sellevoll M.A., Shimamura H., Iwasaki T. & Kanazawa T., 1993. Ocean bottom seismographs used in a crustal study of an area covered wirh flood-basalt off Lofoten, N. Norway. Terra Nova, 5, 76-84.

Mutter C.Z. & Mutter J.C., 1993. Variations in thickness of layer 3 dominate oceanic crustal structure. Tectonophysics, 117, 295-317.

Mutter J.C., Talwani M. & Stoffa P., 1982. Origin of seaward-dipping reflectors in oceanic crust off the Norwegian margin by "subaerial seafloor spreading", Geology, 10, 353-357.

Mutter J.C., Talwani M. & Stoffa P., 1984. Evidence for a thick oceanic crust adjacent to the Norwegian margin, J. Geophys. Res., 89, 483-502.

Myhre A.M., Eldholm O., Faleide J.I., Skogseid J., Gudlaugsson S.T., Planke S., Stuevold L.M. & Vågnes E., 1992. Norway-Svalbard Margin: Structural and Stratigraphical styles. In Poag C.W. & de Graciansky P.C. (eds.), Geologic evolution of Atlantic Continental Rises, Van Nostrand Reinhold, New York, 157-185.

Olafsson I., Sundvor E., Eldholm O. & Grue K., 1992. Møre Margin: Crustal structure from analysis of Expanded Spread Profiles. Mar. Geophys. Res., 14, 137-162.

Pedersen T. & Skogseid J., 1989. Vøring Plateau volcanic margin: Extension, melting and uplift. In Eldholm, O., Thiede, J. Taylor, E. et al., ODP, Sci. Res., 104, College Station TX (Ocean Driling Program), 985-991.

Planke S., 1994. Geophysical response of flood basalts from analysis of wireline logs: ODP Site 642, Vøring volcanic margin. J. Geophys. Res., 99, 9279-9296.

Planke S. & Eldholm O., 1994. Seismic response and construction of seaward dipping wedges of flood basalts: Vøring volcanic margin. J. Geophys. Res., 99, 9263-9278.

Planke S. & Eldholm O., 1993. Seismic properties of seaward dipping wedges of flood basalts: examples from the Vøring volcanic margin. Abstr. Geol. Soc. Am. Annual Meeting, A427.

Planke S. & Flovenz O.G., 1994. Integration of downhole and surface seismic data in flood basalt terrains: implications for seismic imaging and crustal structure. Abstr. Amer. Geophys. Union Fall Meeting.

Roberts D.G., Schnitker D. et al., 1994. Init. Rept. DSDP, 81, US Gov. Printing Office, Washington DC, 1984.

Saemundson K., 1974. Evolution of the axial rifting zone in northern Iceland and the Tjørnes Fracture Zone, Geol. Soc. Am. Bull., 85, 495-504.

Schuepbach M.A. & Vail P.R., 1980. Evolution of outer highs on divergent continental margins. In Continental Tectonics, Nat. Res. Council, Washington D.C., 50-64.

Skogseid J., 1994. Dimensions of the Late Cretaceous-Paleocene Northeast Atlantic rift derived from Cenozoic subsidence, Tectonophysics, 240, 225-247.

Skogseid J. & Eldholm O., this volume. Rifted continental margin off mid-Norway.

Skogseid J. & Eldholm O., 1989. Vøring Plateau continental margin: Seismic interpretation, stratigraphy and vertical movements. In: Eldholm O., Thiede J., Taylor E. et al., Proc. ODP, Sci. Results, 104: College Station (Ocean Drilling Program), 993-1030.

Skogseid J. & Planke S., in press. Seismic reflection and refraction imaging of crustal structure on NE Atlantic margins: Mesozoic continental rifting and Cenozoic volcanic margin formation. Basin Res.

Skogseid J., Pedersen T. & Larsen V.B., 1992a. Vøring Basin: subsidence and tectonic evolution. Norw. Petrol. Soc. Spec. Publ., 1, 55-82.

Skogseid J., Pedersen T., Eldholm O. & Larsen B.T., 1992b. Tectonism and magmatism during NE Atlantic continental break-up: the Vøring Margin. J. Geol. Soc. London, Spec. Publ., 68, 305-320.

Talwani M., 1978. Distribution of basement under the eastern North Atlantic Ocean and Norwegian Sea. Geol. J. Spec. Issue, 5, 25-57.

Talwani M. & Eldholm O., 1977. Evolution of the Norwegian-Greenland Sea. Geol. Soc. Am. Bull., 88, 969-999.

Talwani M. & Eldholm O., 1972. The continental margin off Norway: A geophysical study, Geol. Soc. Am. Bull., 83, 3573-3606.

Talwani M., Ewing J., Sheridan R.E. & Holbrook W.S., this volume. The EDGE experiment and the U.S. East Coast magnetic anomaly.

Thompson R.N. & Gibson S.A., 1991. Subcontinental mantle plumes, hotspots and pre-existing thinspots. J. Geol. Soc. London, 148, 973-977.

Todal A., 1994. Tektono-magmatisk utvikling av Indias vestlige kontinentalmargin. Cand.scient. thesis Univ. Oslo, 133 pp.

White R.S. & McKenzie D., 1989. Magmatism at rift zones: The generation of volcanic continental margins and flood basalts, J. Geophys. Res., 94, 7685-7729.

White R.S., McKenzie D. & O'Nions R.K., 1992. Oceanic crustal thickness from seismic measurements and rare earth element inversions, J. Geophys, Res, 97, 19683-19715.

White R.S., Spence G.D., Fowler S.R., McKenzie D.P., Westbrook G.K. & Bowen A.N., 1987. Magmatism at rifted continental margins, Nature, 330, 439-444.

Zehnder C.M., Mutter J.C. & Buhl P., 1990. Deep seismic and geochemical constraints on the nature of rift-induced magmatism during breakup of the North Atlantic, Tectonophysics, 173, 545-565.

LITHOSPHERE-ASTHENOSPHERE INTERACTIONS BELOW RIFTS

C.E. KEEN and R.R. BOUTILIER
Atlantic Geoscience Centre
Bedford Institute of Oceanography
PO Box 1006
Dartmouth, Nova Scotia
B2Y 4A2   Canada

ABSTRACT.   Quantitative model results of flow within a viscous lower lithosphere and asthenosphere are presented, which result from the thermal and mechanical consequences of extension within the solid upper lithosphere. The flow is calculated using a two-dimensional, time-dependent formulation of the Navier-Stokes equations for an incompressible viscous fluid. The viscosity depends on temperature, pressure and strain-rate. Velocity boundary conditions on the upper surface of the fluid simulate extension in the overlying solid lid. Calculations were made for different viscosity distributions and for different rates and spatial distributions of extension.

Significant small-scale convection develops below the solid upper lithosphere for some model input parameters. The spatial distribution of extension in the overlying solid layer defines the lateral thermal gradients, the primary driving force of the convective flow. The most vigorous flows occur when the rift is narrow and the transition from extended to unextended regions is sharp. The rate of extension is relatively unimportant for the parameter ranges explored here. The strain-rate dependence of viscosity is an important factor in determining the mode of convection.

Small-scale convection will affect both the subsidence history and, in some cases, the volumes of melt produced during extension. The effect on subsidence is potentially significant, as a perturbation to the larger effects of thermal changes within the lithosphere and crustal thinning. Time dependent subsidence is particularly striking on the rift shoulders. Decompression melting of upwelling mantle is shown to deliver significantly more melt to crustal levels due to small-scale convection, and provides a means of producing excess melt volumes without anomalously high mantle temperatures. These results are important for understanding a wide variety of rifted terranes, including rifted continental margins, and they supplement and largely confirm earlier work on rift-generated small scale convection.

Introduction

The role of the asthenosphere in controlling the geological processes of rifted margins and rift basins is not well understood. Processes of deformation and heat transport within the lithosphere have been extensively described and shown to produce most of the first-order properties of rifts (e.g. Bassi, 1991; McKenzie, 1978;

17

E. Banda et al. (eds.), Rifted Ocean-Continent Boundaries, 17–30.
© 1995 *Kluwer Academic Publishers. Printed in the Netherlands.*

18

Royden et al., 1980, Royden and Keen, 1980; Steckler and Watts, 1978). Recently the importance of these lithospheric processes in influencing magma generation and the creation of new basaltic crust has been established (White and McKenzie, 1989). However, it is generally assumed that the asthenosphere rises passively below the thinning lithosphere and that the asthenosphere does not contribute actively to the observed near-surface phenomena.

Several previous studies addressed how the asthenosphere might respond more actively to rifting. Buck (1986) and Keen (1985) showed that small-scale convection in the asthenosphere may occur below rifts and that the convection is driven by the horizontal temperature gradient that develops as a result of deformation of the everlying lithosphere. This rift-induced convection may cause uplift of the rift shoulders (Buck, 1986) and additional thinning of the lower lithosphere (Keen, 1985). Mutter et al. (1988) suggested that small-scale convection could provide a mechanism for generating large quantities of magma and providing the thick igneous crust observed on volcanic rifted margins. The earlier quantitative studies of convection were limited by the physical dimensions of the region of flow and by the assumption of a Newtonian viscosity function in most models.

We have begun numerical modelling of asthenospheric flow which attempts to remove some of the limitations of these previous quantitative models. We present here the method and first results from this model.

Method

The numerical method is based on a two-dimensional finite element formulation of the Navier-Stokes equations for incompressible flow. We solve for the velocity and temperature fields within a fluid substrate underlying a solid plate within which the velocity field is specified (Fig. 1). Model parameters are listed in Table 1.

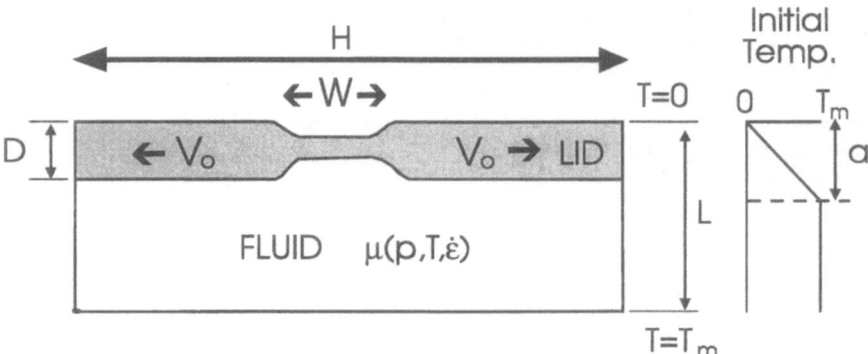

Figure 1. Schematic illustration of the model and the main input parameters. The grey area represents the solid lid in which velocities are specified. Symbols used match those in Table 1.

The upper boundary of the fluid is in no-slip contact with the rigid plate (Fig. 2). The left hand side of the fluid is a symmetry boundary, while the right hand boundary allows for horizontal flow in and out of the model. The lower boundary is free-slip. Constant temperatures are maintained at the top and bottom of the system. The initial conditions specify a linear temperature gradient down to the base of the thermal lithosphere, and a constant temperature below.

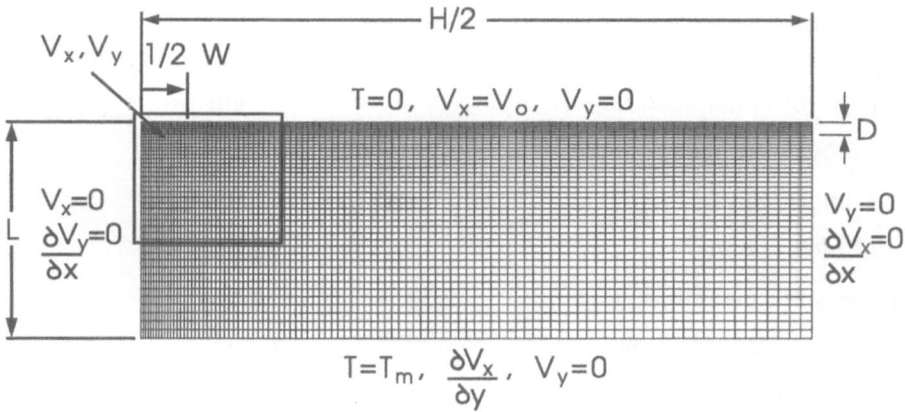

Figure 2. The finite element grid, showing the boundary conditions on the fluid. The shaded upper region of depth D is the solid lid. V_x and V_y are the horizontal and vertical velocities, respectively. The rectangular window on the upper left side of the finite element grid is the region of solution space for which results are shown in subsequent figures.

The specified horizontal and vertical velocities within the overlying solid plate are appropriate for a thinning and rifting lithosphere, assuming a passive rifting mechanism. Each model is characterised by an initial width of the rift zone (W in Fig. 1) and the horizontal velocity outside the rift zone (V_o) at which the solid plates move apart during rifting. At the end of the rift phase the velocities in the solid plate are set to zero and the system is allowed to evolve freely during the post-rift period.

The calculations are performed on an HP 755 computer using a commercially available finite element software package (EMRC Corporation, NISA/Fluid software). A graded 40x102 grid of rectangular, 4-noded elements (Fig. 2) is used. The model is stepped through time and the solutions are found on a fixed Eulerian grid. The velocity boundary conditions must be updated at the end of each time step to simulate a widening rift zone. The use of a fixed grid is problematic as the boundary between the fluid and the overlying solid plate is always horizontal. This limitation is currently being addressed in more complex model descriptions.

The viscosity depends on strain rate and on temperature and pressure, $\mu = A\dot{\varepsilon}^{(1-n)/n}\exp\{(E+PV)/nRT\}$. P, T and $\dot{\varepsilon}$ are pressure, temperature, and strain rate, respectively. This form of viscosity dependence is supported by laboratory measurements on ultrabasic rocks (e.g. Kirby and Kronenberg, 1987); our values of E, V and n were chosen to correspond to those for wet dunite (Chopra and Patterson, 1981; see Table 1). The value of a scaling viscosity, μ_o is a model input variable, where $\mu_o = A\dot{\varepsilon}_m^{(1-n)/n} exp\{(E+P_m V)/nRT_m\}$. This value is the viscosity at the base of our model, at reference temperature, pressure, and strain rate, T_m, P_m, and $\dot{\varepsilon}_m$, respectively.

Table I: Model Physical Properties

thickness of upper solid lid	D	40 km
total model thickness	L	700 km
total model width	H	4200 km
initial Moho depth		35 km
thickness thermal lithosphere	a	125 km
initial asthenosphere temperature	T_m	1350°C
viscosity activation energy	E	444. kJ/*mol*
viscosity activation volume	V	$15x10^{-6}$ m^3/*mol*
viscosity strain rate exponent	n	3.356
mantle density		3300 Mg/m^3
crustal density		2900 Mg/m^3
thermal conductivity		3.1 W/*m*/°C
thermal diffusivity		$8.0x10^{-7}$ m^2/*s*

A variety of solutions have been computed, tracking flow through both syn-rift and post-rift periods. Different values of input parameters have been assessed, including plate separation velocity, width of the rift zone, and scaling viscosity. The input parameters are listed in Table 2. As an illustration of the imposed behaviour within the solid plate, Fig. 3 shows the predicted shape of the Moho at the end of rifting for these models. This shape varies with the width of the rift zone, with the horizontal and vertical velocities in the plate, and with the total rifting time, all of which are model inputs.

We predicted the subsidence versus time across the surface of each model. We did not explicitly include the dynamic topography caused by vertical stresses due to flow, which, as Cordery and Phipps Morgan (1993) point out, is difficult to calculate for

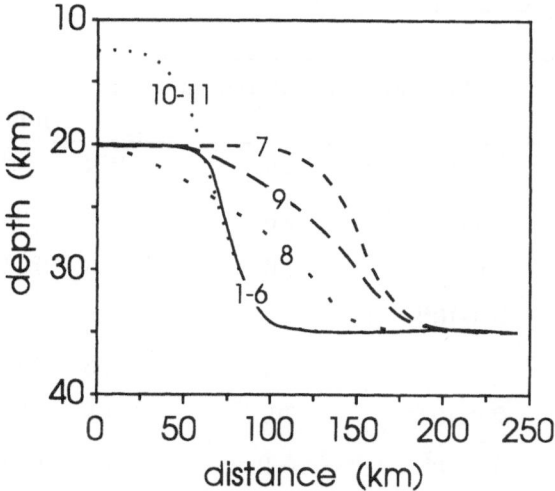

Figure 3. Shape of the base of the Moho after rifting for each of the models listed in Table 2.

these boundary conditions. Instead the isostatic subsidence due to thinning the crust, as predicted by the velocities specified in the solid lid, and due to density variations created by thermal expansion throughout the entire model, was calculated. Airy isostatic compensation was assumed. The isostatic subsidence should be a good first approximation to the complete subsidence which includes the dynamic topography (Jarvis and Peltier, 1982). It differs from the subsidence predicted in kinematic stretching models principally because of the thermal anomalies in the asthenosphere and lower lithosphere which are generated by the flow.

Results

FLOW IN THE ASTHENOSPHERE

A typical result which exemplifies fluid circulation below a rifting lithosphere is shown in Fig. 4 (model 4 in Table 2). The base of the thermal lithosphere is approximated by the 1200°C isotherm and above this depth the thermal lithosphere exhibits a linear gradient, similar to the behaviour of simple kinematic models for lithospheric thinning at rifts. Flow in the asthenosphere starts (Figs. 4A and B) with simple upwelling under the thinning lithosphere. The zone of upwelling tends to become more focused and narrow with time. After about 13 Ma of rifting, a convective instability develops below the rift (Fig. 4C), which persists for tens of Ma into the post-rift phase (Figs. 4D-3F). The convective circulation is time dependent; both the velocities and the shape of the convecting region changes through time (Fig. 4). Generally the circulation is concentrated within the upper 200 km of the asthenosphere where the viscosity is lowest. The maximum mantle velocities are

Table 2: Model Input Parameters

Model No.	V_o mm/yr	scaling viscosity Pa-s	rift duration Ma	β	V_{max} mm/yr
1	10	6.3×10^{21}	3.0	1.8	12.7
2	10	1.1×10^{22}	3.0	1.8	1.05
3	10	2.0×10^{22}	3.0	1.8	0.13
4	1	6.3×10^{21}	30.	1.8	17.9
5	1	1.1×10^{22}	30.	1.8	2.36
6	1	2.0×10^{22}	30.	1.8	0.02
7	10	1.1×10^{22}	6.1	1.8	1.54
8	10	1.1×10^{22}	4.1	1.8	0.10
9	10	1.1×10^{22}	6.3	1.8	0.03
10	1	6.3×10^{21}	42.	2.8	12.6
11	1	2.0×10^{22}	42.	2.8	0.67

β is the maximum amount of extension of the solid lid over the syn-rift period. V_{max} is the plate separation velocity. V_{max} is the maximum velocity of circulation in the fluid at the beginning of the post-rift phase and gives an estimate of the vigour of the flow.

much greater than the velocity of plate separation or of the velocity of passive upwelling.

The viscosity-depth distribution is also shown in Fig. 4. The decrease in viscosity with depth near the base of the lithosphere is due to the increase in temperature , to the maximum, T_m. In the asthenosphere where temperatures are almost constant, the stronger pressure dependence creates a positive viscosity gradient. The lowest viscosities are associated with the high strain rates within the convecting region (Fig. 4). Viscosity minima generally lie between 10^{17} and 10^{19} Pa-s, which are comparable to the viscosities used to characterise the asthenosphere in recent studies of flow below mid-ocean ridges (Chen and Morgan, 1990; Cordery and Phipps Morgan, 1993; Su et al., 1994). In those studies, viscosities between 10^{18} and 10^{19} Pa-s yield predictions of crustal thickness or sea floor topography which are consistent with observations in oceanic regions.

Solutions were also obtained for higher scaling viscosities (Table 2). The overall shape of the circulation and the time dependence of the solutions remain the same

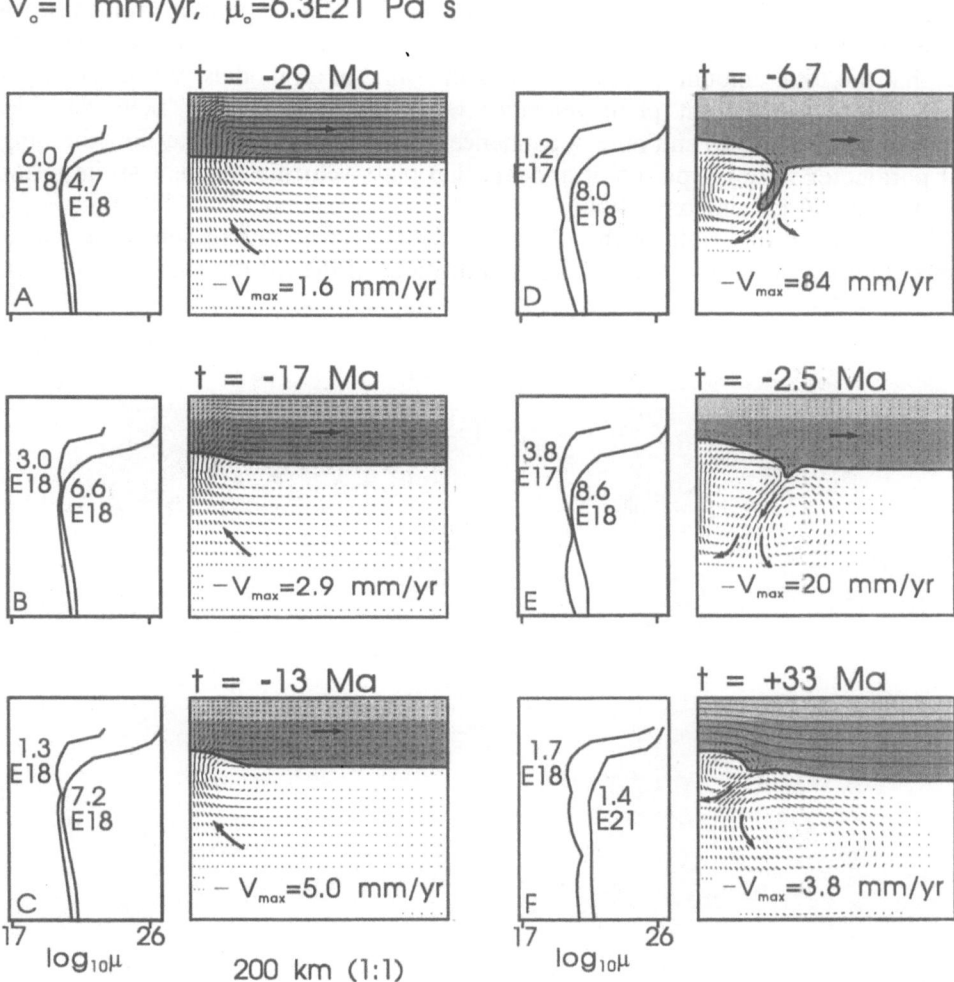

V_o=1 mm/yr, μ_o=6.3E21 Pa s

$t = -29$ Ma

6.0 E18/4.7 E18

A $-V_{max}=1.6$ mm/yr

$t = -6.7$ Ma

1.2 E17/8.0 E18

D $-V_{max}=84$ mm/yr

$t = -17$ Ma

3.0 E18/6.6 E18

B $-V_{max}=2.9$ mm/yr

$t = -2.5$ Ma

3.8 E17/8.6 E18

E $-V_{max}=20$ mm/yr

$t = -13$ Ma

1.3 E18/7.2 E18

C $-V_{max}=5.0$ mm/yr

17   26
$\log_{10}\mu$

$t = +33$ Ma

1.7 E18/1.4 E21

F $-V_{max}=3.8$ mm/yr

17   26
$\log_{10}\mu$

200 km (1:1)

Figure 4. Evolution of model 4 for which input parameters are listed in Table 2. Development of the flow is shown at six times (Figs. 4A to 4F) which are indicated on the figure. A time of t=0 Ma is the end of rifting, and negative and positive times represent syn-rift and post-rift times, respectively. For each time the right-hand box shows the flow velocity vectors as short lines. The length scale of the maximum velocity is indicated in each case. The heavy line is the 1200°C isotherm; for clarity other isotherms are shown only in Fig. 4F. The light grey upper region is the solid lid and the darker grey region is the lower lithosphere. The box for which velocity results are shown is the window indicated in figure 2. The left-hand box shows the log viscosity-depth distribution for two columns, one in the centre of the rift where x=0 and one beyond the rift at x=420 km. Minimum values of viscosity are indicated for each column. In every model, the curve with the minimum viscosity is for the centre of the rift where the strain rates are greatest and temperature is elevated at shallow depths.

but the maximum flow velocity decreases by about a factor of eight when the scaling viscosity is increase by a factor of three.

The effect of fast (10 mm/yr) and slow (1 mm/yr) plate separation velocities was assessed. The duration of rifting was decreased proportionally for the fast solutions, so that the total amount of extension was the same in all models. While the results for these two different plate velocities were not identical, the flow velocities, convective geometries and time dependence were remarkably similar. For the range of parameter values explored, it appears that plate velocities are not an important control on rift-driven convection.

Variations in the width of the rift and in the sharpness of its boundaries has an important effect on flow (Fig. 5). These attributes of the rift are exemplified by the

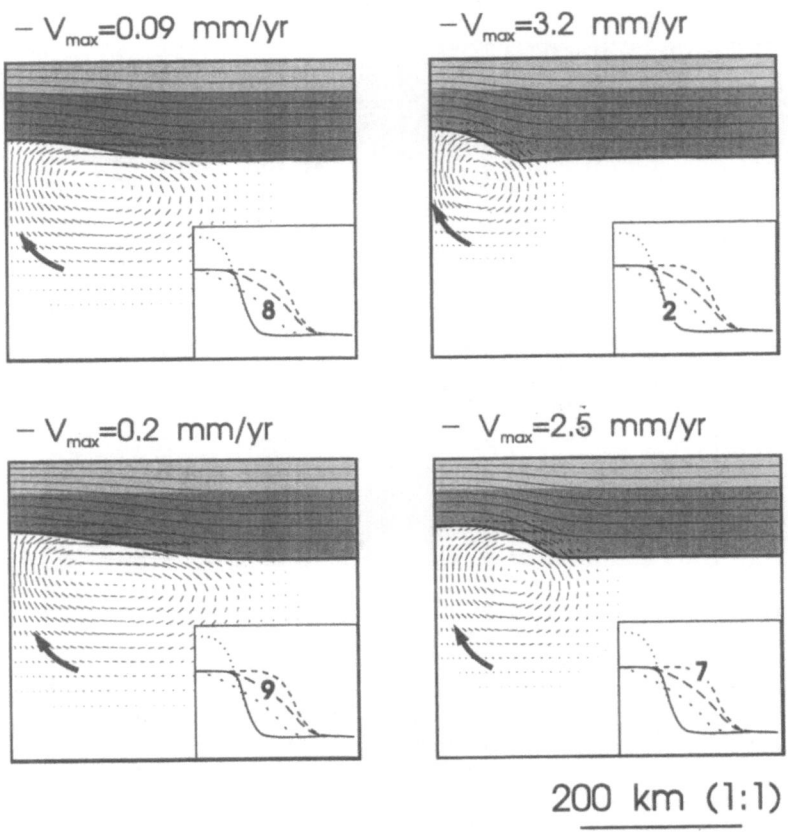

Figure 5. Comparison of various models with different prescribed necking profiles in the solid lid as shown in figure 3 and in the inset for each model. The flow velocity vectors and model properties are shown as described for figure 4. Model numbers are shown on the inset and refer to those listed in Table 2. All results are for a time of 29 Ma after rifting stopped.

Moho deformation predicted at the base of the solid lid (Fig. 3). The four different model results in Fig. 5 show that the most vigourous flow occurs when the rift is narrow and the transition from stretched to unstretched region is sharp (model 2). Widening the rift zone decreases the maximum velocity significantly. Even more striking is the decline in circulation when lithospheric necking is more gradual across the rift (models 8 and 9). These results confirm earlier work (Buck, 1986) suggesting that the vigour of convective flow below a rift may depend critically on lateral thermal gradients in the deep lithosphere and therefore indirectly on the geometry of lithospheric deformation.

VERTICAL MOTIONS AND SURFACE HEAT FLOW

The evolution of the rift basin for a typical model is shown in Fig. 6. Most of the subsidence is identical with that predicted by a simple kinematic model for extension of the lithosphere (e.g. McKenzie, 1978). The model with the highest scaling viscosity, for which flow is least vigourous, best approximates this simple model. In this case, the rift basin is well behaved. There is only minor uplift of the rift shoulders due to lateral heat transport. However, at lower scaling viscosities there are shorter-term changes in the subsidence with time which follow the time-dependence of circulation. At the lowest scaling viscosity, for example, relative uplift occurs over a short period during the syn-rift phase. Also, the rift shoulders exhibit a more complex history, with minor uplift during the 30 Ma long syn-rift period, followed by subsidence during the 33 Ma post-rift period. This reflects the competing effects of lateral heat conduction in the lithosphere, adding heat to this region, and temperature decreases within the underlying asthenosphere associated with a downwelling convective limb (e.g. Fig. 4d). These additional vertical motions are 100 to 200 m in amplitude and may be significant in interpreting rift basin subsidence and stratigraphic histories. Observations of the sediment stratigraphy in deep boreholes drilled on continental margins suggest that deviations from the long-term post-rift thermal subsidence do occur, with periods of 10 to 30 Ma and amplitudes of several hundred metres (e.g Heller et al, 1982).

The surface heat flow shows little departure from that predicted by a simple kinematic model. At the maximum post-rift time of 33 Ma, a model with low scaling viscosity (model 4) exhibits a maximum of 8 % higher heat flow than a comparable model with the highest scaling viscosity (model 6). This slightly higher heat flow is caused by the more vigorous convection, which maintains a thinner thermal boundary layer at the rift.

MELT GENERATION

Decompression melting is predicted to occur at rifts when sufficient lithospheric extension has occurred (under normal thermal conditions $\beta > 2$ or 3 , White and McKenzie, 1989). In most models the amount of melt predicted assumes that the upflow of mantle below a thinning lithosphere occurs passively, with velocities

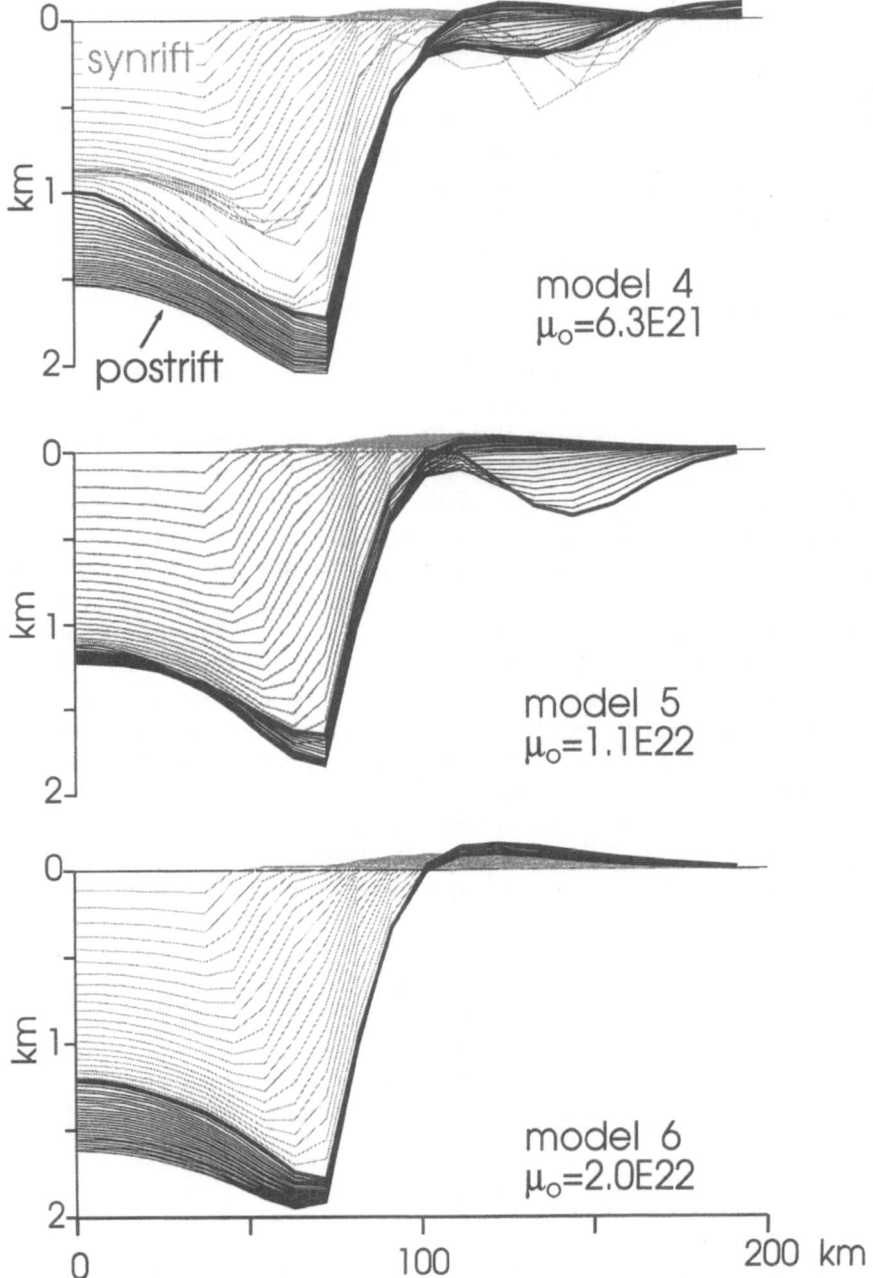

Figure 6. Development of a rift basin with time for models 4 to 6 in Table 2. Each figure shows time lines for the position of basement, spaced 1.32 Ma apart. Maximum post-rift time is 33 Ma. The basin is filled with water. Grey lines and black lines are the syn-rift and the post-rift development, respectively.

dictated by the rate of extension and thinning. Our results suggest that upflow rates may be much higher due to the development of convective instability in the asthenosphere. We have used the melting parameters of McKenzie and Bickle (1988) to estimate the melt volumes produced by lithospheric extension coupled with this asthenopheric flow.

We have divided the mantle into small parcels and tracked the temperature and position of each parcel through time, given the calculated flow velocities. Only the implications of the calculated flow on melt production have been assessed: we do not investigate the effects that melting would have on flow in terms of density and viscosity changes in the mantle. Melting will tend to inhibit convection because the extraction of melt from shallow depths in the asthenosphere reduces the density of the residual mantle and therefore stabilises the density-depth gradient (Su and Buck, 1993). Conversely, melt buoyancy may increase the upflow rate and enhances the melt accreted at crustal levels (Su et al., 1994).

We assume that our temperatures can be considered supra-adiabatic for calculating the melt generated and no additional correction was applied for adiabatic gradient. Each parcel of mantle that crosses the solidus produces melt: melting within each parcel is assumed to occur adiabatically. This is not strictly true because conductive cooling will reduce melt production and our estimate of melt is therefore a maximum. We assume the lateral position at which a parcel yields up its melt is the point at which it reaches its minimum depth. The total melt thickness is the sum over time of all the melt produced below a given point on the surface. All of these simplifying assumptions allow only tentative conclusions at present and we must further assess their effect on the results. The calculations probably provide reasonable first-order predictions of melt volume and are a sensible starting point for further model refinement.

The results are shown here for a model which has been stretched more than those shown earlier to increase the amount of melting (models 10 and 11; Fig. 7). We compared the results for a model with a low scaling viscosity (model 10) which exhibited vigourous convection with results for an identical model but with a high scaling viscosity (model 11) that exhibits laminar flow with no significant convection. There is a striking increase in melt production when small-scale convection is active. The high-viscosity model produces only about 50 m of melt, while the low-viscosity convective flow delivers several kilometers. We can also compare the results of the low-viscosity model with the simple passive flow model described by White and McKenzie (1989) for instantaneous stretching which for this geometry predicts 600 m of melt. The passive model produces more melt than the high-viscosity model (600 m versus 50 m) because there are no heat losses by conduction included in the former. In spite of some heat loss during rifting, the low-viscosity convective model still delivers 3.5 times the amount of melt predicted by the instantaneous passive model.

The distribution of melt volume across the rift shows a gradual decrease which reflects the width of the upwelling and the shape of the convecting zone. Melt

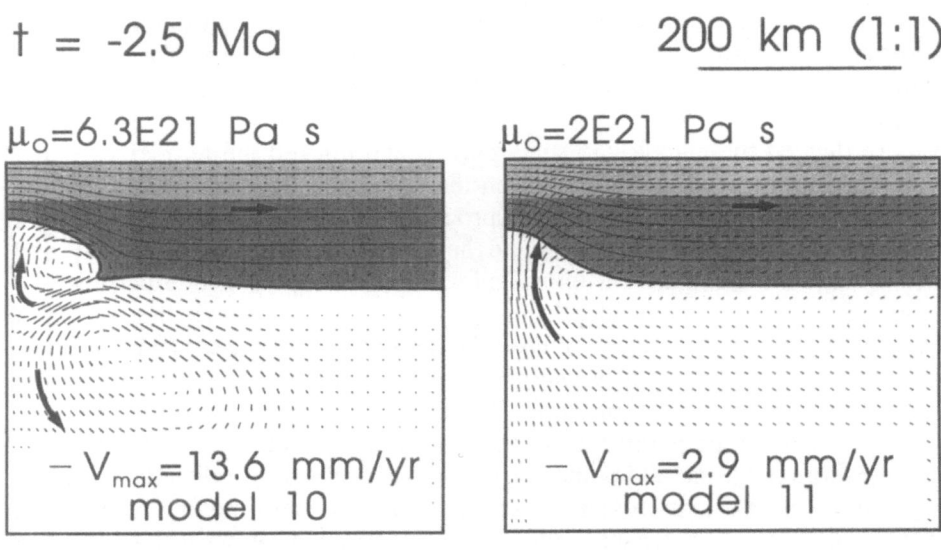

t = -2.5 Ma   200 km (1:1)

μ_o=6.3E21 Pa s μ_o=2E21 Pa s

$-V_{max}$=13.6 mm/yr $-V_{max}$=2.9 mm/yr
model 10   model 11

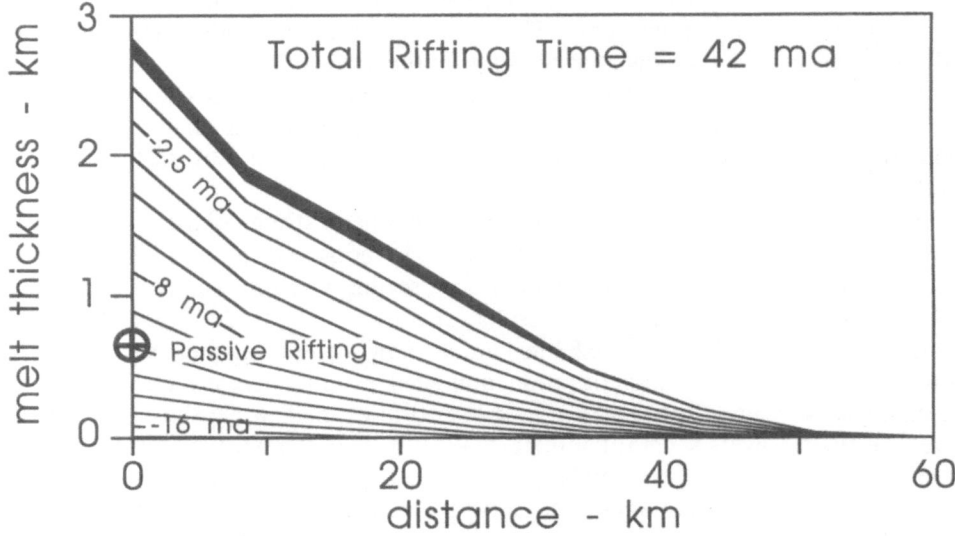

Figure 7. Velocity vectors for models 10 and 11 which are identical except for the scaling viscosity. The low-viscosity model 10 shows vigourous convection while the high-viscosity model shows laminar flow, similar to the assumptions of simple kinematic models. The lower part of this figure shows the time history of melt generation for model 10 (model 11 shows no significant melt thicknesses). The cross represents the predicted melt thickness for the kinematic model of White and McKenzie (1989) for the equivalent amount of lower lithospheric thinning.

volume also decreases with time as convection slows down during the post-rift stage. These results show that melt volumes are enhanced by small-scale convection and that perhaps large melt volumes can be produced without excessively high mantle temperatures.

Conclusions

The models described above are in an early stage of development and our results must be regarded as preliminary. The contrasting results for wide versus narrow, and sharp versus gradual rifts show potential for explaining some of the variations in rifting style provided the viscosities are sufficiently low. These features include perturbations in subsidence history (e.g. Heller et al., 1982) and possibly the large volumes of melt that occur below some rifted margins (e.g. Holbrook and Kelemen, 1993; Mutter et al., 1988). Both sets of observations are at present unexplained.

The present models apply strictly only to continental rifted terrains, and not to rifted continental margins. We intend to develop the modelling strategy further to simulate rupture and the early development of sea floor adjacent to the margin. Continental margins should manifest many of the asthenospheric properties described, perhaps with greater intensity because of the generation of larger thermal gradients between hot oceanic and cool continental lithosphere. However, as the two sides of the rift move farther apart with the creation of new oceanic lithosphere between them, the thermal gradients will be averaged over larger distances and will diminish. Our model results suggest that convection will decline under these circumstances. However, the situation will be complicated by the interaction of the flow centred below the margin with that below the young ridge crest, and the simplest concepts may not apply. Transform margin settings with a different relationship to the adjacent ridge crest may provide a useful contrast to rifted margins in this respect.

ACKNOWLEDGEMENTS. We thank Patrick Potter for his help with the figures. W.R. Buck, R.C. Courtney, I.D. Reid , R. Sabadini, and R. Stephenson gave us constructive comments on the manuscript. Mr. K. Bhatia of EMRC Corporation provided useful comments on the finite element software. Geological Survey of Canada Contribution Number 32794.

References

Bassi, G. 1991. Factors controlling the style of continental rifting: Insights from numerical modelling. Earth and Planet. Sci. Lett., 105, 430-452.
Buck, W.R. 1986. Small-scale convection induced by passive rifting: the cause for uplift of rift shoulders. Earth Planet. Sci. Lett., 77, 362-372.

Chen,Y. and Morgan, W.J. 1990. A non-linear rheology model for mid-ocean ridge axis topography. J. Geophys. Res., 95, 17583-17604.

Chopra, P.X. and Patterson, M.S. 1981. The experimental deformation of Dunite. Tectonophysics, 78, 453-473.

Cordery, M.J. and Phipps Morgan, J. 1993. Convection and Melting at mid-ocean ridges. J. Geophys. Res., 98, 19477-19503.

Heller, P.L., Wentworth, C.M., and Poag, C.W. 1982. Episodic post-rift subsidence of the U.S. Atlantic margin. Geol. Soc. Am. Bull., 93, 379-390.

Holbrook, W.S. and Kelemen, B. 1993. Large igneous province on the United States Atlantic margin and implications for magmatism during continental breakup. Nature, 364, 433-436.

Jarvis, G.T. and Peltier, W.R. 1982. Mantle convection as a boundary layer phenomenon. Geophys. J. R. astr. Soc., 68, 389-427.

Keen, C.E. 1986. The dynamics of rifting: deformation of the lithosphere by active and passive driving forces. Geophys. J. R. astr. Soc., 80, 95-120.

Kirby, S.H. and Kronenberg, A.K. 1987. Rheology of the Lithosphere: selected topics. Rev. Geophys., 25, 1219-1244.

McKenzie, D.P. 1978. Some remarks on the development of sedimentary basins. Earth Planet. Sci. Lett., 40, 25-32.

McKenzie, D.P. and Bickle, M.J. 1988. The volume and composition of melt generated by extension of the lithosphere. J. Petrol., 29, 625-679.

Mutter, J.C., Buck, W.R., and Zehnder, C.M. 1988. Convective partial melting, 1. A model for the formation of thick basaltic sequences during the initiation of spreading. J. Geophys. Res., 93, 1031-1048.

Royden, L. and Keen, C.E. 1980. Rifting processes and thermal evolution of the continental margin of eastern Canada determined from subsidence curves. Earth and Planet. Sci. Lett., 51, 343-361.

Royden. L. Sclater, J.G. and Von Herzen, R.P. 1980. Continental margin subsidence and heat flow, important parameters in formation of petroleum hydrocarbons. Amer. Assoc. Petrol. Geol., 64, 173-187.

Steckler, M.S. and Watts, A.B. 1978. Subsidence of the Atlantic continental margin off New York. Earth and Planet. Sci. Lett., 41, 1-13.

Su, W., Mutter, C.Z., Mutter, J.C., and Buck, W.R. 1994. Some theoretical predictions on the relationship among spreading rate, mantle temperature, and crustal thickness. J. Geophys. Res., 99, 3215-3227.

Su, W. and Buck, W.R. 1993. Buoyancy effects on mantle flow under mid-ocean ridges. J. Geophys. Res., 98, 12191-12205.

White, R.S. and McKenzie, D.P. 1989. Magmatism at rift zones: the generation of volcanic continental margins and flood basalts. J. Geophys. Res., 94, 7685-7729.

FINITE DURATION RIFTING, MELTING AND SUBSIDENCE AT CONTINENTAL MARGINS

JONATHAN W. BOWN & ROBERT S. WHITE
Bullard Laboratories
Madingley Road
Cambridge
CB3 0EZ
U.K.

ABSTRACT. When continental lithosphere is thinned during rifting, basaltic melt is generated by decompression of anhydrous mantle if the geotherm intersects the anhydrous mantle solidus. The quantity of melt generated depends on four principal factors: the degree of lithospheric thinning; the potential temperature of the asthenospheric mantle; the thickness of the lithosphere prior to rifting; and the duration of rifting. Subsidence at rifted continental margins can be affected significantly by melt generation during rifting: mantle melting causes reduced subsidence because both the igneous rock added to the crust and the residual mantle are less dense than the original mantle. At 'volcanic' rifted margins subsidence is also affected by the relative uplift resulting from isostatic compensation of the underlying mantle whose density is reduced by the thermal anomaly caused by the mantle plume. We present results from a uniform pure-shear lithospheric stretching model for melt generation and for subsidence at continental margins rifted at realistic finite rates. Predictions from the model are compared with observations of melt generation and subsidence from the 'non-volcanic' Galicia Bank rifted margin and from the 'volcanic' Rockall Plateau rifted margin in the North Atlantic.

1. Introduction

The general mechanisms causing subsidence in extensional sedimentary basins and at rifted continental margins are well understood. Lithospheric stretching by pure-shear causes thinning of both the crust, which has a lower density than the underlying asthenospheric mantle, and of the lithospheric mantle, which has a higher density than the underlying asthenosphere. Provided the continental crust is of normal thickness prior to stretching, then in order to maintain isostatic equilibrium, subsidence accompanies the stretching event. Subsequently, a further phase of thermal subsidence occurs as the asthenospheric mantle that has welled up beneath the rift cools back to the steady-state thermal profile: the timescale for thermal subsidence, which is exponential in form, is of the order of 50 Ma.

The lithospheric stretching model as an explanation for the generation of extensional sedimentary basins and rifted margins was first postulated by McKenzie (1978), assuming local isostasy and instantaneous, uniform pure-shear extension of the lithosphere. McKenzie's (1978) model has been modified subsequently to accommodate extension at finite rates (Jarvis & McKenzie, 1980). This uniform lithospheric stretching model has been tested in many extensional sedimentary basins and rifted margins around the world. In the North

31

E. Banda et al. (eds.), Rifted Ocean-Continent Boundaries, 31–54.
© 1995 *Kluwer Academic Publishers. Printed in the Netherlands.*

Sea, where the model was first tested rigorously, the amount of stretching measured by crustal thinning and from normal faulting is in good agreement with that derived from subsidence data (Sclater & Christie, 1980; Barton & Wood, 1984; White, 1991). The model also accounts for the subsidence of a number of rifted continental margins. These include the eastern margin of the United States (Steckler & Watts, 1978), the Nova Scotia margin (Royden & Keen, 1980), and the Bay of Biscay (Le Pichon & Sibuet, 1981). Barton & Wood (1984), Watts (1988), Fowler & McKenzie (1989) and Watts & Torné (1992) using a combination of gravity and subsidence data, have shown for several areas of continental extension that the elastic thickness is ~5 km or less, so local isostasy is a good approximation.

Since the uniform stretching model accounts successfully for the main features of extensional sedimentary basins and rifted margins, it provides a good starting point from which to consider the generation of melt during extension and the effect of melt generation on subsidence. Here we present results from a model of uniform lithospheric extension for melt generation and subsidence at continental margins rifted at realistic rates. We first present model results showing the effect on melt generation of rifting at a finite rate. We next present model results highlighting the effect of melt generation on subsidence. Our model results are then compared to observations of melt generation and subsidence from the 'non-volcanic' Galicia Bank rifted margin and from the 'volcanic' Rockall Plateau rifted margin in the North Atlantic.

2. Melt Generation Due to Continental Extension

Decompression of anhydrous mantle rising beneath regions of continental extension results in partial melting if the perturbed geotherm intersects the anhydrous mantle solidus. Small amounts of highly enriched melt may also be generated from layers within the lithosphere metasomatised by small melt fraction melts from the convecting asthenospheric mantle (McKenzie, 1989). The amount of melt generated by decompression of the anhydrous mantle depends on four principal factors: the degree of lithospheric thinning; the potential temperature of the asthenospheric mantle; the initial thickness of the lithosphere; and the duration of extension. The potential temperature of a mantle parcel is the temperature it would have if raised to the surface adiabatically, without melting. The variation of melt production with the degree of lithospheric thinning, with asthenosphere potential temperature and with lithosphere thickness has been modelled by McKenzie & Bickle (1988), using the melting calculations of McKenzie (1984), assuming that the lithosphere extends instantaneously and deforms by uniform pure-shear. McKenzie & Bickle (1988) characterised how the anhydrous mantle melts using parameterizations of all the then published results from melting experiments on anhydrous aluminous lherzolite. Their model results showed that the amount of melt generated by extension at a rift is very sensitive to the potential temperature of the asthenosphere welling up beneath the rift.

In a recent paper we modelled the variation of melt production with rift duration (Bown & White, in press). Our model follows closely from that of McKenzie & Bickle (1988). Melting is described by McKenzie & Bickle's (1988) parameterizations, and in the absence of heat conduction our melting calculations, which are based on the determination of local instantaneous melt production rates (Watson & McKenzie, 1991), are equivalent to those used by McKenzie & Bickle (1988). Similarly, the lithosphere deforms by uniform pure-shear during extension (McKenzie, 1978; Jarvis & McKenzie, 1980). Below we present

Variable	Meaning	Value used	Units
a	Steady-state thickness of lithosphere; plate thickness	125×10^3	m
c_P	Specific heat capacity at constant pressure	1.2×10^3	J kg^{-1} °C^{-1}
	Entropy change on melting	400	J kg^{-1} °C^{-1}
g	Acceleration due to gravity	9.81	m s^{-2}
h	$= \frac{g\alpha(T+273)}{c_P}$, adiabatic temperature gradient (McKenzie & Bickle, 1988)		°C m^{-1}
$S(t)$	Subsidence due to lithospheric extension		m
t	Time since start of rifting		s
t_c	Pre-rift thickness of continental crust	34.7×10^3	m
$T(z,t)$	Real temperature		°C
$\overline{T}$	Normal real temperature at the base of the plate		°C
v	Vertical velocity of the mantle		m s^{-1}
x	Horizontal co-ordinate		m
X	Proportion of melt by weight		none
z	Vertical co-ordinate, measured upwards from the base of the plate		m
α	Thermal expansion coefficient of matrix	3.2×10^{-5}	°C^{-1}
	Thermal expansion coefficient of melt (Dane, 1941)	6.8×10^{-5}	°C^{-1}
β	Stretching factor (extensional strain)		none
κ	Thermal diffusivity of the mantle	8.05×10^{-7}	m^{2}s^{-1}
ρ_a	$= \rho_s \left(1 - \alpha \overline{T}\right)$, normal density of asthenosphere		kg m^{-3}
ρ_c	Density of continental crustal material at s.t.p.	2800	kg m^{-3}
ρ_m	Density of igneous addition to the crust	3000	kg m^{-3}
$\Delta\hat{\rho}$	Density decrease from undepleted lherzolite to depleted harzburgite	65	kg m^{-3}
ρ_r	$= \rho_s - \frac{\Delta\hat{\rho}}{0.25}X \quad X < 0.25$ $= \rho_s - \Delta\hat{\rho} \quad X \geq 0.25$, density of mantle residue at s.t.p.		kg m^{-3}
ρ_s	Density of undepleted mantle at s.t.p.	3300	kg m^{-3}
ρ_w	Density of seawater	1030	kg m^{-3}

Table 1: Values and definitions of model parameters used.

model results from our one-dimensional model in which heat conduction occurs only in the vertical direction, and from our two-dimensional model in which heat conduction also occurs horizontally from extended to adjacent unextended lithosphere. Values and definitions of the model parameters are given in Table 1.

In both one- and two-dimensional models the temperatures are determined in two stages. First, temperatures are calculated ignoring the thermal consequences of melting. These temperatures are then corrected for the absorption of latent heat of fusion by the melting process. We ignore the advection of heat with the melt and the latent heat released by solidification of the melt, since this heat is released in the crust, not in the mantle source region and its effect on melt generation is negligible.

In the one-dimensional model, the temperatures during a finite period of extension, ignoring the loss of latent heat of fusion, are found by solving the advection-diffusion equation,

$$\frac{\partial T}{\partial t} = \kappa \frac{\partial^2 T}{\partial z^2} - v\left(\frac{\partial T}{\partial z} + h\right) \tag{1}$$

for the physical system shown in Figure 1. T is temperature, t time, z the vertical co-ordinate, v vertical velocity, κ the thermal diffusivity of the mantle and h the adiabatic temperature gradient. As the material within the plate is stretched, asthenospheric mantle of constant potential temperature wells up across the plane $z = 0$ to replace material lost by horizontal flow within the plate. All model calculations are done for extension at a constant strain rate and v is determined in the same way as that used by Jarvis & McKenzie (1980). Velocities within the deforming plate are not affected by the melt extraction process. The initial temperature distribution comprises a steady-state conductive temperature profile in the plate and a uniform potential temperature in the underlying asthenosphere; at this instant the thickness of the lithosphere is the same as that of the plate. Temperatures are corrected for the absorption of latent heat of fusion using the same approach as that of Watson & McKenzie (1991) but with a value of $400\,\mathrm{J\,kg^{-1}\,{}^\circ C^{-1}}$ for the entropy of fusion rather than the value of $250\,\mathrm{J\,kg^{-1}\,{}^\circ C^{-1}}$ that they used.

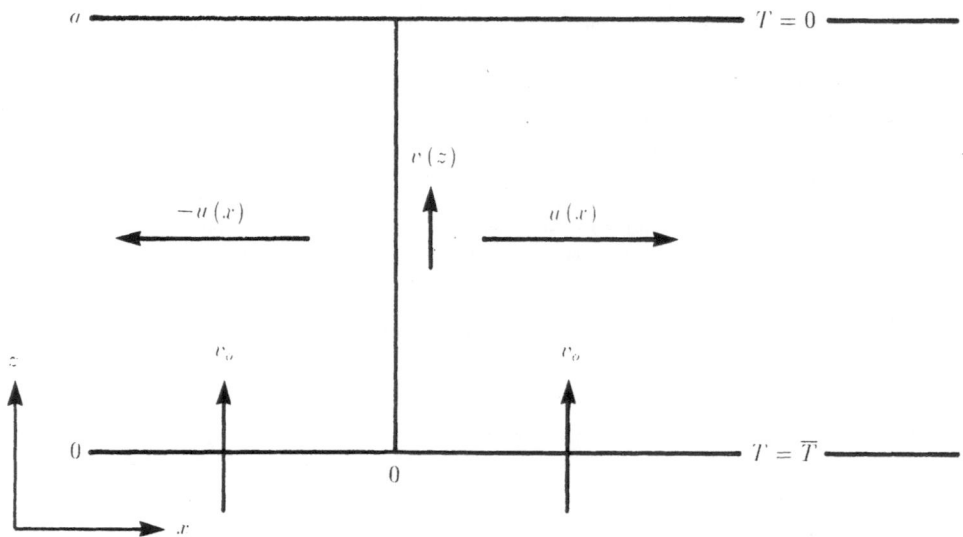

Figure 1: Physical system used to calculate the temperature profile during continental extension (Jarvis & McKenzie, 1980).

In our two-dimensional model the distribution of extension across a symmetric rift is described by a linear variation across the rift at the start of rifting of the final β experienced by each vertical section. Temperatures are calculated allowing for the horizontal conduction of heat. Melting is calculated in the same way as in the one-dimensional model.

The significant effect on melt generation of vertical heat conduction during a prolonged period of continental extension was first demonstrated by Pedersen & Ro (1992), also using a one-dimensional model for decompression melting of anhydrous mantle during uniform pure-shear lithospheric extension. The primary difference between our models and their model is that we calculate melting using McKenzie & Bickle's (1988) parameterizations rather than Ringwood's (1975) older description of the melting of a model mantle of pyrolite containing 0.1 % H_2O. In addition we use a value for the entropy of fusion of $400\,\mathrm{J\,kg^{-1}{}^\circ C^{-1}}$ which is approximately twice that used by Pedersen & Ro (1992), and closer to values determined from laboratory experiment (Fukuyama, 1985) and from the average slope of the mantle solidus (McKenzie, 1984). We also present total melt thicknesses calculated by our models rather than thicknesses relative to the amount of melt generated by instantaneous rifting.

We show below results from our models that illustrate the variation of melt production with the degree of lithospheric extension (β), with asthenosphere potential temperature and with rift duration, with particular emphasis on the last. All model results presented here were calculated using a value of 125 km for the initial thickness of the lithosphere (Parsons & Sclater, 1977).

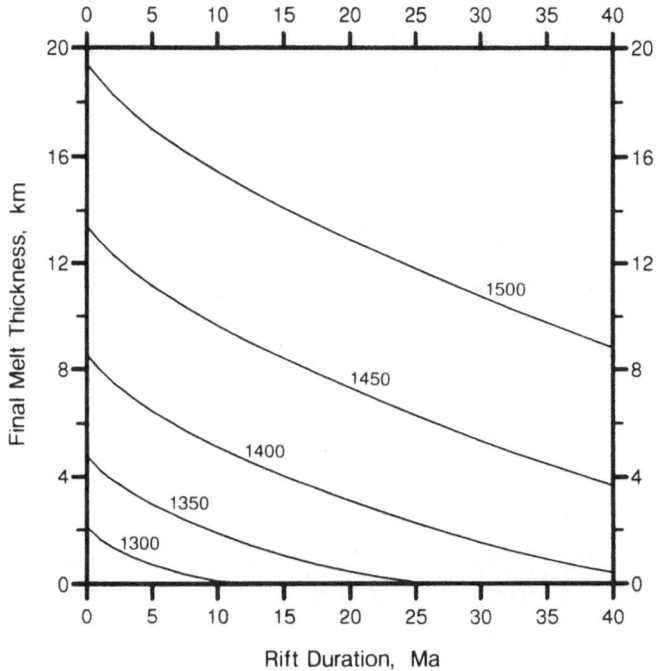

Figure 2: Variation of final thickness of melt with rift duration for a stretching factor of 5. Curves are shown for cases where the continental lithosphere is initially 125 km thick, and in thermal equilibrium with asthenospheric mantle at potential temperatures of 1300, 1350, 1400, 1450 and 1500°C.

The variation of melt thickness with rift duration is illustrated in Figure 2 by results from our one-dimensional model for extension by a β of 5 (a value typical of stretched continental crust near the oceanward edge of rifted continental margins) and asthenosphere potential temperatures between 1300°C (the normal potential temperature of the asthenosphere; Bown & White, 1994) and 1500°C. For all asthenosphere potential temperatures, the melt thickness decreases considerably as the duration of rifting increases. A rift duration of 15 Ma is sufficient to reduce from 2.1 km to zero the thickness of melt generated above normal asthenosphere with a potential temperature of 1300°C, to halve from 8.5 km to

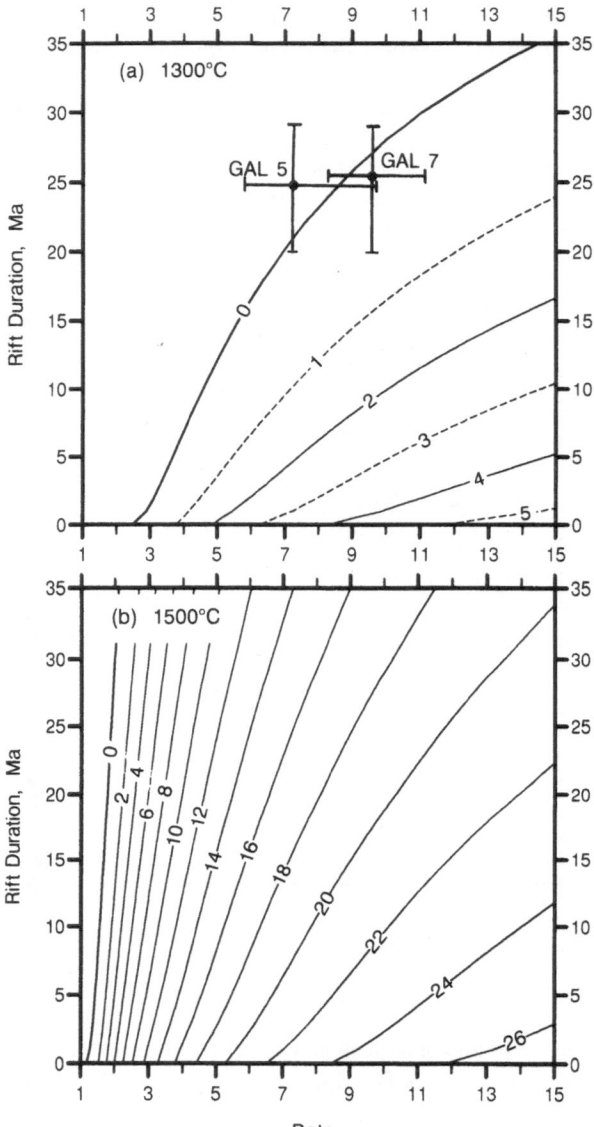

Figure 3: Variation of final melt thickness in kilometres with the degree of lithospheric extension (β) and with rift duration, for continental lithosphere initially 125 km thick and in thermal equilibrium with underlying asthenospheric mantle at potential temperatures of (a) 1300°C, and (b) 1500°C. The crosses on (a) show our estimates of the rift duration and the stretching factor near the continent-ocean transition at the Galicia Bank 'non-volcanic' margin: GAL 5 from seismic velocity model of Horsefield (1992); GAL 7 from seismic velocity model of Whitmarsh *et al.* (1993).

4.0 km the thickness of melt generated by extension above asthenosphere with a potential temperature of 1400°C, and to reduce by a quarter from 19.3 km to 14.1 km the thickness of melt generated by extension above asthenosphere with a potential temperature of 1500°C. The amount of melt generated beneath continental rifts depends strongly on the duration of extension because most melt is generated near the top of the upwelling asthenosphere (if indeed any is generated), where during a protracted rift episode conductive heat loss is greatest. Final melt thicknesses calculated from our one-dimensional model for a wide

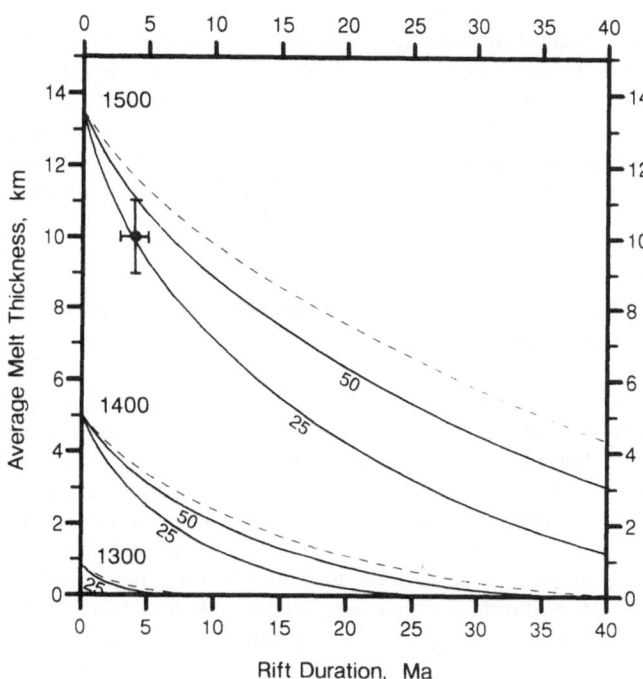

Figure 4: Variation with rift duration of the average melt thickness at the end of rifting, calculated with horizontal heat conduction for rifts with initial half-widths of 25 and 50 km, and without horizontal heat conduction for a rift with an initial half-width of 25 km, for asthenosphere potential temperatures of 1300, 1400 and 1500°C. The curves calculated without horizontal heat conduction are shown dashed. All curves are calculated for continental lithosphere which is initially 125 km thick and in thermal equilibrium with the asthenosphere. The horizontal distribution of extension within the rift is defined by a linear function in the material before rifting occurs. The final stretching factor experienced by columns in the rift increases linearly from 1 at the edge of the rift to 5 in the centre of the rift. The decrease in average melt thickness which can be attributed to horizontal heat conduction is the difference between the curves calculated with and without horizontal heat conduction. The cross shows our estimate of the average melt thickness and rift duration at the Rockall Plateau 'volcanic' rifted margin.

range of β and for rift durations up to 35 Ma are shown in Figure 3 for asthenosphere potential temperatures of 1300°C and 1500°C. Clearly heat loss by vertical conduction from the upwelling mantle during a protracted period of continental extension cannot be ignored when calculating the amount of melt generated.

Figures 2 and 3 also demonstrate the strong dependence of the amount of melt generated on the potential temperature of the asthenosphere welling up beneath the rift. Rifting above asthenosphere with a potential temperature of 1400°C, only 100°C hotter than normal, increases by a factor of four the total melt produced by instantaneous rifting.

The effect on melt generation of horizontal heat conduction from extended to adjacent unextended continental lithosphere is illustrated in Figure 4 by results from our two-dimensional model. The effect of horizontal heat conduction is most important in narrow, rapidly formed rifts. In such rifts the decrease in melt production due to horizontal heat conduction is approximately the same as that due to vertical heat conduction. In wider rifts, or those formed less rapidly, the effect of horizontal heat conduction on the total amount of melt generated is much less and is considerably smaller than that due to vertical heat conduction.

3. The Effect of Melt Generation and Mantle Plumes on the Subsidence Caused by Continental Extension

Melt generation may reduce considerably the subsidence at continental rifts. Melt generation reduces subsidence because mantle rock melts partially to produce igneous and residual rocks which are less dense than the original mantle (White & McKenzie, 1989). The one-dimensional model used here to calculate the effect of melt generation on subsidence caused by continental extension follows directly from the one-dimensional melting model. It is described fully by Bown (1994). We calculate subsidence in two steps. First, the subsidence caused by uniform pure-shear extension of the lithosphere is determined in the absence of melting (McKenzie, 1978; Jarvis & McKenzie, 1980). This first value is then corrected for the relative uplift caused by melting and by isostatic compensation of hot plume-derived material beneath the plate. Since the density changes caused by melting are permanent, so is the resulting uplift. Values and definitions of model parameter values are given in Table 1.

Assuming local isostasy, the water-loaded subsidence of the top of the lithosphere in the absence of melt generation $S(t)$ is given by (McKenzie, 1978; Jarvis & McKenzie, 1980),

$$S(t) = A \left(1 - \frac{1}{\beta} \right) - BQ(t) \tag{2}$$

where

$$A = \frac{t_c \left(\rho_s - \rho_c \right)}{\left(\rho_a - \rho_w \right)}$$

$$B = \frac{\alpha \rho_s}{\left(\rho_a - \rho_w \right)}$$

and

$$Q(t) = \int_0^a \left[T(z, t) - T(z, \infty) \right] dz$$

where β is the extension factor, t_c the pre-rift thickness of the continental crust, ρ_s the mantle density, ρ_c the crustal density, ρ_a the density of normal asthenosphere, ρ_w the density of seawater and α the thermal expansion coefficent. The first term on the right-hand side of Equation 2 is equal to the subsidence that would occur if the crustal portion of the lithosphere were thinned by a factor β, in the absence of any thinning of the lithosphere as a whole. The second term on the right-hand side accounts for the uplift or subsidence caused by changes in the thermal profile during and after extension. assuming that the plate is composed entirely of mantle material.

The effect of melt generation on subsidence is calculated directly from the one-dimensional melting model. We assume that melt is extracted and solidifies at crustal levels instantaneously. Models of melt flow in the mantle indicate that melt is extracted rapidly from the mantle in less than a few million years (McKenzie, 1985). We use a constant value of 3000 kg m^{-3} for the density of the new igneous rock added to the crust (Herzberg et al., 1983; White & McKenzie, 1989). The density of the mantle is reduced by melt extraction because melting reduces the Fe/Mg ratio and the proportion of the dense aluminous phase in the residue. Oxburgh & Parmentier (1977) calculated that there is a reduction in density of 65 kg m^{-3} between undepleted garnet lherzolite and depleted harzburgite, formed after extraction of 25 weight % of basaltic melt. Like Klein & Langmuir (1987) we assume that the density of the residue changes linearly with the degree of melting up to 25 weight % melting at which point the total density change is 65 kg m^{-3}. Thereafter the density of the residue is assumed not to change.

We calculate the uplift caused by melting by integrating over time the rates of uplift, allowing for the thinning by later extension of the new igneous addition to the crust and of the depleted mantle. Uplift rates are calculated from the distribution with depth of the local instantaneous melt production rate, assuming local isostasy. Implicit in our calculations of uplift are the assumptions that melt migrates only vertically and that the depleted mantle remains in situ. In calculating the uplift caused by melting we also assume that the new igneous rocks and the depleted mantle have the same total thickness as the original mantle from which they were derived. This assumption affects the model results by a negligible amount and is also made in models for the subsidence caused by thermal relaxation of the ocean floor and of continental rifts (Parsons & Sclater, 1977; McKenzie, 1978). Whilst the absorption of latent heat by the melting process affects melt generation significantly, its effect on subsidence is negligible and we ignore it.

If all the basaltic melt generated during extension is added to the lower crust, our subsidence calculations determine the position of the top of the pre-existing continental crust. If, however, some of the basaltic melt is extruded as lava flows then our subsidence calculations determine the position of the top of the lavas if the amount of sediment intercalated with the lavas is negligible. Drilling results from the seaward-dipping basaltic lava sequences beneath the 'volcanic' rifted margins of the northern North Atlantic indicate that the thickness of intercalated sediment is small (e.g. Roberts, Schnitker, et al., 1984; Eldholm, Thiede, Taylor, et al., 1987; 1989; ODP Leg 152 Shipboard Party, 1994).

In order to model the subsidence at 'volcanic' rifted margins, such as those in the northern North Atlantic, it is also necessary to consider the buoyancy of the anomalously hot plume-derived material beneath the plate. Mantle plumes emplace anomalously hot mantle beneath the base of the lithosphere over regions up to 2000 km in diameter; isostatic compensation of the hot plume-derived material can uplift oceanic lithosphere by 1–2 km (e.g. Cape Verde hotspot, Courtney & White, 1986; Hawaii hotspot, von Herzen et al., 1982).

Here we model this elevation by the uplift resulting from isostatic compensation of a layer of asthenospheric mantle beneath the plate with a constant potential temperature that is higher than normal. This uplift was called dynamic uplift by White & McKenzie (1989). We assume that active convection within mantle plumes does not thin the lithosphere appreciably (Sleep, 1994). At present, the interaction between extending lithosphere and the head of a mantle plume is poorly constrained (Arndt & Christensen, 1992). We assume that during rifting the depth to the base of the hot plume-derived material remains constant while plume material wells up into the space created by thinner lithosphere beneath the rift. In our model the relative uplift caused by the layer of plume-derived material beneath the deforming plate is therefore unaffected by rifting. The extra plume material is assumed to be supplied by lateral flow beneath the plate.

The modelled effect of melt generation and of mantle plumes on the syn-rift subsidence caused by instantaneous extension of continental lithosphere is illustrated in Figure 5. Whilst instantaneous rifting does not occur in the Earth, the results shown in Figure 5 do illustrate how the subsidence due to continental extension is changed by melt generation and by the presence of mantle plumes. The model results shown in Figure 5 are similar to those of White & McKenzie (1989). As the amount of melt generated increases with stretching factor and with the temperature of the asthenosphere welling up beneath the rift, so does the relative uplift caused by melt generation. For the model parameters used to calculate Figure 5 the uplift caused by the addition of new igneous rock to the crust is $\sim 1\frac{1}{2}$ times that caused by mantle depletion. Using a value for the density of the igneous addition to the crust that is either $100\,\mathrm{kg\,m^{-3}}$ greater than or $100\,\mathrm{kg\,m^{-3}}$ less than the value of $3000\,\mathrm{kg\,m^{-3}}$ used to calculate the curves in Figure 5 changes the uplift due to igneous addition by 30–40%. Increasing the density of the igneous addition reduces uplift, while decreasing the density increases uplift. Similarly, increasing the density changes in the mantle due to depletion by 50% increases the uplift due to mantle depletion by approximately 50%. In addition to the uplift resulting from isostatic compensation of the hot plume-derived material beneath the plate (dynamic uplift), the subsidence due to lithospheric thinning (determined from Equation 2) is also suppressed by the welling up of hotter than normal plume-derived material to fill the space left by the thinning of the

---

Figure 5: [Next page.] Variation with stretching factor of syn-rift water-loaded subsidence caused by instantaneous rifting of continental lithosphere, initially 125 km thick and thermally equilibrated with a potential temperature of 1300°C at its base. At the start of rifting a layer of hot plume-derived mantle 75 km thick is emplaced beneath the plate, causing uplift of its top. The curve of long dashes shows the water-loaded subsidence due to lithospheric thinning. The curve of intermediate dashes shows the water-unloaded uplift caused by the addition of new igneous rock to the crust. The curve of short dashes shows the water-unloaded uplift caused by the melting of the mantle to a less dense solid residue. The thin solid curve shows the water-unloaded dynamic uplift caused by the layer of plume-derived material beneath the plate. The thick solid curve is the total subsidence, which is water-loaded below sea-level, and unloaded above. The lower panel shows the corresponding variation with stretching factor of melt thickness. Potential temperatures of plume-derived mantle of (a) 1300°C, (b) 1400°C, and (c) 1500°C.

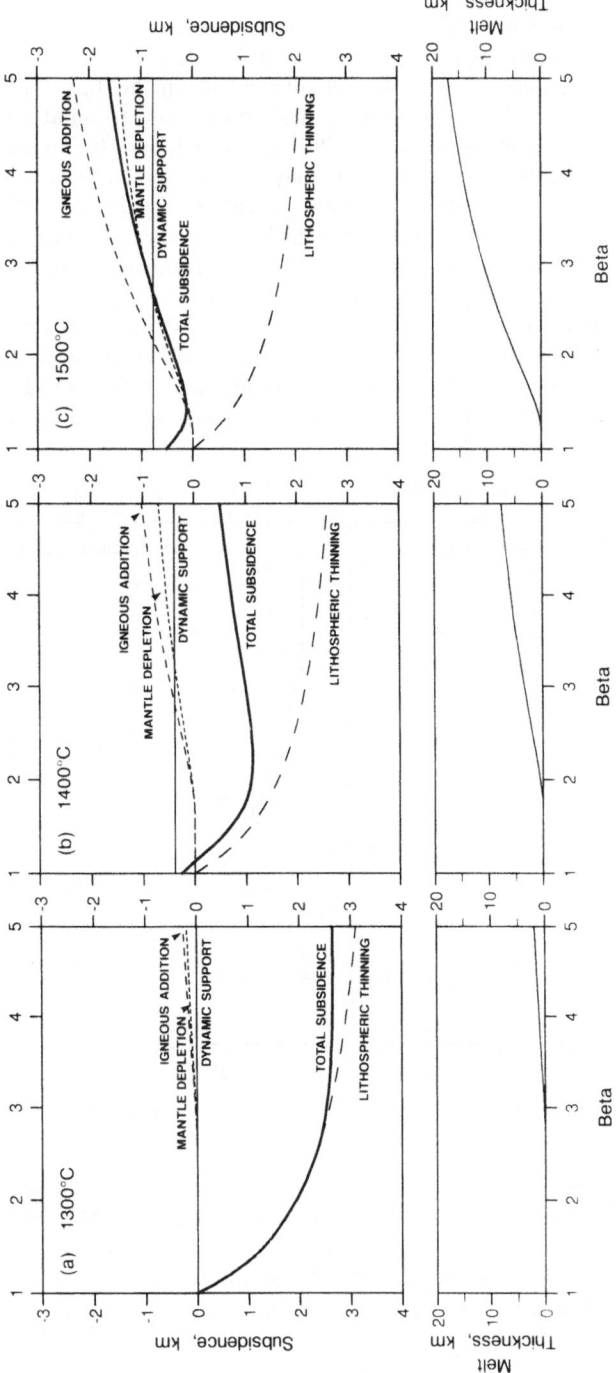

42

lithosphere. The magnitude of the dynamic uplift varies linearly with the thickness of the plume layer beneath the plate and with the temperature anomaly associated with it.

In the Earth, continental extension occurs at a finite rate. Since the amount of melt generated by continental extension is strongly dependent on the duration of rifting, so too is the effect on subsidence of melt generation. The effect of extension at a finite rate on the relative uplift caused by melt generation is illustrated in Figure 6. Figure 6 shows water-loaded subsidence curves for instantaneous rifting and for rifting over a period of 25 Ma. In both cases the curves were calculated for the extension of continental lithosphere initially 125 km thick by a β of 5. The lithosphere which has a normal temperature gradient is underlain either by normal potential temperature asthenosphere or by a 75 km thick layer of plume-derived material with potential temperatures 100°C or 200°C hotter than normal. In each case, the uplift caused by melt generation when rifting occurs over an interval of 25 Ma is much less than the uplift caused when rifting occurs instantaneously because much less total melt is generated by slow rifting. The upward kick of the subsidence curve during finite duration rifting reflects the generation of melt towards the end of rifting.

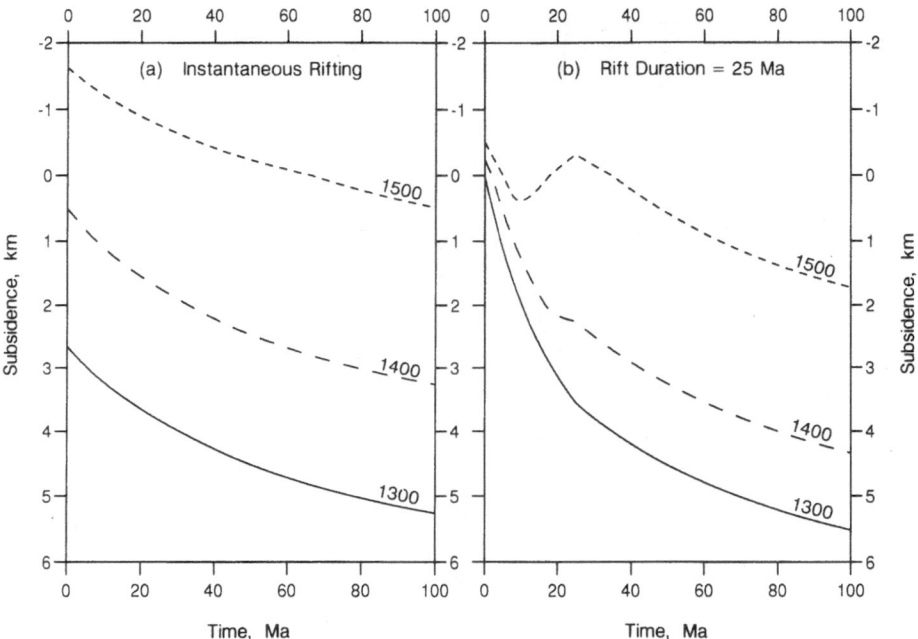

Figure 6: Water-loaded subsidence caused by stretching by a factor of 5 of continental lithosphere, initially 125 km thick and thermally equilibrated with a potential tempera-ture of 1300°C at its base. A 75 km thick layer of plume-derived mantle with a potential temperatures of 1300°C (solid curve), 1400°C (curve of long dashes) and 1500°C (curve of short dashes) is then inserted at the base of the lithosphere at the onset of rifting. (a) instantaneous rifting; (b) rifting over 25 Ma.

The model results presented in Figure 6 were calculated assuming that the temperature of the layer of plume-derived material does not change with time. Mantle plumes are not, however, steady-state features; results from numerical models and laboratory experiments indicate that the temperature and flow rate associated with starting plume-heads are likely to be considerably greater than the subsequent flow (White & M°Kenzie, 1989; Griffiths & Campbell, 1990; Loper, 1991). These results are consistent with the observed decrease in the crustal thickening, and hence in the rate of melt production, along the trace of the Réunion mantle plume over the 30 Ma following the generation of the Deccan flood basalts (White, 1993). The dynamic uplift above a mantle plume is, therefore, likely to vary with time. The dynamic uplift of a rift will also be affected by continental drift. As a rift moves away from an underlying mantle plume, the isostatic uplift of the rift caused by underlying hot mantle will decay to zero.

4. Comparison of Model Calculations With Observations From Two North Atlantic Continental Margins

Two types of rifted continental margin have been identified in the North Atlantic: 'volcanic' margins such as the Rockall Plateau and Vøring Plateau margins, which exhibit voluminous syn-rift magmatism; and 'non-volcanic' margins such as the Biscay and Galicia Bank margins, with only minor magmatism (White, 1992a). The 'volcanic' margins were formed by continental rifting above hot asthenosphere fed by the Iceland mantle plume (White & M°Kenzie, 1989; Hill, 1991; White, 1992b; Fram & Lesher, 1993; White & M°Kenzie, in press), while the 'non-volcanic' margins were formed by continental rifting over a prolonged period above asthenosphere with a normal potential temperature (White & M°Kenzie, 1989; White, 1992b; Bown & White, in press).

4.1. MELTING AND SUBSIDENCE ON THE 'NON-VOLCANIC' GALICIA BANK MARGIN

Seismic velocity models of the Galicia Bank margin (Figure 7a) provide no evidence for the addition of significant quantities of melt to the continental crust, even though the crust has been thinned to only 3–4 km (Horsefield, 1992; Whitmarsh et al., 1993). Furthermore, no basaltic rocks have been found above stretched continental crust (Boillot, Winterer, Meyer et al., 1987; Boillot, Winterer, et al., 1988). Assuming that the continental crust was initially 30 km thick, the crust has been thinned by a β of 7–10. Seismic stratigraphy calibrated by drilling results indicates that the main phase of rifting at the Galicia Bank margin lasted from the early Valanginian (141–138 Ma; Harland et al., 1990) to the late Aptian (118–112 Ma; Harland et al., 1990) a period of some 20–29 Ma (ODP Leg 103 Shipboard Scientific Party, 1987). The one-dimensional melting model predicts that less than 0.5 km of melt would be generated by rifting over such periods above asthenosphere of normal potential temperature (Figure 3a), compared to 3.5–4.5 km of melt if extension occurred instantaneously. The scarcity of syn-rift magmatism at the Galicia Bank margin, therefore, reflects conductive heat loss during the prolonged period of continental rifting prior to break-up. The model calculations in Figure 3a, however, take no account of the horizontal heat conduction from extended to adjacent unextended lithosphere, which would reduce further the amount of melt generated.

The nature of the oceanic crust adjacent to the Galicia Bank margin provides further

44

evidence for significant conductive cooling of the upwelling asthenosphere during continental rifting prior to break-up. The first formed oceanic crust adjacent to Galicia Bank is anomalously thin, at only 3–4 km (Whitmarsh *et al.*, 1993; in prep.), but increases to a normal thickness within ~15 km of the ocean-continent transition (OCT; at 40 km on Figure 7a), after which it remains a constant thickness. The initially thin oceanic crust was

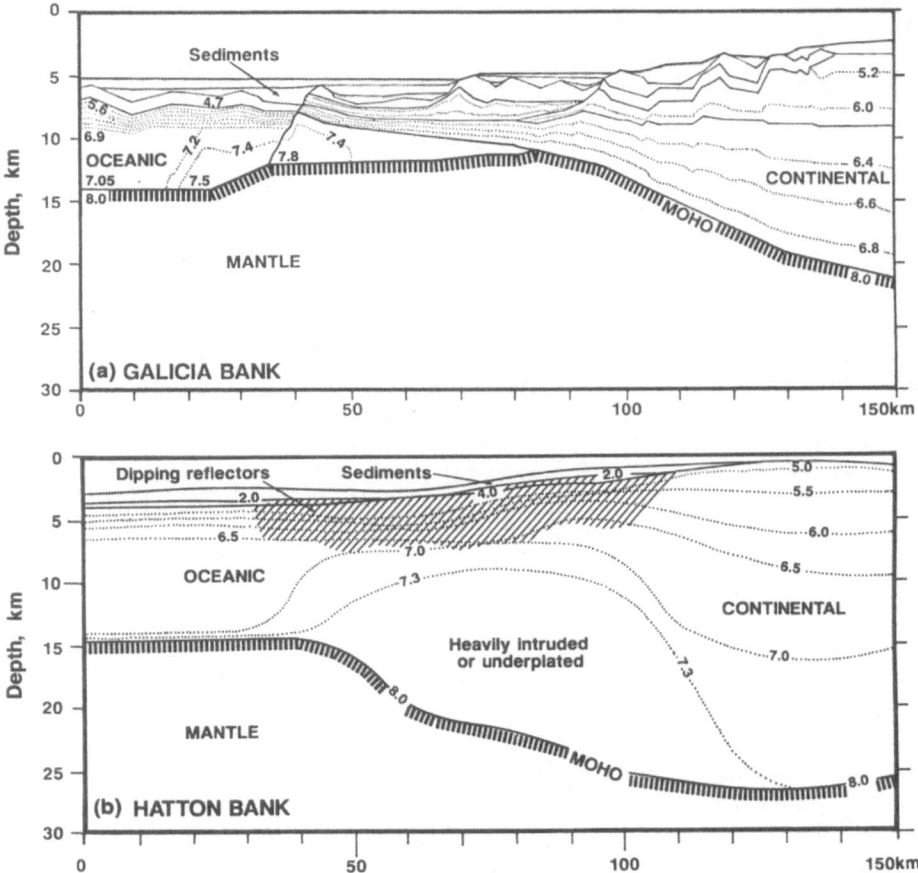

Figure 7: Cross sections at the same scale showing the deep structure from wide-angle seismic experiments and normal-incidence seismic reflection profiles across (a) the 'non-volcanic' Galicia Bank margin (oceanic structure between 0–40 km from Whitmarsh *et al.* (in prep.); continental structure between 40-150 km from Horsefield, 1992), and (b) the 'volcanic' Rockall Plateau margin (Fowler *et al.*, 1989; Morgan *et al.*, 1989; Spence *et al.*, 1989; White, 1992a). In both profiles dense stipple shows sediment. In (b) diagonal shading shows the extent of extrusive basalts generating seaward-dipping reflectors.

probably formed from mantle cooled significantly by conduction as it welled up beneath the slowly stretching continental lithosphere. Continued seafloor spreading then allowed fresh mantle with normal temperatures to well up beneath the spreading centre. The extreme thinness of the oceanic and continental crust in the vicinity of the Galicia Bank OCT is also indicated by the occurence near the OCT of a ridge of partially serpentinized mantle peridotite, which was emplaced towards the end of continental extension (Boillot *et al.*, 1989).

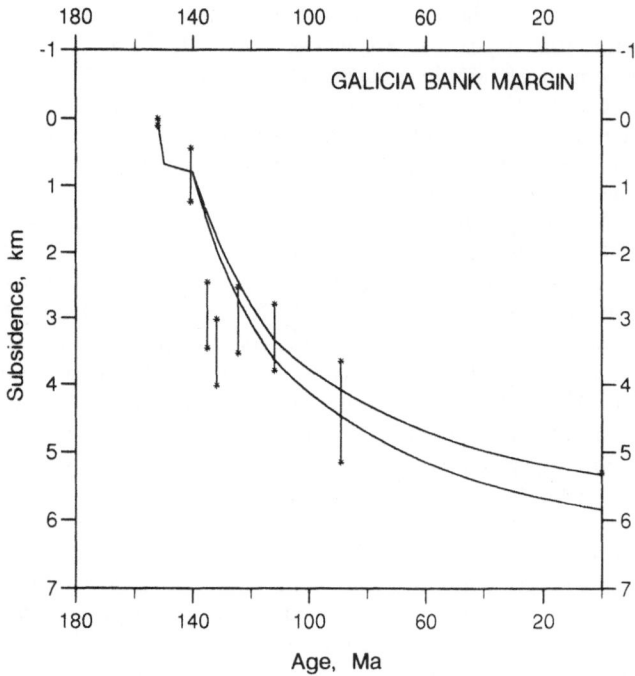

Figure 8: Model subsidence curves superimposed on water-loaded basement subsidence determined from the sediments drilled at ODP Sites 638, 639 and 641 on the Galicia Bank rifted margin (ODP Leg 103 Shipboard Scientific Party, 1987). The backscripping process employed to determine the water-loaded basement subsidence in the vicinity of the ODP sites is described in the text. Palaeobathymetry estimates are from Moullade *et al.* (1988). The vertical dimension of each bar is equal to the difference between Moullade *et al.*'s (1988) upper and lower estimates of palaeobathymetry. The theoretical subsidence curves are calculated from the uniform lithospheric stretching model outlined in the text. The continental lithosphere is initially 125 km thick and in thermal equilibrium with normal asthenosphere with a potential temperature of 1300°C. The lithosphere is extended by a β of 1.2 between 152–150 Ma, and by β of 3 or 4 during the main phase of rifting between 140–112 Ma. During both rift episodes extension occurs at a constant rate.

During the main phase of rifting the Galicia Bank margin subsided rapidly. The margin, however, remained relatively starved of sediment and deep water sediments were deposited soon after the start of rifting (Moullade et al., 1988). In order to compare the subsidence at the Galicia Bank margin with model subsidence curves it is necessary to determine the water-loaded subsidence history of the basement. Water-loaded subsidence of stretched continental crust near the Galicia Bank OCT was determined from the stratigraphic sequence drilled at Ocean Drilling Program (ODP) Sites 638, 639 and 641 (ODP Leg 103 Shipboard Scientific Party, 1987) and from the palaeobathymetry estimates of Moullade et al. (1988). This backstripping process is done by converting the sediment load at any time, allowing for compaction, into a water load and adding on the depth of water beneath which the sediments were deposited (Steckler & Watts, 1978; Sclater & Christie, 1980). Backstripping was done assuming local isostasy and using Sclater & Christie's (1980) and Schmoker & Halley's (1982) descriptions of how sediment density varies with depth for different sediment types. No corrections for eustatic changes in sea-level were applied, due to the uncertain nature of the correction (Wooler et al., 1992). Absolute ages within the stratigraphic sequence recorded at ODP Sites 638, 639 and 641 were determined using the timescale of Harland et al. (1990). The resulting water-loaded subsidence history is shown in Figure 8. Because the Galicia margin has been relatively starved of sediment the back-stripped water-loaded subsidence is governed largely by the estimates of palaeobathymetry, which are inherently uncertain. In the vicinity of ODP Sites 638, 639 and 641 it is, however, clear that rifting was accompanied by rapid water-loaded subsidence of 2–3 km (Figure 8).

The main phase of continental rifting lasted from the early Valanginian (141–138 Ma) to the late Aptian (118–112 Ma) (ODP Leg 103 Shipboard Scientific Party, 1987). The deposition of ~500 m of Tithonian (152–146 Ma) carbonates in the vicinity of ODP Sites 638, 639 and 641, however, probably reflects a small amount of continental extension earlier in the late Jurassic (Moullade et al., 1988). The basement subsidence during the late Jurassic can be accounted for by lithospheric stretching at the start of the Tithonian by a β of ~1.2 (Figure 8). The subsidence during the main phase of rifting can then be accounted for by further stretching by β of 3–4 (Figure 8). The large misfit between the model curves and the backstripped basement subsidence in the Hauterivian (135–132 Ma) is probably due to extension at a non-uniform rate and to error in Moullade et al.'s (1988) palaeobathymetry estimates. Stretching factors of 3–4 for the main phase of rifting are consistent with the β value of 3.4 estimated from the geometry of normal faulting (Moullade et al., 1988). The total β of 4–5 is consistent with the crustal thickness of 6–7 km beneath ODP Sites 638, 639 and 641, determined from a coincident wide-angle seismic profile (Figure 7a; Horsefield, 1992). Observations of basement subsidence, extension on normal faults and crustal thinning are, therefore, all consistent with the model of lithospheric extension.

4.2. MELTING AND SUBSIDENCE ON THE 'VOLCANIC' ROCKALL MARGIN

The 'volcanic' continental margin of Rockall Plateau formed during the early Tertiary when Greenland and northwest Europe rifted apart above the thermal anomaly surrounding the Iceland mantle plume (White & McKenzie, 1989 and in press; Hill, 1991; White, 1992b; Fram & Lesher, 1993). Igneous activity is manifest in the upper crust by thick sequences of convex-upwards seaward-dipping reflectors imaged on normal-incidence seismic reflection profiles, caused by basaltic lava flows (Roberts et al., 1984; White et al., 1987). Wide-angle seismic profiles indicate that huge volumes of new igneous material were also added to the lower crust (Figure 7b; Fowler *et al.*, 1989; Morgan *et al.*, 1989; Spence *et al.*, 1989). The new igneous material underplated beneath, or intruded into the lower continental crust beneath the seaward-dipping reflectors exhibits unusually high seismic velocities, typically greater than $7.3\,\mathrm{km\,s^{-1}}$. Such high velocities are consistent with the high MgO content of the melt generated by decompression of abnormally hot mantle (White & McKenzie, 1989). Lead isotope data from basalts drilled from the seaward-dipping reflector sequence suggest that the basalts were erupted through continental basement (Morten & Taylor, 1987; Merriman *et al.*, 1988). Furthermore, the first seafloor spreading magnetic anomaly adjacent to the western margin of Rockall Plateau (anomaly 24) is west of the seaward feather-edge of the dipping reflectors (Roberts *et al.*, 1984). These observations suggest that the bulk of the melt added to the crust at the Rockall Plateau margin was generated and emplaced during continental extension prior to break-up of the northern North Atlantic.

The total volume of melt added to continental margins can be estimated from seismic velocity models of the crust determined from wide-angle profiles. Due to several uncertainties such estimates may be in error by factors of up to two. The main uncertainty is in the ratio of igneous rock to pre-existing continental crust in the intrusive section, since the lower crust cannot be sampled directly. The second major uncertainty is that large volumes of melt may flow large distances away from, and along 'volcanic' rifts (White, 1992b). Bearing in mind these caveats we estimate the total cross-sectional area of melt along a profile across the Rockall Plateau margin (Figure 7b) to be $\sim900\,\mathrm{km^2}$. This cross-sectional area corresponds to an average thickness of $\sim10\,\mathrm{km}$ across the rift, which is approximately 90 km wide.

The bulk of the basalts associated with the break-up of the northern North Atlantic were erupted over a very short interval of only 2–3 Ma at most (Eldholm *et al.*, 1989; White & McKenzie, 1989). It is, however, impossible to know over what interval the intrusives in the lower continental crust were added. If rifting occurred within an interval of 3–5 Ma then our melting model reproduces the total amount of melt generated beneath the Rockall Plateau margin provided the potential temperature of the asthenosphere was about 1500°C (Figure 4). This temperature is some 200°C hotter than normal.

Oceanic crust formed immediately after break-up has a thickness of $11\pm0.5\,\mathrm{km}$ (Fowler *et al.*, 1989; Morgan *et al.*, 1989), which can be accounted for by seafloor spreading above asthenosphere with a potential temperature of 1360–1400°C (Bown & White, 1994). This temperature is only some 80°C hotter than normal. Hence, following continental break-up the temperature of the material supplied by the Iceland mantle plume to the base of the lithosphere appears to have dropped by about 100°C. Such a drop is consistent with the suggestion that the voluminous magmatism at the time of break-up was caused by the initiation of a new mantle plume supplying hotter material than the subsequent flow (White & McKenzie, 1989; Griffiths & Campbell, 1990; Loper, 1991). If, however, the rising central core of the Iceland plume lay beneath the rift, then the material welling up at the spreading centres on either side of the plume would have already been depleted considerably by partial melting within the rising core of the plume (White *et al.*, in press). The oceanic crust would then be thinner than expected for single-stage melting of asthenosphere with a potential temperature of 1500°C even though its temperature was still high, because the material welling up beneath the spreading centre had already undergone an earlier stage of melting.

Drilling on the Rockall Plateau margin has shown that the basaltic lava flows, constituting the seaward-dipping reflector sequence, were erupted in shallow-water marine or terrestrial

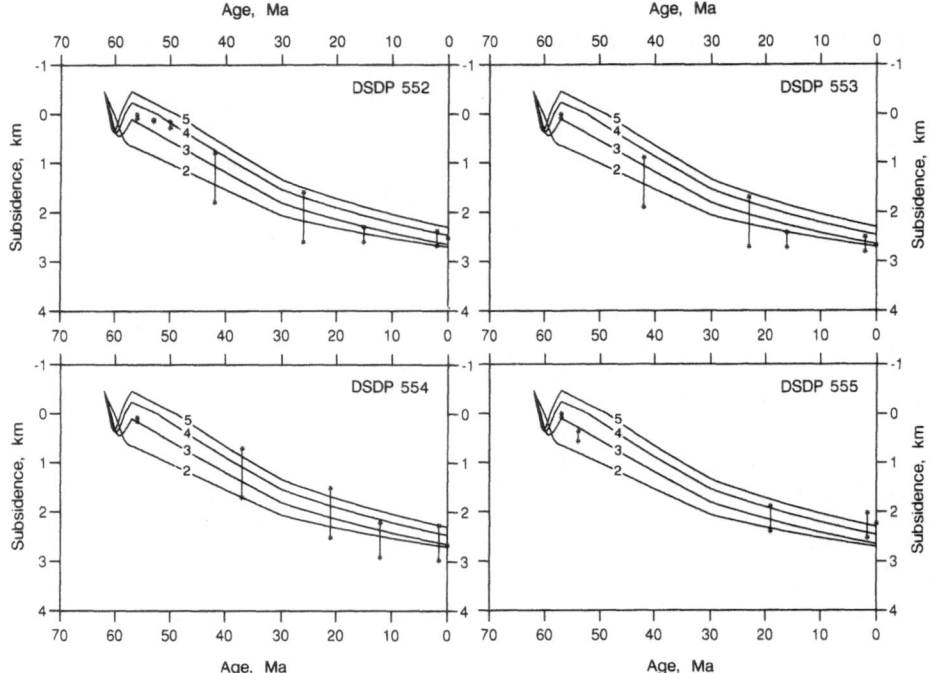

Figure 9: Theoretical subsidence curves superimposed on the water-loaded subsidence of the top of the basaltic lava flows at DSDP Sites 552–555 on the 'volcanic' rifted margin of Rockall Plateau. Backstripping was done in the same way as for the data from Galicia Bank. Palaeobathymetry estimates are from Murray (1984). The vertical dimension of each bar represents the difference between Murray's (1984) upper and lower estimates of the palaeobathymetry. The theoretical subsidence curves are calculated from the uniform lithospheric stretching model outlined in the text, for β of 2–5 (labelled on curves). Rifting lasts from 62–57 Ma and occurs at a constant strain rate. The continental lithosphere is initially 125 km thick and in thermal equilibrium with normal asthenosphere with a potential temperature of 1300°C. At the start of rifting a layer of hot plume-derived material 75 km thick and with a potential temperature of 1480°C is emplaced beneath the plate, causing uplift of its top by 0.5 km. The temperature in the plume-derived layer beneath the plate decays linearly to 1300°C by 30 Ma ago.

environments (Roberts, Schnitker, *et al.*, 1984). On the conjugate East Greenland margin, the plateau basalts generated during continental break-up flowed from offshore to onshore where they were intercalated with fluvial and lacustrine sediments (Larsen, 1984). The rift offshore must, therefore, have been elevated with respect to the hinterland when the basalts were extruded.

The water-loaded subsidence of the top of the basaltic lava flows was determined at Deep Sea Drilling Project (DSDP) Sites 552–555 on the Rockall Plateau margin (Roberts,

Schnitker, *et al.*, 1984), using the palaeobathymetry estimates of Murray (1984). Backstripping was done in the same way as for the ODP data from Galicia Bank. Like Galicia Bank, the Rockall Plateau rifted margin is relatively starved of sediment and the backstripped water-loaded subsidence is governed largely by the estimates of palaeobathymetry. Site 555 is located above the flat-lying basalts on the Plateau, Sites 552 and 553 are located above the seaward-dipping reflector sequence, and Site 554 is located above the outer high marking the ocearward termination of seaward-dipping reflectors (Roberts, Schnitker, *et al.*, 1984). It is, therefore, likely that the lithosphere was extended most beneath Site 554 and least beneath Site 555. The backstripped subsidence curves from all four sites are, however, remarkably similar (Figure 9). At all four sites the present day water-loaded subsidence of the top of the basalts is ~2.5 km.

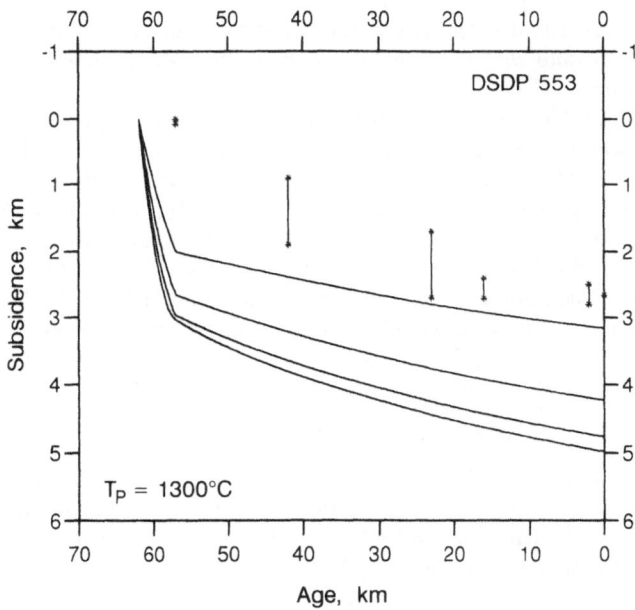

Figure 10: Theoretical subsidence curves superimposed on the water-loaded subsidence of the top of the basaltic lava flows at DSDP Site 553 on the 'volcanic' rifted margin of Rockall Plateau. The water-loaded subsidence of the top of the basalts is the same as shown in Figure 9b. The vertical dimension of each bar represents the difference between the upper and lower estimates of the palaeobathymetry. The theoretical subsidence curves are calculated from the uniform lithospheric stretching model outlined in the text, for β of 2–5 (labelled on curves) and for continental lithosphere that is initially 125 km thick and in thermal equilibrium with normal asthenosphere at a potential temperature of 1300°C. Rifting lasts from 62–57 Ma and occurs at a constant strain rate. Note that these theoretical subsidence curves of extension above normal potential temperature asthenosphere are unable to reproduce either the absolute depth, or the slope of the water-loaded subsidence determined from the backstripping.

The water-loaded subsidence curves from DSDP Sites 552–555 cannot be fitted adequately by model curves calculated for extension above asthenosphere of normal potential temperature (Figure 10). The model curves for normal potential temperature asthenosphere do not reproduce the start of thermal subsidence near sea-level of thermal subsidence, or the steep slope of the post-rift subsidence. The observed subsidence is, however, fitted well by rifting above a layer of plume-derived material 75 km thick with an initial potential temperature of 1480°C (Figure 9). The uplift caused by melt generation and by isostatic compensation of the hot plume-derived material accounts for the eruption close to sea-level of the lavas produced near the end of rifting. We account for the steep slope of the post-rift subsidence by the reduction in dynamic uplift caused by reducing the temperature anomaly of the plume layer back to that of normal asthenosphere between the start of rifting and 30 Ma ago (Figure 9). The post-rift subsidence is too fast to be accounted for purely by thermal relaxation of the plate. A decrease in the temperature of the asthenosphere beneath the plate is consistent with drift of the rift away from the sphere of influence of the Iceland mantle plume and with a gradual decrease in the heat and mass transported by the Iceland mantle plume.

This subsidence derived value of 1480°C for the potential temperature of the asthenospheric material welling up beneath the rift is indistinguishable from the value required to account for the total volume of melt generated during rifting. Furthermore, the model fits the observations well for a wide range of β (Figure 9). The similarity of the model subsidence curves for β of 2–5 arises from the generation of large amounts of melt during rifting, which when added to the crust, buffer its thickness. Rifting above anomalously hot asthenosphere can, therefore, account for the similar shape of the subsidence curves from regions across the margin which are likely to have been stretched by very different amounts.

5. Summary

Model results presented above showed that the amount of melt generated by decompression of anhydrous mantle welling up beneath regions of continental extension is strongly dependent on the duration of rifting, in addition to the degree of lithospheric thinning and to the temperature of the asthenospheric mantle. We have also demonstrated that the subsidence of regions of continental extension can be affected significantly by melt generation and by the dynamic support of mantle plumes.

Observations of melt generation, subsidence, crustal thinning and rift duration at the 'non-volcanic' Galicia Bank and the 'volcanic' Rockall Plateau rifted continental margins in the North Atlantic can be accounted for by our model for melting and subsidence at rifted margins. The sparsity of syn-rift magmatism at 'non-volcanic' margins is due to rifting over a prolonged period above asthenospheric mantle of normal potential temperature, whilst the huge volume of melt emplaced at 'volcanic' margins reflects rapid rifting above asthenospheric mantle some 200°C hotter than normal. This model also accounts for the observed rapid syn-rift subsidence at 'non-volcanic' margins, and the observation from 'volcanic' margins that the basaltic lava flows were deposited close to or above sea-level during rifting.

ACKNOWLEDGMENTS. We are grateful to Dan McKenzie and Nicky White for their advice, and to Norman Sleep and Rob Govers for helpful reviews. This work forms part of a wider study of magmatism associated with lithospheric rifting, supported in part by grants to RSW from the Natural Environment Research Council. JWB was supported in part by a studentship from Shell International Petroleum Co. Ltd.. Department of Earth Sciences, Cambridge, contribution number 3951.

References

Arndt, N.T., & U. Christensen, 1992. The role of lithospheric mantle in continental flood volcanism: Thermal and geochemical constraints, J. Geophys. Res., 97, 10967–10981.

Barton, P., & R. Wood, 1984. Tectonic evolution of the North Sea basin: crustal stretching and subsidence, Geophys. J. R. astr. Soc., 79, 987–1022.

Boillot, G., E.L. Winterer, A.W. Meyer, et al., 1987. Proc. Ocean Drill. Program, Init. Repts., 103, College Station, TX (Ocean Drilling Program).

Boillot, G., E.L. Winterer, et al., 1988. Proc. Ocean Drill. Program. Sci. Results, 103, College Station, TX (Ocean Drilling Program).

Boillot, G., G. Féraud, M. Recq, & J. Girardeau, 1989. Undercrusting by serpentinite beneath rifted margins, Nature, 341, 523–525.

Bown, J.W., 1994. Melting and subsidence at rifts, Ph.D. dissertation (unpublished), University of Cambridge, pp. 238.

Bown, J.W., & R.S. White, 1994. Variation with spreading rate of oceanic crustal thickness and geochemistry, Earth Planet. Sci. Lett., 121, 435–449.

Bown, J.W., & R.S. White, 1995. The effect of finite extension rate on melt generation at rifted continental margins, J. Geophys. Res., 100, in press.

Courtney, R.C., & R.S. White, 1986. Anomalous heat flow and geoid across the Cape Verde Rise: Evidence for dynamic support from a thermal plume in the mantle. Geophys. J. R. astr. Soc., 87, 815–867, and microfiche GJ 87/1.

Dane, E.D., 1941. Densities of molten rocks and minerals. Am. J. Sci., 239, 809–818.

Eldholm, O., J. Thiede, E. Taylor, et al., 1987. Proc. Ocean Drill. Program. Init. Repts., 104, College Station, TX (Ocean Drilling Program).

Eldholm, O., J. Thiede, E. Taylor, et al., 1989. Proc. Ocean Drill. Program. Sci. Results, 104, College Station, TX (Ocean Drilling Program).

Eldholm, O., J. Thiede & E. Taylor, 1989. Evolution of the Vøring volcanic margin, Proc. Ocean Drill. Program, Sci. Results, 104, 1033–1065.

Fowler, S., & D. McKenzie, 1989. Gravity studies of the Rockall and Exmouth Plateaux using SEASAT altimetry, Basin Research, 2, 27–34.

Fowler, S.R., R.S. White, G.D. Spence & G.K. Westbrook, 1989. The Hatton Bank continental margin, II, Deep structure from two-ship expanding spread profiles. Geophys. J., 96, 295–309.

Fram, M.S, & C.E. Lesher, 1993. Geochemical constraints on mantle melting during creation of the North Atlantic basin, Nature, 363, 712–715.

Fukuyama, H., 1985. Heat of fusion of basaltic magma. Earth Planet. Sci. Lett., 73, 407–414.

Griffiths, R.W., & I.H. Campbell, 1990. Stirring and structure in mantle starting plumes, Earth Planet. Sci. Lett., 99, 66–78.

Harland, W.B., R.L. Armstrong, A.V. Cox, L. Craig, A.G. Smith & D.G. Smith, 1990. A Geological Timescale 1989, Cambridge University Press, Cambridge.

Herzberg, C.T., W.S. Fyfe & M.J. Carr, 1983. Density constraints on the formation of the continental Moho and crust, Contr. Miner. Petrol., 84, 1–5.

Hill, R.I., 1991. Starting plumes and continental break-up, Earth Planet. Sci. Lett., 104, 398–416.

Horsefield, S.J., 1992. Crustal structure across the continent-ocean boundary, Ph.D. dissertation (unpublished), University of Cambridge, pp. 215.

Jarvis, G.T., & D.P. McKenzie, 1980. Sedimentary basin formation with finite extension rates, Earth Planet. Sci. Lett., 48, 42–52.

Klein, E.M., & C.H. Langmuir, 1987. Global correlations of ocean ridge basalt chemistry with axial depth and crustal thickness, J. Geophys. Res., 92, 8089–8115.

Larsen, H.C., 1984. Geology of the East Greenland shelf, in: Petroleum Geology of the North European Margin, Spencer, A.M., et al., eds., pp. 329–339, Graham & Trotman.

Le Pichon, X., & J.-C. Sibuet, 1981. Passive margins: a model of formation, J. Geophys. Res., 86, 3708–3720.

Loper, D.E., 1991. Mantle plumes, Tectonophys., 187, 373–384.

McKenzie, D., 1978. Some remarks on the development of sedimentary basins, Earth Planet. Sci. Lett., 40, 25–32.

McKenzie, D., 1984. The generation and compaction of partially molten rock, J. Petrology, 25, 713–65.

McKenzie, D., 1985. The extraction of magma from the crust and mantle, Earth Planet. Sci. Lett., 74, 81–91.

McKenzie, D., 1989. Some remarks on the movement of small melt fraction in the mantle, Earth Planet. Sci. Lett., 95, 53–72.

McKenzie, D., & M.J. Bickle, 1988. The volume and composition of melt generated by extension of the lithosphere, J. Petrology, 29, 625–679.

Merriman, R.J., P.N. Taylor & A.C. Morton, 1988. Petrochemistry and isotope geochemistry of early Palaeogene basalts forming the dipping reflector sequences SW of Rockall Plateau, NE Atlantic, in: Early Tertiary Volcanism and the Opening of the NE Atlantic, Spec. Publ. Geol. Soc. Lond., 39, Morton, A.C., & L.M. Parson, eds., pp. 123–134.

Morgan, J.V., P.J. Barton & R.S. White, 1989. The Hatton Bank continental margin, III, Structure from wide-angle OBS and multichannel seismic refraction profiles, Geophys. J. Int., 98, 367–384.

Morton, A.C., & P.N. Taylor, 1987. Lead isotope evidence for the structure of the Rockall dipping-reflector passive margin, Nature, 326, 381–383.

Moullade, M., M.-F. Brunet & G. Boillot, 1988. Subsidence and deepening of the Galicia margin: The paleoenvironmental control, Proc. Ocean Drill. Program, Sci. Results, 103.

Murray, J.W., 1984. Paleogene and Neogene benthic foraminiferas from Rockall Plateau, Initial Rep. Deep Sea Drill. Proj., 81, 503–534.

ODP Leg 103 Shipboard Scientific Party, 1987. Introduction, objectives, and principal results: ocean drilling program Leg 103, West Galicia margin, Proc. Ocean Drill. Program, Init. Repts., 103, 3–17.

ODP Leg 152 Shipboard Party, 1994. Drilling unearths "Fire and Ice" at Southeast Greenland Margin, EOS Trans. Am. Geophys. Un., 75, 401–406.

Oxburgh, E.R., & E.M. Parmentier, 1977. Compositional and density stratification in the oceanic lithosphere – causes and consequences, J. Geol. Soc. Lond., 133, 343–354.

Parsons, B., & J.G. Sclater, 1977. An analysis of the variation of ocean floor bathymetry and heat flow with age, J. Geophys. Res., 82, 803–827.

Pedersen, T., & H.E. Ro, 1992. Finite duration extension and decompression melting, Earth Planet. Sci. Lett., 113, 15–22.

Ringwood, A.E., 1975. Composition and Petrology of the Earth's Mantle. McGraw-Hill, New York, N.Y..

Roberts, D.G., D. Schnitker, et al., 1984. Initial Rep. Deep Sea Drill. Proj., 81. Washington (U.S. Government Printing Office), 1984.

Roberts, D.G., J. Backman, A.C. Morton, J.W. Murray & J.B. Keene, 1984. Evolution of volcanic rifted margins: Synthesis of Leg 81 results on the west margin of Rockall Plateau, Initial Rep. Deep Sea Drill. Proj., 81, 913–923.

Royden, L., & C.E. Keen, 1980. Rifting process and thermal evolution of the continental margin of eastern Canada determined from subsidence data. Earth Planet. Sci. Lett., 51, 343–361.

Schmoker, J.W., & R.B. Halley, 1982. Carbonate porosity versus depth. Am. Assoc. Petroleum Geologists Bull., 66, 2561–2570.

Sleep, N.H., 1994. Lithospheric thinning by midplate mantle plumes and the thermal history of hot plume material ponded at sublithospheric depths. J. Geophys. Res., 99, 9327–9343.

Sclater, J.G., & P.A.F. Christie, 1980. Continental stretching: An explanation of the post-mid-Cretaceous subsidence of the central North Sea Basin. J. Geophys. Res., 85, 3711–3739.

Spence, G.D., R.S. White, G.K. Westbrook & S.R. Fowler, 1989. The Hatton Bank continental margin, I, Shallow structure from two-ship expanding spread seismic profiles, Geophys. J., 96, 273–294.

Steckler, M.S., & A.B. Watts, 1978. Subsidence of the Atlantic-type continental margin off New York. Earth Planet. Sci. Lett., 41, 1–13.

von Herzen, R.P., R.S. Detrick, S.T. Crough, D. Epp & U. Fehn, 1982. Thermal origin of the Hawaiian swell: Heat flow evidence and thermal models. J. Geophys. Res., 87, 6711–6723.

Watson, S., & D. McKenzie, 1991. Melt generation by plumes: A study of Hawaiian volcanism. J. Petrology, 32, 501–537.

Watts, A.B., 1988. Gravity anomalies, crustal structure and flexure of the lithosphere at the Baltimore Canyon Trough, Earth Planet. Sci. Lett., 89, 21–238.

Watts, A.B., & M. Torné, 1992. Crustal structure and the mechanical properties of extended continental lithosphere in the Valencia trough (western Mediterranean). J. Geol. Soc. Lond., 149, 813–827.

White, N., 1991. Does the uniform stretching model work in the North Sea?, in: The Structure and Evolution of the North Sea, Blundell, D.J., & A.D. Gibbs, eds., Oxford Scientific Publications, Clarendon Press, Oxford, 217–235.

White, R.S., 1992a. Crustal structure and magmatism of North Atlantic continental margins, J. Geol. Soc. Lond., 149, 841–854.

White, R.S., 1992b. Magmatism during and after continental break-up, in: Magmatism and the Causes of Continental Break-up, Spec. Publ. Geol. Soc. Lond., 68, Storey, B.C., T. Alabaster & R.J. Pankhurst, eds., pp. 1–16.

White, R.S., 1993. Melt production rates in mantle plumes. Phil. Trans. Roy. Soc., London, Series A, 342, 137–153.

White, R., & D. McKenzie, 1989. Magmatism at rift zones: The generation of volcanic continental margins and flood basalts, J. Geophys. Res., 94, 7685–7729.

White, R.S., & D. McKenzie, 1995. Mantle plumes and flood basalts, J. Geophys. Res., 100, in press.

White, R.S, J.W. Bown & J.R. Smallwood, 1995. The shape of the Iceland mantle plume, J. Geol. Soc. Lond., in press.

White, R.S, G.K. Westbrook, A.N. Bowen, S.R. Fowler, G.D. Spence, C. Prescott, P.J. Barton, M. Joppen, J. Morgan & M.H.P. Bott, 1987. Hatton Bank (northwest U.K.) continental margin structure, Geophys. J. R. astr. Soc., 89, 265–272.

Whitmarsh, R.B., L.M. Pinheiro, P.R. Miles, M. Recq & J-.C. Sibuet, 1993. Thin crust at the western Iberia ocean-continent transition and ophiolites, Tectonics, 12, 1230–1239.

Whitmarsh, R.B., R.S. White, J.-C. Sibuet, S.J. Horsefield & M. Recq, in prep.. The ocean-continent boundary off the western continental margin of Iberia – III. Crustal structure across the Galicia Bank Margin, Geophys. J. Int..

Wooler, D.A., A. Smith & N. White, 1992. Measuring lithospheric stretching on the margins of Tethys, J. Geol. Soc. Lond., 149, 517–532.

STRESSES ASSOCIATED WITH CONTINENTAL BREAK-UP

M. H. P. BOTT
Department of Geological Sciences
University of Durham
South Road
Durham DH1 3LE
United Kingdom

ABSTRACT. The structural setting of passive margins indicates major extensional stress during the rifting stage. Local gravitational loading stresses which develop as the lithosphere thins are not large enough to have much influence on the tectonic development. A source of sufficiently large renewable extensional stress to cause lithospheric failure is the hot, low-density upper mantle of the type associated with a plume. Widespread extensional stress may also result from subduction on opposite sides of a large continental plate. These two sources of extensional stress may combine to initiate and cause break-up. The abrupt change to a more subdued stress regime at the onset of seafloor spreading can be explained by abrupt weakening of the lithosphere at the newly forming plate boundary.

1. Introduction

Two inferences about the state of stress during passive margin development can be drawn from the observed structure. First, extreme stretching of the continental crust during the rift stage indicates persistent horizontal extensional stress large enough to cause failure and finite stretching of the continental lithosphere. Second, the cessation of horizontal tectonics at the time of the post-rift unconformity indicates an abrupt change to a more subdued stress system during post-rift development of the margin.

Worldwide observations of stress (Zoback, 1992) indicate that present-day low-lying continental regions are subjected to thrust or strike-slip stress regimes, rather than to tension. This suggests that normal continental regions are not now extensionally unstable. Present-day continental regions displaying active extension are mainly restricted to regions of high topography which are isostatically supported by anomalously low density upper mantle (hot spot or thinned lithosphere) or by thickened crust.

This paper briefly discusses possible sources of stress affecting the continental lithosphere at the time of break-up. A fuller account has been given by Bott (1992) where technical details of the finite element modelling, omitted from this account, are given together with a fuller list of references. The models shown in Figs 1 to 3 are presented in a different form from the earlier paper, and it is newly demonstrated that tension due to an upper mantle hot spot can cause whole lithospheric failure.

E. Banda et al. (eds.), Rifted Ocean-Continent Boundaries, 55–63.
© 1995 *Kluwer Academic Publishers. Printed in the Netherlands.*

2. Stress in the lithosphere

Several types of stress system affect the lithosphere. These include stress due to plate boundary forces, thermal stress, membrane stress, and stresses due to gravitational loading which may include lithostatic stress, bending stress and buoyancy stress. However, bending stress, thermal stress and membrane stress cannot be renewed on a short enough time scale to give rise to ongoing finite deformation such as occurs during the rifting stage at margins. The only two types of stress system obviously capable of causing such finite deformation are buoyancy stresses and stresses due to plate boundary forces.

Gravitational loading normally involves both surface topographic loading and subsurface loading in isostatic balance with each other. Where the surface and subsurface loading mirror-image each other, as in simple local Airy isostasy, then buoyancy stresses occur but there are no bending stresses. The differential stress in a vertical plane is approximately equal to the load, with anomalously low subsurface densities accompanied by elevated topography producing horizontal differential extensional stress, and an anomalously dense substratum accompanied by topographical depression producing compression. When there is asymmetry between surface and subsurface loading, as in flexural isostasy, then both bending and buoyancy stresses occur. Buoyancy stresses, which concentrate in the strong regions of the lithosphere, have the important property that they occur whether the causitive subsurface load is within the lithosphere or within deeper viscous parts of the underlying weaker mantle. Furthermore for loads wide in comparison to their depth, the deeper the subsurface load, the larger the resulting buoyancy stresses in the overlying strong parts of the lithosphere (Bott, 1991). For this reason, density anomalies of thermal origin in the upper mantle are a powerful source of lithospheric stress.

Plate boundary forces are also indirectly caused by the anomalous density structure beneath the plate boundary, such as the dense subducting slab or low density upwelling asthenosphere beneath ocean ridges. The resulting buoyancy stress system is intersected by a zone or plane of weakness which cuts right across the strong lithosphere at the plate boundary. In order to minimize the stress differences in the weak region, supplementary boundary tractions act on edge of the adjacent plates and cause redistribution of stress. This has the effect of strongly reducing the deviatoric stresses in the vicinity of the plate boundary and at the same time producing deviatoric stress of opposite polarity within the interiors of the adjacent plates. Thus the plate boundary forces and the associated plate interior stresses originate from buoyancy forces (Bott, 1991).

The conditions under which whole failure of the continental lithosphere can occur have been investigated by Kusznir and Park (1987) for realistic profiles of rheology. The strength of the lithosphere is best expressed in terms of the tectonic force required to cause such failure. This is the difference between the horizontal and vertical principal stresses integrated with respect to depth across the lithosphere. Kusznir and Park (1987) estimated that a tensile tectonic force of about 3×10^{12} N/m would cause the continental lithosphere to fail where the heat flow is above average.

3. Local buoyancy stresses produced by lithospheric thinning

As the lithosphere is stretched during the rifting stage, the crust thins producing a region of anomalously high density with negative buoyancy. This gives rise to compressive buoyancy

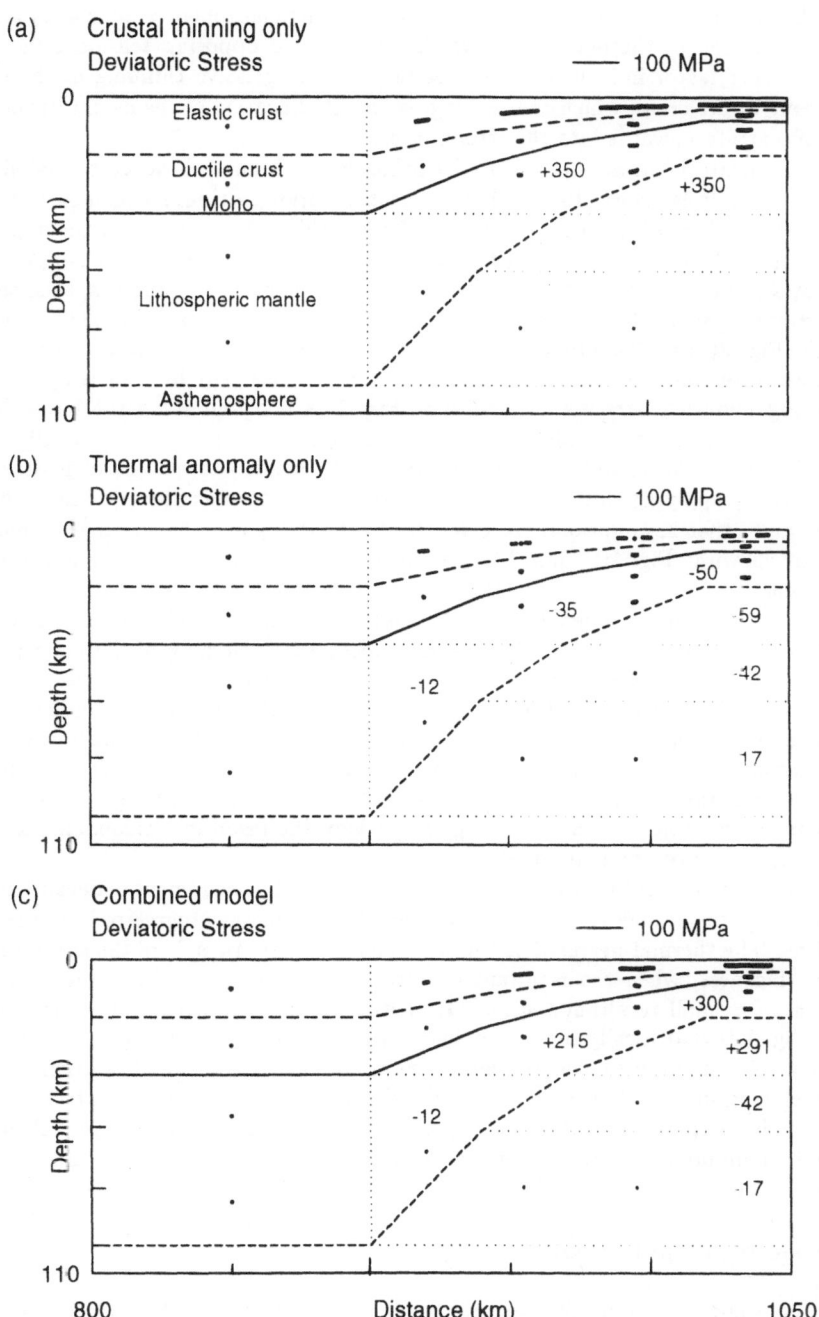

Fig. 1. Models showing deviatoric principal stresses (solid line =compression, broken line with dot at centre =tension) for (a) crustal thinning, (b) thermal anomaly due to lithospheric thinning, and (c) both crustal thinning and thermal effects. The models are all symmetrical about the right edge and only the axial region down to 110 km is shown. Selected values of anomalous density due to crustal thinning and thermal effects are shown in kg/m³. Surface loading appropriate to local Airy isostasy is incorporated but not shown.

stress in the strong regions. In contrast, the upwelling asthenosphere produces an anomalous buoyant region of thermal origin, which gives rise to opposing tensional buoyancy stress. A further factor influencing the stresses is the progressive thinning of the strong stress-bearing upper crust, which causes progressive amplification of the deviatoric stresses as they concentrate upwards into the strong layer.

The local deviatoric stresses produced by lithospheric thinning have been modelled by finite element analysis (Bott, 1992). The lithosphere prior to thinning is assumed to be 100 km thick, with a 40 km thick crust and 20 km thick elastic layer at the top. The lithosphere is stretched by a factor of five at the maximum. This relatively simple model demonstrates principles which should carry over to more complicated models, such as with a second strong layer just below the Moho. The models have been run for 1000 time steps of 500 y each (Fig. 3(c) for 2500 time steps) to approach viscoelastic relaxation during which stress concentrates into the elastic layer and flexural isostatic equilibrium is approached. The resulting buoyancy stresses are shown in Fig. 1, which represents the left half of a symmetrical model which extends for 2100 km horizontally and 400 km vertically. Only the central part of the model extending down to 110 km is shown. Local isostasy has been assumed by applying surface loads equal and opposite to the subsurface load. Elastic properties and other assumptions are given by Bott (1992) but the assumed anomalous densities are shown in Figs 1(a) and (b). Zero deviatoric stress in the out-of-plane direction has been assumed.

Fig. 1(a) isolates the buoyancy stresses due to crustal thinning. The anomalously dense region of upper mantle resulting from a fivefold maximum stretching of the crust and the associated surface depression, represented by a negative surface load, give rise to a horizontal deviatoric compression which reaches a maximum value of 196 MPa in the thinned region of maximum thinning of the elastic layer. Fig. 1(b) shows the isolated effect of the hot, low density thermal anomaly caused by the lithospheric thinning. This produces a maximum horizontal deviatoric tension of about 98 MPa in the elastic layer. These two effects oppose each other, as shown in Fig. 1(c) where the resultant maximum horizontal deviatoric stress is a compression of 98 MPa.

The combined model of Fig. 1(c) thus shows that the local development of density anomalies produces a horizontal deviatoric compression which increases progressively as the lithosphere thins. The thermal perturbation may be larger or the lithosphere thicker in relation to crustal thickness, when the resultant compressional stress would be reduced or there might even be a small resultant tension. However, the main conclusion from the models shown in Fig. 1 is that locally developed buoyancy stresses as the lithosphere is stretched are unlikely to contribute to the stretching process and probably would normally tend to oppose stretching in a small way. These locally developed stresses produce a tectonic force which is small compared to the value required to cause lithospheric failure. They thus appear to be unimportant in the tectonic development of the margin, and we must seek the main cause elsewhere.

4. Tension from upper mantle hot spot

A prominent source of strong lithospheric tension is provided by a low density region in the upper mantle resulting from anomalously high temperatures (hot spot) such as may arise from a mantle plume. Such a region gives rise to isostatic uplift at the surface and to large near-horizontal deviatoric tension. The prominent occurrence of extensional stress

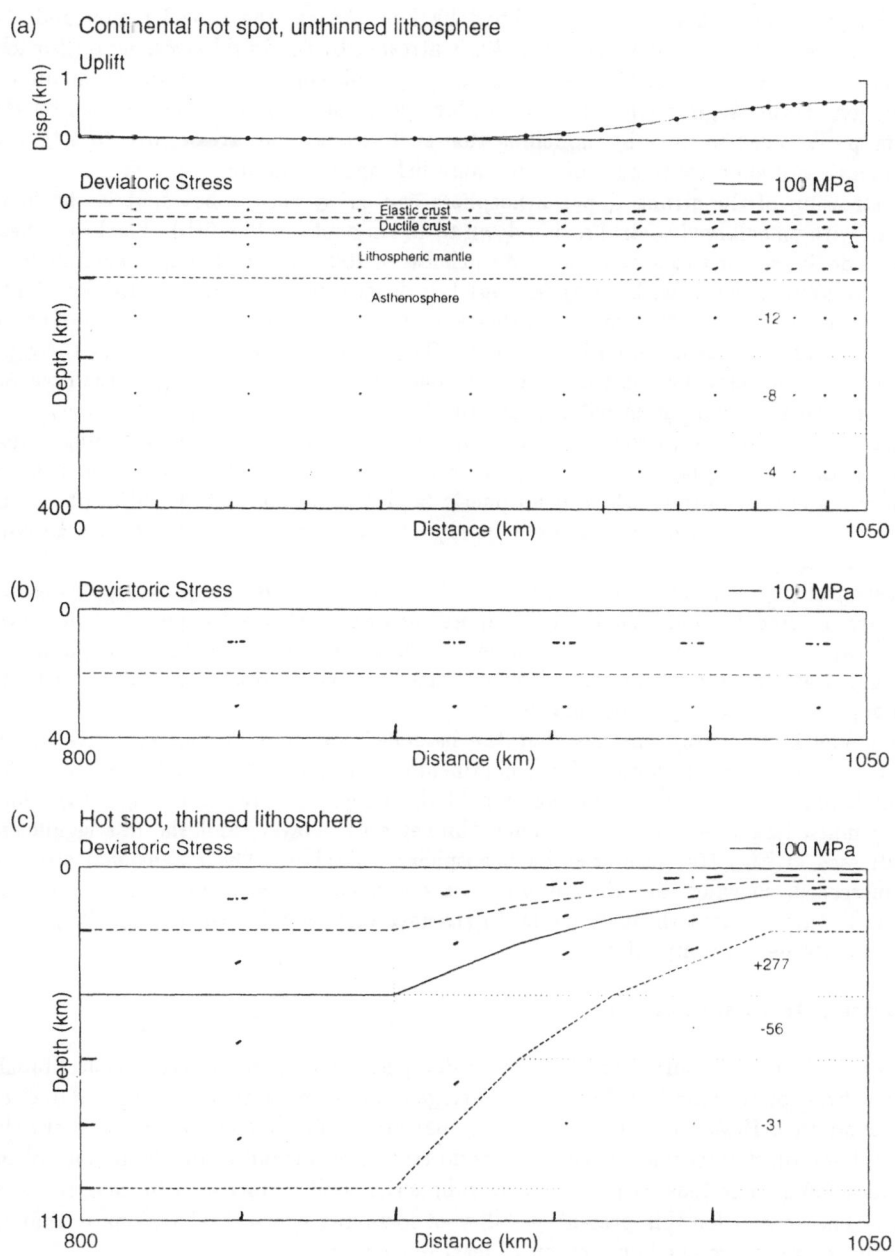

Fig. 2. Deviatoric principal stresses produced by a sub-lithospheric hot spot: (a) shows stresses due to the anomalous density field in kg/m³ for unthinned lithosphere, with flexural isostatic profile above; (b) is an enlargement of the axial region of the symmetrical model; (c) shows the effect on the hot spot stresses of the lithospheric thinning as shown in Fig. 1(c).

(normal regime) in such regions, exemplified by East Africa, the Baikal region and the Basin and Range Province of western U.S.A., is attested by the world stress map (Zoback, 1992) and in the development of extensional faulting. This contrasts with the predominant compression (thrust regime) characteristic of low lying continental regions as seen on the stress map. We need to show by modelling that such extensional stresses due to a typical upper mantle hot spot are large enough to cause lithospheric failure.

The stressing of the strong upper lithosphere caused by such a hot spot is shown in Fig. 2(a), with an enlarged central region down to 40 km depth in Fig. 2(b). The lithosphere in this model is assumed to have a uniform thickness of 100 km. The hot spot extends from 100 km to 400 km depth with an anomalous low density of -12 kg/m^3 at 100 km depth decreasing to -4 kg/m^3 at 400 km depth, corresponding to a temperature anomaly of about 120 K at 100 km decreasing to 40 K at 400 km. The model is symmetrical about the right edge, the hot spot being 300 km wide. The hot, buoyant region and the associated isostatic uplift of the surface (maximum 650 m) give rise to a maximum near-horizontal deviatoric principal stress of 75 MPa above the centre of the hot spot. This stress extending across the 20 km thick strong layer is equivalent to a tectonic force of 3.0×10^{12} N/m. This is just large enough to initiate whole lithospheric failure provided that the lithosphere has above average temperatures, which might be expected as a result of early igneous activity produced in the plume head.

Continental plateaus such as East Africa and the Basin-and-Range province of western U.S.A. are uplifted by 2 km, suggesting a more substantial thermal anomaly than shown in Fig. 2(a) and corresponding larger tensions. Thus the observed extensional failure in such regions would be expected to arise from the buoyancy stresses associated with the underlying hot, low density upper mantle.

The model in Fig. 2(c) assumes that the modelled hot spot of Fig. 2(a) has caused extreme stretching and thinning of the continental lithosphere by a factor of five. The stresses displayed result from superposition of the buoyancy stresses of Fig. 2(a) which are now much larger because of the much thinner strong layer, and the less significant local stresses of Fig. 1(c) produced by lithospheric thinning. The extensional stress is much increased, showing that the thinning process, once started, should accelerate until complete in this constant rheology model (if strength increases due to cooling, the process may terminate before completion).

5. Tension from subduction pull

The only other obvious source of renewable tension possibly able to cause continental break-up arises from plate boundary forces at convergent plate margins, especially subduction plate boundaries. However, in the present-day pattern of plates, compressive or strike slip stress regimes affect most unelevated continental regions as a result of the dominance of the ridge push force. The tension produced by subduction pull is only seen in local back-arc regions. The reason for this present stability of low-lying continental regions is that all major plates are at present bounded on at least one side by an ocean ridge.

The disposition of plates may have been more favourable to tensional regimes at certain periods in the past when subduction was occurring on opposite sides of a large essentially continental plate such as Pangaea prior to Mesozoic break-up. In this situation, the subduction pull acting on opposite sides of the supercontinental region may overcome the dominance of ridge push to cause plate-wide extensional stress. Subduction pull occurs

61

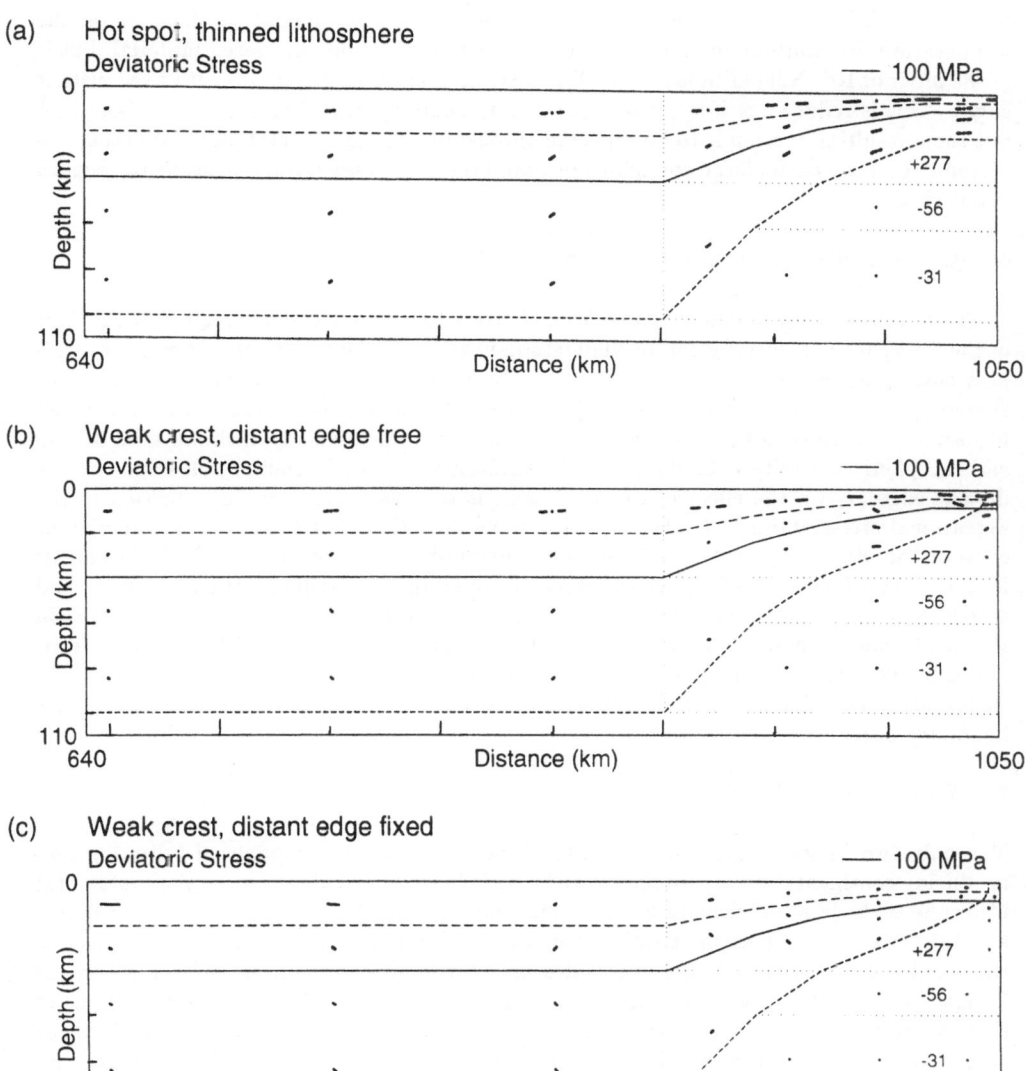

Fig. 3. Models shown the effect of weakening the crestal region: (a) shows the unweakened model of Fig. 2(c); (b) shows the effect of weakening the crestal region to the surface with free boundary conditions at the distant edge; and (c) shows the same model with the distant edge of the elastic layer constrained to zero horizontal displacement.

when the subduction fault is unlocked, with low friction. Its magnitude depends on the temperature distribution and length of the subduction slab and associated material, but is of the order of 10^{13} N/m as modelled by Whitaker et al.(1992) using plates with fixed distant edges. This is reduced as a result of resistance to subduction and transmitted ridge push, but some resulting tension is to be expected throughout the continental plate. Whether the tectonic force would be large enough to initiate break-up in unelevated continental regions is not clear.

6.   Change of stress regime at break-up

The abrupt change from strong extensional stress to a subdued stress regime at the time of the post rift unconformity can be understood in terms of the formation of a zone of weakness cutting across the lithosphere at the newly developed plate boundary where seafloor spreading starts. This is modelled in Fig. 3. Fig. 3(a) shows the model of strong thinned lithosphere identical to Fig. 2(c). Fig. 3(b) shows a modification in which material having asthenospheric viscosity extends up to the surface, with free boundary conditions at the distant edge (x=0); the effect of the weak zone is to cause a substantial reduction in the extensional stress. Model 3(c) has the further modification that the nodes on the distant edge of the elastic layer at x=0 have been constrained to zero horizontal displacement; here the extensional stresses in the vicinity of the margin are reduced almost to zero, and a compressional stress occurs within the adjacent plate interior. In reality, the model in Fig. 3(b) underestimates the resistance at the distant edge, and that in Fig. 3(c) overestimates it. The true situation lies between these two extremes. The main point is to explain theoretically the observation that a substantial drop in tension occurs when the new plate boundary becomes active.

7.   Discussion and conclusions

The only two known sources of renewable lithospheric tension capable of initiating and producing continental break- up appear to be as follows: (1) local tension in the continental lithosphere above a low density upper mantle hot spot of the type produced by a plume; and (2) plate interior tension resulting from subduction pull which would be most effective when subduction occurs on opposite sides of a predominantly continental plate such as Pangaea prior to break-up. Whether or not break-up can occur just in response to one of these sources of tension is unclear. In the present day compressional stress regime, it is not obvious whether a crack developing in an uplifted hot spot region such as East Africa could propagate laterally through the normal continental lithosphere. Furthermore, subduction pull by itself might not develop a large enough tectonic force to initiate break-up.

It is suggested that the most favourable scenario for continental break-up occurs when both sources of tension occur contemporaneously. Thus a continental interior region may be extensionally unstable as a result of subduction pull. Break-up could then be initiated if a plume produces a hot spot beneath such lithosphere, and it could readily propagate laterally to produce a new plate boundary. Magma arising from the hot spot may help to weaken the lithosphere. Crustal lines of weakness may control the geometry of break-up. When the lithosphere weakens abruptly at the onset of seafloor spreading, the stress regime becomes subdued in response.

8. References

Bott, M. H. P., 1991. Sublithospheric loading and plate-boundary forces. *Phil. Trans. R. Soc. Lond.* A, **337**, 83-93.

Bott, M. H. P., 1992. The stress regime associated with continental break-up. In Storey, B. C., Alabaster, T. and Pankhurst, R. J. (eds), *Magmatism and the Causes of Continental Break-up*, Geological Society Special Publication No. 68, 125-136.

Kusznir, N. J. and Park, R. G., 1987. The extensional strength of the continental lithosphere: its dependence on geothermal gradient, and crustal composition and thickness. In Coward, M. P., Dewey, J. F. and Hancock, P. L. (eds), *Continental Extensional Tectonics*, Geological Society Special Publication No. 28, 35-52.

Whittaker, A., Bott, M. H. P. and Waghorn, G. D., 1992. Stresses and plate boundary forces associated with subduction plate margins. *J. Geophys. Res.*, **97**, 11,933-11,944.

Zoback, M. L., 1992. First- and second-order patterns of stress in the lithosphere: the world stress map project. *J. Geophys. Res.*, **97**, 11,703-11,728.

GRAVITY ANOMALIES AND THE THERMAL AND MECHANICAL STRUCTURE OF RIFTED CONTINENTAL MARGINS.

A. B. WATTS and C. MARR,
Department of Earth Sciences,
University of Oxford,
Parks Road,
Oxford OX1 3PR,
U.K.

ABSTRACT. Simple models for the free-air gravity "edge effect" anomaly at rifted continental margins have been constructed in which the lithosphere responds to sediment loading by flexure. Two main types of edge effect have been recognised: a long-wavelength, high-amplitude "single" associated with high rigidity, strong, margins and a short-wavelength, low-amplitude "double" associated with low rigidity, weak, margins. Within the weak margin category two additional types have been distinguished: the onshore and offshore dipping "doubles" which indicate relatively strong and weak lithosphere respectively. A "case history" study of the onshore dipping double at the East Coast, USA margin, where the deep crustal structure is constrained by seismic refraction data, suggests that this type of edge effect is caused by sediment loading of stretched continental lithosphere which is mechanically weaker than adjacent oceanic lithosphere. Why stretched lithosphere is weak is enigmatic but, the lack of evidence of any hot-spot activity suggests that it is caused by rheological effects related to crustal composition, rather than by re-heating of the lithosphere following rifting. Despite a region of weak lithosphere, there has been a substantial tectonic subsidence at the margin which is difficult to explain if the crust has been thickened by magmatic material on its surface or at its base. Application of the simple models to double and single edge effects along the African margin suggest that it is highly segmented as regards its strength, unlike the East Coast, USA margin. The margin offshore S. Morocco, Somalia, Kenya, S. Mozambique, SE and SW Africa, Namibia, Gabon and Nigeria are examples of weak margins whereas those offshore Liberia, Senegal, Sierra-Leone, N. Mozambique and S. Angola are strong. Weak margins correlate with regions that have been influenced by hot-spot activity, the weakest examples (i.e. offshore dipping double) correspond to segments where hot-spot traces are parallel to the local trend of the margin, the stronger ones (onshore dipping doubles) where traces are at a high angle to it. The strongest margins (single) occur in regions that have not been influenced by any significant hot-spot activity. Thermal re-heating of the lithosphere by mantle hot-spots may therefore be a contributing factor to the apparent weakness of stretched continental lithosphere.

1. Introduction

The free-air gravity anomaly or "edge effect" associated with rifted continental margins is one of the most distinctive features of the oceanic gravity field. Previous studies attributed it to the gravity effect of the thinning of the crust from continental to oceanic regions. Worzel (1968), for example, used an Airy model to predict the geometry of the thinning and showed how the edge effect would change if the ocean/continent transition was moved landward or seaward of the shelf break. Talwani and Eldholm (1973) used the Airy model to correct the free-air anomaly for the gravity effect of the transition and showed that many margins are characterised by an "outer high" which has a steep gradient on the landward side and a "tail" on the oceanward side. Rabinowitz and LaBrecque (1977) showed that the high persisted even

65

E. Banda et al. (eds.), Rifted Ocean-Continent Boundaries, 65–94.
© 1995 Kluwer Academic Publishers. Printed in the Netherlands.

when the sediments at margins were also compensated according to the Airy model. Both Talwani and Eldholm (1973) and Rabinowitz and LaBrecque (1977) interpreted the outer high as the result of density differences in the crust and used it - along with other geophysical observations such as magnetic anomaly patterns and P-wave velocity-depth data - as one criteria for locating the boundary between the continental and oceanic crust.

The sediments which accumulate at a rifted margin represent a load on the surface of the lithosphere which should flex under their weight (Watts and Ryan, 1976). The Airy model, which assumes that the lithosphere deforms locally rather than regionally to applied loads, may not therefore be applicable at margins. Walcott (1972) and Cochran (1973), for example, used a thin elastic plate to model the edge effect at the Mississippi and Amazon river deltas. They considered the displacement of water by sediments as a mass excess and the flexural response as a mass deficiency due to the subsidence of relatively low density crust into denser mantle. By comparing the combined gravity effect of the sediment load, its compensation, and the anomaly due to the crustal configuration prior to loading to the observed free-air gravity anomaly, Walcott (1972) and Cochran (1973) estimated that the elastic thickness, Te, of the lithosphere, which is determined by the flexural rigidity (Table 1), was relatively high (20-30 km) suggesting some role for flexure at rifted margins.

In order to quantitatively access the role of flexure, Karner and Watts (1982) and Diament et al. (1986) used spectral techniques of analysing the relationship between gravity and topography. Karner and Watts (1982) showed that the outer high (Talwani and Eldholm, 1973; Rabinowitz and LaBrecque, 1977) could be explained by flexure. These workers also pointed out that although flexure describes the mechanism of compensation better than Airy, the Te determined by spectral techniques only represents the average response of the crust and lithosphere to sediment loading during margin evolution. Although both Karner and Watts (1982) and Diament et al. (1986) found some difficulties with the application of spectral techniques to rifted margins (e.g. complications in mirror imaging, unknown crustal structure, lateral heterogeneity in the basement and sediment densities), they were able to confirm the role of flexure, rather than Airy, in their development. They could not, however, determine whether Te at a margin increases with age following a rifting event, although they did find that the average Te was smallest for young margins and largest for older ones.

Studies of the edge effect at margins where there are seismic constraints on the crustal and upper mantle structure have been used to evaluate the dependence of Te on age since rifting. At the East Coast, USA margin, for example, Watts (1988) suggested that Te is low (about 5 km) and has not changed significantly since rifting. Similar Te values have been reported from the Exmouth and Rockall Plateau margins (Fowler and McKenzie, 1989) and the North Sea basin (Barton and Wood, 1984). Such low values are difficult to reconcile with stratigraphic data which suggest that rift-type basins generally widen with time (Watts et al., 1982). White and McKenzie (1988) argued, however, that the widening could be explained by a low Te *provided* that it is combined with a depth-dependent stretching model (Rowley and Sahagian, 1986; Royden and Keen, 1980) in which the total amount of extension in the crust and lithosphere is the same but, the region of mantle thinning extends over a broader region than the crustal thinning. More difficult to explain is a) evidence that some continental rifts (e.g. East Africa, (Weissel and Karner, 1989)) and rifted margins (e.g. New Zealand, Newfoundland) have relatively high Te and, b) oceanic flexure studies (Watts, 1978) which suggest that Te, and hence lithospheric strength, should increase following a rifting event.

A problem with these previous studies is that they have focused on the edge effect in the vicinity of the shelf break. They have therefore ignored the complete system of gravity highs and lows that may flank the edge effect because of flexure. In this paper, we present the results of some simple models to illustrate the relationship between the gravity anomaly pattern and the thermal and mechanical properties of the lithosphere at rifted margins. A "case history" study of the East Coast, USA margin, is used to "ground truth" the relationship between one type of gravity pattern and the spatial and temporal variations in flexural strength across a margin. These relationships are then applied to a new continent-wide compilation of gravity anomaly data from the African region. The overall objectives of the study are to better understand a) the role of flexure in contributing to free-air gravity anomaly edge effect, b) the long-term mechanical properties of the lithosphere and, c) the origin of weak zones within rifted continental margins.

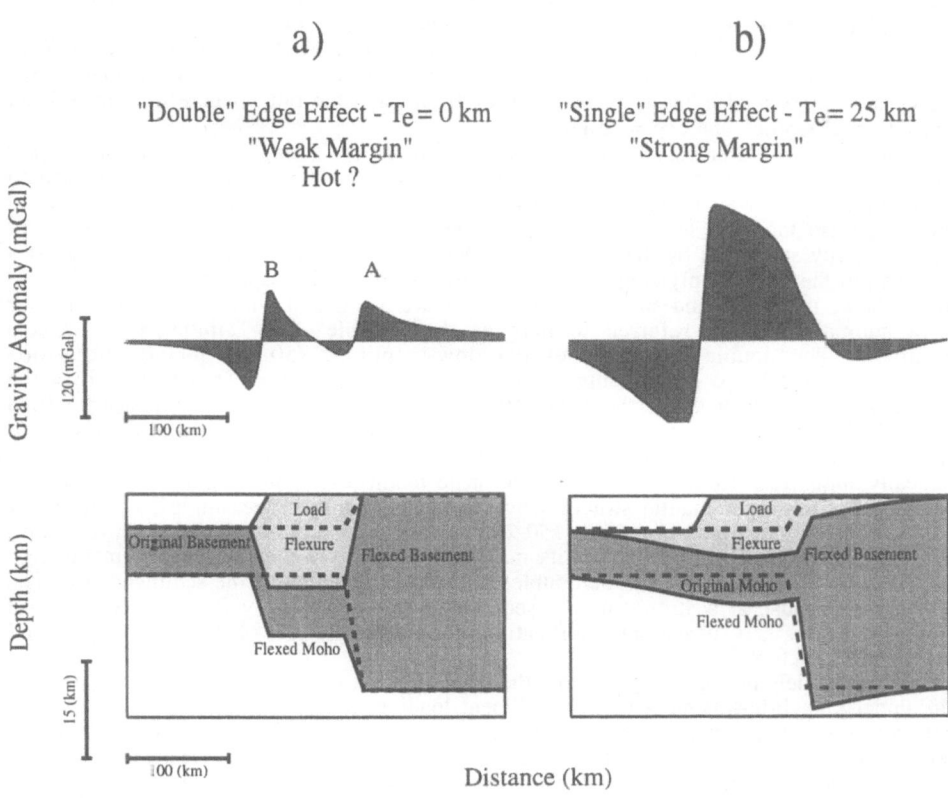

a) "Double" Edge Effect - T_e = 0 km
"Weak Margin"
Hot ?

b) "Single" Edge Effect - T_e= 25 km
"Strong Margin"

Fig. 1 Simple models for the flexure of the lithosphere caused by sediment loading of a rifted continental margin. Two "end-member" models are shown: a) a relatively weak, low Te, margin. and b) a relatively strong, high Te, margin. The dashed line shows the original crustal structure prior to sediment loading and the solid lines show the modified crustal structure. The shaded profiles above the crustal model show the free-air gravity anomaly edge effect, assuming a density of water, sediment, crust and mantle of 1030, 2500, 2800 and 3330 kg m^{-3} respectively. The weak, low Te, margin is associated with a "double" edge effect whereas the strong, high Te margin is associated with a "single". A = onshore high and B = offshore high.

2. Simple models of Sediment Loading.

It is conceptually useful when modelling the gravity anomaly at rifted margins (Watts, 1988) to consider the edge effect as the result of two competing contributions: one derived from "rifting" and the other from "sedimentation". The rift anomaly is associated with the crustal and upper mantle structure at the time of rifting and, in the case of the stretching model (McKenzie, 1978), is given by the combined gravity effect of the uplifted mantle and the water filled subsidence. The positive effect of the deeper mantle uplift occurs over a broader region than the negative effect due to the water so the rift anomaly comprises of a positive-negative "couple" with a high located over the unstretched crust and a low over the stretched crust. The sedimentation anomaly, on the other hand, consists of a positive component due to the displacement of water by denser sediment and a negative due to their compensation. Since the sediment load is shallower than the compensation, the high dominates and so the sedimentation anomaly is made up of a single high with flanking lows.

Fig. 1 shows the edge effect that would be expected for a rifted margin, initially in Airy isostatic equilibrium, that has been loaded by sediments. Two cases have been considered: one in which sediments load a "weak" margin (Te = 0 km) and the other where they have been emplaced on a "strong" one (Te = 25 km)

We caution that by referring to rifted margins as "weak" or "strong", we do not mean their actual strength since this requires knowledge of the maximum stresses that are needed to cause failure. Rather, we use the terms as a means to describe the ability of the lithosphere to support sediment loads for long periods of time. Thus a weak margin is one where sediment loads are mainly supported by the buoyancy of the underlying fluid substratum whereas at a strong margin they are mainly supported by the strength of the lithosphere.

Fig. 1 shows that weak and strong rifted margins are associated with a distinctive free-air gravity anomaly pattern; referred to here as the "double" and "single" edge effects respectively. The double comprises of two low-amplitude (50-100 mGal) and short-wavelength (100-150 km) gravity "highs". The onshore high (A, Fig. 1) is located over the initial shelf break whereas the offshore high (B, Fig. 1) coincides with the position of the final shelf break. The two highs are separated by a gravity "low" which is confined to the region of the thickest sediments. The double arises because at a weak margin the sediments are more or less locally supported. There is therefore little or no modification of the gravity effect of the initial crustal and upper mantle structure by sediment loading. The "single", on the other hand, comprises of a high-amplitude (150-200 mGal) and long-wavelength (200-300 km) gravity high. The gravity high reaches its maximum value over the thickest sediments and, unlike the two highs that make up the double, delineates the region of the sediment load. The single arises because at a strong margin sediments are supported by flexure over a broad region and so there is a substantial modification of the gravity effect of the initial crustal and upper mantle structure.

The simple models in Fig. 1 suggest that the edge effect is a strong function of the strength of the underlying lithosphere. Although sediment loading modifies the gravity effect of the initial crustal and upper mantle structure at both weak and strong rifted margins, the strength of the lithosphere determines how the anomaly is modified and whether or not, for example, there is any record in the edge effect of the gravity effect of the initial continental margin. Thus, whether the edge effect resembles a double or single is a useful "rule of thumb" indicator that might indicate the strength of the lithosphere at a rifted margin.

3. The Edge Effect, Tectonic Subsidence and the Mechanism of Sediment Accumulation

Tectonics in the form of thermal contraction, along with global changes in sea-level, is widely believed to be (Sleep, 1971) one of the principal factors which determines the accommodation space for sediments to infill rift-type basins. It is therefore important to consider the contribution of changes in the tectonic subsidence through time as well as the different mechanisms by which sediments infill a marine basin when discussing the edge effect anomaly.

Backstripping studies (e.g., Steckler and Watts, 1978) have shown that the tectonic subsidence history of rifted margins can, in many cases, be quite well explained by a uniform stretching model (McKenzie, 1978) or its modifications for depth dependent extension (Royden and Keen, 1980) and lateral heat flow and flexure (Cochran, 1981). According to the McKenzie model, there is an initial subsidence due to crustal and lithospheric thinning which is followed by a thermal subsidence as crust and lithosphere cool. We have assumed in Fig. 1 that the rift anomaly is given by the initial crustal and upper mantle structure and therefore that it does not change with time following a rifting event. Fortunately, when calculating the edge effect it is not necessary to consider the increments to the rift anomaly that are caused by this cooling process.

Consider, for example, the total edge effect gravity anomaly after the nth time step, T_n. We then have:

$$T_n = T_{n-1} + (R_n - R_{n-1}) + S_n \qquad (1)$$

where Rn is the rift anomaly after the nth step. S_n is the sedimentation anomaly produced by the new sediment load after the nth step. At the first time step (n=1)

$$T_1 = R_1 + S_1 \qquad (2)$$

because T_0 and R_0 reflect the previous time step to which there are no contributions. At the second time step (n=2),

$$T_2 = T_1 + (R_2 - R_1) + S_2$$

Substituting from (2) gives

$$T_2 = R_1 + S_1 + (R_2 - R_1) + S_2$$

or

$$T_2 = R_2 + S_1 + S_2$$

therefore the total anomaly, Tn, due to all the time steps can be written as

$$T_n = R_n + \sum_{n=1}^{\infty} S_n \qquad (3)$$

Thus, (3) shows that it is only the final rift anomaly, R_n corresponding to the time since rifting of the final sediment load that needs to be considered in calculating the total edge effect. The sedimentation anomaly, however, has to be calculated at each time step because of changes in load geometry and the resulting lithospheric response.

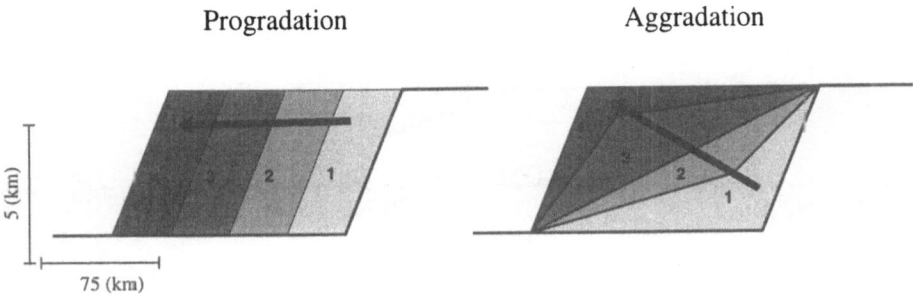

Fig. 2 *Simple models of progradation and aggradation at a rifted continental margin. The figure shows the geometry of 4 successive loads.*

The mechanisms by which sediments are transported into a rift-type basin depend on a number of factors including the rate of sediment supply and the topography, lithology and climate of potential source areas. Although sediments in many rift-type basins have accumulated upwards by aggradation, some clastic and carbonate margins (Sarg, 1988) show a tendency to build outwards by progradation (Fig. 2). The case of progradation is best illustrated, perhaps, at large river delta systems where it typically involves the transportation of sediments directly from source areas by rivers, across the shoreline and into a marine environment. For aggradation, the transport mechanism is more complex. Sediment supply is

70

not limited to rivers but, can occur parallel to the shoreline by processes such as long-shore drift.

Fig. 3 shows the edge effect that would result from the progradation and aggradation of sediments on a weak ($T_e = 0$ km) and strong ($T_e = 25$ km) margin. In the modelling, we assume that a sufficient accommodation space has been provided so as to allow a load up to 5 km high over a horizontal distance of 100 km to accumulate by progradation or aggradation. Over the four incremental time steps assumed, the final sediment load geometry produced by both mechanisms is the same. In the case of progradation, each step has a lateral extent of 25 km, and fills the accommodation space to the full depth of 5 km. For the aggradational case, however, each step covers the total lateral extent of 100 km, but it does not fill the whole accommodation space.

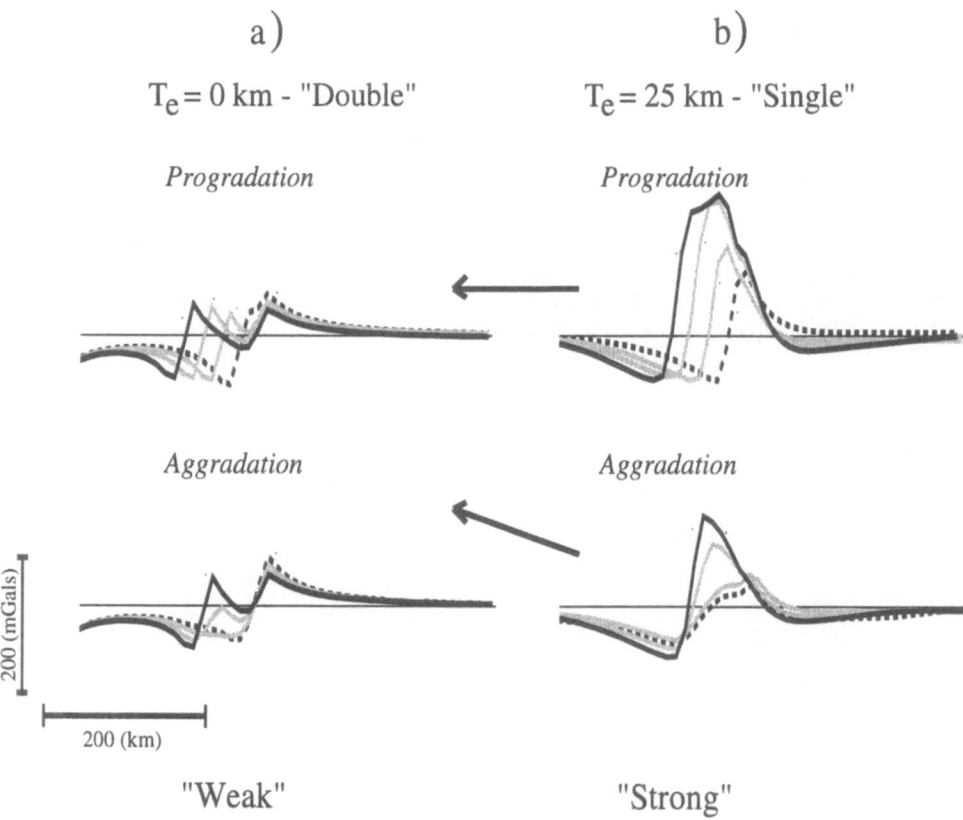

Fig. 3 Simple models for the edge effect associated with the transport of sediment by aggradation and progradation onto rifted margins of different flexural strength. The geometry of the sediment loads is shown in Fig. 2 and the edge effect has been calculated at each load step. Dashed line = Edge effect after deposition of the first sediment load. Solid

line = Edge effect after deposition of the final load. a) weak, low T_e (0 km) and b) strong, high T_e (25 km) rifted margin.

Since the models in Fig. 3 assume that the accommodation space at the continental margin has already been established prior to sediment loading and, that the incremental sediment loads are of equal cross-sectional area, then R_n in equation (4) is given by the initial rift anomaly, R_0. The figure shows that the edge effect is independent of the mode of sediment accumulation for both strong and weak margins but, the incremental steps show variations that depend on whether sediments accumulate by progradation or aggradation. The weak margin (Fig. 3a) shows a double edge effect for both types of sediment accumulation, since by the final step the transfer of mass between the initial margin and the new shelf break has been completed. In the case of progradation, the shelf break is always apparent which causes the double anomaly to persist throughout sedimentation. Additionally the onshore high maintains its position over the initial margin while the offshore high migrates seaward with the sediment load. In the case of aggradation, however, the double does not totally develop until the sediment profile forms a new shelf break. Prior to this, the margin has a slope rather than a shelf profile and so the offshore high is not developed. The strong margin (Fig. 3b) shows a single edge effect throughout its development and sediment build up for both modes of sediment accumulation. Progradation over a strong margin results in an increase in both the amplitude and wavelength of the gravity anomaly highs. In the aggradational model only the amplitude increases with the maximum wavelength having already been achieved by the initial step. Therefore, for a single edge effect the amplitude of the high directly reflects the lateral extent of the sediment, whereas for a double it is the position of the two gravity highs which mark the limits of the sediment load.

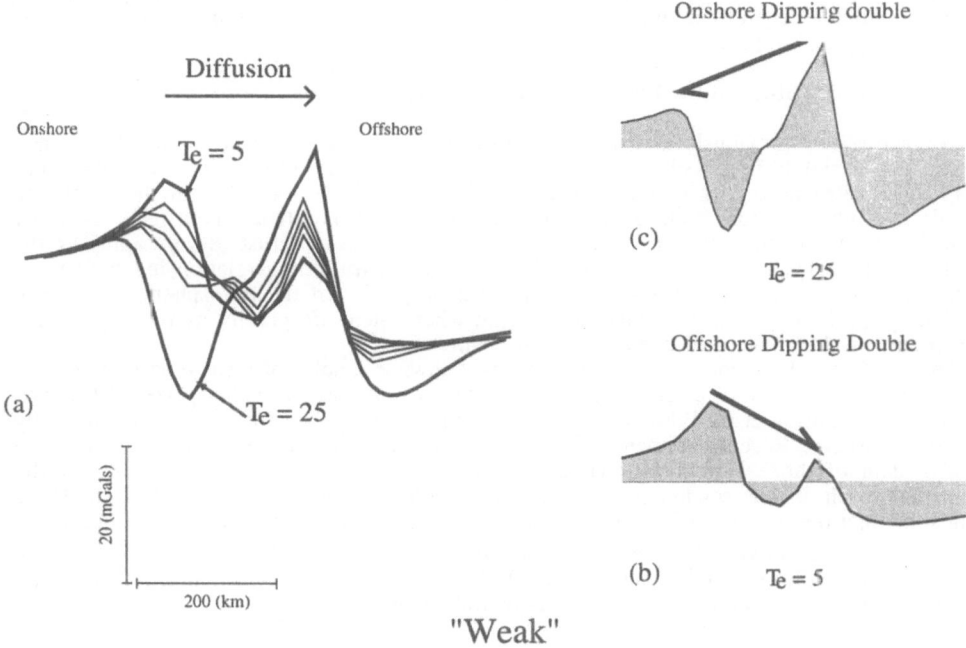

Fig. 4 Simple models for the edge effect associated with the transport of sediments by diffusion at a rifted margin. a) calculated edge effect for $5 < T_e < 25$ km. b) calculated edge effect for Te = 5 km showing the "offshore dipping double" and c) calculated edge effect for Te = 25 km showing the "onshore dipping double".

Several studies suggest that the transport of sediment from sub-aerial to marine environments is best described by some sort of diffusion process. According to this model, a balance exists between the amount of material removed and the amount deposited. Kenyon and Turcotte (1985) applied a diffusion model to describe the changes that occur in the shape of the depositional front through time at large river delta systems while Sinclair et al. (1991) and Flemings and Jordan (1990) applied a diffusive model to describe the transfer of material from thrust/fold belts into flanking foreland basins. Both models assume a depositional front which is exponential in form and migrates in a seaward direction. The main difference between them is that Kenyon and Turcotte (1985) assumed a sediment supply whereas Sinclair et al. (1991) and Flemings and Jordan (1990) computed the amount of material removed and deposited from the slope of the topography profile.

Fig. 4 shows the gravity effect of a diffuse sediment load that builds out on a margin with a T_e ranging from 5 to 25 km. A diffusion model produces both the double and single type of edge effects. The main difference from the aggradation and progradation models is that the double - with its pattern of onshore and offshore highs - persists for higher values of T_e. For example, the T_e = 25 km case still has the appearance of a double (Fig. 4), unlike either the aggradation or progradation cases which show a well developed single for this value of Te. The diffusion model eventually produces a single but, not until T_e exceeds a value of about 40 km.

A characteristic feature of the double in the diffusive model is that the relative amplitude of the onshore and offshore highs appears to depend on T_e. Fig. 4 shows, for example, that when the slope between the two gravity highs dips offshore T_e is low. As T_e is increased, the slope becomes horizontal, and then begins to dip onshore. The form of the edge effect is the same, however, in both cases and resembles the double of the non-diffusive models in Figs. 2 and 3. As is the case for the double edge effect in the non-diffusive models, the dipping doubles also reflect relatively weak lithosphere. We therefore consider these features as a subset of the double edge effect and refer to them here as the "offshore dipping" double and the "onshore dipping" double.

4. Case History Study of the East Coast, USA Rift-Type Margin

We have shown previously that irrespective of differences in the mechanisms of sediment deposition, it should be possible to use the edge effect to infer information about the long-term mechanical properties of the lithosphere. The models developed so far, however, are simple and ignore factors such as lateral changes in T_e and modifications to the crust other than flexure due to the addition or subtraction of material to the crust. Before applying the models more generally, it is useful therefore to "ground truth" the relationship between a particular type of edge effect and the mechanical properties of the lithosphere by forward modelling, preferably at a well surveyed margin where there are gravity as well as seismic constraints on the crustal and upper mantle structure.

The East Coast, USA margin is one of the best known examples of a rift-type margin. The margin shows evidence of detrital "syn-rift" rift-basins both onshore (Manspeizer, 1985) and offshore (Hutchinson et al. 1986) which are overlain by a thick "post-rift" sequence of gently dipping Mesozoic-Recent sediments (Schlee et al. 1976). Seismic reflection and refraction profile data offshore New Jersey (LASE, 1986; Diebold et al, 1988) indicate that the sediments reach thicknesses in excess of 12 km beneath the Baltimore Canyon Trough. These data show that the thickest sediments are underlain by a 8-10 km thick crust. The thin crust is associated with relatively high P wave velocities (>7.1 km s^{-1} which have recently been interpreted by Holbrook and Kelemen (1993) as indicating that large amounts of magmatic material have been added to the crust during or following rifting.

4.1 ESTIMATION OF T_e FROM SEISMIC AND GRAVITY DATA

As Fig. 5 shows, the East Coast, USA margin is associated with a well developed free-air gravity anomaly edge effect. The edge effect consists of a low-amplitude and short-wavelength gravity high at the shelf break which is flanked by gravity lows. The seaward low is of larger amplitude and longer wavelength than the landward low. There is evidence that the landward low is bounded by a small-amplitude high at the New Jersey shore, although

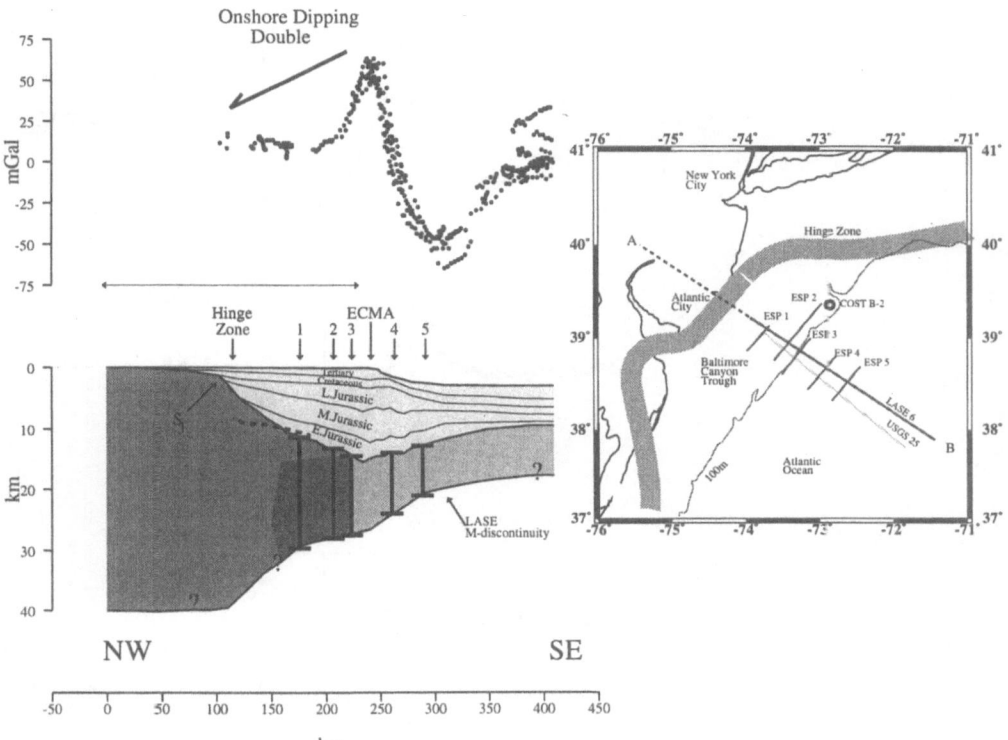

Fig. 5 Free-air gravity anomaly and crustal structure along the LASE profile of the East Coast, USA rifted margin offshore New Jersey. The gravity data is based on (Watts, 1988) and the crustal structure on the (LASE, 1986). The margin is associated with a well developed edge effect which shows a small-amplitude, short-wavelength high over the outer shelf and well developed lows over the continental rise and the middle shelf. There is evidence for a small-amplitude high in the vicinity of the shoreline. The edge effect is characteristic of an onshore dipping double.

identification of this high is complicated by the presence of the Appalachian gravity high. This high has been interpreted (Karner and Watts, 1983) as the result of buried loads that were emplaced in the crust during the Appalachian orogeny. The shoreline high, the presence of a gravity "low" over the thickest sediments and, the small-amplitude high at the shelf break suggest that the edge effect is an example of an onshore dipping double, rather than a single, and hence reflects relatively weak lithosphere.

In order to better understand the relationship between an onshore dipping double and the mechanical properties of the underlying lithosphere, we follow the approach of Watts (1988) who considered the edge effect as the sum of the gravity effects of three main processes: rifting, sedimentation and erosion. If seismic refraction data are available, then the "rifting anomaly" can be computed by isostatically restoring the crust and mantle in the absence of sediments. The "sediment anomaly" can then be computed by estimating the sediment load from the difference between the present-day topography and the top of the restored crust and

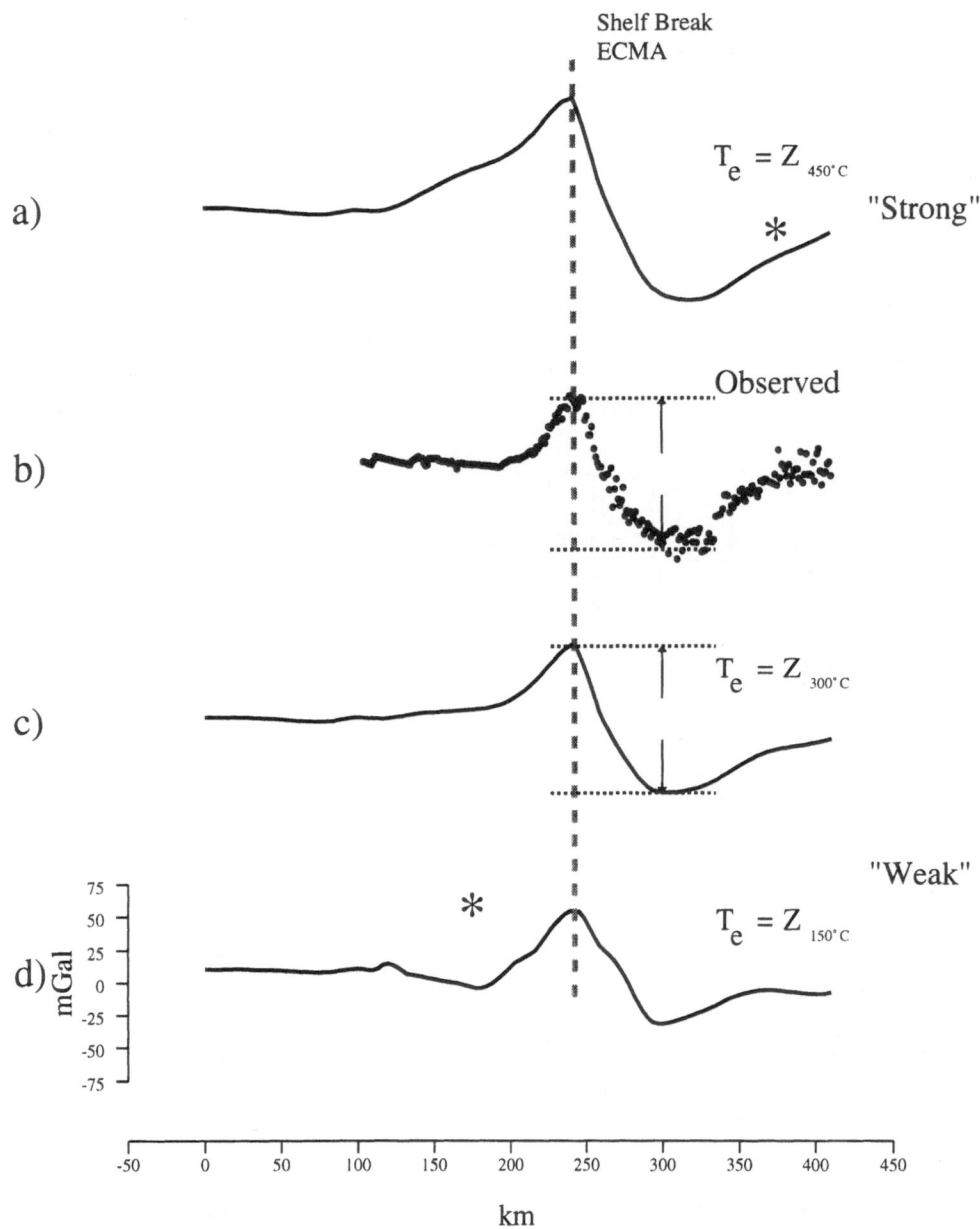

Fig. 6 Comparison of the observed edge effect to calculations obtained by flexurally backstripping individual sedimentary layers along the LASE profile. The calculations assume Te is given by the depth of the a) 450°, c) 300° and d) 150° oceanic isotherms. The asterisks indicate those portions of the calculated profiles that fit the observed data particularly well.

then calculating the response of the crust and upper mantle to this load using the elastic plate model. Finally, the erosion anomaly can be computed by estimating the amount of bulge material that has been removed so as to produce the present-day topography of the margin. By comparing the sum of these processes to the observed edge effect, it is possible (e.g., Watts, 1989) to constrain T_e.

Unfortunately, methods based only on seismic refraction data yield a single estimate of T_e because it assumes that the margin sediments load the lithosphere during one time interval. In order to determine whether T_e varies with time, it is necessary to calculate the contribution to the sedimentation anomaly of each sediment load as it is emplaced. This can be carried out by flexural backstripping seismic reflection profile data layer by layer (e.g. Fig. 6). Like Airy, flexural backstripping yields the depth that basement would be in the absence of sediment and water loading (i.e., the tectonic subsidence/uplift) and hence, by isostatic balancing, the depth to the Moho. The main limitation is that a value of T_e needs to be specified for each backstripped layer

Oceanic flexure studies (e.g., Watts 1978) suggest that T_e increases with age or the lithosphere at the time of loading and is given approximately by the depth to the 450° oceanic isotherm. There is evidence (Calmant and Cazenave, 1986) that regions such as the Pacific superswell (McNutt and Fischer, 1987) follow a lower isotherm because the lithosphere there has been re-heated by one or more hot-spots. Also, fracture zones appear to be associated with a higher isotherm (Wessel and Haxby, 1990) due, perhaps, to differences in crustal thickness and, hence composition, along the ridge axis. Since stretched crust was once close to a ridge then it is reasonable, despite these variations, to suppose that Te of stretched lithosphere may increase following a rifting event. Fig. 6 shows the edge effect for three cases where the sediments are backstripped according to whether Te follows the 150°C, 300°C or 450°C oceanic isotherm. Although no one isotherm explains the edge effect, the calculated profiles explain portions of the observed data. For example, seaward of the shelf break the edge effect is quite well described by the calculated profile based on a 450° isotherm whereas landward of the shelf break the 150°C isotherm fits better. Thus, a good fit to all the features of the edge effect might be for a model in which there are lateral changes in the flexural strength of the lithosphere across the margin.

Fig. 7 shows that a model in which sediments load weak crustal landward of the shelf break and stronger lithosphere seaward of the break explains the edge effect quite well. There is a particularly good fit to the amplitude and wavelength of the high in the region of the shelf break. More importantly, the two lows that flank the high are also fit. The best overall fit is for a T_e profile with a weak zone under the shelf that is flanked by stronger zones either side. If, for example, we had assumed that the lithosphere underlying the shelf was of similar strength to the oceanic lithosphere - then Fig. 7 shows (thick dashed line) that the anomaly over the shelf would be too high. The shelf break also corresponds to the location of the crest of the East Coast Magnetic Anomaly (ECMA, Figs. 6 and 7) which previous studies suggest (e.g., Grow, 1981) is the location or the ocean/continent boundary. An onshore dipping double appears therefore to characterise a region where sediments have loaded weak stretched continental lithosphere which abuts stronger oceanic lithosphere.

Fig. 7 Free-air gravity anomaly and crustal structure associated with the best fitting T_e model along the LASE profile (Fig. 6). ECMA = East Coast Magnetic Anomaly. ESP = Expanding Spread Profile. a) Calculated (dashed and solid lines) and observed (solid dots) free-air gravity anomaly. Heavy dashed line = "Strong/Strong" case. Grey dashed line = "Weak/Weak" case. b) Crustal model. Heavy dashed line shows initial (i.e. backstripped) crustal structure - prior to sediment loading. The best fit to the observed gravity data is for amodel in which T_e (Bodine et al., 1981) varies laterally across the margin. In particular, sediments landward of the ECMA have loaded relatively weak ($T_e = Z_{150°}$, Z = depth) lithosphere while sediments seaward of the ECMA have loaded stronger (Te = $Z_{450°}$) lithosphere.

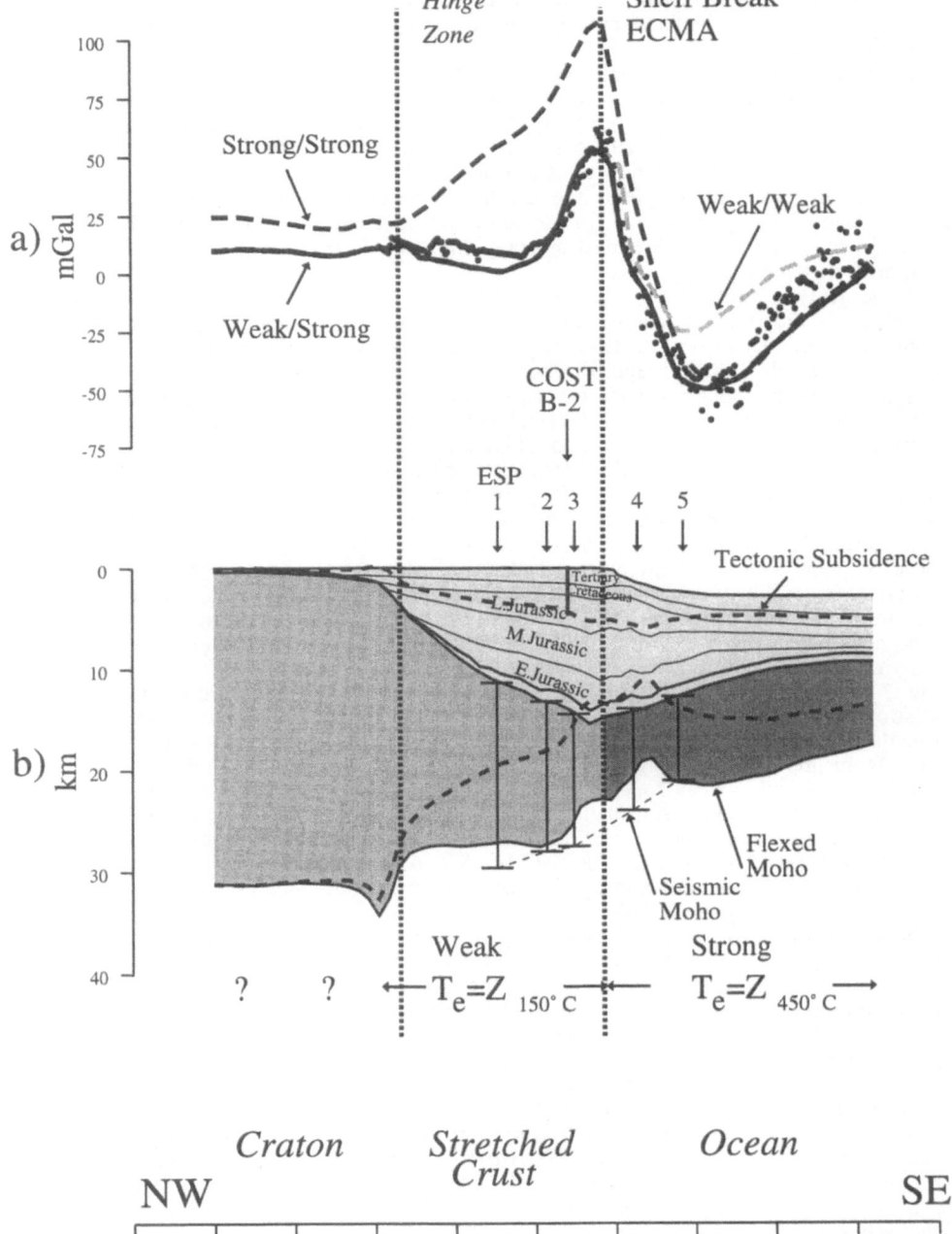

a) mGal

Hinge Zone

Shelf Break
ECMA

Strong/Strong

Weak/Weak

Weak/Strong

COST
B-2

ESP
1   2 3   4   5

b) km

Tectonic Subsidence

Tertiary
Cretaceous

L.Jurassic

M.Jurassic

E.Jurassic

Flexed
Moho

Seismic
Moho

?   ?

Weak
$T_e = Z_{150°\,C}$

Strong
$T_e = Z_{450°\,C}$

Craton   *Stretched Crust*   *Ocean*

NW   SE

-50   0   50   100   150   200   250   300   350   400   450

km

4.2 MAGMATISM

A consequence of determining the T_e structure of a rifted margin is that the depth to the Moho can be "forward modelled" and compared to the Moho as defined by seismic refraction data. Flexural backstripping, for example, reveals the crustal and upper mantle structure *at the time of rifting.* Subsequent loading by sediments modifies this Moho which, because it is based on backstripping, is referred to (Watts and Torné 1992) as the "backstrip Moho". Since the backstrip Moho is the depth that Moho should be on the basis of the thermal and loading history of the margin, differences between the backstrip and the seismically constrained Moho - or, as it is referred here, the "seismic Moho" - indicates whether material has been added to the thinned and loaded crust (e.g. by magmatism) or subtracted from (e.g., by tectonic erosion) it following rifting. This is important to resolve, especially as it may relate to current controversies concerning the amount of igneous material associated with rifted margins (Holbrook and Kelemen, 1993).

Fig. 8 shows a detailed comparison of the backstrip and seismically constrained Moho along a portion of the LASE profile of the Baltimore Canyon Trough. The figure shows that the East Coast USA margin is a case where the seismic Moho is deeper than the backstrip Moho by about 2 km suggesting that some material may indeed have been added to the base of the crust at this margin following rifting

The amount of additional material is difficult to assess but, the cross-sectional area of the region between the backstrip and seismic Moho on the LASE profile is about $2 \times 160 = 320$ km^2. If it is assumed that a seaward dipping reflector sequence on USGS Line 25 beneath the outer shelf at about 9 and 10 s two-way travel time (Poag 1985) also represents magmatic material then, the total area of the added material could be as much as 640 km^2. Although this estimate appears high, it is less than the 1600 km^2 estimated by Holbrock and Kelemen (1993). The reason for this is that these workers only used seismic refraction data to estimate the amount of added material; assigning *all* the material with a P wave velocity higher than 7.1 km s[-1] as newly added magmatic material. Fig. 8 shows that when the seismic velocities are superimposed on the flexure model then it seems that only velocities in excess of 7.3 km s[-1] actually represent the newly added material: velocities in the range 7.1-7.3 km s[-1] being representative of the pre-existing (stretched) crust. This interpretation is in accord with that of White et al. (1987) at the Hatton Bank margin who attribute velocities of 7.0-7.3 km s[-1] to pre-existing continental crust and velocities of > 7.3 km s[-1] to the newly added magmatic material. Unfortunately there is no evidence at the East Coast, USA margin that velocities of 7.0-7.3 km s[-1] are in fact, representative of pre-existing continental crust because of the lack of reliable velocity-depth data beneath onshore regions.

One problem with using the flexural backstripping technique is that it requires that the T_e structure of the margin is known. Although gravity modelling constrains T_e, it cannot distinguish, for example, between weak lithosphere and the Airy model (Te = 0 km) because of uncertainties in densities and other factors.

An alternative approach is to backstrip stratigraphic data from deep stratigraphic wells. The advantage of this approach, which determines the path of the tectonic subsidence/uplift curve through time, is that the extension factor, ß, and hence the amount of crustal and lithospheric thinning can be estimated by best fitting thermal models *directly* to the backstrip data. A disadvantage is that the tectonic subsidence is only determined at a single well site. This is not too critical along the LASE profile, however, because one of the wells, the COST B-2, is located in the outer shelf where the estimate of material that has been added to the crust appears from flexural backstripping to be a maximum.

Fig. 9 shows an attempt to estimate ß by backstripping bio-stratigraphic data at the COST B-2 well - assuming a model (McKenzie, 1978) in which the COST and lithosphere were stretched by an equal amount at the time of rifting. The figure shows that ß depends critically on the paleobathymetry, porosity-depth and sea-level that are assumed in the backstripping. For example, if maximum water depth estimates, the Watts and Steckler (1979) sea-level curve, and a smooth porosity-depth curve with initial porosity of 55% had been used then the best fit value of ß is 3.0 which corresponds to a mean crustal thickness of 10.4 km and a depth to the backstrip Moho of 25.2 km, assuming an unstretched crustal thickness of 31.2 km.

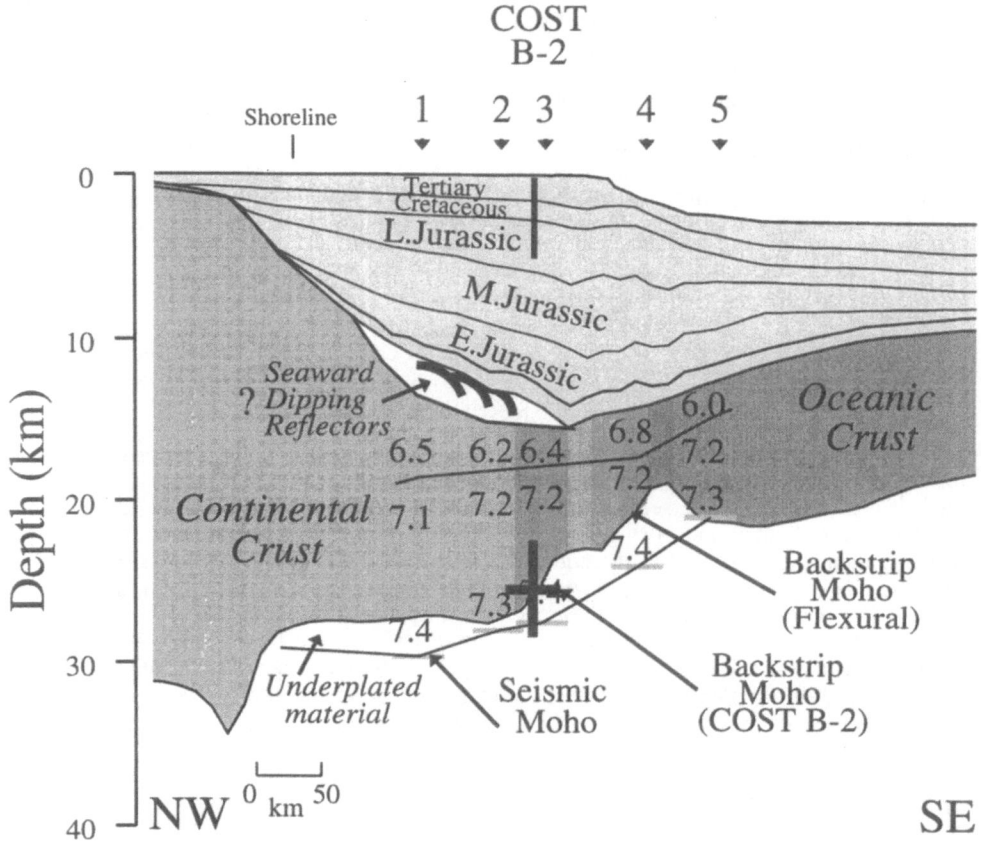

Fig. 8 Crustal structure along a portion of the LASE profile between 50 and 350 km (Fig. 7). Numbers 1-5 refer to location of ESP mid-points. P-wave velocities at the ESP mid-points are shown in km s⁻¹ and are based on (LASE, 1986). Grey vertical and horizontal bar = estimate of depth to backstrip Moho based on backstripping bio-stratigraphic data at the COST B-2 well (Fig. 9). The region between the backstrip Moho and the seismic Moho is interpreted as material that has been added to the crust following initial rifting. The extent of the seaward dipping reflectors is based on (Poag,. 1985).

If, on the other hand the minimum water depth and the Pitman (1978) sea-level curve had been used, the corresponding ß and depth to backstrip Moho would be 2.4 and 4.4 and 27.8 and 21.9 km respectively for the same unstretched crustal thickness. These variations in the backstripping parameters imply a depth to the backstrip Moho as shallow as 21.9 km or as deep as 27.8 km.

A potential problem with both types of backstripping is that they only reveal the *total* tectonic subsidence/uplift in a rift-type basin. Any uplift that results from the addition of material to the base of the crust (McKenzie, 1984), for example, will therefore be included - along with the subsidence/uplift due to stretching - in the backstrip curves derived from seismic or well data. Assuming isostatic balance before and after material is added then the uplift, h, can be written:

$$h = r\frac{(\rho_m - \rho_c)}{(\rho_c - \rho_w)} \qquad (4)$$

where r is the amount of added material and ρ_w, ρ_c and ρ_m are the densities of sea-water, crust and mantle respectively. Using r = 2 km the estimate based on flexural backstripping), ρ_w =1030, ρ_c =2800 and ρ_m = 3330 kg m^{-3} gives h = 0.6 km.

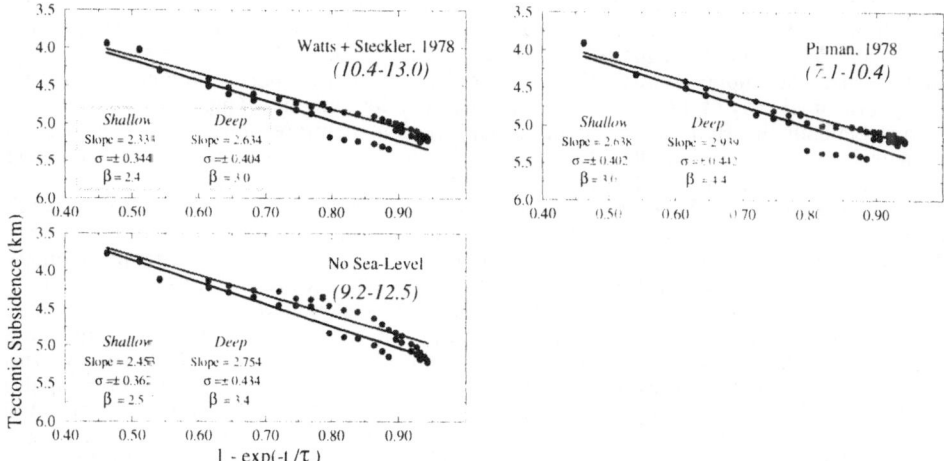

Fig. 9 Plot of tectonic subsidence vs. $(1 - e^{(-t/\tau)})$ for the COST B-2 well. τ = thermal time constant = 62.5 Ma, t = age (Ma) since rifting. The tectonic subsidence curves have been calculated for different assumptions concerning the water depth of deposition and sea-level changes through time. Heavy dots = shallow water depth. Grey dots = deep water depth. Sea-level curves are based (Pitman, 1978) (right panel) and (Watts and Steckler, 1979) (upper left panel). The lower left panel shows the tectonic subsidence for no sea-level change. β = stretching factor. σ = standard deviation (km) between the tectonic subsidence and the best fit straight line. The numbers in brackets in the upper right of each panel indicate the range of thickness of the stretched crust based on the best fit estimates of β.

The calculations in (4) ignore the thermal effects of underplating stretched crust during a rifting event. These effects could be modelled using a depth-dependant extension model (Royden and Keen, 1980; Rowley and Sahagian, 1986) in which the mantle is stretched by a greater amount than the crust. However, these models require that the amounts of crust and mantle stretching be specified.

A different approach is to consider the observational constraints at rifted margins such as the North Atlantic which are known to have been influenced by deep mantle plumes. Turner and Scrutton (1993), for example, have shown from wells in the margin north of Scotland a rapid uplift when this region was close to the Iceland plume. Other studies in the Porcupine Seabight (Hall and White, 1994) have documented a rapid subsidence. In contrast, the tectonic subsidence at the COST B-2 well, like other wells along the East Coast, USA margin, shows a smooth exponential decrease through time. The main exception is the tectonic subsidence at a well near Sable Island which shows a relative uplift followed by a steep subsidence, but this has been attributed (Watts and Steckler, 1979) to vertical movements due to movement of the Argo salt. The oldest strata in the COST B-2 well are Upper Jurassic in age so that it is possible that the effects of magmatism were recorded in the early development of the margin. If this is the case, then magmatism would have little effect on our estimates of β since the only data that was used to estimate this parameter is from the well itself. Magmatism could, however, decrease the amplitude of the tectonic subsidence/uplift, which implies that the amount of stretching inferred from flexural backstripping may have been underestimated. The close agreement between the backstrip Moho derived from the well and seismic reflection profile data together with the fact that the tectonic subsidence at the COST B-2 well is already close to the maximum that can be produced by thermal models suggest to us, however, that magmatic processes have not significantly affected the tectonic subsidence/uplift history of the Baltimore Canyon Trough region.

Although it is not possible from these considerations to accurately determine the amount of newly added magmatic material, backstripping suggest thicknesses in the range of 0 to 5 km. These estimates are much less than the 20 to 25 km reported by (White, 1992) from the Rockall Bank and offshore Norway where the Iceland plume has influenced margin development. However, they are not significantly different from what would be expected from melting models (e.g. McKenzie and Bickle, 1988) for adiabatic decompression of the asthenosphere. For example, if the lithosphere has been stretched by a ß factor of 2.4 to 4.4 - as the backstripping implies - then a thickness of about 2-5 km of melt is predicted, assuming a lithospheric thickness of 125 km and a "normal" mantle potential temperature of 1330°C. The crust beneath the Baltimore Canyon Trough region therefore appears to have had no more material added to it than would be expected from decompression models; in agreement with the lack of evidence there of significant hot-spot activity.

There is evidence that the onshore dipping double, and hence zone of relatively weak lithosphere, extends away from the Baltimore Canyon Trough along much of the East Coast, USA margin. Fig. 10 shows, for example, ensemble averages of 5 free-air gravity anomaly profiles from 5 segments of the margin between Nova Scotia and Florida. Ensemble averaging (e.g., McKenzie and Bowin, 1976) is a useful technique to isolate those features, such as the edge effect, which are common to each profile while at the same time eliminating effects such as changes in the density of the basement which are not. A difficulty with applying the technique to the East Coast, USA margin is that the surface (i.e., topographic) and sub-surface loads of the Appalachians are associated with gravity highs which also trend sub-parallel to the coastline and therefore may also be common to each profile. These highs can, however, be identified and separated from the edge effect anomalies. As Fig. 10 shows the East Coast, USA is associated with a small-amplitude offshore high which is limited spatially to the region of the shelf break in slope. Moreover, the offshore high is bounded by flanking lows. There is evidence on a number of profiles for a small-amplitude onshore high. These features are characteristic of an onshore dipping double and since they appear on widely separated profiles they suggest to us that weak stretched continental crust may extend along much of the length of the margin. At present we do not know why stretched continental crust is mechanically weak over such large horizontal distances but, the lack of evidence of hot-spot activity during margin development suggest that thermal re-heating of the lithosphere following rifting is probably not involved.

5. Application of the Simple Models to the African Margin

The East Coast, USA margin "case history" study of an onshore dipping double illustrates how the edge effect may be used to better understand the crustal structure and the evolution of the margin. The study also implies that it should be possible to qualitatively compare edge effects observed over different rifted margins and, more importantly, arrange these margins into an order with respect to their thermal and mechanical properties.

To illustrate this point, we have analysed the edge effect along the African margin. This margin was chosen because there is relatively little seismic data to constrain the actual T_e structure and part of the margin is conjugate to the East Coast, USA. Another reason is the availability of new gravity anomaly data compiled at the University of Leeds as part of its African Gravity Project (AGP). This data set comprises of all the available "point" gravity measurements from academic, national government and commercial sources (Fig. 11 - inset). Both land and marine data have been verified for inter-survey compatibility, rogue points, tilt, and bias. The corrected point data were then smoothed over 2.5 x 2.5 minute "squares" and gridded at 5 x 5 minute interval using a minimum curvature technique (e.g., Wessel and Smith, 1993).

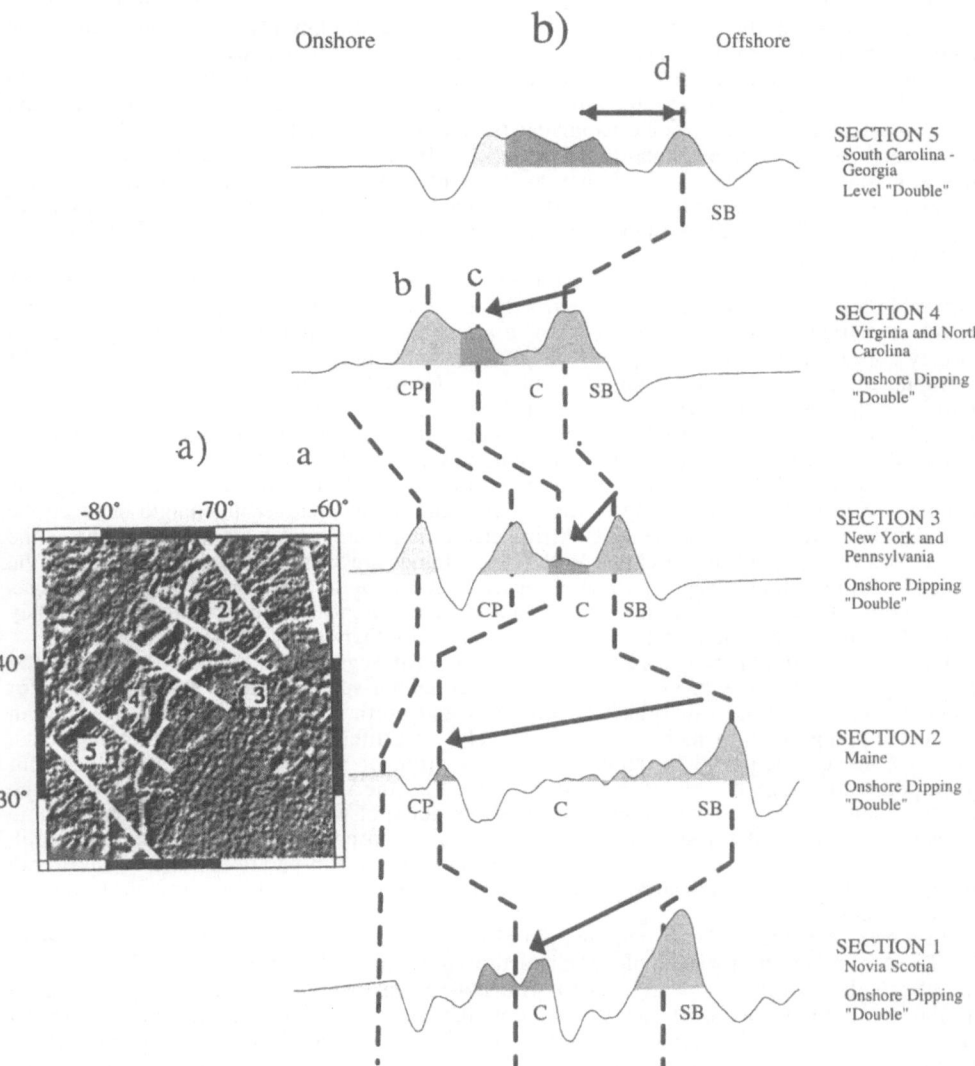

Fig. 10 Ensemble averaged free-air gravity anomaly profiles from S segments of the East Coast, USA margin. **a)** Shaded relief gravity anomaly map (Bouguer onshore, Free-Air offshore) of North America based on the DNAG data set. The grey shades represent the intensity of the slope in the gravity field in the direction of the artificial sun (azimuth: 135 degrees). The numbers 1-5 refer to segments within which gravity and topography profiles have been ensemble averaged in order to determine the "typical" edge effect profile across the East Coast, USA margin. CP = Onshore limit of coastal plain. SB = Shelf break. C = Coastline. a = Gravity high due to surface topography. b = Gravity high due to buried (sub-surface) topography. c = Onshore high. d = Offshore high. **b)** Ensemble averaged profiles.

For the purposes of analysis, the margin was divided up into a number of "segments" (Fig. 11) each of which were defined on the basis of the similarity of their free-air gravity anomaly patterns. Within each segment, 10 profiles were projected orthogonal to the local trend of the margin, along which the gridded gravity data was sampled at 5 km intervals. Similar ensemble averaging techniques to those applied to the East Coast, USA margin were used to define the form of the edge effect in each segment. As Fig. 12 shows, the African margin exhibits both the double and single types of edge effect. The double is recognised by its low amplitude and short-wavelength features. Double anomalies were observed over the margins offshore Libya, Morocco, Nigeria, Gabon-Cameroon, Angola, S. Africa, S. Mozambique, Tanzania, Kenya and Somalia. The single, on the other hand, is recognised by its large amplitude, long-wavelength positive peak. Single edge effect anomalies are observed offshore the margins of Algeria-Tunisia, Senegal, Mauritania, Sierra-Leone, Liberia, Ivory Coast, Ghana, Namibia, and N. Mozambique.

The edge effect cannot always be isolated from ensemble averaged gravity profiles. A particular problem along the African margin is features such as fracture zone ridge and trough topography which may be common to a number of profiles within each segment and so may appear along with the edge effect in an ensemble profile. The transform margins, for example, offshore the Gulf of Guinea and Ivory Coast trend more or less parallel to the margin so that the fracture zone edge effect (Louden and Forsyth, 1976) may contaminate the margin edge effect. Other problems are due to density structures in the crust - although we believe that ensemble averaging in this case tend to removes these features.

The strength of the lithosphere inferred from the edge effect along the African margin is summarised in Fig. 13a. A striking feature of the figure is the contrast in strength between the west and east Africa margins: the west Africa margin is generally of high strength whereas the East Africa margin is of low strength. Within these trends are variations which suggest that the African margin is highly segmented with regard to its strength. While these strength changes may, like the East Coast, USA margin, be caused by rheological effects, we believe that they more likely reflect the modification of the lithosphere by thermal effects following rifting.

Fig. 13b shows that there is a correlation between weak segments of the African margin and those regions that have been influenced by mantle hot-spots (Demets et al., 1990). For example, the S. Morocco margin between 25°N and north of 29°N is characterised by an offshore dipping double which indicates relatively weak lithosphere. The flanking segments, however, are characterised by singles indicating stronger lithosphere. The weak segment corresponds to a region which has been influenced by the same hot-spot that generated the Canary Islands. We speculate, therefore, that this hot-spot may have re-heated the oceanic lithosphere such that it appears weaker than it should be based on its age. According to Holik and Rabinowitz (1992), the trace of the Canary hot-spot is sub-parallel to the S. Moroccan margin and therefore the reheating, and hence weakening, may have been particularly intense at this margin.

Other weak regions of the African margin also appear to correlate with the hot-spot traces. The SE Africa margin, for example, is characterised by an offshore dipping double.

Although the margin is not associated with hot-spot activity at the present day, it has a similar trend to the trace of the Bouvet hot-spot which plate kinematic models suggest subjected the margin to a prolonged period of re-heating during 40-80 Ma. The SW Africa margin, on the other hand, is characterized by an onshore dipping double. This margin therefore appears as a relatively weak margin even though it trends across the traces of individual hot-spots. We attribute this weakness to the fact that the margin appears to have been influenced by both the Vema (20-40 Ma) and Discovery (60-80 Ma) hot-spots.

The margin to the north of the SW Africa margin, offshore N. Namibia-Angola, also appears to have been influenced by hot-spots and yet, is associated with a single edge effect. One explanation of this is that the hot-spots involved (Tristan and Gough) only influenced the margin early on its evolution (100-120 Ma), and so the margin has had sufficient time to cool and regain its strength. Similarly, the coastline from Senegal round to Nigeria, is characterised by a number of single edge effects. There is notable lack of hot-spot activity along this segment of margin suggesting that the single reflects sediment loading on strong continental lithosphere which has not been weakened by thermal events subsequent to rifting. A particularily well developed single occurs offshore N. Mozambique-Tanzania which we also attribute to the lack of hot-spot activity.

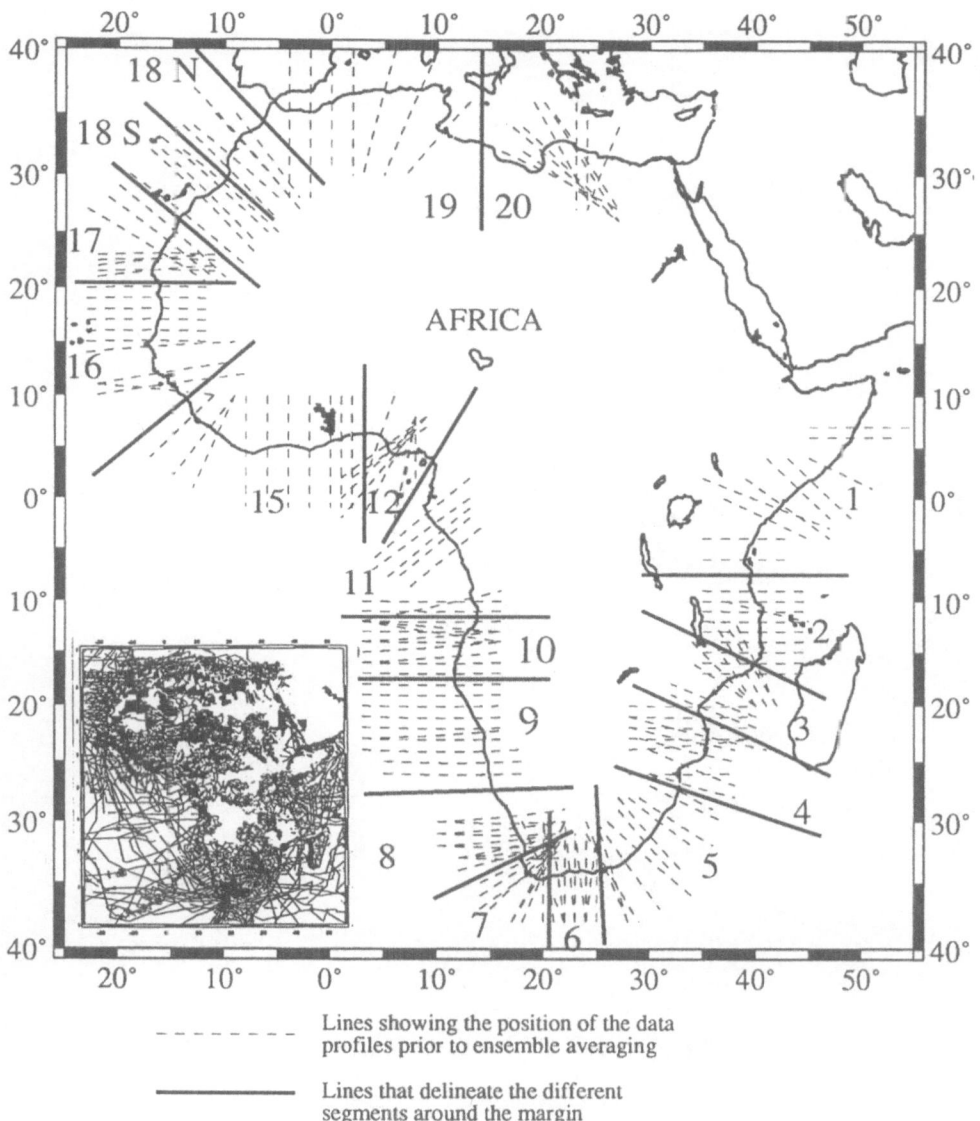

Fig. 11 Map showing the location of the free-air gravity anomaly profiles that were used to construct the ensemble averages of the African margin. The profiles were constructed from a 5 x 5 minute grid of individual measurements which have been smoothed over 2.5 x 2.5 minute 'squares'. The distribution of the smoothed values is shown in the inset.

84

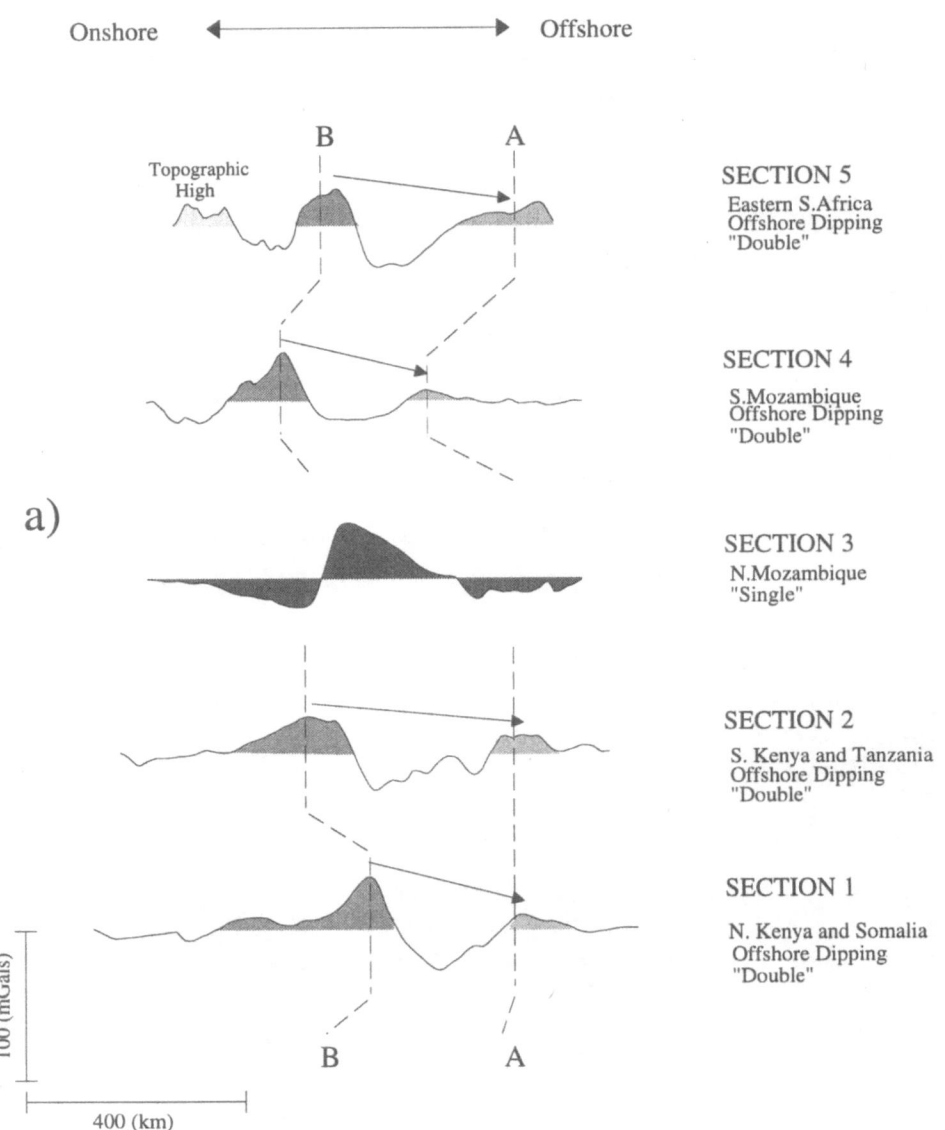

Onshore ◄————————► Offshore

Topographic High

B   A

SECTION 5
Eastern S.Africa
Offshore Dipping
"Double"

SECTION 4
S.Mozambique
Offshore Dipping
"Double"

a)

SECTION 3
N.Mozambique
"Single"

SECTION 2
S. Kenya and Tanzania
Offshore Dipping
"Double"

SECTION 1
N. Kenya and Somalia
Offshore Dipping
"Double"

100 (mGals)

B   A

400 (km)

*Fig. 12 Ensemble average profiles of free-air gravity anomaly across the African margin. **a)** Sections 1-5 (Fig. 11) and **b)** Sections 11-18N (fig 11). The arrows indicate the double edge effect. A = offshore high and B = onshore high (e.g. Fig., 1).*

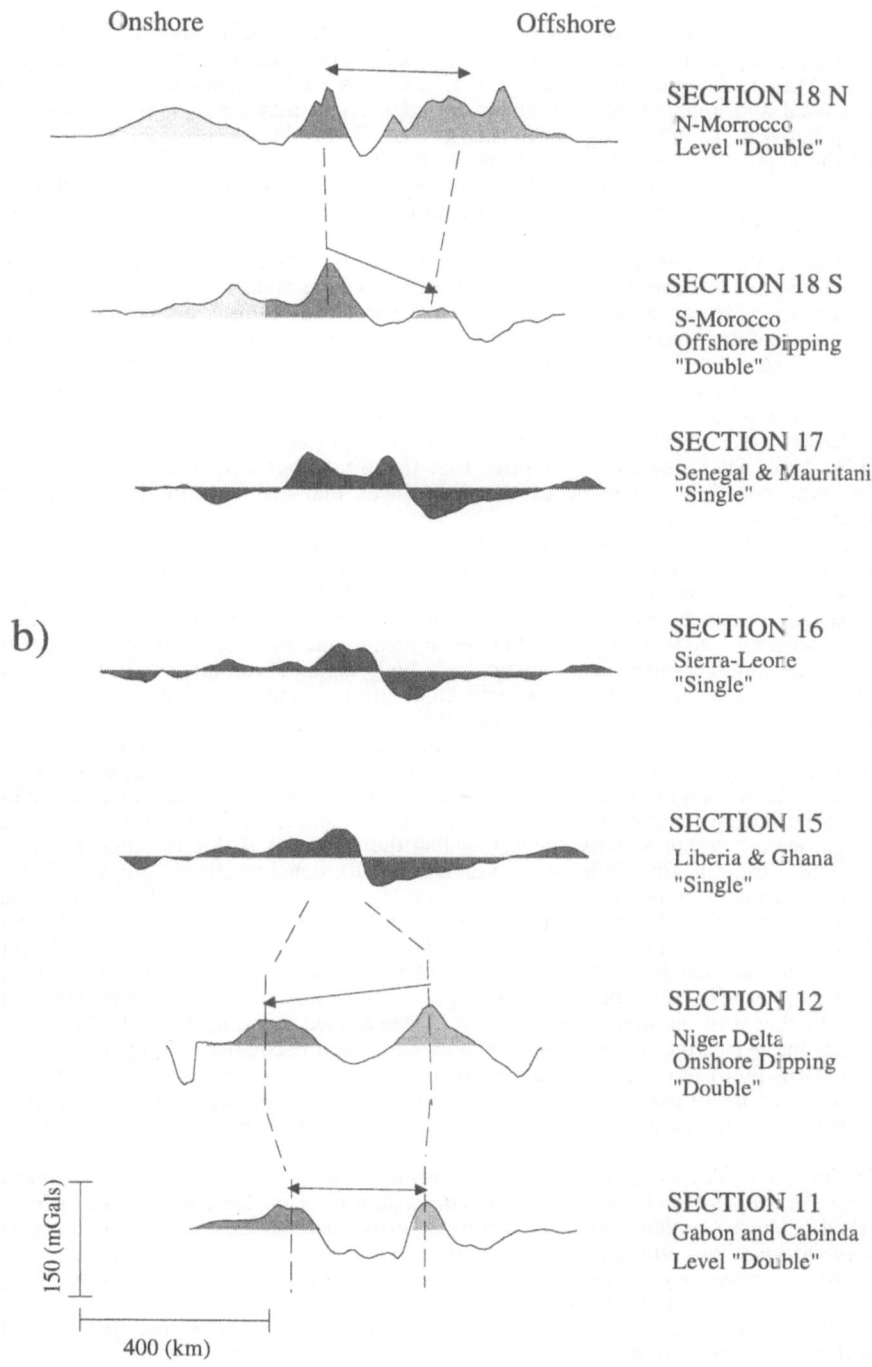

Onshore   Offshore

SECTION 18 N
N-Morrocco
Level "Double"

SECTION 18 S
S-Morocco
Offshore Dipping
"Double"

SECTION 17
Senegal & Mauritania
"Single"

b)

SECTION 16
Sierra-Leone
"Single"

SECTION 15
Liberia & Ghana
"Single"

SECTION 12
Niger Delta
Onshore Dipping
"Double"

SECTION 11
Gabon and Cabinda
Level "Double"

150 (mGals)

400 (km)

6. Discussion

We have shown that the free-air gravity anomaly "edge effect" can be used to constrain the elastic thickness, T_e, at rifted continental margins and hence, the long-term strength of the lithosphere. At those margins such as the East Coast, USA while seismic constraints exits on the deep crustal and upper mantle structure, the edge effect may be used to determine the relationship between T_e and the age since rifting. Once T_e has been estimated, it can be used to predict the depth that the Moho should be on the basis of the loading history of a rifted margin. Hence, by comparing this depth to the Moho determined by seismic refraction, we can estimate whether material has been added to or subtracted it from the crust following rifting.

The analysis in this paper suggest that the East Coast, USA margin is quite uniform in its long-term strength - in contrast to the African margin which appears highly segmented. At the African margin, strong regions occur alongside weak ones. This gravity "segmentation" cannot be described by a particular wavelength: margins are characterized as strong and weak over distances that range from a few hundred to several hundred km. A consequence is that it is likely there will be a significant asymmetry in the strength of the lithosphere between conjugate margin pairs. The edge effect at the Florida margin, for example, is an onshore dipping double (Fig. 10) whereas its conjugate at the Senegal and Mauritania margins is a single (Fig. 12b). Such asymmetry implies that flexure, rather than the tectonic style of extension, may contribute to some of the differences that are seen in the stratigraphic development of conjugate margins pairs.

We believe that in the case of the African margin, the segmentation is probably the result of the influence of hot-spots that - depending on their longevity - have weakened the lithosphere sufficiently to produce the double type of edge effect. The offshore dipping double appears to be a case of pronounced weakening due, perhaps, to a hot-spot trace that trends parallel to the margin whereas onshore dipping doubles indicate one or more hot-spot traces that because of plate motions intersect the margin at a high angle. The single edge effect, on the other hand, has been interpreted as indicating cold, strong, cratonic regions which have not been influenced by hot-spots in recent geological time.

Previous authors have argued, however, that independent of hot-spot activity, extended continental crust may be fundamentally weak. Watts (1988), for example, suggested that at the East Coast, USA margin the lithosphere is weak and, apparently, has remained so for long periods of time during margin evolution. There is no evidence, however, along this margin for any significant hot-spot activity suggesting that the weakness is due to other rheological effects. Yield Strength Envelope (e.g., Goetze, 1978) considerations suggest that the continental lithosphere consists of a strong upper layer, a weak intermediate layer and a strong lower layer. The thickness of the weak intermediate layer is determined by the rheological properties of crustal and upper mantle rocks but, assuming a quartz-rich upper layer and an olivine-rich lower layer then, strain rate, porosity, crustal thickness and the continental geotherm have all been shown to play a role. Because T_e is determined by the mechanical thickness of the lithosphere, the low T_e determined at the East Coast, USA margin suggests that the two strong layers may somehow have been decoupled during rifting. Burov and Diament (in prep) have recently argued that mechanical de-coupling of the lithosphere can occur in response to extension. They showed that the onset of mechanical decoupling is greatly enhanced by increases in crustal thickness and for thicknesses exceeding 35 km can

Fig. 13 Summary diagram showing the relationship between strength variations along the African margin and hot-spot tracks. a) Strength variations. Unfilled columns (i.e. white) = weak. Filled columns (i.e. black/grey) = "strong". Partly filled columns correspond to margin sections of intermediate strength. Note that the African margin is highly segmented with relatively rigid regions (e.g. sections 15-17) juxtaposed to weak ones (e.g., sections 18S, 11-12). b) Absolute motion of the African plate based on (Demets et al., 1990). The motion is shown at 10 Ma interval. Large filled circle = present day position of the hot-spot. Note that although the African margin is not associated with significant hot-spot activity at the present day, they were in the past. Hot-spot activity has been particularly intense along the S. Morocco margin (Canary hot-spot, the SE Africa margin (Bouvet) and the Gabon and Cabinda margins (Cameroon + St Helena).

a)

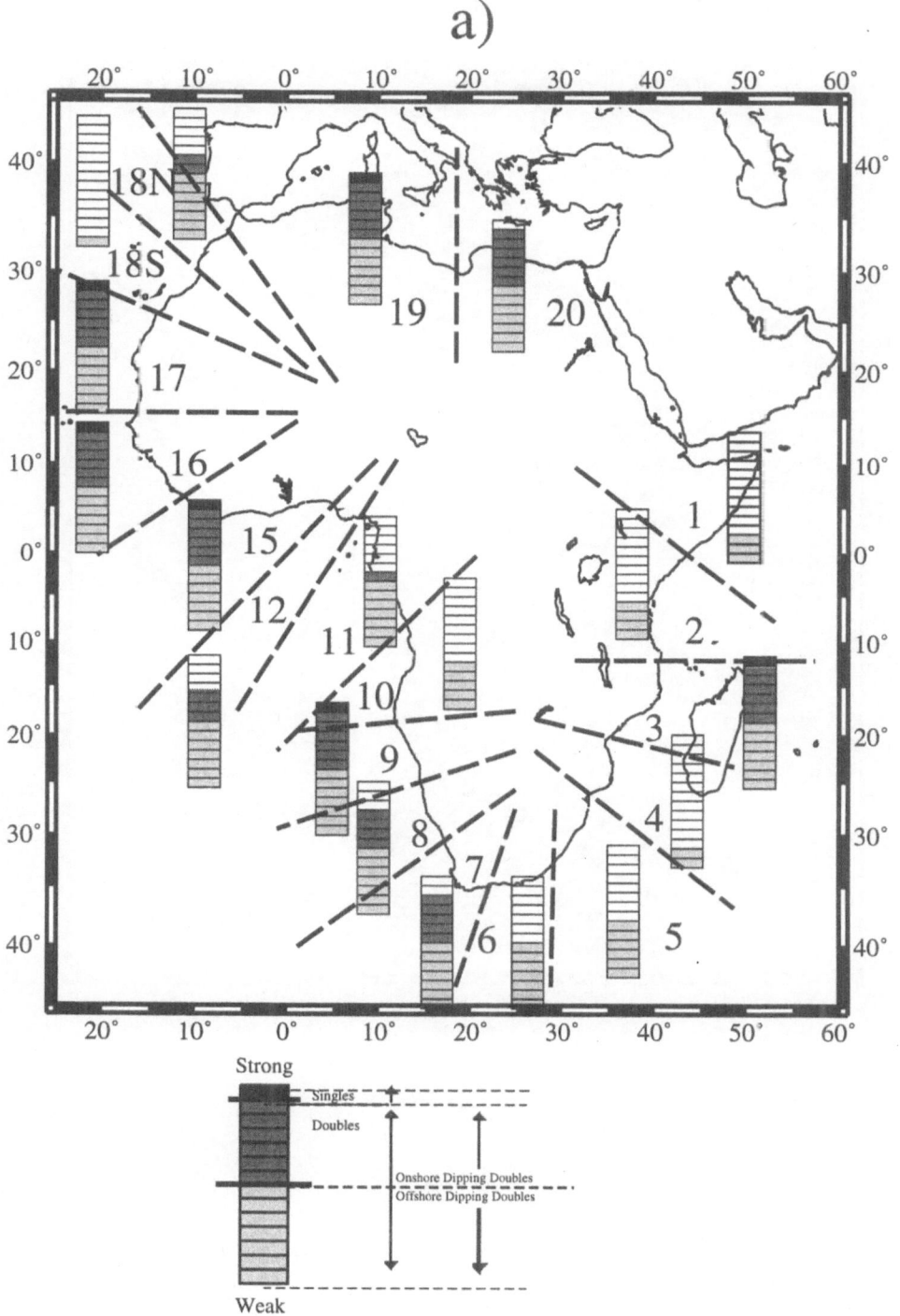

b)

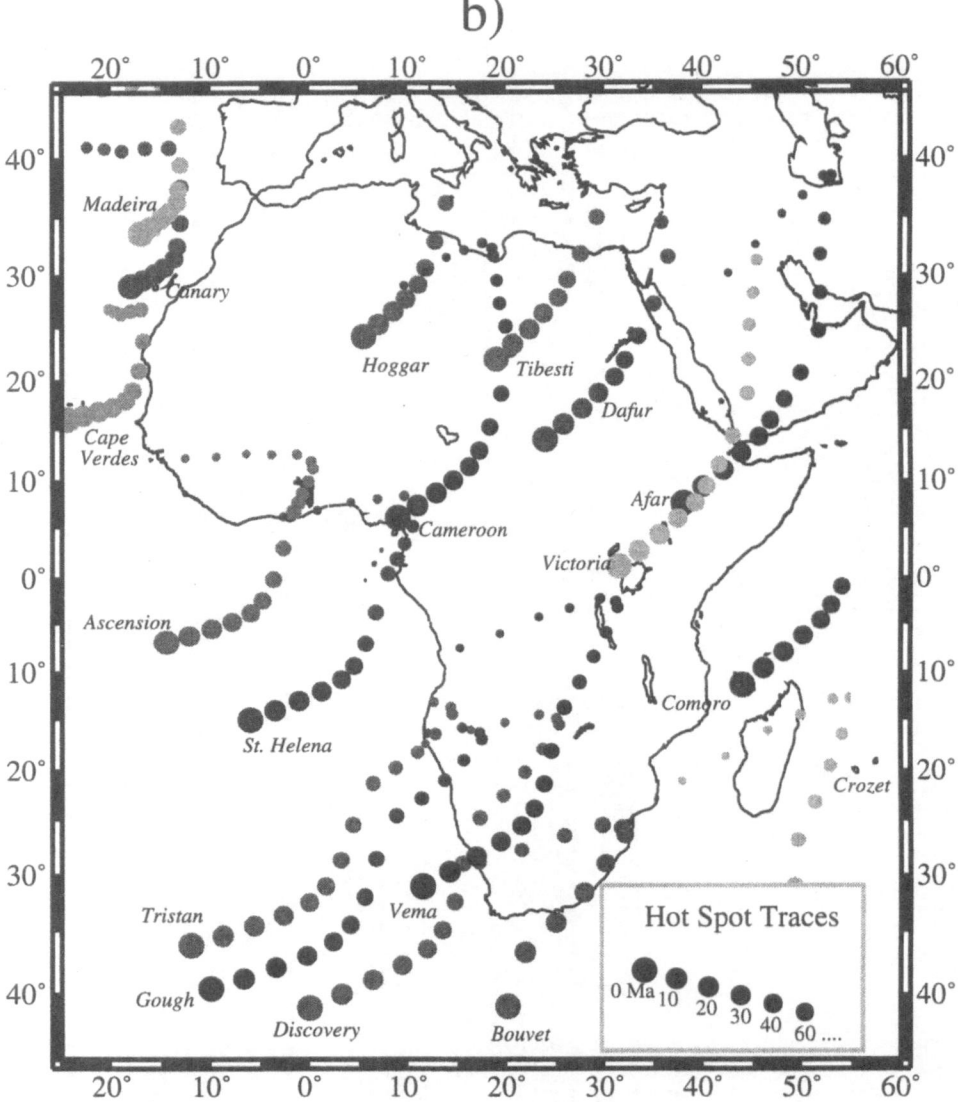

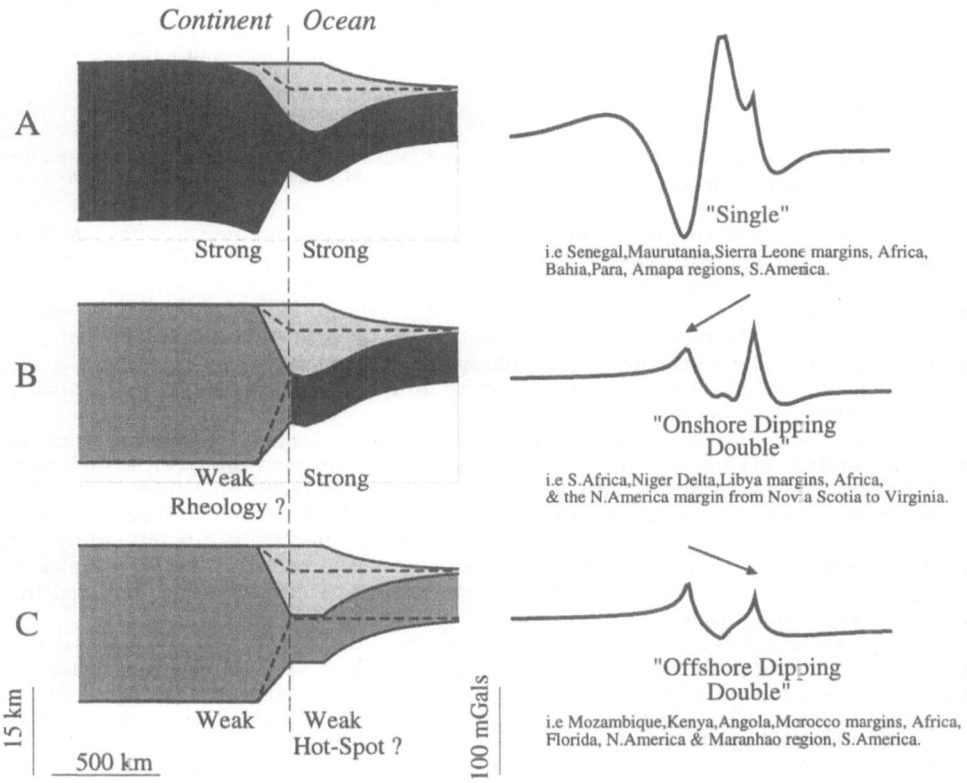

Fig. 14 Summary of the main "types" of rifted continental margins according to the free-air gravity anomaly edge effect. The calculated flexure and edge effect are based on the same initial crust/mantle structure and sediment load. Type A = "Single" edge effect which corresponds to sediment loading of strong continental crust that abuts strong oceanic crust. Type B = "Onshore Dipping Double" edge effect corresponds to sediment loading of weak continental crust that abuts strong oceanic crust. Type C = "Offshore Dipping Double" edge effect corresponds to weak continental crust that abuts weak oceanic crust.

occur quite abruptly. Although the crustal thickness is not well known beneath the East Coast, USA margin, landward of the hinge zone, it is likely that it is somewhat thicker than normal because of the proximity of the margin to the previously stabilised Appalachian orgenic belt.

Irrespective of the actual cause, the low T_e at the East Coast, USA margin suggests that hot-spots may not explain all the segmentation in strength of the lithosphere along the African margin. The offshore dipping double may, for example, be the only reliable indicator of hot-spot activity because it requires that *both* the oceanic and continental are weak (Fig. 14). The onshore dipping double, on the other hand, only requires a relatively weak continent and this may arise because of rheological or other effects.

Despite these difficulties, the edge effect is, we believe, generally diagnostic of the strength of the lithosphere at rifted margins (Fig. 14). Offshore dipping doubles appear to indicate margins with both weak continent and ocean: the weak ocean, at least, being explained by hot-spot activity. Onshore dipping doubles, however, indicate only weak continent which abuts strong or normal ocean. The cause of the weakness is not clear but, it could be indicative of a rheological control. Finally, a single indicate old cratons and in this sense the T_e reflects the background strength of these regions. Whether T_e of the cratons depends on thermal age, however, is still controversial, although it is likely that both composition and present day geothermal gradients play some role.

7. Conclusions

This study of gravity anomalies and lithospheric flexure at rifted continental margins allow the following conclusions to be drawn:

1. Two types of free-air gravity anomaly edge effects occur at rifted margins: the long-wavelength, high-amplitude "single" edge effect associated with high T_e. strong margins, and the short-wavelength, low amplitude "double" edge effect associated with low Te, weak margins. Within the weak margin category two other anomaly types have been recognised: "onshore dipping" doubles which are indicative of stronger lithosphere and "offshore dipping" doubles which indicate weaker lithosphere. These relationships between the edge effect and the strength of the lithosphere appear to be independent of the mechanism by which sediments are transported to a margin.

2. A "case history" study of the East Coast, USA margin - where seismic reflection and refraction profile data are available - show that an onshore dipping double corresponds to the case where sediments have loaded relatively weak stretched continental lithosphere which abuts stronger oceanic and (?) continental lithosphere. Why stretched crust is weak is enigmatic but, the lack of evidence of any hot-spot activity suggest that it is caused by some form of rheological effect (e.g., crustal thickness, composition) which has de-coupled the strong uppermost part of the crust from any support it might otherwise have received from the underlying strong mantle.

3. One consequence of a weak zone at the East Coast, USA margin is that sediment loading has contributed a large part to the total sediment accumulation. The tectonic subsidence (i.e., that part of the sediment accumulation not caused by sediment loading) is also large, however, and close to the maximum that can be produced by thermal models. This suggests that the margin has not had much material added to it following rifting (e.g., by magmatism) since such material would cause uplift and require even larger amounts of thermal subsidence in order to explain the same amount of tectonic subsidence.

4. Application of the simple models to the African margin suggest that it is highly segmented as regards its strength. This segmentation appears to be due to thermal re-heating of the lithosphere following rifting. In particular, weak margins (e.g. doubles) correspond to those that have been thermally reactivated by hot-spot activity in recent geological time, the weakest examples (e.g., offshore dipping double) correspond with segments where the hot-spot trails are parallel to the margin and where the reheating has been quite intense, the stronger ones (e.g., onshore dipping doubles) are associated with hot-spot tracks that are perpendicular to the margin. The strongest margins (e.g. singles) do not appear to have been recently exposed to any thermal source.

5. These studies of the edge effect suggest that conjugate margin pairs maybe highly asymmetric as regards their long-term strength. Thus, a weak margin (e.g., East Coast, USA) may appear alongside a strong one (e.g., West Africa). The cause of such asymmetry is not clear but, rheological effects and thermal re-heating may both play a role. Irrespective of the cause, the thermal and mechanical properties of the lithosphere could explain some of the differences that are seen between conjugate margin pairs, especially in their stratigraphic development.

Acknowledgements

We are grateful to J. D. Fairhead (University of Leeds) for provision of the African Gravity data and to M. McNutt and an anonymous reviewer for their constructive comments on the manuscript..

References

Barton, P. J., and R. J. Wood, 1984. Tectonic evolution of the North Sea Basin: Crustal stretching and subsidence, Geophys. J. R. Astr. Soc., 79, 98-1022.

Bodine, J. H., M. S. Steckler, and A. B. Watts, 1981. Observations of flexure and the rheology of the oceanic lithosphere, J. Geophys. Res., 86, 3695-3707.

Calmant, S., and A. Cazenave, 1986. The effective elastic lithosphere under the Cook-Austral and Society islands, Earth Planet. Sci. Lett., 77, 187-202.

Cochran, J. R., 1973. Gravity and magnetic investigations in the Guiana Basin, Western Equatorial Atlantic,, Geol. Soc. Am. Bull., 84, 3249-3268.

Cochran, J. R., 1981. Simple models of diffuse extension and the pre-seafloor spreading development of the continental margin of the northwestern Gulf of Aden, Proceedings of the 26th International Cong. Symposium on Continental Margins, Oceanologica Acta, 154-165.

Demets, C., R. G. Gordon, D. F. Argus, and S. Stein, 1990. Current plate motions , Geophys. J. Int., 101, 425-478.

Diament, M., J. C. Sibuet, and A. Hadaoui, 1986. Isostasy of the Northern Bay of Biscay continental margin, Geophys. J. R. Astr. Soc., 86, 893-907.

Diebold, J. B., P. L Stoffa, and LASE study Group, 1988. A large aperture seismic experiment in the Baltimore Canyon Trough, in The Geology of North America, The Atlantic Continental Margin, U.S., vol. 1-2, edited by R. E Sheridan. pp. 387-398. Geol. Soc. Amer., Washington.

Flemings, P. B., and T. E. Jordan, 1990. Stratigraphic modelling of foreland basins: Interpreting thrust deformation and lithosphere rheology. Geology, 18, 430-434.

Fowler, S., and D. McKenzie, 1989. Gravity studies of the Rockall and Exmouth Plateaux using SEASAT, Bas. Res., 2, 27-34.

Goetze, C., 1978. The mechanisms of creep in olivine, Phil. Trans. R. Soc., 288, 99-119.

Grow, J. A., Structure of the Atlantic Margin of the United States, in Geology of Passive Continental Margins, vol. Education Course Notes # 19. edited by A. W. Bally, pp. 3-1-3-41, Amer. Assoc. Pet. Geol., Tulsa, Oklahoma.

Hall, B. D., and N. White, 1994. Origin of anomalous Tertiary subsidence adjacent to North Atlantic continental margins, Mar. and Pet. Geol., 11, 702-714.

Holbrook, W S., and P. B. Kelemen, 1993. Large Igneous Province on the US Atlantic Margin and Implications for Magmatism During Continental Break-up, Nature, 364, 433-436.

Holik, J. S., and P. D. Rabinowitz, 1992. Structural and Tectonic Evolution of Oceanic Crust within the Jurassic Quiet Zone, Offshore Morocco, in Geology and Geophysics of Continental Margins, vol. Memoir 53, edited by J. S Walkins. F. Zhiqiang and K. J. McMillen, pp. 259-282, Amer. Assoc. Pet. Geol., Tulsa, Oklahoma.

Hutchinson, D., K. D. Klitgord, and R. S. Detrick, 1986. Rift basins of the Long Island Platform, Geol. Soc. Amer. Bull., 97, 688-702.

Karner, G. D., and A. B. Watts, 1982. On isostasy at Atlantic-type continental margins, J. Geophys. Res., 87, 2923-2948.

Karner, G. D., and A. B. Watts, 1983. Gravity anomalies and flexure of the lithosphere at mountain ranges, J. Geophys. Res., 88, 10,449-10,477.

Kenyon, P. M., and D. L. Turcotte, 1985. Morphology of a Delta Prograding by Bulk Sediment
Transport, Geol. Soc. Amer. Bull., 96, 1457-1465.

LASE study Group, 1986. The Structure of the US East Coast Passive Margin from Large Aperture Seismic Experiments (LASE), Marine Petrol. Geol., 3, 234-242.

Louden, K. E., and D. W. Forsyth, 1976. Thermal conduction across fracture zones and the gravitational edge effect, J. Geophys. Res., 81, 4869-4874.

Manspeizer, W., 1985. Early Mesozoic History of the Atlantic Passive Margins, in Geological Evolution of the United States Atlantic margin, edited by C. W. Poack, pp 1-24, Van Nostrand Reinhold Company, New York.

McKenzie, D., 1978. Some remarks on the developments of sedimentary basins. Earth Planet. Sci, Lett., 40, 25-32.

McKenzie, D., 1984. A possible mechanism for epiorogenic uplift, Nature, 307, 616-618.

McKenzie, D. P., and M. Bickle, 1988. The volume and composition of melt generated by extension of the lithosphere, J. Petrology, 29, 625-679.

McKenzie, D. P., and C. O. Bowin, 1976. The relationship between bathymetry and gravity in the Atlantic Ocean, J. Geophys. Res., 81, 1903-1915.

McNutt, M. K., and K. M. Fischer, 1987. The south Pacific superswell, in seamounts, islands, and atolls, edited by B. H. Keating, P. Fryer, R. Batiza and G. W. Boehlert, pp. 25-34, American Geophysical Union.

Pitman, W. C., 1978. The relationship between Eustasy and Stratigraphic Sequences of Passive Margins, Geol. Soc. Amer. Bull., 89, 1389-1403.

Poag, C. W., 1985. Depositional History and Stratigraphic Reference Section for Central Baltimore Canyon Trough, in Geologic evolution of the U.S. Atlantic Margin, New York, edited by C. W. Poag, pp. 217-264, Van Nostrand Reinhold Company, New York.

Rabinowitz, P. D., and J. L. LaBrecque, 1977. The isostatic gravity anomaly: A key to the evolution of the ocean-continent boundary, Earth Planet. Sci. Lett., 35, 145-150, 1977.

Rowley, D. B., and D. Sahagian, 1986. Depth-dependent stretching: a different approach, Geology, 14, 32-35.

Royden, L., and C. E. Keen, 1980. Rifting Process and Thermal Evolution of the continental Margin of Eastern Canada determined from Subsidence curves, Earth Planet. Sci. Lett., 51, 343-361.

Sarg, J. F., 1988. Carbonate sequence stratigraphy, in Sea-level changes; An Integrated Approach, vol. Spec. Pub. No. 42, edited by B. S. H. C.K. Wilgus, C. G. St. C. Kendall, H.W. Posamentier, C.A. Ross and J.C. Van Wagoner, pp. 155-182, Soc. Econ. Paleon. Min.

Schlee, J., J. C. Behrendt, J. A. Grow, J. M. Robb, R. E. Maattick, P. T. Taylor, and B. A. Lawson, 1976. Regional Geologic Framework off Northeastern United States, Amer. Assoc. Pet. Geol. Bull., 60, 926-951.

Sinclair, H. D., B. J. Coakley, P. A. Allen, and A. B. Watts, 1991. Simulation of foreland basin stratigraphy using a diffusion model of mountain belt uplift and erosion: an example from the central Alps, Switzerland, Tectonics, 10, 599-620.

Sleep, N. H., 1971. Thermal effects of the formation of Atlantic continental margins by continental break-up, Geophys. J. Roy. Astr. Soc., 24, 325-350, 1971

Steckler, M. S., and A. B. Watts, 1978. Subsidence of the Atlantic-type continental margin off New York, Earth Planet. Sci. Lett., 41, 1-13.

Talwani, M., and O. Eldholm, 1973. The boundary between continental and oceanic crust at the margin of rifted continents, Nature, 241, 325-330.

Turner, J. D., and R. A. Scrutton, 1993. Subsidence patterns in western margin basins: evidence from the Faeroe-Shetland basin, in Petroleum Geology of Northwest Europe: Proceedings of the 4th conference, edited by J. R. Parker, pp. 975-983, Geol. Soc. London.

Walcott, R., 1972. Gravity, flexure, and the growth of sedimentary basins at a continental edge, Geol. Soc. Am. Bull., 83, 1845-1848.

Watts, A. B., 1978. An analysis of isostasy in the world's oceans: 1. Hawaiian-Emperor Seamount Chain, J. Geophys. Res., 83, 5989-6004.

Watts, A. B., 1988. Gravity anomalies, crustal structure and flexure of the lithosphere at the Baltimore Canyon Trough, Earth Planet. Sci. Lett., 89, 221-238.

Watts, A. B., G. D. Karner, and M. S. Steckler, 1982. Lithospheric flexure and the evolution of sedimentary basins,. The Evolution of Sedimentary Basins, vol. 305A, 249-281 pp., Phil. Trans. Roy. Soc. London.

Watts, A. B., and W. B. F. Ryan, 1976. Flexure of the lithosphere and continental margin basins, Tectonophysics, 36, 25-44.

Watts, A. B., and M. S. Steckler, 1979. Subsidence and Eustasy at the continental margin of eastern North America, Deep Drilling Results in the Atlantic Ocean: Continental Margins and Paleoenvironment, vol. Amer. Geophysical Union, 218-234 pp., Maurice Ewing Series vol. 3

Watts, A. B, and M. Torne, 1992. Crustal Structure and the Mechanical Properties of Extended Continental Lithosphere in the Valencia trough (Western Mediterranean), Journal Geol. Soc. London 149, 813-827.

Weissel, J. K., and G. D. Karner, 1989. Flexural uplift of Rift Flanks due to Mechanical Unloading of the Lithosphere during Extension, J. Geophys. Res., 94, 13,919-13,950.

Wessel, P., and W. F. Haxby, 1990. Thermal stresses, differential subsidence and flexure at oceanic fracture zones, J. Geophys. Res., 95, 375-391, 1990.

Wessel, P., and W. H. F. Smith, 1991. Free software helps map and display data, EOS Trams, Amer. Union, 72, 441-446.

White, N., and D. P. McKenzie, Formation of the "Steer's Head" Geometry of Sedimentary Basins by Differential Stretching of the Crust and Mantle, Geol., 16, 250-253.

White, R. S., 1992. Crustal Structure and Magmatism of North Atlantic Continental Margins, J. Geol. Soc. London, 149, 841-854.

White, R. S., G. K. Westbrook, A. N. Bowen, S. R. Fowler, G. D. Spence, C. Prescott, P. J. Barton, M. Joppen, J. Morgan, and M. H. P. Bott, 1987. Hatton Bank (Northwest UK) continental margin structure, Geophys. J. R. Astr. Soc., 89, 265-267.

Worzel, J. L., 1968. Advances in marine geophysical research of continental margins, Can. J. Earth Sci., 5, 963-983.

Table 1. Thermal and Mechanical Parameters assumed in calculations

Youngs Modulus = E = 100 GPa
Poisson's Ratio = σ = 0.25

$$T_e = \sqrt[3]{\frac{12D(1-\sigma^2)}{E}}$$

where = D = flexural rigidity of the Lithosphere
Thermal Thickness of the Lithosphere = a =125 km
Density of Lithosphere at 0^oC = ρ_{lo}= 3330 kg m^{-3}
Density of water = ρ_w = 1030 kg m^{-3}
Coefficient of Volume Expansion = α = 3.28 x 10^{-5} $^oC^{-1}$
Temperature of mantle = T_1 = 1303°C

Slope of best fit line of Tectonic Subsidence vs $(1-e^{(-t/\tau)})$ = E_o r
where

$$E_0 = \frac{4a\rho_{lo}\alpha T_1}{\pi^2(\rho_{lo}-\rho_w)} \qquad r = \left(\frac{\beta}{\pi}\right)\sin\left(\frac{\pi}{\beta}\right)$$

NATURE OF THIN CRUST ACROSS THE SOUTHWEST GREENLAND MARGIN AND ITS BEARING ON THE LOCATION OF THE OCEAN-CONTINENT BOUNDARY[1]

S.P. SRIVASTAVA and W.R. ROEST[2]
Atlantic Geoscience Centre,
Geological Survey of Canada,
Bedford Institute of Oceanography,
P.O Box 1006, Dartmouth, N.S., Canada B2Y 4A2

ABSTRACT. The detailed seismic refraction and a limited amount of deep reflection measurements across the southwest Greenland margin show presence of a wide zone of thin upper crust overlying a serpentinized mantle. The nature of the crust in this zone, however, remains in dispute. Interpretation that this is thinned continental crust is in conflict with the interpretation based on magnetic data which suggest it to be oceanic. The magnetic data from this region have, therefore, been re-examined here and it is shown that even though the magnetic anomalies are small in amplitudes and variable in shape over a short distance, they can reasonably be correlated with synthetic seafloor spreading anomalies. It is possible for these anomalies to be caused by injection of volcanic material through continental crust, but their overall continuity and linear character, their resemblance to anomalies formed in the southern Labrador Sea, their symmetry across the extinct ridge, and their correlation with seafloor spreading models, strongly argue for their formation by seafloor spreading. Such an interpretation is also consistent with the plate kinematic motions derived for the North American and Eurasian Plates, and shows that the Labrador Sea essentially started to form along a northwest continuation of the Mid-Atlantic Ridge at chron 33 time.

The correlation of magnetic anomalies in the present model shows a drastic change in the half rate of spreading at chron 30 from 5.8 mm/y before to a mean value of 12.0 mm/y after. It is suggested that the decrease in amplitude of magnetic anomalies arises from fragmentation of the oceanic crust formed at such low spreading rates. The change in the rate of spreading correlates well with a change in basement topography, from rough during slow, to smooth during faster spreading and the occurrence of thin crust during slow spreading and a normal thickness during faster spreading. These changes in crustal properties are remarkably similar to those observed across the central Labrador Sea, where the half spreading rate changed from 10 mm/y to 3.5 mm/y before the cessation of spreading. Here a clear division in basement morphology and crustal thickness is observed between the crusts formed at these two rates. The crust formed at 10 mm/y half spreading rate exhibits smoothly undulating basement with slightly less than normal crustal thickness, while the crust formed at 3.5 mm/y half spreading rate show evidence of intense normal faulting, with many faults showing large offsets and extending to lower crust and Moho depths. Refraction results together with gravity modelling show the crust to be abnormally thin overlying serpentinized upper mantle. The similarities between the crustal structures formed at the central Labrador Sea and that formed across the SW Greenland margin support the suggestion that the thin crust across SW Greenland margin is oceanic and was formed during slow seafloor spreading. Furthermore, the magnetic modelling suggest that the ocean-continent boundary lies fairly close to the bottom of the continental slope in this region. Comparison of this crust with crust formed across several other continental margins show great similarities suggesting that they

---

[1]Geological Survey of Canada Contribution no. 37994

[2]Geophysics, Geological Survey of Canada, 1 Observatory Crescent, Ottawa, Ont., K1A 0Y3.

95

E. Banda et al. (eds.), Rifted Ocean-Continent Boundaries, 95–120.
© 1995 *Kluwer Academic Publishers. Printed in the Netherlands.*

also were formed at slow spreading rates.

Introduction

The Labrador Sea, located between the coasts of Labrador and Greenland is a relatively small ocean basin. The vast amount of geological and geophysical data collected in this region (for details see Bell, 1989) has been used to decipher its history of evolution by Srivastava (1978); Srivastava et al. (1981) and Roest and Srivastava (1989a). They have shown that the region was formed by seafloor spreading with rifting starting between Greenland and Labrador some time in the early Cretaceous and culminating in the formation of seafloor magnetic anomalies starting at chron 34 in the southern Labrador Sea and at chron 33 in its northern part (Fig. 1).

In spite of the large amount of geophysical and geological data, a serious disagreement exists about the extent of oceanic crust in the basin and the location of the ocean-continent boundaries. This has been a subject of much debate for several years (e.g. Grant, 1984) and more so recently (Chalmers et al., 1993). The existence of continental crust, extensively intruded by igneous material has been suggested by Chalmers (1991) for the northwest margin of Greenland. Detailed seismic refraction measurements carried out across the SW Greenland margin (CL, Fig. 1) have been interpreted in terms of thinned continental crust overlying serpentinite mantle over a wide zone (80 to 100 km) by Chian and Louden (1994). Deep seismic reflection measurements carried out in close vicinity (line 90-3 in Fig. 1) to the refraction measurements show the possibility that the region could be continental (Keen et al., 1994). This is because of the disturbance and faulting seen, not only in the acoustic basement but also in the deeper layers of the sediments lying between basement blocks beneath the continental rise, suggesting that the regions may contain syn-rift sediments (Keen et al., 1994).

The interpretations favouring thinning of the continental crust imply the existence of about 100 km of such crust across these margins. This may be in direct conflict with the numerical computations of Bassi et al., (1993) which suggest that the lithosphere will rupture long before such extreme thinning can occur over such long distances. The purpose of this paper is to re-examine the magnetic measurements across the SW Greenland margin in light of the additional geophysical data which now exist there and to decipher the nature of the crust across it. Recent deep reflection measurements made across the extinct spreading centre in the central part of the Labrador Sea (line 90-2, Fig. 1) are specially useful when examining the SW Greenland margin as they show for the first time that extensional faulting seems to play a significant role in forming abnormally thin oceanic crust at very slow spreading rates (Srivastava and Keen, 1994).

THE PROBLEM AND PREVIOUS OBSERVATIONS

For the sake of clarity, a brief review of existing interpretations of the SW Greenland margin is given below before examining other possibilities for the nature of this thin crust.

Chalmers Model: The complexity of the subsurface structures present under the Labrador Sea was presented by Hinz et al., (1979) based on a number of seismic reflection profiles collected by the Bundesanstalt fur Geowissenschaften und Rohstoffe (BGR) in 1977. One of the lines from this survey, BGR-12 across northwest Greenland margin (Fig. 1), was reprocessed and interpreted by Chalmers (1991). He showed the presence of rotated and faulted basement blocks overlain by layered sediments which lie parallel to the down throw sides of the faulted blocks, in a region where magnetic anomaly 33 was tentatively identified by Roest and Srivastava (1989a). Farther oceanwards, the seismic data showed the presence of a seaward dipping layer, which Chalmers interpreted as arising from a contact between continental crust below and volcanic material on top.

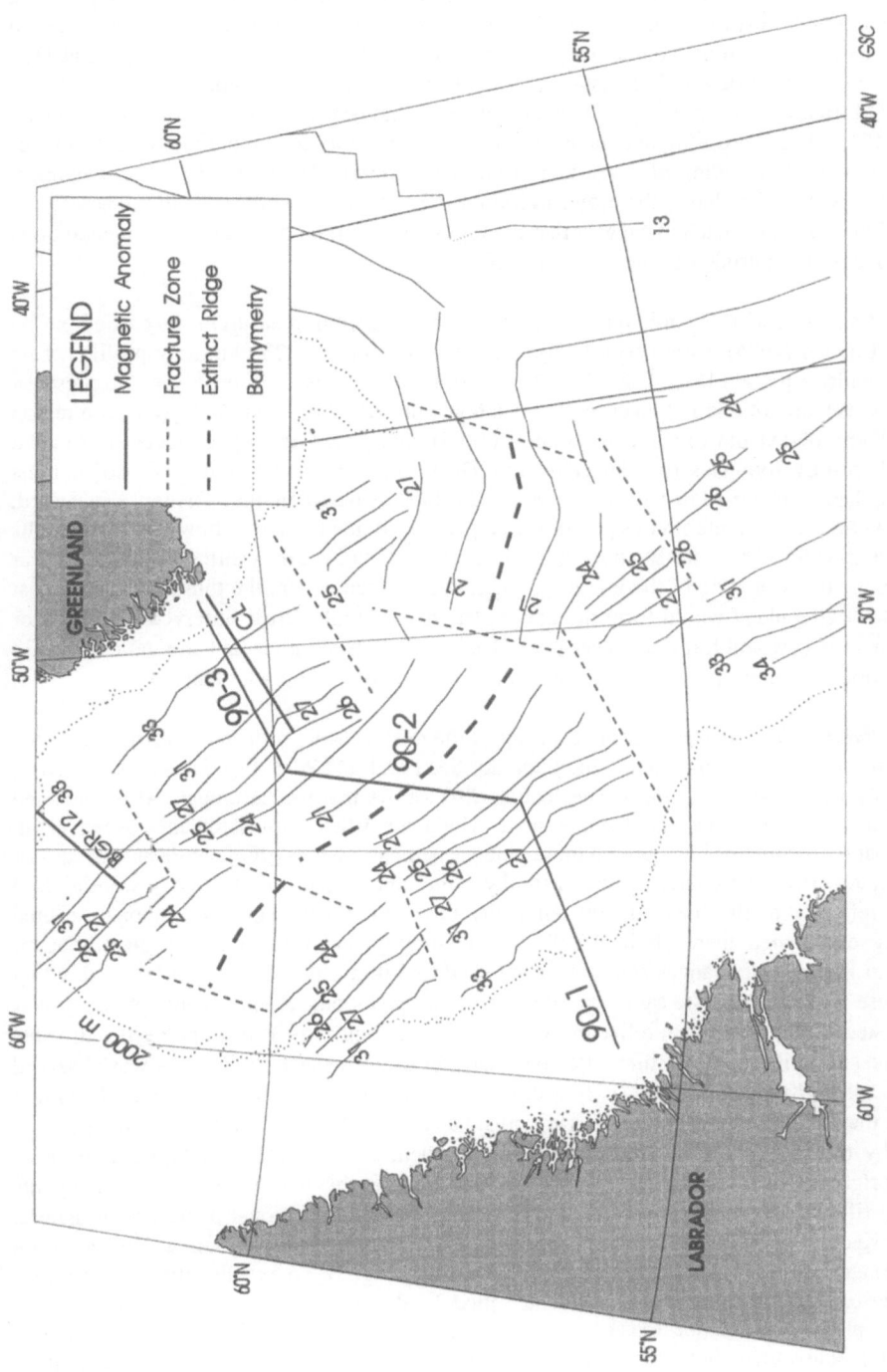

Fig. 1. Map showing the location of multichannel seismic reflection lines 90-1, 90-2, 90-3 and BGR-12. Also shown are the magnetic isochrons, fracture zones and the extinct ridge derived from the map of Roest and Srivastava (1989b). CL- refers to location of the Chian and Louden (1994) refraction line.

Chalmers, therefore, interpreted the magnetic anomalies landward of chron 27 as arising from a thinned continental crust overlain by reversely magnetised volcanic material. This interpretation has now been extended to the southern margins of Greenland and Labrador (Chalmers and Holt Laursen, 1995), where Roest and Srivastava (1989a) had identified the presence of chron 27 to 33. One of the difficulties, as shown by Chalmers and Holt Laursen (1995), in accepting magnetic anomalies 27 to 33 as spreading anomalies, has been the difficulty in simulating these anomalies based on the rate of spreading of Roest and Srivastava (1989a). Chalmers (1991) and Chalmers and Holt Laursen (1995) have, therefore, proposed the absence of chrons 28 to 33 across the margins of the Labrador Sea and suggest that the region is underlain by thinned continental crust which is extensively intruded by igneous material.

Chian and Louden Model: Seismic refraction results obtained from measurements carried out by Chian and Louden (1994) using ocean bottom seismometers along a 230 km long profile across the SW Greenland margin (Fig. 1 and Fig. 4) are shown in Figure 2. These show the presence of a thin low velocity (4.6 km/s) layer overlying a high velocity (7.0-7.6 km/s) layer over a region 80 to 100 km in extent. Chian and Louden (1994) interpreted the upper layer as thinned continental crust overlying a serpentinized mantle. The results also show thinning of a 30 km crust below Greenland to about 3 km across the continental slope. A basement high lies farther seaward, which Chian and Louden interpret as part of the serpentinized mantle. They show that their results are consistent with a two dimensional gravity modelling along this profile. A simple shear mechanism for the formation of the margin is suggested by them where the thin continental crust interpreted off Greenland would form the upper plate and a similar crust observed off Labrador (Chian et al., 1994) would form the lower plate. Their model also implies that anomalies 28 to 33 are not seafloor anomalies, but they do not explain what caused these anomalies.

Keen et al. Model : The interpreted line drawing of the deep seismic reflection profiles collected along sections of conjugate margins of the Labrador Sea (90-1 and 90-3, Fig. 1) as given by Keen et al., (1994) are shown in Figure 3. The section 90-1 shows that the basement (B) is dissected by numerous faults, forming small depressions which are filled with sediments. Keen et al. proposed that the sediments lying within these grabens perhaps are syn-rift in nature as they appear to be separated from the overlying sediments by a prominent reflector (U). They were able to relate this reflector to the break up unconformity under the shelf using the regional seismic stratigraphy established there (Bell, 1989). The continuity and reflectivity of this reflector, according to Keen et al. changes seaward of the shelf and they show it as U_1 to U_3. According to them these may not have the same age or tectonic interpretations. A similar interpretation was then suggested for line 90-3 with reflectors W and W_1 separating syn-rift from post-rift sediments based on the similarity of the seismic reflective character of this reflector with what was observed along line 90-1 (Keen et al. 1994). If sediments under these reflectors are indeed syn-rift sediments, then it implies that the underlying crust has to be continental in nature similar to that proposed by others (Chalmers, 1991; Chian and Louden, 1994). On the other hand, if these reflectors originate from some other tectonic process, it is possible that the underlying sediments are not syn-rift and therefore the crust underlying them may be transitional or oceanic in nature. Therefore, Keen et al. give two scenarios; one where the continent-ocean boundary lies at chron 33 (COB 1) and the other where it lies at chron 27 (COB 2, Fig. 3). To examine these possibilities further they carried out gravity and subsidence model calculations along these sections. These calculations also suggested the existence of thin crust across both margins with high density material lying below them, but were not conclusive enough to distinguish between the two models.

The answer to the problem whether the thin crust is oceanic or continental perhaps lies in the identification of the magnetic anomalies and their association with the basement features as seen

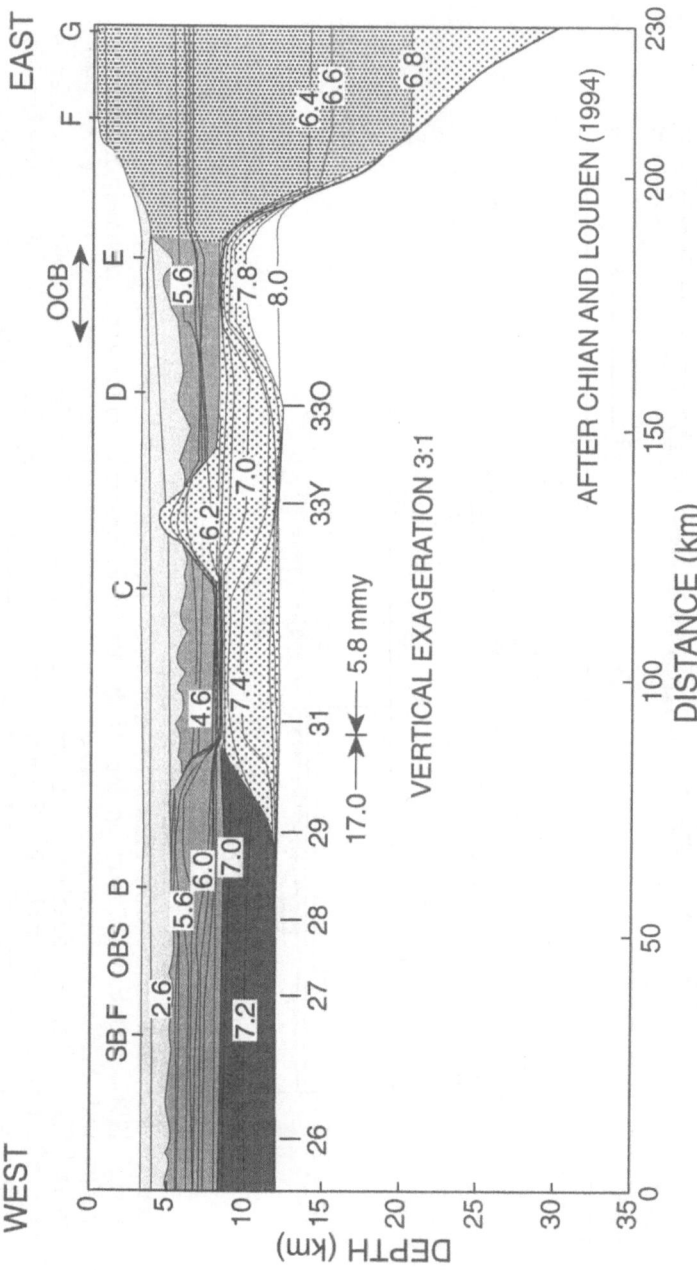

Fig.2. Two dimensional velocity model showing iso-contours of V_p (in km/s) across SW Greenland margin as given by Chian and Louden (1994). Also shown are locations of ocean bottom seismometer and sonobuoy F. of magnetic isochrons (number 27 etc.) as identified here and the location where a drastic change in spreading rate is obtained. The interpreted position of the ocean-continent boundary is shown.

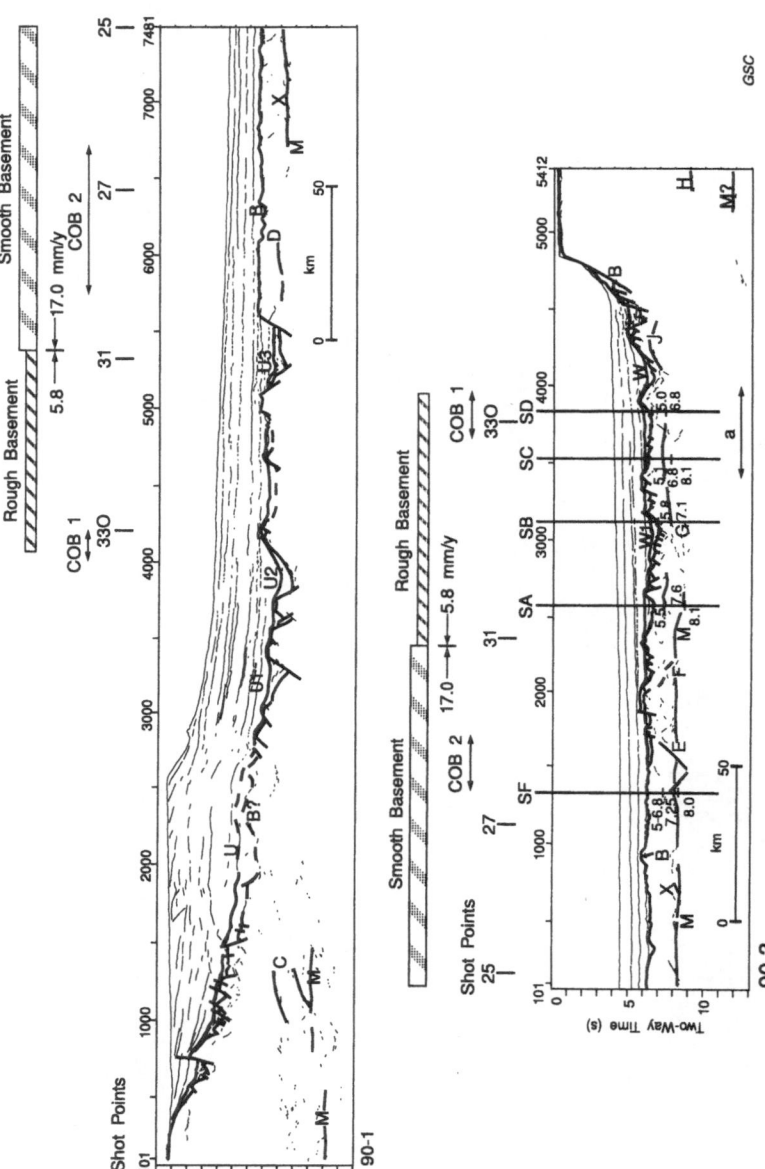

Fig. 3. The interpreted line drawing of MCS lines 90-1 and 90-3 (location see Fig.1) showing the main feature as given by Keen et al., (1994). Numbers (27 etc) on top of the line drawings refer to magnetic anomaly numbers as identified here together with half rate of spreading. Letters SF etc designate position of sonobuoy lines where they cross line 90-3 (location see Fig.4). Velocity distribution (in km/s) below these sonobuoy lines are shown. Letters refer to M-moho; X-dipping reflectors at moho level, E F strong dipping events; U, U1, U2, U3, W W1 -unconformities, B - basement, G J sub-basement reflectors. OCB1 and OCB2 are the two alternative positions of continent ocean boundary as given by Keen et al., (1994). a- refers to the enlarged section of the record shown in Fig. 9.

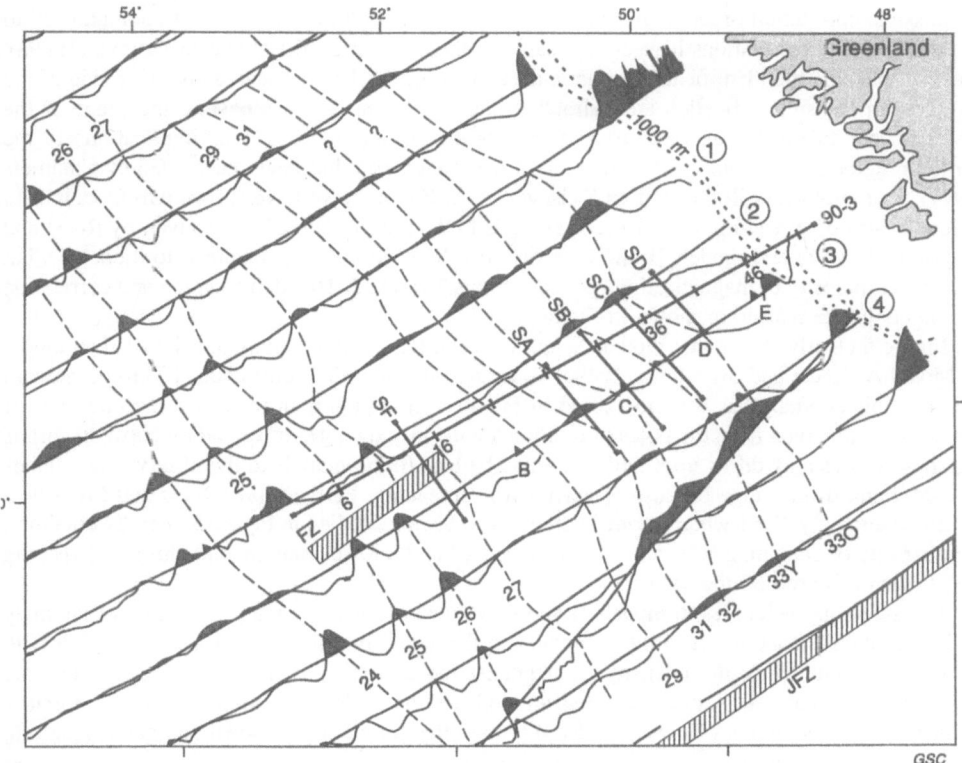

Fig. 4. Plot of magnetic anomalies along tracks, with their correlation forming isochron lines. Circled numbers refer to tracks shown in Fig.5. Also shown are the positions of MCS line 90-3, sonobuoy lines (SF etc) and the ocean bottom seismometers B to E shown by solid triangles. JFZ - Julianhaab Fracture Zone from Roest and Srivastava (1989b).

in the reflection and refraction data.

Magnetic Anomalies

Identification of magnetic anomalies throughout the Labrador Sea was originally carried out by Srivastava (1978) and was subsequently modified by Roest and Srivastava (1989a) to take into account the misidentification of anomaly 25 at places in the earlier identification of Srivastava (1978). This resulted in a slightly different identification of older anomalies and hence in the poles of rotation for these anomalies. The details of their identification were also published on a large scale map (Roest and Srivastava, 1989b). A portion of the resulting lineations extracted from this map is shown in Figure 1. We will examine a portion of this data across the southwest Greenland margin where details of both the seismic refraction and reflection measurements are available.

Figure 4, extracted from Roest and Srivastava's (1989b) chart of the Labrador Sea, is a plot of the magnetic anomalies along ships tracks across the southwest Greenland margin. It also shows the possible correlation of anomalies between ship tracks. With the exception of anomalies 29 to 33Y (young) the correlations here are the same as those given by Roest and Srivastava (1989a). We will focus on the identification of anomalies along tracks 1 to 4, located on either side of the region where seismic reflection and refraction data exist (Fig. 5). It is beyond the scope of the present paper to discuss all data used in the correlation and discuss their implications to the northern region as well, where identification of these anomalies has been questioned by Chalmers (1991). This problem will be dealt with elsewhere. In Figure 5, we have shown two models with different rates of spreading. Model A corresponds to the rates of spreading as given by Roest and Srivastava (1989a) and models B and C as obtained here. Model C is identical to model B , but with lower intensity of magnetisation between chrons 29 and 33. The identification and correlation for the conjugate margin is shown in Figure 6.

Figure 5 clearly shows that models B and C match the observed anomalies better than model A. Model A (Roest and Srivastava, 1989a) generates anomaly 33O (old) about 10 km oceanward to where this anomaly has been identified in Figure 5. It is important to realize that the rates of spreading as obtained from the poles of rotation are the average rates of spreading between chrons and these values may differ from those obtained by best fitting models to the observed anomalies on each flank of the ridge because of asymmetric spreading. Furthermore, Roest and Srivastava identified anomaly 30/31 where anomaly 29 has now been identified in Figure 5, thereby obtaining a higher rate of spreading between chrons 27 and 29 and a lower rate beyond chron 31, keeping the position of chron 33 the same.

The spreading velocities in models B and C were constrained by the small width of anomaly 30-31 and the distance between the anomalies in the actual profiles. To get a closer match with the observed anomalies, the intensity of magnetisation between chrons 29 to 33 was decreased (model C, Figs. 5 and 6). A lower magnetisation in regions of slow spreading rates could possibly result from fragmentation of the crust. Even though the correlation between anomalies resulting from model C is favoured here, not much difference was observed between model C and another model based on a uniform rate of spreading of 12 mm/y between chrons 26 and 30/31 except for a larger distance between chrons 26 and 27 than observed from the data. What is most important to note here is that all models that provide a reasonable fit imply a significant change in the rate of spreading in the vicinity of anomaly 30. This observation correlates well, as we will show later, with changes in the seismic reflection and refraction measurements.

Anomaly 33Y shows up as a small amplitude anomaly along most, tracks contrary to that shown in model C. To examine if this can be caused by basement topography, calculations for model C were carried out along the Chian and Louden (1994) seismic line (Fig. 7). We compared these with the observations along a track coincident with their profile (track 3, Fig. 4). Figure 7 shows that this anomaly is indeed largely influenced by basement topography. Significant variations in the basement topography explain why this anomaly is not present along every track in this region (Fig. 4).

In modelling the magnetic anomaly at the foot of the slope, where a large negative anomaly is observed, two possibilities were considered: One, that the region is entirely underlain by thinned continental crust; Two, that part or all of it contains negatively magnetised volcanic rocks as can be expected during pre-chron 33O time with a junction between continental and volcanic rocks near OBS E (Fig. 7), where the crust thins to 3 km (Fig. 2). The results are shown in Figure 7 as calculated anomalies B and A respectively. It can be concluded from these calculations that the large negative anomaly is not entirely caused by continental crust and that some negatively magnetised volcanic rocks are needed for this purpose. The answer perhaps lies between the two models. In spite of the simplicity of the model the fit between the model and observations is remarkably good, suggesting that the thin crust observed along this profile is most likely oceanic in origin.

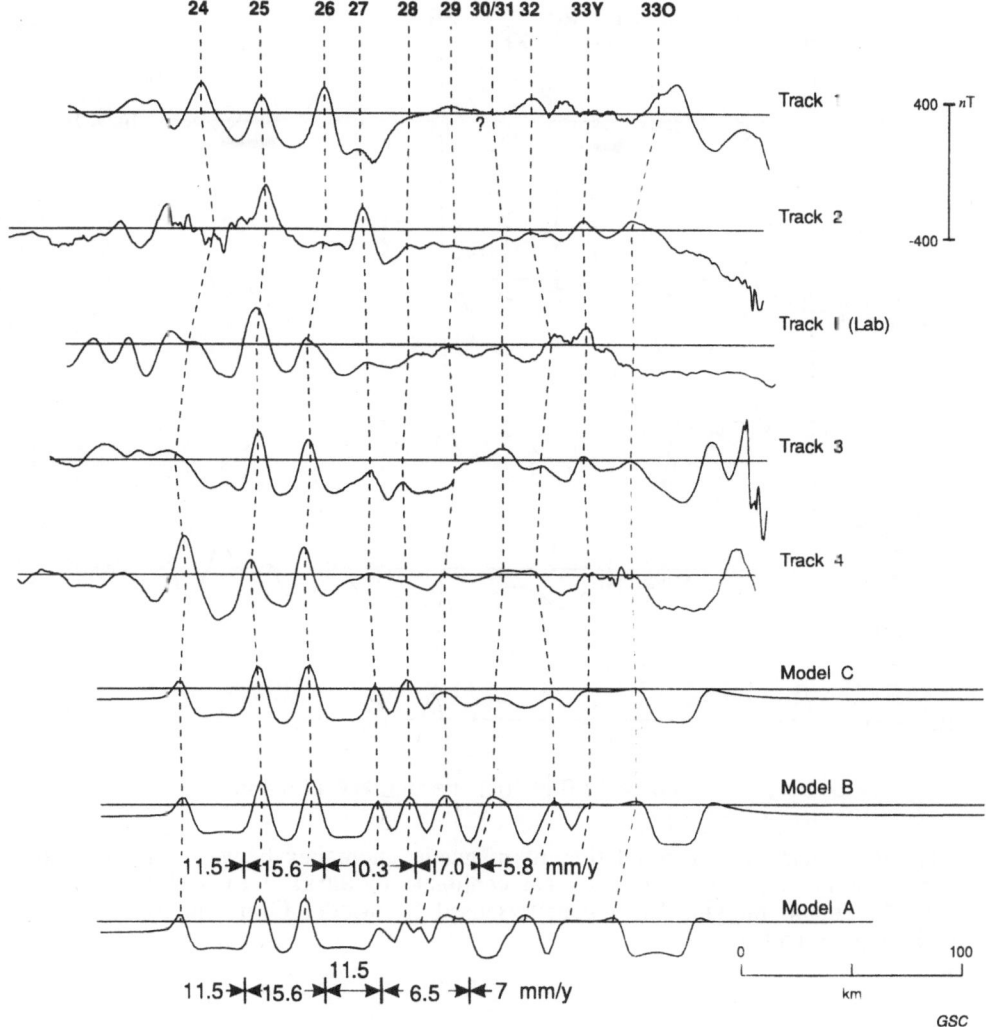

Fig. 5. Identification and correlation of magnetic anomalies, between tracks 1 to 4 (Fig.4) across the SW Greenland margin and track I conjugate to track 2 across the Labrador margin, based on the models as shown. The half spreading rates for the three models are shown below them. Model C and B have the same rates of spreading but different intensities of magnetisation. Model A corresponds to the rate of spreading as given by Roest and Srivastava (1989a). Calculations were done using I = 63° and D = -28° for remanent magnetisation and I = 71° and D = -34° for the induced magnetisation. Intensity of magnetisation for all bodies 1 km thick was 5 A/m except for bodies corresponding to chrons 25 and 26 which had values of 7 A/m. Intensity of magnetisation used for model C are shown in Fig.7.

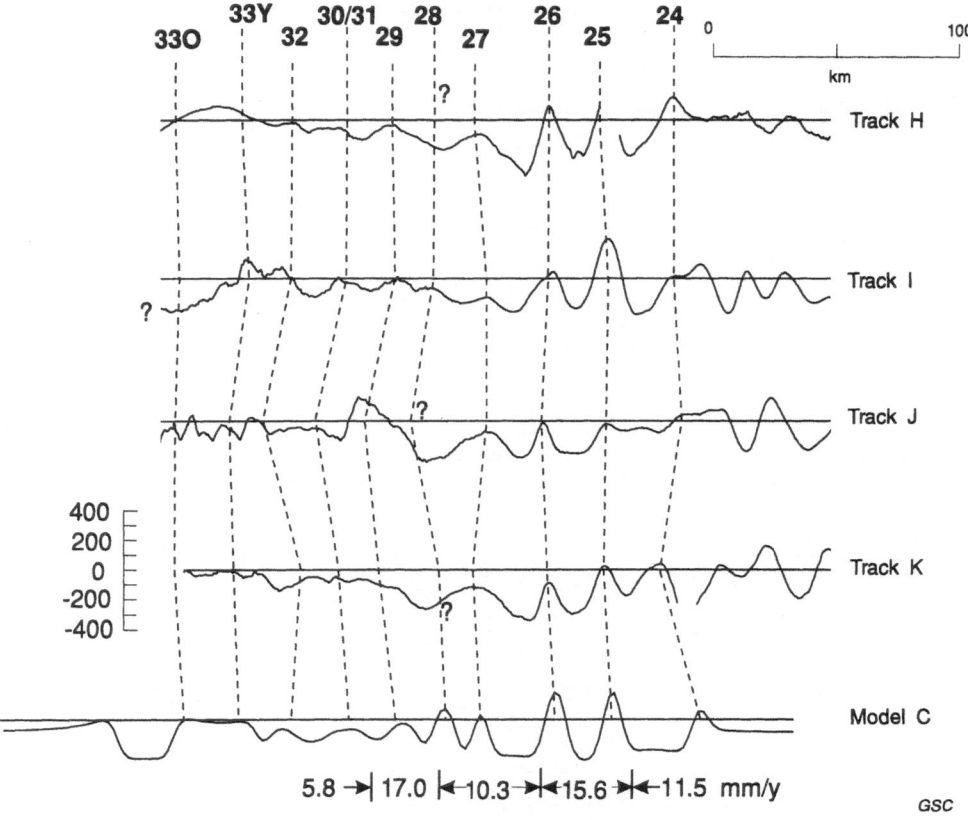

Fig. 6. Identification and correlation of magnetic anomalies between tracks H to K across the Labrador margin which are conjugate to tracks 1 to 4 (Fig.4) across the SW Greenland margin. The parameters used for Model C are the same as given in Fig. 5 and 7.

Comparison With Other Data

The magnetic modelling results show that the rate of spreading changed drastically at about chron 30 time. Such a decrease in spreading rate coincides remarkably well with the change in the basement topography as well as with the change in crustal thickness (Fig. 2). Not much is known about what kind of variations in crustal properties can be expected at such low spreading rates, except that theoretical calculations suggest some decrease in crustal thickness (Reid and Jackson, 1981; Su et al., 1994; Bown and White, 1994). In this regard the deep multichannel seismic reflection measurements carried out across the central part of the Labrador Sea (line 90-2, Fig. 1) are of much help. As a number of inferences derived from these measurements have been used here, we will briefly summarize their findings.

DEEP MCS MEASUREMENTS ACROSS THE CENTRAL LABRADOR SEA

Figure 8 shows the interpretation and line drawing of profile 90-2 (Fig. 1) as given by Srivastava

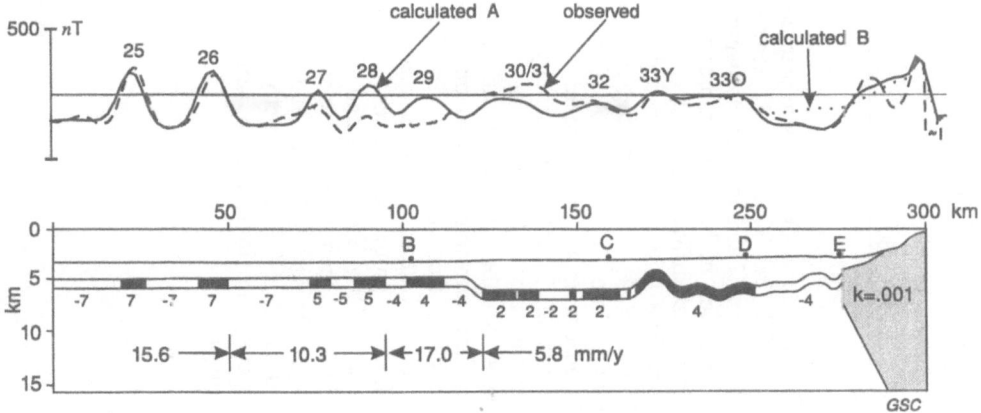

Fig. 7. Correlation between calculated and observed magnetic anomalies along the seismic refraction profile of Chian and Louden (1994) where the source of the magnetic anomalies have been assumed to lie within a 1 km thick layer with the intensities of magnetisation (in A/m) as shown below them wherever possible. Black blocks are normally magnetised. Susceptibility (k) value used for the continental crust is shown. Numbers with arrows signify the rate of spreading.

and Keen (1994). The entire line is shown in three sections, here called zones; each showing a region formed at a different rate of spreading. Zones A and C cover the sections formed between chrons 25 and 21, when the rate of spreading was about 10 mm/y, while zone B shows the region between chrons 21, when the spreading rate had started to decrease before the ridge became extinct at about chron 13. This zone was formed at a mean half rate of spreading of about 3.5 mm/y. The data show a clear division between two types of crust formed at different rates of spreading. The oldest crust in zones A and C, between chrons 21 and 25, exhibits a smoothly undulating basement with only minor normal faulting. In contrast, the younger crustal region (Zone B) displays evidence of intense normal faulting of the crust, suggesting extension. The geometry of the basement resembles that of rotated blocks bounded by normal faults, usually observed across passive continental margins which undergo extension. Some of the blocks contain reflective packages arising from a mix of volcanic and sedimentary material and which may also have been subjected to normal faulting and tilting. Their reflective characteristics suggest that they were formed during and after the formation of the crust in this region (Srivastava and Keen, 1994).

Integration of these data with previous refraction measurements (Osler and Louden, 1992) and gravity modelling suggest that the regions on both flanks (between chrons 21 and 25) of the extinct ridge exhibit slightly thinner than normal crust (4.8 km or less versus a normal thickness of about 7 km) and a much thinner crust of about 3 km in the central region between chrons 21 (Srivastava and Keen, 1994). The fault geometry along the profile shows that the central region underwent about 70% extension where the mean half spreading rate was about 3.5 mm/y against about 15% extension where the half spreading rate was about 10 mm/y. Because intense faulting is seen even at the beginning of zone B rather than only near the median valley of the extinct ridge where the spreading rates were probably the lowest prior to its extinction, it was concluded that extension can be a significant contributing factor in giving rise to thin crust at slow spreading rates (Srivastava and Keen, 1994).

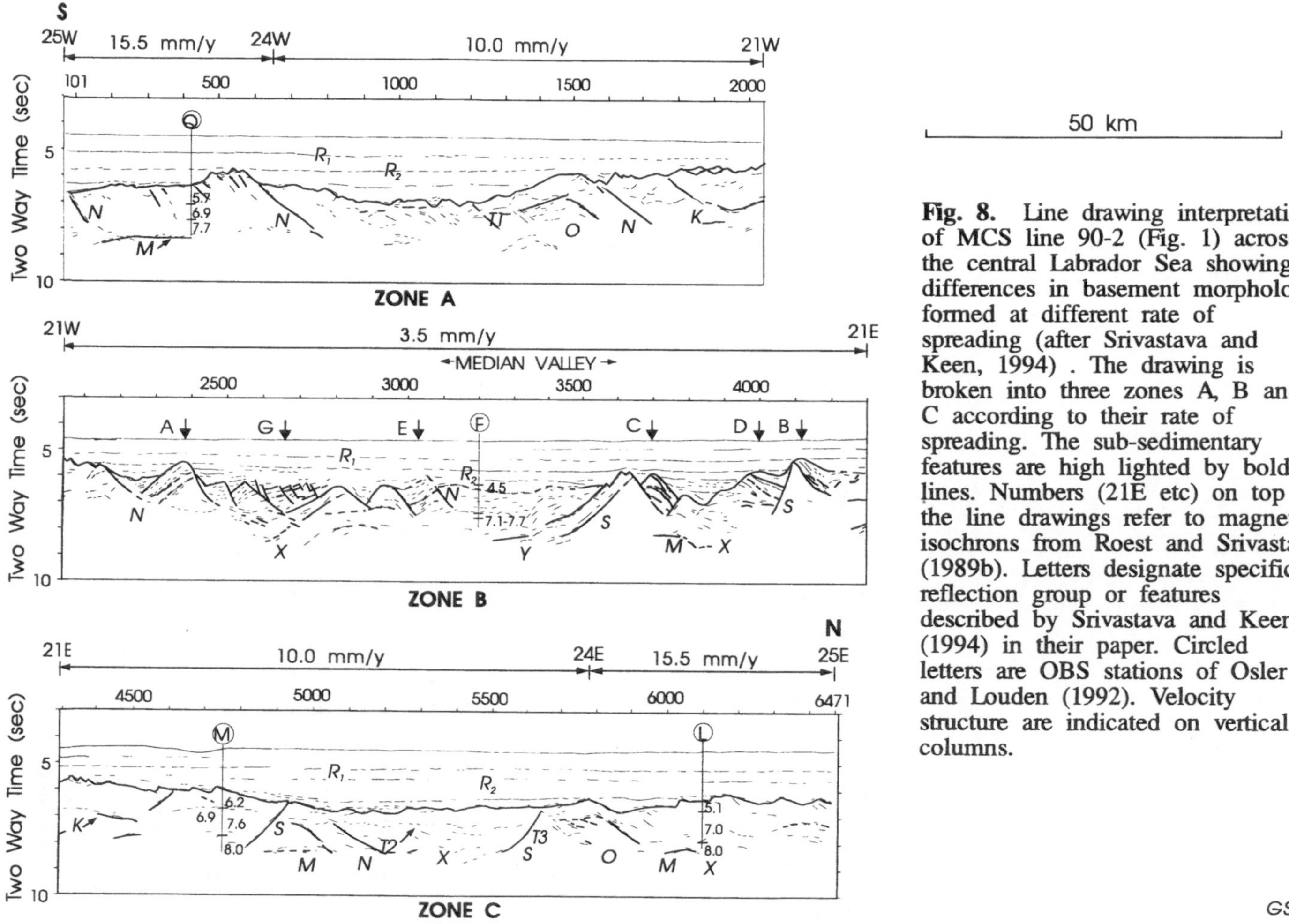

106

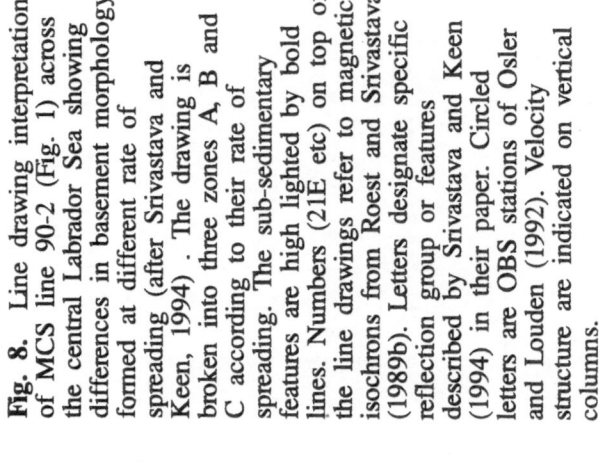

Fig. 8. Line drawing interpretation of MCS line 90-2 (Fig. 1) across the central Labrador Sea showing differences in basement morphology formed at different rate of spreading (after Srivastava and Keen, 1994) . The drawing is broken into three zones A, B and C according to their rate of spreading. The sub-sedimentary features are high lighted by bold lines. Numbers (21E etc) on top of the line drawings refer to magnetic isochrons from Roest and Srivastava (1989b). Letters designate specific reflection group or features described by Srivastava and Keen (1994) in their paper. Circled letters are OBS stations of Osler and Louden (1992). Velocity structure are indicated on vertical columns.

GSC

COMPARISON WITH SEISMIC REFLECTION AND REFRACTION MEASUREMENTS

If we regard the identifications of magnetic anomalies and their rates of spreading as described above and shown in Figs. 4 to 7 essentially correct, some interesting questions arise about how these relate to the features as interpreted by Keen et al. (1994) in the seismic reflection profile and to the velocity distributions as obtained by Chian and Louden (1994) across the southwest Greenland margin.

The spreading rates as calculated for model C, Figure 5, are shown along reflection profiles 90-1 and 90-3 in Figure 3 and also along the refraction profile, Figure 2. These figures show existence of rougher basement topography in the regions of slow spreading rate (5.8 mm/y) compared to where spreading rate was higher (17.0 mm/y). Here basement is broken up into small and large faulted blocks with relief of up to half a second. A much smoother topography is observed in regions formed at spreading rates higher than 10 mm/y (Fig. 3). By analogy with the observations from the central Labrador Sea (Fig. 8, Srivastava and Keen, 1994) where the spreading was comparable (3.5 mm/y) to the present instance (5.8 mm/y), one would expect a lot of faulting and disturbance not only in the basement but also in deeper regions of the basins formed between the faulted blocks, resembling characteristics usually seen in syn-rift sediments across rifted margins. This would suggest that the reflectors U_1 and W, W_1 observed below the rise (Fig. 3) perhaps are not related to break up unconformities in these regions and signify the end of the period during which spreading was low.

A similar influence of the rate of spreading is also seen on crustal thickness when a comparison is made with the refraction results of Chian and Louden (1994). We have regarded the velocity value of 7.6 km/s and higher as signifying the crust mantle boundary as done elsewhere (e.g Whitmarsh et al., 1993). Based on this criterion we find the existence of thinner crust (3.3 km) within the region formed at a spreading rate of about 6 mm/y (Fig. 2) similar to what is observed across the central Labrador Sea (2 to 4 km. Osler and Louden, 1992) where spreading rate was even lower (3.5 mm/y). Therefore, the degree of thinning of the crust may be different across the SW Greenland margin from that observed across the central Labrador Sea. Thinning of the crust from a normal value of about 6 km at higher spreading rates to about 3-4 km when the rate of spreading decreased to 5.8 mm/y in the present instance fits rather well with such a scenario. We don't know if the degree of thinning as observed here is what one would expect and we will come back to this question later.

Furthermore, considerable similarities exist in the velocity distribution as observed by Chian and Louden here and those observed across the central Labrador Sea by Osler and Louden (1992). The presence of low velocities (4-5 km/s) as observed in the upper crust in this region could arise from faulting and alteration of oceanic crust similar to those observed in the central Labrador Sea. The high velocity layer (7.0 - 7.6 km/s) between chrons 31 and 33O underneath would then correspond to the altered upper mantle material like serpentine or gabbroic material. We agree with Chian and Louden's (1994) interpretation of this being altered upper mantle material rather than gabbroic material for the reasons given by them. One, for a crystallized melt to have a velocity as high as 7.6 km/s, a high asthenospheric temperature is required, which for a stretching factor of 5, as they calculated, would produce a minimum crustal thickness of 20 km. This is not observed here. Two, such thick volcanic material would produce dipping reflectors, usually observed across volcanic margins and this is not observed here (Fig. 3) either. Thirdly, the V_p and V_s velocities observed in this region though not conclusive enough to draw firm distinction between the two type of materials, are supportive for this material being serpentinized upper mantle. Furthermore, these velocity values are very similar to those observed across the central Labrador Sea, zone B (Fig. 8, Osler and Louden, 1992; 1994). In this region Osler and Louden (1992) relate the high velocity in the lower crust to altered serpentinized upper mantle too. Thus we see a great similarity in the velocity distributions between the SW Greenland margin and the central Labrador Sea. The variability in the velocity values as obtained across Greenland margin could arise from the variability in the degree of alteration of upper mantle from hydrothermal circulation. Like in the central Labrador Sea, presence of faults in the crust would be the obvious conduits for hydrothermal circulation to take place to greater depths in the region. The basement highs as seen in Figure 2 between C and D and also in Figure 3 at shot point 4000, would then largely contain serpentinized peridotite upper mantle material. This is also supported by the drilling results off Iberia (ODP Drilling Leg 149, 1993), where serpentinized peridotites have been drilled at the basement highs exhibiting similar velocity distributions.

The hydrothermal circulation would also lower the velocity in the upper crust. It is difficult to reconcile with the idea that this layer is altered continental crust as Chian and Louden (1994) suggest, considering that normal continental crustal velocity values are observed farther to the west where the entire crust has thinned substantially more (Fig. 2).

The velocity distribution obtained across SW Greenland margin also bears great similarity with that observed across fossil fracture zones in the North Atlantic (Detrick et al. 1993). If we consider the observed bending of the magnetic lineations in the region where refraction and reflection measurements were carried out (Fig. 4), the existence of a fracture zone in this region can not be ruled out completely. A small fracture zone has been postulated to offset anomalies 25 to 27 in this region (Fig. 4). If this fracture zone continues to the east then it must lie between the reflection and refraction profiles as demanded by flow lines, and may not influence the reflection and refraction results.

There are several observations which argue against the possibility for the presence of a major fracture zone in this region. One, the extension of fracture zone at chrons 25 to 27 to older regions is not supported on the Labrador side, where a dense network of gravity and magnetic data exist. Two, similar velocity distributions have been obtained across other non-volcanic rifted margins (Iberia Abyssal Plain, Whitmarsh et al., 1990, 1993; Tagus Abyssal Plain, Pinheiro et al., 1992; and south Newfoundland margin, Reid, 1994) where no fracture zones are present. And, lastly, there seems to be a remarkable correlation between the rate of spreading and the reflection and refraction characteristics of the crust, not only here but also elsewhere in the Labrador Sea, which strongly argues that both the changes in basement morphology and in velocity distribution are caused by changes in the rate of spreading.

Interpretation of a thin oceanic crust underlain by serpentinite or peridotite upper mantle material also fits with some other observation made from the seismic reflection data. Two very prominent reflectors are observed below basement in the region of thin crust and these were identified as G and J (Fig. 3) by Keen et al. (1994). Figure 9 shows a portion of the line, where this reflector is clearly visible on both sides of a basement high but not underneath it. Based on an OCB located either at chron 27 or at 33O, Keen et al. suggested several possibilities for these reflectors: A detachment zone for normal faults in continental basement, a suggestion which was also favoured by Chian and Louden (1994); or the top of a serpentinite mantle below very thin oceanic crust. If our interpretation of magnetic anomalies is correct, the first possibility can be ruled out. The seismic refraction line of Chian and Louden (1994) is at an angle with the seismic reflection line (Fig. 4) and therefore a direct correlation between the two is difficult. Nonetheless, reflectors G and J seem to lie at a depth where changes in velocity values are observed by Chian and Louden.

To examine the correlation between the refraction and reflection results in more detail, the refraction measurements made across this line using expendable sonobuoys (Fig. 4) many years ago (Srivastava and Woodside, 1979) and analyzed by Stergiopolous (1984) have been plotted along this line. All of the sonobuoys, except for sonobuoy SD, were reversed, giving a fairly accurate velocity distribution along this line. The results were constrained using ray tracing techniques. The results, Figure 3, show that the reflector G is coincident with the velocity changes. Here velocity of 5.0 to 5.8 km/s for the crustal layer changes to velocity of 6.8 to 7.6 km/s for the

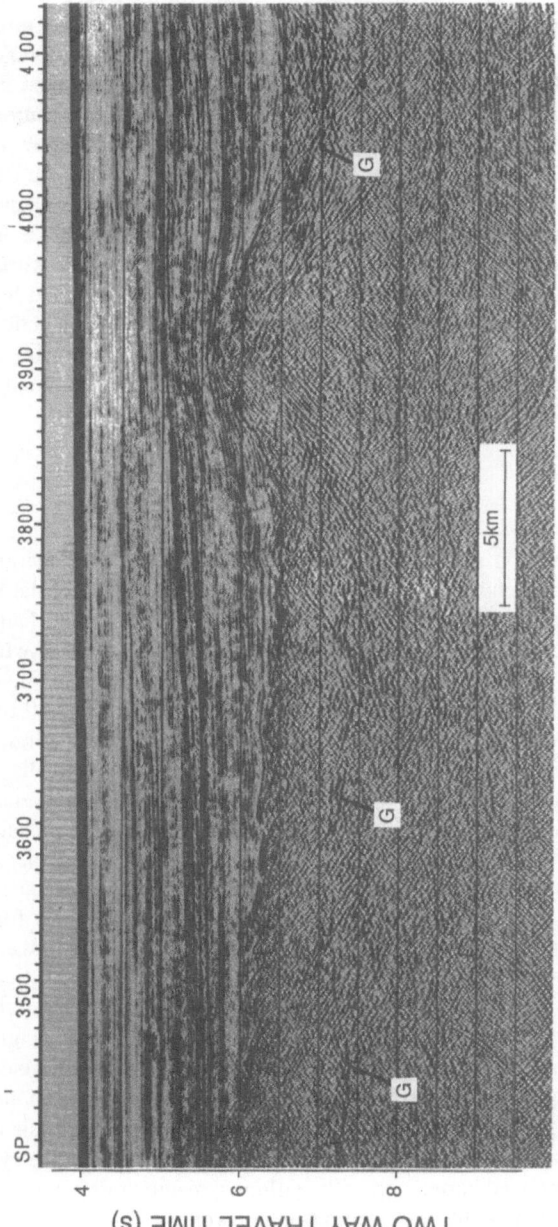

Fig. 9 Section (a) from Fig. 3 of original MCS line 90-3 showing the prominent reflector G seen below basement and the basement high interpreted as a peridotite ridge by Chian and Louden (1994)

upper mantle below. The large variation in the velocity values for the upper mantle perhaps is an indication of the degree of alteration in the upper mantle material. These results are in fair agreement with those obtained by Chian and Louden (1994) using OBS and would suggest that this reflector marks the boundary between the crust and upper mantle at most place along this profile. This would then support our and Chian ad Louden's (1994) interpretations that the two basement highs observed in Figures 2 and 3 should largely contain upper mantle material. At the western end of the profile, where this reflector lies at a shallow depth below the basement, it may mark a boundary between the upper and lower crust. It is not certain if the nature of reflector J as observed farther landward below a basement high (Fig. 3) also represents a similar situation because of lack of sonobuoy measurements in this location. However, it approximately lies in the region where Chian and Louden's (1994) results show the continental crust to have thinned the most (Fig. 2) and therefore, it is possible that it corresponds to a similar junction between low velocity mantle material below and low velocity fractured continental crustal rocks above; a situation similar to that observed across the Galicia margin where a prominent S reflector is observed, but only in continental domain (Boillot et al., 1988).

Discussion

Seismic reflection and refraction measurements across the SW Greenland margin have clearly established the existence of an unusually thin crust. The primary objective of the work described here is to examine whether this thin crust could be oceanic in nature, rather than continental as interpreted by Chian and Louden (1994). There are several problems associated with a continental interpretation: It conflicts with theoretical calculations of Bassi et al. (1993), predicting that continental crust will rupture long before it is stretched to form a thin crust 80 to 100 km long. It also fails to account satisfactorily for the observed marine magnetic anomalies, which display symmetry across the extinct spreading axis. The assumption that these anomalies are caused by the injection of large quantities of volcanic material into a thinned continental crust, as suggested for the northwest Greenland margin (Chalmers, 1991; Chalmers and Holt Laursen, 1995), would imply the presence of mafic material rather than altered mantle material under the thin crust. However, a number of arguments mentioned earlier support the presence of serpentinite mantle rather than mafic material. In addition, an intrusion model can not account for the observed symmetry in the magnetic patterns across the extinct ridge (Roest and Srivastava, 1989a).

Admittedly, the magnetic anomalies closest to the margins of the Labrador Sea are weak and their character is variable from track to track. The influence of track to track variability is greatly reduced when the magnetic observations are interpolated onto a regular grid. Figure 10 shows a shaded relief representation of the gridded magnetic observations. This figure clearly shows the existence of two, margin parallel, positive anomalies in crust older than anomaly 27. These anomalies are interpreted as anomalies 31-33 in Figure 4. We argue on the basis of their overall continuity and linear character, their resemblance to anomalies further south, the symmetry across the extinct ridge, and the presented correlation with seafloor spreading models, that anomalies 28 - 33 in the Labrador Sea are compatible with geomagnetic reversals.

The relatively small amplitude of these anomalies may then be the result of a very slow rate of spreading (about 6 mm/y). The compilation of Jackson and Reid (1983) show a systematic variation in the amplitude of magnetic anomalies with spreading rates. They attributed this to greater irregularity in the distribution of magnetic polarities in the crust at slow spreading. At very slow spreading rates, crustal accretion becomes an episodic rather than a continuous process, resulting in fragmentation, as the deep reflection data here and across the central Labrador Sea illustrate. This will give rise to inhomogeneities in the magnetic source rocks, effectively resulting in a reduced intensity of magnetisation and giving rise to smaller amplitude magnetic anomalies

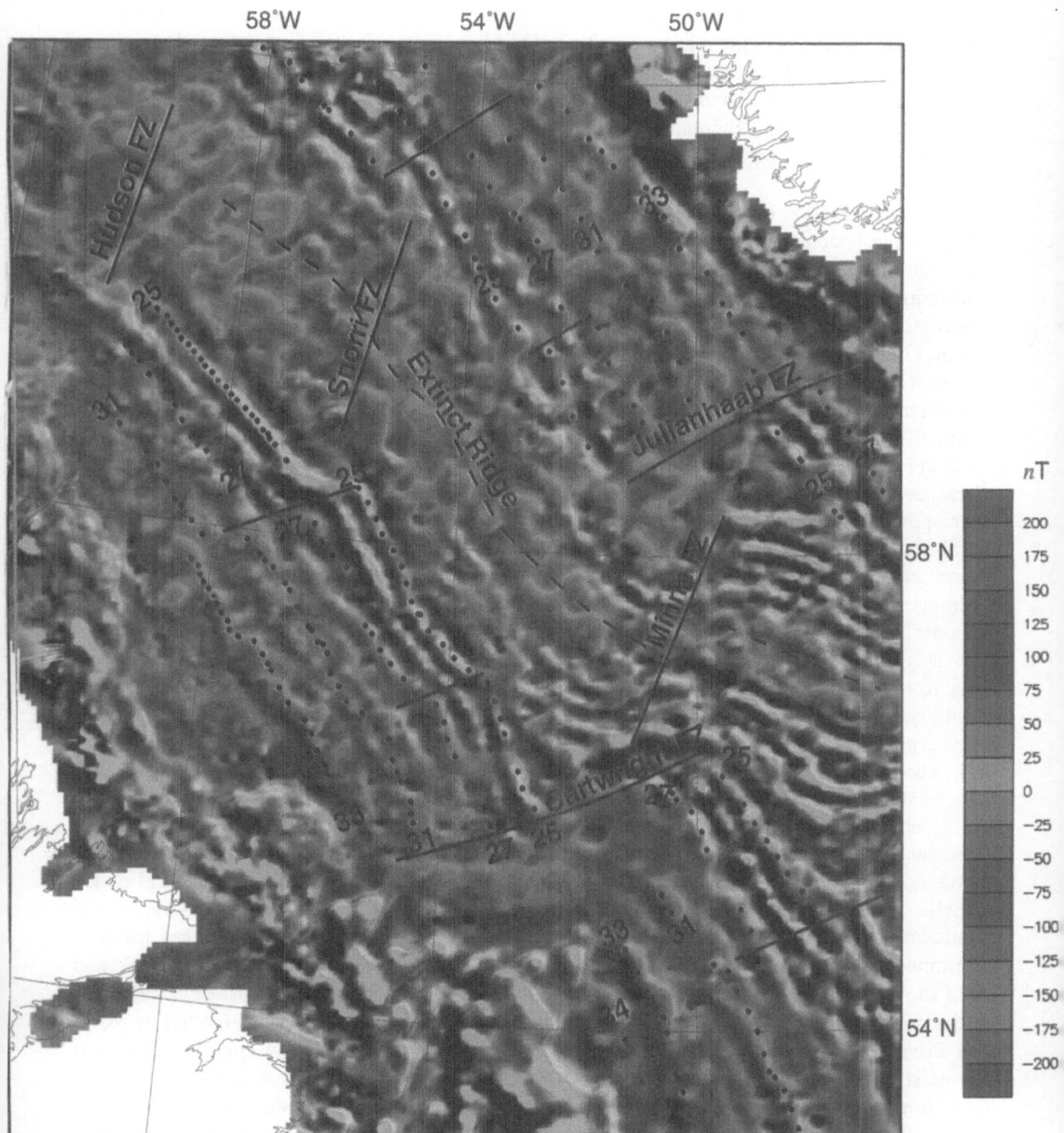

Fig. 10. Colour shaded relief representation of the magnetic anomalies in the Labrador Sea. The black dots represent the location where these anomalies have been identified from track data. Continuous black lines mark the location of fracture zones. Note the symmetry of the positive anomalies off the Labrador and Greenland Coasts, and their similarity to anomalies 31-33 south of the Cartwright Fracture Zone. Anomalies 27 to 25 form a sequence of well defined, narrow, positive anomalies.

with variable shapes (cf. Schouten et al., 1982). In addition, the anomalies here are contaminated by temporal variations in the magnetic field, including geomagnetic storms and diurnal variations, which can be particularly strong in the auroral zone. High frequency noise is easily recognizable (see, for example, tracks 1 and 4 in Figure 5 and tracks I and J in Fig. 6), but the slower, diurnal, variations may give rise to anomalies that are similar in wavelength and amplitude to the seafloor spreading anomalies and hard to eliminate without base station information (e.g. Jackson et al., 1979).

The interpretation of these anomalies as seafloor spreading anomalies formed at a slow rate of spreading is consistent with the plate kinematic motions derived for the North American and Eurasian Plates (Fig. 11). It should be stressed that the position of Greenland at chron 33 time is constrained by the entire plate configuration in the North Atlantic and Arctic Oceans. The relative positions of the North American and Eurasian plates require Greenland to occupy a position at chron 33 time close to the one given in Figure 11, in order to avoid a large overlap between eastern Greenland and northern Europe. Thus, Figure 11 shows that the Labrador Sea essentially formed along a continuation of the Mid-Atlantic Ridge at chron 33 time. On the contrary, if we assume that the region between chrons 33 and 27, north of an assumed plate boundary (marked APB in Fig. 11) separating the North American-Greenland plate system from the North American-Eurasian plate system including the Rockall Plateau, was undergoing extension, while a corresponding region to the south of this boundary was formed by seafloor spreading, it raises a number of problems. One, extension to the north and seafloor spreading to the south of APB from chron 34 to chron 27 would create strike slip motion across this boundary, making it a very prominent feature. No evidence for its existence can be seen in the gravity (Srivastava, 1986; Woodside, 1989), magnetic (Fig. 10) or the limited amount of seismic reflection observations (Srivastava, 1978) in this region. Furthermore, extension north of APB during this time would have to take place at about the same rate as seafloor spreading farther south, and to a total width of in excess of 200 km; this would be difficult to achieve without rupturing the lithosphere. The same argument applies if seafloor spreading had started at chron 31 instead, north of CFZ and JFZ, as has been implied by Chian and Louden (1994). In this case the amount of extension needed to remove the overlap between Greenland and Rockall Plateau (Fig. 11) would be of the order of 140 km. Such an overlap would become half if we assume that the regions between chron 31-33, on Greenland and Labrador sides, are thinned continental crust and have separated from each other along a detachment surface like the reflectors G and J (Fig. 3) as suggested by Chian and Louden (1994). On the other hand, if the Labrador Sea was opening by seafloor spreading, as the continuation of anomalies 31 and 33 north of APB (Fig. 10) suggests, there is no need for such a prominent boundary to exist. Finally, the Cartwright and Julianhaab Fracture Zones (CFZ and JFZ) show up as major features in both magnetic (Fig. 10) and gravity (Woodside, 1989) maps of the regions. They extend all the way to the ocean continent boundary, as interpreted by Roest and Srivastava (1989a). The fact that their geophysical character does not change along their strikes strongly suggests that they were entirely formed as fossil traces of an oceanic transform fault. A model where they are partly transform margin and partly fracture zone is hard to conceive.

Another observation which also supports initiation of seafloor spreading at chron 33 time is the offset in anomalies 33 to 25 across CFZ and JFZ (Fig. 11). If seafloor spreading had started at chron 27 time, the question arises why this and younger anomalies got offset towards the Labrador side across these fracture zones. The reconstructions suggest that at chron 27 time the ridge could have easily continued to the north without any offset across these fracture zones. The offsets in the ridge to the Labrador side was not created because of the existence of thinnest crust under chron 27. The refraction results clearly show a thicker crust there, compared to that under chrons 31-33. We have shown reconstructions of the Labrador Sea at chrons 31 and 33 times in Fig.11 to demonstrate the reason for this offset in all anomalies starting at chron 33 time. As can be seen, anomaly 33 had to offset on the Labrador side as the thinnest crust existed west of the

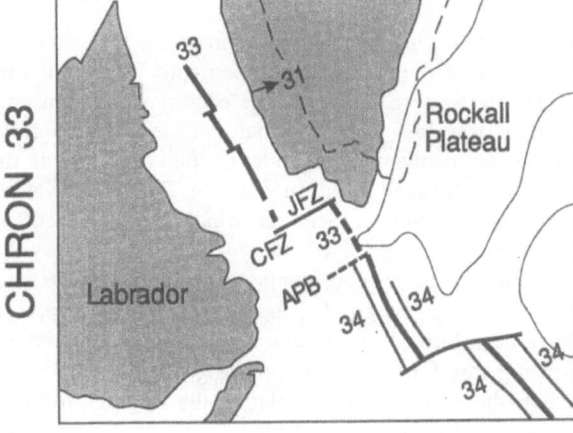

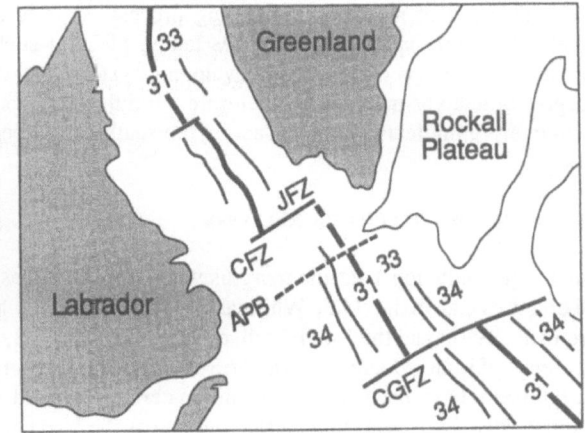

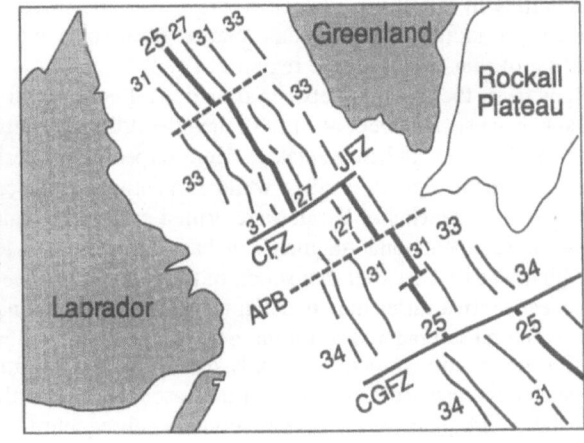

Fig. 11.
Reconstructions of the Labrador Sea at chrons 25, 31 and 33 times similar to those given in Srivastava and Roest (1989). Also shown is the position of Greenland at chron 31 time by dashed line in chron 33 reconstruction. APB - an assumed plate boundary between the Eurasian and the Grenland plates. CGFZ - Charlie Gibbs Fracture Zone, CFZ - Cartwright Fracture Zone, JFZ - Julianhaab Fracture Zone.

GSC

Greenland position at that time, to start seafloor spreading there. This offset thus perpetuated in all anomalies generated subsequently.

Thus, formation of oceanic crust at a slow spreading rate between chrons 31-33 remains the most applicable model at present. The change from rough basement topography during the time of slow spreading to smooth topography at a faster spreading rate; the occurrence of thin crust during the time when the rate of spreading was low and a normal thickness during faster spreading; and the presence of low velocity upper mantle material under the thin crust, giving rise to a prominent reflector G in seismic reflection profiles, all fit this model well. These observations are similar to those observed across the central Labrador Sea, for which the slow seafloor spreading origin is well established (Srivastava and Keen, 1994) and which also shows very low amplitude magnetic anomalies that are difficult to relate to geomagnetic reversals. Therefore, even though magnetic anomalies across SW Greenland margin by themselves may not present conclusive evidence for this region to have been formed by seafloor spreading, its combination with other observations provides a convincing case for a seafloor spreading origin.

It is realised that some caution needs to be exercised when relating the findings from the central part of the Labrador Sea to those observed across the margins as the two relate to some what different systems. The central Labrador Sea is a case where spreading was slowly decreasing while the present instance deals with the start of a seafloor spreading process, soon after the opening of an ocean basin. Nonetheless, seafloor spreading processes taking place at such low rates will have a great deal in common, whether it is a dying or a developing system. The difference between them may lie in the degree of tectonism. We have, therefore, used the findings from the central Labrador Sea as a guide in interpreting features we see across the southwest Greenland margin.

APPLICABILITY OF SLOW SPREADING MODEL TO OTHER REGIONS

The margins of the Labrador Sea are not the only non-volcanic margins where thin crust has been observed. Other margins are the Iberia Abyssal Plain (IAP, Whitmarsh et al., 1990 and 1993), Tagus Abyssal Plain (TAP, Pinheiro et al., 1992) and the Newfoundland Basin margin (NB, Reid, 1994). All of these margins show presence of high velocity material (7.3-7.6 km/s) underlying thin crust. In all of these cases strong arguments have been made that this material is altered upper mantle. In the IAP it outcrops, forming ridges. Recent drilling carried out across the IAP confirms this interpretation, as peridotite and serpentinite were drilled at some of the basement highs (ODP Drilling Leg 149, 1993). The similarities in the crustal velocities present across these widely separated margins and the presence of peridotite ridges in some places, raise some interesting questions about the mode of formation of thin crust in these regions.

If we regard that peridotite ridges mark the junction between oceanic and continental crusts as the drilling results off Galicia Bank suggest (Boillot et al., 1988), then the drilling results from Leg 149 in IAP (ODP Drilling Leg 149, 1993) are rather anomalous. Here serpentinized peridotite was drilled at the most oceanward basement high (Site 897) while serpentinized breccia and peridotite interlayered with sediments and fragments of basalt were drilled at the next landward basement high (Site 899). Presence of sediments and fragments of basalt at this high led the shipboard party to suggest that the high was formed after deposition of the mass flow. The uplift of this high after the formation of oceanic crust raises doubts about the idea that this is indeed a transition zone. The results from the central Labrador Sea, on the other hand, clearly show that during very slow spreading extension seems to take place, not only during the emplacement of crust but also subsequently. Under such circumstances, formation of the basement high in the IAP where serpentinized breccia was drilled, took place in the oceanic domain subsequent to crustal emplacement during very slow spreading. Srivastava et al. (1990) indeed indicate a slow spreading rate of 5 mm/y between the IAP and Newfoundland, prior to the formation of the J-anomaly ridge.

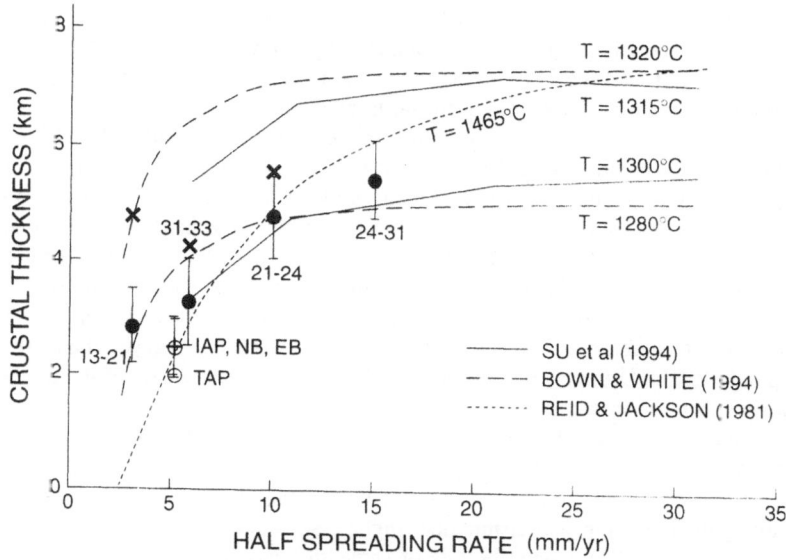

Fig. 12 A plot of crustal thickness against half spreading rates. The solid dots refer to the crustal thicknesses from the Labrador Sea between different chrons, open circles are crustal thicknesses from IAP - Iberia Abyssal Plain (Whitmarsh et al., 1993), TAP - Tagus Abyssal Plain (Pinheiro et al., 1992), NB - Newfoundland Basin Margin (Reid, 1994) and EB - Eurasian Basin (Jackson et al., 1982). Also shown are the theoretical estimates of Su et al., (1994), Bown and White (1994) and Reid and Jackson (1981). The crosses refer to the average crustal thicknesses when corrected for observed extension.

To examine further the similarities in the crustal properties across these margins, a comparison of their crustal thickness with the rate of seafloor spreading was made and the results are shown in Figure 12. Also included here, for comparison, are the results from two regions which undoubtedly were formed by seafloor spreading, the central part of the Labrador Sea (Srivastava and Keen, 1994) and the Eurasian Basin (EB, Jackson et al., 1982). The extreme values of the crustal thickness, as obtained from reflection and refraction measurements at different chrons formed at different rate of spreading in the Labrador Sea, are supplemented by the values obtained from combined study of gravity and refraction measurements where ever possible. Therefore, some of the values (e.g for chron 24-31) can only be considered as crude approximations as they lack detailed refraction measurements for the entire duration considered. Nonetheless, the results show a remarkable decrease in the thickness of the crust formed at low spreading rates (chron 31 to 33 and 13 to 21) against that formed at faster spreading rate (chron 24 to 31).

In calculating the crustal thickness in all these regions it was regarded that 7.3 - 7.6 km/s velocity observed in these regions marks the location of crust - mantle interface. Further, it was assumed that, since Newfoundland and Iberia margins are conjugate to each other (Srivastava et al., 1990), they were formed at the same rate of spreading. Assuming that the separation between Iberia and North America took place at chron M25 (Srivastava et al., 1990), results in obtaining a half rate of spreading as 5 mm/y between chron M0 and M25, based on the poles of rotation for chron M0 and M25 (Srivastava et al., 1990). We have used this value in plotting the crustal thicknesses at these places in Figure 12. We realise that this may not be completely right in view of different rate of spreading (10 mm/y) as obtained by Whitmarsh et al. (1990) and Pinheiro et al. (1992) for the TAP and the IAP respectively. It is interesting to note that at rate of spreading around 6 mm/y all regions show similar crustal thicknesses.

Also shown in Figure 12 are the theoretical curves derived by Reid and Jackson, (1981); Su et al., (1994) and Bown and White, (1994) for steady state systems. Su et al. (1994) consider that upwelling beneath the spreading centre is influenced by the buoyancy of the melt, while Reid and Jackson consider an isoviscous mantle whose upwelling is solely guided by plate spreading. Bown and White (1994) consider that upwelling beneath ridges is controlled by corner flow between the diverging plates. Reid and Jackson (1981) calculations are perhaps an oversimplification of the actual situation as was also pointed out by others. In spite of their different approaches, they all show a significant decrease in crustal thickness at spreading rates lower than 20 mm/y. The observed points for the Labrador Sea, strictly speaking, do not follow any of the predicted curves but do show a decrease in crustal thickness with decrease in the rate of spreading.

The question then arises if this decrease in crustal thickness is solely due to a reduction in melt generation, as the theoretical models suggest. This does not seem to be the case as suggested by Srivastava and Keen (1994). By measuring the horizontal offsets on the normal faults in the basement along the Labrador Sea seismic line (Fig. 8), Srivastava and Keen (1994) found that the central region (chron 13-21), where the rate of spreading was about 3.5 mm/y, underwent about 70% extension while the adjacent regions (chrons 21-24), where the spreading rate was about 10 mm/y, underwent only 15% extension. A similar calculation for the region between chrons 31-33 showed a 30% extension. If we apply these corrections to the crustal thickness values in Figure 12, shown by crosses, we find that the crustal thicknesses do not show much variations with changes in the rate of spreading. However, these calculations are based on the assumption that all extension took place post crustal emplacement which, as the results from the central Labrador Sea show, is not completely right. In spite of this approximation, the results show that most of the variations in crustal thickness as seen in Figure 12 can be explained by the increase in extension with decline in the spreading rate. As some of the extension could be synchronous with crustal emplacement, the estimates shown in Figure 12 are maxima. If these are reduced then some of the thinning may be due to decline in magma generation as the theoretical calculations show.

Nonetheless, the similarity in the crustal thickness values at low rates of spreading from very different regions shown in Figure 12 is rather interesting and suggests that the formation of thin crust at slow spreading rates is not just restricted to margin regions only but is equally true at spreading centres (e.g. Eurasian Basin).

This raises an interesting question, why peridotite ridges are mainly observed across the margins formed at slow spreading rates, like IAP and SW Greenland. The answer may lie in the large extension which these regions have undergone prior to and during seafloor spreading. The results presented here show that in most of these instances these ridges lie more in the oceanic than in continental domain. As such, their occurrences are related both to a decrease in melt generation, as the results of Fig. 12 suggest, and to excessive thinning of the continental and oceanic crusts in these regions.

The magnetic calculations across SW Greenland margin, Fig. 7, constrained by refraction results necessitated the presence of a negatively magnetised body lying adjacent to a thick continental body. The calculations, however, do not rule out the presence of a thin layer of continental crust with low susceptibility under the negatively magnetised volcanic body, which would imply that the OCB lies between chron 33O and the west Greenland shelf break. A positive magnetic anomaly (33Y) overlies the basement high (Fig. 6). The calculation, though not conclusive, suggests that this basement high may not be entirely composed of serpentinized mantle material and may contain some volcanic material on top. This is because serpentinized/peridotite have been found to be very variable in their remanent magnetisation (between 0 and 10 A/m,

Dunlop et al., 1982), while others have found it to contain low remanent magnetisation (Stokking et al., 1992).

Conclusions

The following conclusions can be made from the combined analysis of geophysical data across the SW Greenland margin:

1. A re-examination of magnetic data from this region together with those across the conjugate margin off Labrador confirms earlier interpretations that the magnetic anomalies in these regions were formed by seafloor spreading. Because these anomalies are small in amplitude and variable in shape, we can not entirely rule out that they could have been caused by injection of volcanic material through continental crust. However, their overall continuity and linear character, their resemblance to anomalies formed in the southern Labrador Sea as part of the North American and Eurasian plate system, their symmetry across the extinct ridge, and their correlation with seafloor spreading models, strongly argue for their formation by seafloor spreading.
2. The model calculations show that a drastic change in the rate of spreading took place at chron 30, contrary to earlier calculations. This change in the rate of spreading is found to be coincident with the change in basement topography from rough during the low spreading rate to smooth during faster spreading rates and the occurrence of thin crust during the slow spreading to a normal thickness during faster spreading.
3. It is suggested that the small amplitudes of these anomalies is caused by fragmentation of the crust during slow spreading as suggested by the reflection data.
4. The interpretation that these anomalies were formed at slow rate of spreading is consistent with the plate kinematic motions derived for the North American and Eurasian Plates, and shows that the Labrador Sea essentially was formed along a continuation of the Mid-Atlantic Ridge at chron 33 time.
5. Based on these observations and magnetic modelling it is suggested that the thin crust observed across the SW Greenland margin is most likely oceanic and that a very narrow zone of transitional crust exists landward of it, before true continental crust is encountered below the continental shelf.
6. Comparison of this crust with crust formed across other continental margins shows great similarities, suggesting that they were also formed at slow spreading rate. The drilling results in Iberia Abyssal Plain fit well with a slow spreading model.

Acknowledgements

We thank Ian Reid and Jean-Claude Sibuet for their helpful comments on an earlier version of this manuscript. The manuscript has benefited from comments and observations of Dale Sawyer and an anonymous referee. We are grateful to Jacob Verhoef for allowing us to use the new gridded data from the Labrador Sea in this paper. Drafting of the figures done by digital cartography section of the Atlantic Geoscience Centre and by drafting and illustration section of the Bedford Institute of Oceanography are acknowledged.

References

Bassi, G., Keen, C.E., and Potter, P., 1993. Contrasting styles of rifting: models and examples from the eastern Canadian margin. Tectonics. **12**, 639-655.

Bell, J.S., (Coordinator)1989. East Coast Basin Atlas Series, Labrador Sea. Atlantic Geoscience Centre, Geological Survey Of Canada, Dartmouth, Nova Scotia.

Boillot, G., Winterer, E.L., et al., 1988. Proceedings of the Ocean Drilling Program, Scientific Results. **103**, 858 pp. Ocean Drilling Program, College Station, Texas.

Bown, J.W. and R.S. White, 1994. Variation with spreading rate of oceanic crustal thickness and geochemistry. Earth and Planetary Science Letters. **121**, 435-449.

Chalmers, J.A., 1991. New evidence on the structure of the Labrador Sea/Greenland continental margin. J. Geological Soc. London. **148**, 899-908.

Chalmers, J.A., Pulvertaft, T.C.R., Christiansen, F.G., Larsen, H.C., Laursen, K.H. and Ottesen, T.G., 1993. The southern west Greenland continental margin: rifting history, basin development, and petroleum potential. Petroleum Geology of Northwest Europe: Proceedings of the 4th conference. ed. J.R. Parker. Geological Soc. London. 915-931.

Chalmers,J.A., and Holt Laursen, K., 1995. Labrador Sea: The extent of continental and oceanic crust and the timing of the onset of seafloor spreading. Marine and Petroleum Geology.(In Press).

Chian, D., and Louden, K.E., 1994. The continent-ocean transition across the southwest Greenland margin. J. Geophys. Res. **99**, 9117-9135.

Chian, D., Louden, K.E., Reid, I., and Keen, C.E, 1994. The structure of the conjugate margins of the Labrador Sea based on coincident MCS and wide angle seismic profiles. Abstract European Geophysical Society, Annales Geophysicae, **12**, C36.

Detrick, R.S., White, R.S., and Purdy, G.M., 1993. Crustal structure of North Atlantic fracture zones. Reviews of Geophysics, **31**, 439-458.

Dunlop, D.J. and Prevot, M., 1982. Magnetic properties and opaque mineralogy of drilled submarine intrusive rocks. Geophysical Journal of the Royal Astronomical Society of London, **69**, 763-802.

Grant, A.C., 1980. Probems with plate tectonics: Labrador Sea. Bull. Canadian Petroleum Geolgy, **28**, 252-278.

Hinz, K., Schluter, H.-U., Grant, A.C., Srivastava, S.P., Umpleby, D., and Woodside, J., 1979. Geophysical transects of the Labrador Sea: Labrador to southwest Greenland. Tectonophysics, **59**, 151-183.

Jackson, R.H., Keen, C.E., Falconer, R.K.H., and Appleton, K.P., 1979. New geophysical evidence for sea-floor spreading in central Baffin Bay, Canadian Journal of Earth Sciences, **16**, 2122-2135.

Jackson, R.H., Reid, I. and Falconer, R.K.H., 1982. Crustal structure near the Arctic Mid-Ocean Ridge. Journal of Geophysical Research **87**, 1773-1783.

Jackson, R.H. and Reid, I., 1983. Oceanic magnetic anomaly amplitudes: Variations with seafloor spreading rate and possible implications. Earth and Planetary Science Letters **63**, 368-378.

Keen, C.E., Potter, P., and Srivastava,S.P., 1994.Deep seismic reflection data across the conjugate margins of the Labrador Sea. Canadian Journal of Earth Sciences, **31**, 192-205.

ODP Leg 149 Shipboard Scientific Party, 1993. ODP drills the west Iberia Rifted margin. EOS, **74**, 454-455.

Osler, J.C., and Louden, K.E., 1992. Crustal structure of an extinct rift axis in the Labrador Sea: Preliminary results from a seismic refraction survey. Earth and Planetary Sciences, **108**, 243-258.

Osler, J.C., and Louden, K.E., 1995. The extinct spreading centre in the Labrador Sea: I-Crustal structure from a 2-D seismic refraction velocity model. Journal of Geophysical Research. In Press

Pinheiro, J.C., Whitmarsh, R.B., and Miles, P.R., 1992. The ocean-continent boundary off the western continental margin of Iberia- II. Crustal Structure in the Tagus abyssal plain. Geophysics Journal International, **109**, 106-124.

Reid, I, and Jackson, R.H., 1981. Oceanic spreading rate and crustal thickness. Marine Geophys. Researches 5, 165-172.

Reid, I., 1994. Crustal structure of a nonvolcanic rifted margin east of Newfoundland. Journal of Geophysical Research. 99, 15,161-15,180.

Roest, W.R., and Srivastava, S.P., 1989a. Sea-floor spreading in the Labrador Sea: A new reconstruction. Geology, 17, 1000-1003

Roest, W.R. and Srivastava, S.P., 1989b. Sea floor spreading history I Labrador Sea. Magnetic anomalies along track. Scale 1:2,000000. in J.S. Bell (Coordinator), East Coast Basin Atlas Series: Labrador Sea. Geological Survey of Canada, Atlantic Geoscience Centre, Dartmouth, N.S., 86.

Schouten, H., Denham, C., and Smith, W., 1982. On the quality of marine magnetic anomaly sources and sea-floor spreading topography, Royal Astronomical Society Geophysical Journal, 70, 245-259.

Srivastava, S.P., 1978. Evolution of the Labrador Sea and its bearing on the early evolution of the North Atlantic. Royal Astronomical Society Geophysical Journal, 52, 313-357.

Srivastava, S.P., 1986. Geophysical maps and geological sections of the Labrador Sea. Geological Survey of Canada Paper 85-16.

Srivastava, S.P., and Keen, C.E., 1994. A deep seismic reflection profile across the extinct mid-Labrador Sea spreading centre. Tectonics. (In Press).

Srivastava, S.P., and Roest, W.R., 1989. Sea floor spreading history II-VI Labrador Sea in J.S. Bell (Coordinator), East Coast Basin Atlas Series: Labrador Sea. Geological Survey of Canada, Atlantic Geoscience Centre, Dartmouth, N.S., 100-109.

Srivastava, S.P., and Woodside, J.M., 1979. Report of Cruise No. 79-013, CSS Hudson, June 18 -July 10, 1979. 33p

Srivastava, S.P., Falconer, R.K,H., and MacLean, B., 19981. Labrador Sea, Davis Strait, Baffin Bay: Geology and Geophysics - A review. Geology of the North Atlantic Borderlands. Canada. Soc. Petrol. Geol. Memoir 7, J.Wm. Kerr, A.J. Fergusson and L.C. Machan, 333-398.

Srivastava, S.P., Roest, W.R., Kovacs, L.C., Oakey, G., Levesque, S., Verhoef, J., and Macnab, R., 1990. Motion of Iberia since the Late Jurassic: Results from detaied aeromagnetic measurements in the Newfoundand Basin. Tectonophysics. 184, 229-260.

Stergiopolous, A.B., 1984. Geophysical crustal studies off the southwest Greenland margin. M.Sc thesis. Dalhousie University, Halifax, Nova Scotia. 250p.

Stokking, L.B., Merill, D.L., Haston, R.B., Ali, J.R., and Saboda, K.L., 1992. Rock magnetic studies of serpentinite seamounts in Mariana and Izu-Bonin region, in Fryer, P., Pearce, J.A. and Stokking, L.B., eds., Proceedings of Ocean Drilling Program, Scientific Results, College Station, Texas, Government Printing Office, Washington, D.C. 125, 561-579.

Su, W., Mutter, C.Z., Mutter, J.C. and Buck, R., 1994. Some theoretical predictions on the relationship among spreading rate, mantle temperature and crustal thickness. Journal of Geophysical Research 99, 3215-3227.

White, R.S., McKenzie, D. and Onion, R.K., 1992. Oceanic crustal thickness from seismic measurements and rare earth element inversions. Journal of Geophysical Research 97, 199,683-199,715.

Whitmarsh, R.B., Miles, P.R., and Mauffret, A., 1990. The ocean-continent boundary off the western continental margin of Iberia-I. Crustal Structure at 40° 30'N. Geophysical Journal International. 103, 509-531.

Whitmarsh, R.B., Pinheiro, L.M., Miles, P.R., Recq, M. and Sibuet, J.-C., 1993. Thin crust at the western Iberia ocean-continent transition and ophiolites. Tectonics. **12**, 1230-1239.

Woodside, J.,1989. Gravity anomaly. Scale 1:2,000000. *in* J.S. Bell (Coordinator), East Coast Basin Atlas Series: Labrador Sea. Geological Survey of Canada, Atlantic Geoscience Centre, Dartmouth, N.S., 94.

ANISOTROPY MEASUREMENTS AND THE DEEP STRUCTURE OF A PASSIVE MARGIN: SOUTHWESTERN GREENLAND

CARBONELL, R.[1], SPEECE, M.A[2], CLEMENT, W.P.[3], SMITHSON, S.B.[3]
[1] CSIC - Institute of Earth Sciences "Jaume Almera",
    Marti i Franques s/n, 08028 Barcelona, Spain
[2] Geophysical Engineering Department, Montana Tech,
    W. Park Street, Butte 597'1 Montana, U.S.A.
[3] Department of Geology and Geophysics, University of Wyoming,
    P.O. Box 3006, Laramie 82071, Wyoming, U.S.A.

ABSTRACT. The non-uniqueness of the laboratory measured physical properties (P-wave velocities and densities) can be partly overcome by the inclusion of anisotropy. Seismic anisotropy estimation can constitute an additional constraint in the determination of the mineralogical composition of the deep crust from indirect seismic measurements. Resolution analysis studies reveal that P- and S-wave velocity measurements derived from wide-angle seismic reflection/refraction data are usually characterized by large error estimates (i.e., very low resolution). These error estimates render physically meaningless Poisson's ratio profiles. Also the duality in S-wave velocities due to anisotropy complicates the estimation of Poisson's ratio. The almost general use of single component (vertical) seismic instruments and low resolution of refraction/wide-angle reflection experiments (mostly due to spatial under sampling) have prevented the use of anisotropy as a constraint for the determination of lower crustal composition in favor of estimates on Poisson's ratio. Therefore, we suggest that anisotropy estimates can place new and more relevant constraints on the different rock types present in the deep crust. In order to assess indirect seismic anisotropy measurements we employed three component, densely space large aperture (0-250 km offset, at 100-150 m spacing) seismic recordings acquired along the south west coast of Greenland utilizing REFTEK PASSCAL instruments deployed by the University of Wyoming. With the aid of an inversion scheme that uses reflected and converted energy we determined P- and S-wave velocity-depth functions for the passive margin of southwestern Greenland. The inversion suggests a P-wave velocity structure characterized by two gradient zones: a relatively high gradient from the surface to approximately 5 km depth where velocities exceed 6.0 km/s, followed by a low gradient to the base of the crust where velocities reach 7.0 km/s. A high P-wave velocity layer (7.2-7.4 km/s) can be identified between 6-8 km above the Moho. Gravity modeling suggests relatively high densites 3.0-3.1 kg/m^3 for this layer. Independent analysis of the radically and transversely polarized horizontal components revealed average velocities of 4.9 ± 0.1 km/s and 4.5 ± 0.1 km/s respectively suggesting a seismically anisotropic crust. A time delay of 0.25 s between the radial and the transverse horizontal components of the S$_i$S phase is observed at offsets of 70 km. The radically polarized S-wave is parallel to the southwest coast of Greenland. From the S-wave analysis, the ocean-continent transitional crust is clearly seismically anisotropic above a high velocity layer in the lower continental crust. The density and velocity values suggested for this high velocity structure above the Moho, and the anisotropy measured just above it seem to favor an accretion of hot, mafic mantle material (underplating) at the base of the crust during a rifting episode. Possibly, magmatic underplating during Late Cretaceous rifting of the Labrador Sea heated the preexisting lower crust promoting plastic flow and enabling alignment of anisotropic minerals to produce the seismic anisotropy.

121

E. Banda et al. (eds.), Rifted Ocean-Continent Boundaries, 121–146.
© 1995 Kluwer Academic Publishers. Printed in the Netherlands.

1. Introduction

Over the last decade numerous studies have provided a large wealth of knowledge on the structure, nature and development of margins [Talwani et al., 1979; LASE Study group, 1986; Sheridan et al., 1993]. However, there are still ambiguities in the constitution of lower crustal structures due to the indirect nature of the geology and geophysical studies. Geology and geophysics have indicated intra- and lower-crustal volcanic flows in passive margins. These have also been sampled at various locations by the ODP program and imaged by deep seismic reflection seismology in the form of seaward dipping reflectors (SDR). These volcanic flows are associated with volumetrically large structures characterized by high P-wave velocities ranging approximately 7.0-7.4 km/s and densities derived from gravity modeling which are within 3.0-3.1 kg/m^3. These findings have been interpreted as a result of underplating processes. However, serpentinite generated by hydrothermal alteration of the upper most mantle (undercrusting) [Boillot et al., 1989] has also been sampled by the ODP program in areas close to continental margins. Serpentinite presents a broad spectra of velocity and density values depending on its mineralogy. In general, its physical properties (density, P- and S- wave velocities) overlap those of the lower continental crust with the result that neither seismic refraction nor the gravity data can be used to distinguish definitely between lower continental crust and serpentinized upper mantle material [Christensen, N.I., 1978; Boillot et al., 1989]. The main difference in these deep geological structures are the emplacement mechanisms: underplating, and undercrusting. Underplating is related to emplacing mantel material at the bottom of the crust at high temperatures, in other words, upward magma crystallization at the crust-mantle boundary during the rifting process [McKenzie and Bikle, 1988]. Undercrusting is related to the generation of serpentinite by hydrothermal alteration of the upper mantle. Although the depth to which water may reach in the lithosphere is not known, there seems to be evidence for the presence of water at lower crustal depths. In continental lithosphere, for example, Hyndman and Klemperer, [1989] suggested that a possible cause for the reduction of the average crustal velocities was high pore pressure due to fluids (water). Moreover, the physics involved in fluid flow through porous media can explain lower crustal reflectivity [Suetnova et al., 1994].

There are many indications of seismic anisotropy in the lithosphere. The presence of anisotropy is well documented [Babuska and Cara, 1991]. Pn velocity anisotropy has been observed both beneath the oceans and the continents [Bamford, 1977], (Pn normally describes the velocity distribution immediately below the Moho). Crampin [1967] observed high mode surface wave anomalies for continental paths which can be used as a diagnostic of anisotropic layering [Crampin, 1975]. Long range seismic refraction/wide-angle reflection profiles are also consistent with P-wave azimuthal anisotropy: in southern Germany [Fuchs, 1983]; in the crust of the Basin and Range (Nevada, U.S.A.) [Carbonell and Smithson, 1989]; and in southwestern Spain [Diaz, et al., 1993]. In the upper mantle, anisotropy is generally assumed to be related with the lattice preferred orientation (LPO) of the mantle materials [Nicolas and Christensen, 1987]. The origin of this LPO can be attributed to multiple physical processes, e. g., present day stresses pervading the subcontinental mantle [Nicolas and Christensen, 1987]; past deformations of the lithosphere [Silver and

Chang, 1988]; and passive motion of the lithosphere over a stationary asthenosphere [Hirn, 1977]. The most likely causes for seismic anisotropy in the crust are: large scale structural differences in the geological structures due to present day tectonics [Carbonell and Smithson, 1991], fluid filled microcracks [Nur and Simmons, 1969; Crampin, 1984] and the alignment of anisotropic minerals such as plagioclase, biotite, amphibolites, and pyroxenes [Babuska, 1981; Ji and Mainprice, 1988; Mainprice and Nicolas, 1989; Siegesmund and Kruhl, 1991; Siegesmund and Vollbrecht, 1991]. This mineral alignment suggests that anisotropy can be used as an additional constraint to describe the mineralogical composition of the deep crust.

The combined use of the near-vertical, normal incidence seismic reflection data with wide-angle record sections has provided invaluable indirect information on the physical properties of the deep crust. Information on the P- and S- wave velocity depth functions [Holbrook, 1992] and, Poisson's ratios [Holbrook, 1987, 1988; Hawman et al, 1990, Hawman and Phinney, 1991], is obtained through the detailed analysis of the travel times of the phases and the seismic wave field characteristics such us frequency content, reflectivity character, etc. [Berzon, 1965]. By comparing the physical properties determined in field experiments to those measured in laboratory using rock samples, we hope to estimate the gross composition of the deep crust. However, such data cannot unambiguously resolve the petrologic nature of this largely inaccessible region of the lithosphere. The large overlap or non-uniqueness observed in the laboratory measurements makes it very difficult to assign rock types to the velocities determined from seismic experiments. Futhermore, this non-uniqueness is increased by the low resolution of the velocity depth functions derived from

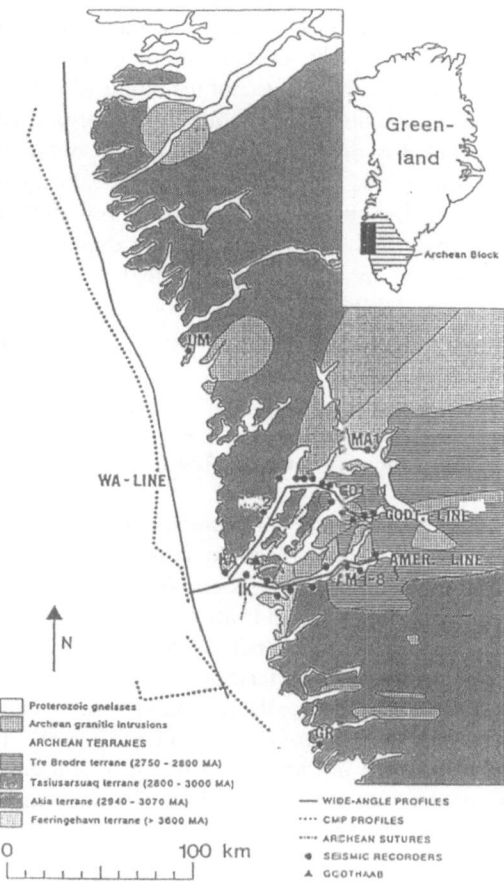

Fig. 1. Location map of shots and receiver stations of the seismic experiment in southwestern Greenland. Solid lines: positions of source array for wide-angle recording; thick-dotted lines: positions of the CMP lines. Three component seismic recorders (REFTEK instruments) located along the coast (UM, KA, IK, GR) and in two fjords, Gothaabfjord and Ameralik (GO 1 to 11, Am 1 to 8 and MA 1 and 2) recorded densely spaced air-gun shots (100-150 m) from profiles WA, GO, and AM. The map also shows the general geology of the area studied in southwestern Greenland. Compilation of published maps from the Geological survey of Greenland [1982], Nutman et al. [1989], and Friend and Nutman [1991].

seismic refraction/wide-angle reflection measurements determined by forward modeling and/or inversion schemes. Thus, the estimates of Poisson's ratio are also characterized by large error bars [Hawman et al., 1990; Hawman and Phinney, 1992]. This large uncertainty significantly decreases the utility of Poisson's ratio to constrain the nature of the lower crust.

In the present manuscript we study whether the laboratory measured anisotropy can be used in conjunction with other physical properties, (e. g., P-wave velocity and density) to characterize different rock types. In order to be able to address indirect anisotropy measurements we need well resolved velocity depth profiles. Therefore, we use an inversion scheme that utilized seismic P and S reflected and P-to-S and/or S-to-P converted energy to determine P- and S-wave velocity depth functions in southwestern Greenland. The 10% anisotropy measured suggests that the high velocity (7.2-7.4 km/s) and relatively high density (3.0-3.1 kg/m^3) layer located at the base of the crust of this passive margin is a result of underplating processes rather than serpentinized upper mantle due to undercrusting.

2. Geology and Geophysics in the Labrador Sea

The geographical configuration of the narrow Labrador Basin makes it an ideal natural laboratory within which to conduct geological and geophysical studies addressing the tectonic processes which enables the development of the rifted basin and the geologic structures that resulted from them.. The asymmetric margins of the Labrador Sea have been the subject of a number of geophysical studies mostly seeking to compare the crustal structures and their tectonic development in order to provide a physically reasonable, conceptual model for the generation and evolution of rifted basins. Our study covers part of the Archean block of SW Greenland (Fig. 1). This is a rifted part of the Nain province that also outcrops in Labrador, Canada [Hoffman, 1989]. The Archean rocks in southwestern Greenland consists of Gneissic, Granitic, and supracrustal complexes with ages ranging between 3800 and 2750 Ma [McGregor, 1973, 1979; Bridgewater, et al., 1976; Brown et a., 1981; McGregor et al., 1986]. Highly deformed, thin mylonitic zones (sutures) occur at the boundaries of the complexes suggesting that four Archean subterranes assembled at about 2750-2650 Ma [Nutman et al., 1989]. After this period the Nain province remained tectonically stable until the opening of the Labrador Sea which started in the late Cretaceous. The rifting expanded to the north in the early Paleocene and ended in the Davis Strait and Baffin Bay during the early Oligocene [Hinz et al., 1979; Srivastava et al., 1980; Johnson et al., 1982; Srivastava, 1983; Srivastava and Tapscott, 1986]. Between 60° to 62° N Latitude Hinz et al., [1979] observed a steeply dipping continental fault which they attribute to normal faulting. Normal faulting can be inferred from ocean bathymetry and gravity modeling [Speece, 1992]. Gravity and magnetic profiles across the Labrador Sea [Geological Survey of Canada 1988a, b; Woodside and Verhoef, 1989] support the asymmetry of the margins indicating steeper gradients and higher amplitude anomalies in the Greenland side than in the Labrador margin [Keen et al., 1994]. This is often associated with a steeply dipping basement and a thin or even missing sequence of shelf sediments [Woodside and Verhoef, 1989]. Gravity studies of Godthäbsfjord and Ameralik region reveal a positive seaward gradient of the long-wavelength Bouguer anomaly of approximately 2 mGal/km, suggesting a 15-20° slope of the crust-mantel boundary along

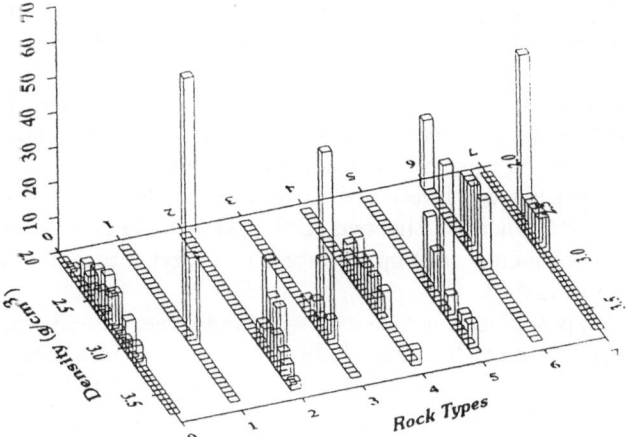

Fig. 2. Statistical distribution of laboratory measured densities for 8 rock types: (0) serpentinite, (1) amphibolite, (2) eclogite, (3) dunite, (4) granulite, (5) peridotite, (6) anorthosite, (7) gneiss. The laboratory data are from Christensen, [1989].

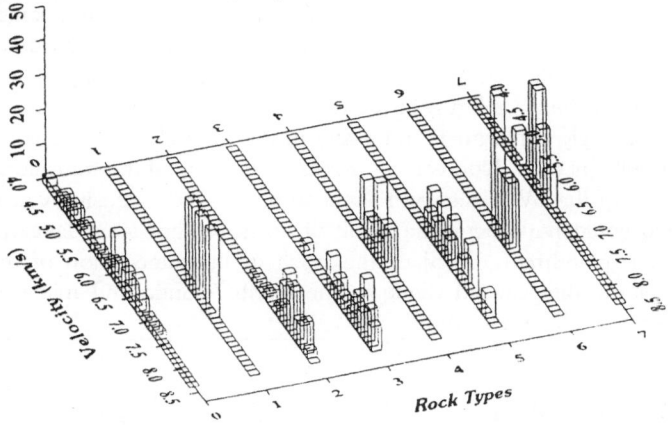

Fig. 3. Statistical distribution of laboratory measured P-wave velocities for 8 rock types (0) serpentinite, (1) amphibolite, (2) eclogite, (3) dunite, (4) granulite, (5) peridotite, (6) anorthosite, (7) gneiss. The laboratory data are from Christensen, [1989].

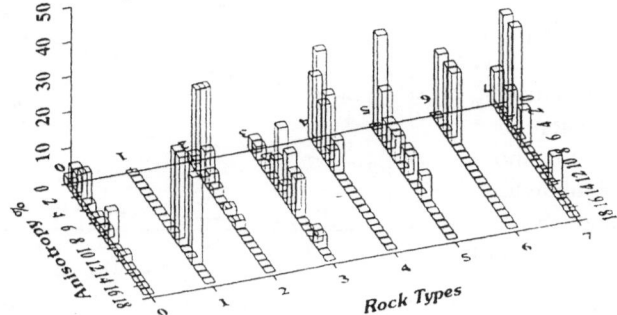

Fig. 4. Statistical distribution of laboratory measured S-wave anisotropy for 8 rock types (0) serpentinite, (1) amphibolite, (2) eclogite, (3) dunite, (4) granulite, (5) peridotite, (6) anorthosite, (7) gneiss. The laboratory data are from Christensen, [1989].

126

the shelf [Speece, 1992]. Seismic reflection and refraction data across the Labrador Basin provide additional evidence for the asymmetry of the margins by imaging a thick sedimentary wedge in the Labrador margin which is non-existent in the Greenland side [Hinz et al., 1979, Keen et al, 1994]. A large number of high amplitude diffractions indicate the existence of a steeply dipping normal faults in the acoustic basement [Hinz et a., 1979, Keen et al., 1994]. As in many other rifted margins the location of the Continent-Ocean Boundary is a controversial issue. Roest and Srivastava [1989] locate it close to the continental slope (chron 33) whereas Chalmers [1991] places it farther into the Labrador Sea, extending the thin continental crust up to chron 27. A careful discussion on this topic can be found in Keen et al. [1994].

The apparent lack of exposed volcanism along the Labrador Sea margins suggests a non volcanic rifting event [White and McKenzie, 1989]. However, up to 8 km thick basaltic formations are found on Disko Island and on the conjugate margin on Baffin Sea [Denham, 1974; Clarke and Pederson, 1976]. These volcanics were extruded after the initial opening of the Labrador Sea during the spreading of the Davis Strait and Baffin Bay approximately at about 61-58 Ma [Clark and Upton, 1971]. Hinz et al., [1979] suggested that the flat basement and the intra-basement structures observed by the multichannel seismic surveys were related to younger basaltic flows overlaying pyroclastic or sedimentary sequences which were deposited on top of older oceanic crust. Refraction/wide-angle velocity determinations [Hinz et al., 1979; Stergiopoulos, 1984] revealed high velocity zones (P-wave velocities higher than 7.2 km/s) in the lowermost crust on each side of the conjugate margin, which can imply a magmatic rift event in the Labrador Sea. Chian and Louden, [1992] found no evidence for high velocity zones at the base of the transitional crust across the southern limits of the SW Greenland Archean block. The extensive volcanism and the crustal thickening could have been the result of an active hot spot developed under the Davis Strait alone [Hyndman, 1975] or the result of the intersection of the rift with an asthenospheric plume that caused rifting in the North Atlantic and migrated underneath Greenland [White and McKenzie, 1989].

3. Poisson's Ratio vs Anisotropy

One of the justifications for refraction/wide-angle measurements is that we hope to estimate the nature and composition of the deep crust by comparing the velocity determined in field experiments to those measured in the laboratory. However, the P- and S-wave seismic velocities of the deep crust are not only a function of the mineralogical composition but they are also determined by other factors: confining pressure,

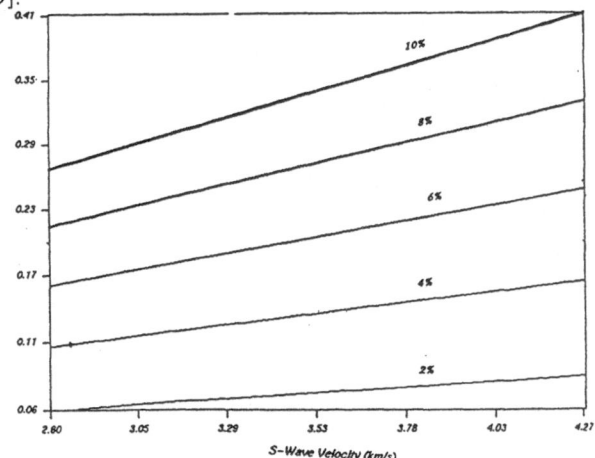

Fig. 5. Uncertainty in S-wave velocity due to 2, 4, 6, 8, and 10 % anisotropy. (a) 2%, (b) 4%, (c) 6%, (d) 8%, (e) 10%. We calculated the uncertainty by $\Delta\beta = \beta_{max} \dfrac{2a}{(a+2)}$ derived from equation (1).

temperature, anisotropy, and pore fluid pressure. Unfortunately, although, we can in some cases correct for these factors by theoretical relationships, the correlation between the seismic wave velocity and composition is sufficiently non-unique that such comparisons are limited to only very rough estimates of crustal composition. The statistical distributions of laboratory measurements of physical properties -density, P-wave velocities and anisotropy percentage- (Fig. 2, 3, and 4) of 8 different rock types candidates for lower crustal rocks - Anorthosite, Amphibolite, Gneiss, Dunite, Serpentinite, Granulite, Peridotite, Eclogite- illustrates the overlap, non-uniqueness, in the laboratory estimates. These rock types are characterized by relatively high P-Wave velocities. Serpentinite has a very broad velocity and density spectra since its elastic constants are strongly influenced by accessory and serpentinite mineralogy. Nevertheless, serpentinite has been purposed as an explanation for lower crustal reflectivity in highly stretched continental rifted areas [Boillot et al., 1989, Boillot et al., 1990]. Average values of S-wave anisotropy range from 4 to 6% (Fig. 4) except for amphibolite and dunite which range approximately from 8 to 12% (Fig. 4) [Holbrook et a., 1992].

Indirect seismic velocity determinations result in average measurements (i.e. averaged along the ray path). Therefore, for anisotropic structures seismically derived velocities will have an uncertainty due to the amount or degree of anisotropy. The anisotropy percentage (a) is usually calculated by:

$$a = \frac{\beta_{max} - \beta_{min}}{\left(\frac{\beta_{max} + \beta_{min}}{2}\right)} \quad (1)$$

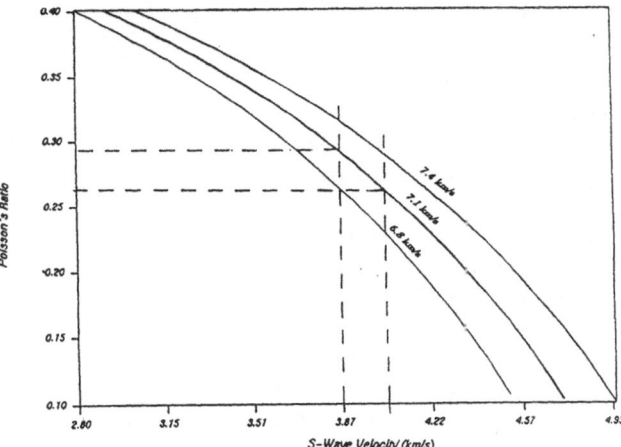

Fig. 6. Uncertainty in Poisson's ratio due to an uncertainty in S-wave velocity of 0.2. This uncertainty was estimated using a constant P-wave velocity of 7.2 km/s and equation (2).

where β_{max} and β_{min} are the maximum and minimum S-wave velocities respectively. For a given anisotropy percentage we can use equation (1) to calculate the change in S-wave velocity. The S-wave velocity range (uncertainty) induced by an anisotropy percentage of 2, 4, 6, 8 and 10% is illustrated in Fig. 5. This uncertainty increases with the S-wave velocity. Taking into consideration lower crustal S-wave velocities of 3.5-4.0 km/s [Holbrook et al., 1992] and an average anisotropy of 4-6% a mathematically reasonable uncertainty for S-wave velocities is ±0.2 km/s due to anisotropy.

The S-wave information is normally used to partly overcome the non-uniqueness of the P-wave velocity values [Christensen and Fountain, 1975]. The relationship between P- and S-wave velocity for a given rock or mineral can be expressed in terms of Poisson's ratio, which increases with increasing P- wave velocity or decreasing S-wave velocity. Poisson's ratio varies from about 0.21 to 0.33 in common rock types and is particularly sensitive to quartz content; whereas most of the rock-forming minerals have Poisson's ratio around

0.25-0.30, quartz has a value of 0.08. There are few studies concerning the influence of anisotropy [Carbonell and Smithson, 1991] and fluid pressure [Hyndman and Klemperer, 1989] in the indirect measurements of physical properties. This anisotropy and fluid pressure are factors that increase the non-uniqueness of the seismic velocities and Poisson's ratio. The presence of anisotropy, for example, implies a dependence of the seismic velocities, and as a consequence of the Poisson's ratio estimates on the direction of wav·· propagation. Poisson's ratio (γ) are usually calculated by

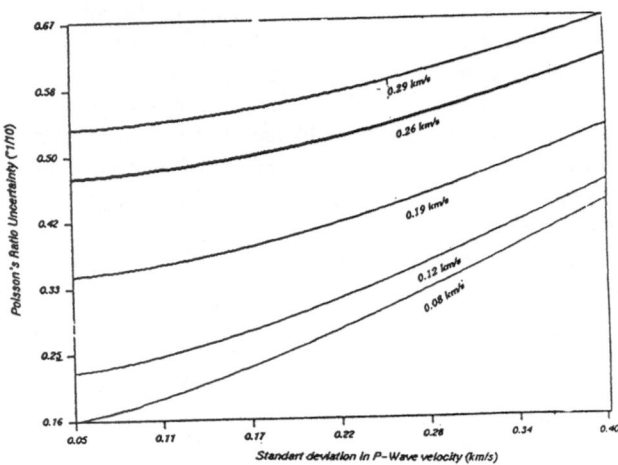

$$\gamma = \frac{\left(\dfrac{\alpha}{\beta}\right)^2 - 2}{2\left[\left(\dfrac{\alpha}{\beta}\right)^2 - 1\right]} \quad (2)$$

where α and β are the P- and S-wave velocities, respectively. Although, equation (2) was derived considering isotropic rocks we can use (2) to calculate the uncertainty of the Poisson's ratio due to an uncertainty of ±0.2 km/s in the S-wave

Fig. 7. *Uncertainty in Poisson's ratio -equation (3)- for different values of P-wave uncertainty σ_α. $\sigma_\beta = 0.2$ km/s .*

velocity. This can be calculated for different P-wave velocities (Fig. 6). For typical lower crustal velocities (P wave velocities of 7.0-7.4 km/s and S-wave velocities of 3.5-4.0 km/s) the average uncertainty in Poisson's ratio is approximately ±0.025. In seismic velocity determinations both P-and S-velocities are characterized by an uncertainty or error estimate, these P and S error estimates will propagate into the Poisson's ratio calculation. Using equation (2) and following Hawman et al., [1990], the uncertainties of the Poisson's ratio can be calculated by,

$$\sigma_\gamma^2 = \sigma_\alpha^2 \frac{\beta^4}{\alpha^6\left(\dfrac{\beta^2}{\alpha^2}-1\right)^4} + \sigma_\beta^2 \frac{\beta^2}{\alpha^4\left(\dfrac{\beta^2}{\alpha^2}-1\right)^4}, \quad (3)$$

where α and β are the P- and the S-wave velocities, respectively, and σ_α and σ_β are uncertainties in α and β respectively. Forward modeling solutions do not supply the formal resolution estimates that the inversion schemes provide. In inversion schemes these σ_α and σ_β can be estimated as half the measure of the spread of the resolution kernels of the velocity solutions [Aki and Richards, 1980; Menke, 1984; Hawman, 1988; Hawman et al.,

1990]. Representative estimates of σ_α and σ_β approximately of ±0.1 to ±0.3 km/s [Holbrook et al., 1987, 1988; Hawman et al., 1990, Holbrook et al., 1992]. The mean σ_γ approach 0.05-0.07 for σ_β of: 0.08, 012, 019, 9.26, and 0.29 km/s, using P- and S-wave velocity of 7.0 and 4.0 km/s, respectively, and σ_α ranging from 0.05-0.4C km/s. In field experiments, Hawman et al., [1990] and Hawman and Phinney, [1992] obtained error estimates for Poisson's ratio that range from 0.05 to 0.12 and from 0.05 to 0.20 in the Basin and Range and for the Great valley and Allegheny plateau of eastern Pennsylvania, respectively.

4. Acquisition, Processing and Analysis of the Seismic Data

4.1 ACQUISITION

During the spring and summer of 1987 and 1989 the University of Wyoming conducted geophysical experiments on the southwestern coast of Greenland. We performed a gravity survey and acquired a seismic wide-angle data set. Over three thousand shots were fired by an airgun array (6000 cu. i.) towed along the coast and inland along the Ameralik and Godhaab fjords in order to better understand the transitional continental-oceanic crust of the Labrador Sea. For the entire seismic experiment 15 three-component PASSCAL REFTEK 73A digital seismic recording instruments were deployed at 35 locations (Fig. 1). These instruments recorded the seismic energy released by the air gun array of five air guns (19.7 l each) with a peak pressure of 140 bars fired at 15 m depth with a spacing of 100 to 150 m between shots. The large number of shots allowed the recording of densely spaced, generally unaliased, seismic data which permitted a reliable correlation of the arrivals.

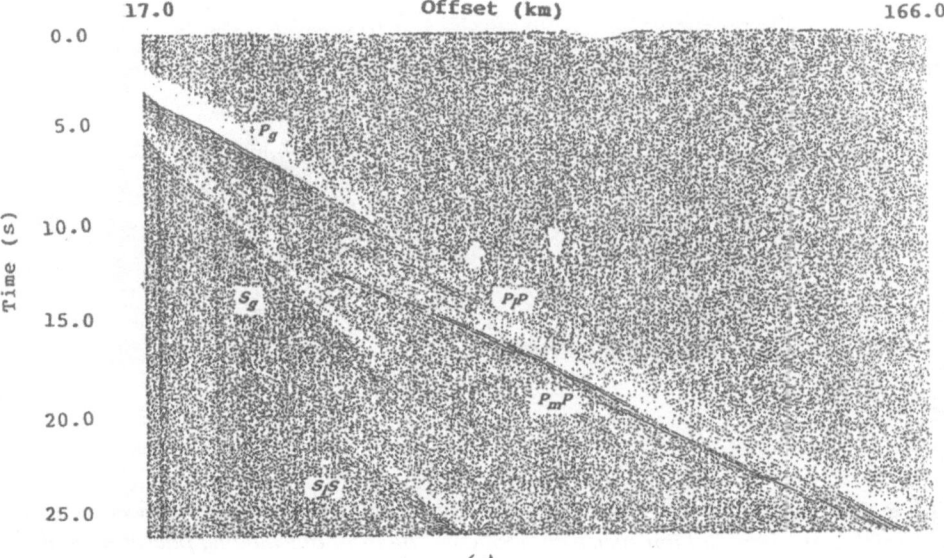

(a)

130

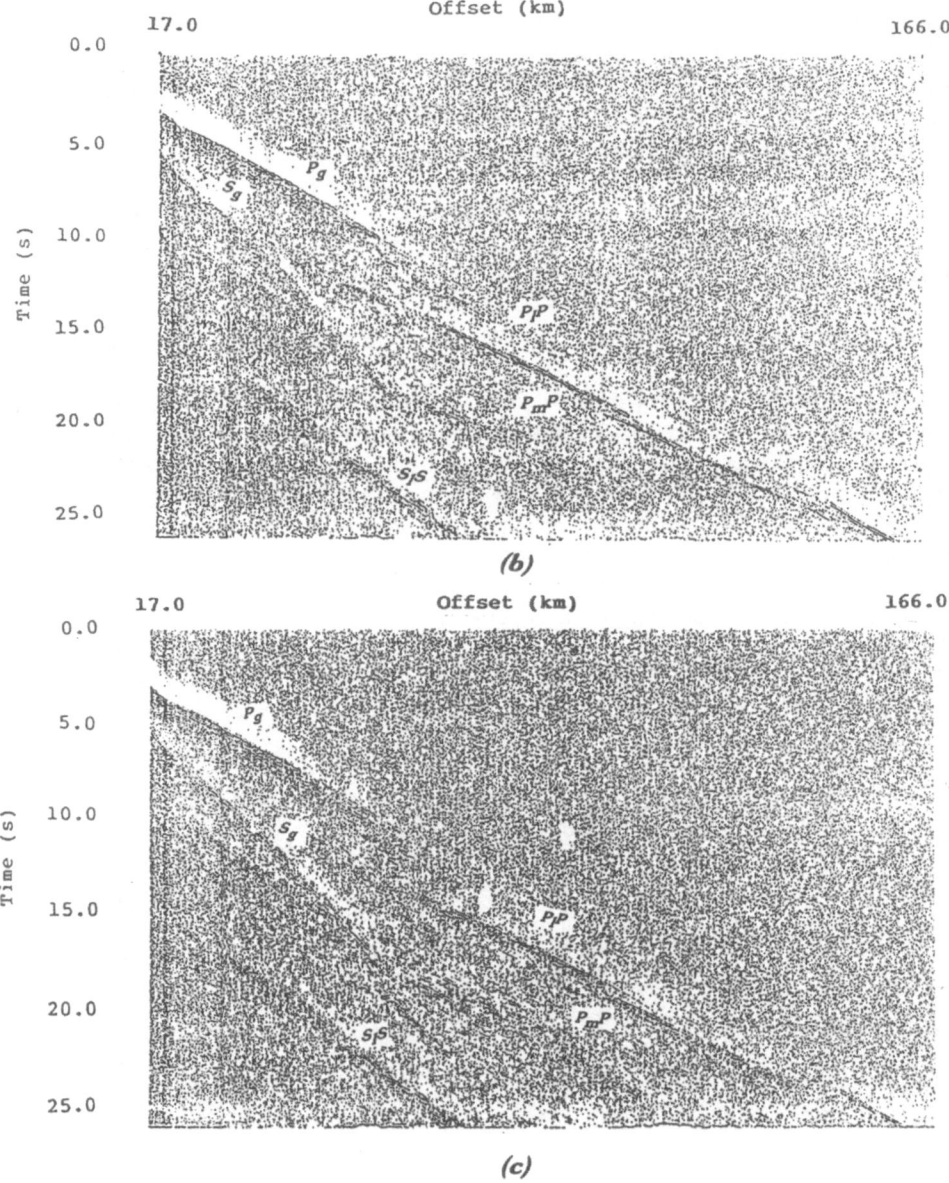

Fig. 8. *Time-offset record sections from the Kangak receiver gather. (a) vertical component, (b) radial component, (c) transverse component. The most outstanding phases a labeled.*

especially for phases with low signal-to-noise ratio. All the instruments recorded three components, the strong S-wave energy identified in the vertical and horizontal components was probably generated at the water-basement interface by conversion from P wave energy. Gohl, [1991], and Gohl and Smithson [1993] presented the results based on the analysis, modeling, and one- and two-dimensional inversion of the vertical component data. In the present study we analyze the horizontal component recordings of station KA (Fig. 1). It should be pointed out that the recordings correspond to receiver gathers of shots fired at sea. Therefore, we avoided source coupling differences. The fact that all the traces in the receiver gather were recorded by the same geophone (emplaced directly on bedrock) prevented differences due to receiver coupling. The dominant frequency range was from 7 to 9 Hz. To enhance the correlation of the phases we performed some basic processing which consisted of: 1) static corrections to remove the undulations on the arrivals caused by the sea floor topography; 2) seismic deconvolution to decrease the source-generated noise (water reverberations, air-bubble) and to temporally sharpen the phases; and 3) band pass filtering, between 4 and 13 Hz to remove the low frequency water-wave energy and high frequency ambient noise.

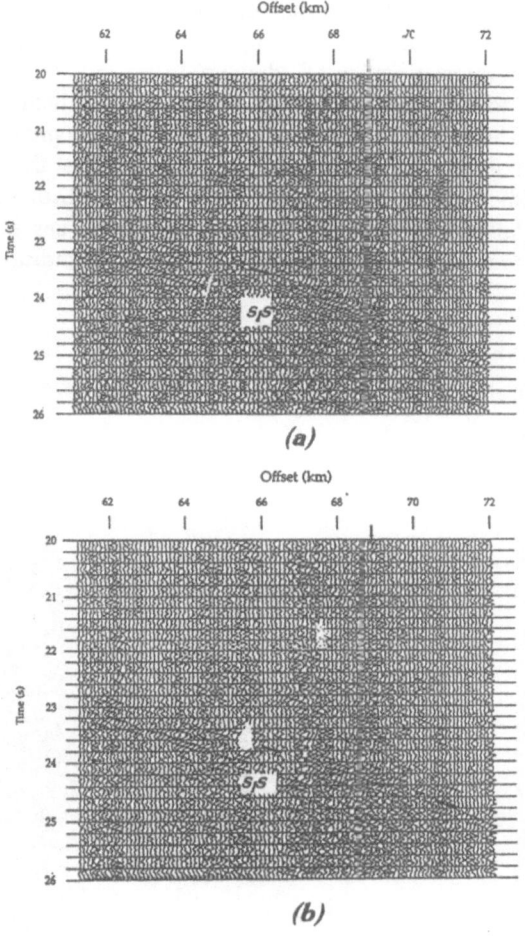

Fig. 9. *Enlarged time-offsets sections around SS phases (a) radial component, (b) transverse component. The arrows mark the SS phase. A delay of approximately 0.25s can be observed. The arrow along the offset axis indicates the trace location for the particle motion diagram show in Fig. 12.*

4.2 ANALYSIS, AND SPECIALIZED PROCESSING

The horizontal component receiver gathers were then rotated into radial and transverse components using the azimuth readings of the ship track. The resulting image shows very distinctive arrivals which seem to mimic the phases observed in the vertical component recordings (Fig. 8a, b, c). For example the most prominent phases are: the P_g direct arrival; the P_lP, an intra-crustal reflected P phase, the P_mP or Moho reflection, and with higher dip (slower apparent velocity) the S-wave phases, an S_g and a S_lS. Note in Fig. 8b and 8c numerous moderately dipping, lower amplitude phases, (an approximately linear fabric). Its moderate dip indicates a velocity intermediate between the P- and the S-wave velocities. This suggest P-to-S, or S-to-P converted energy. This energy has been identified because of the close trace spacing. Up to here most of the interpretation parallels the interpretation of the vertical components [Gohl and Smithson, 1993]. However, a careful analysis of the linear events reveals that the S_lS phase has a different dip in the radial and transverse receiver gathers (compare Fig. 9a, and b) indicating slightly different apparent velocities.

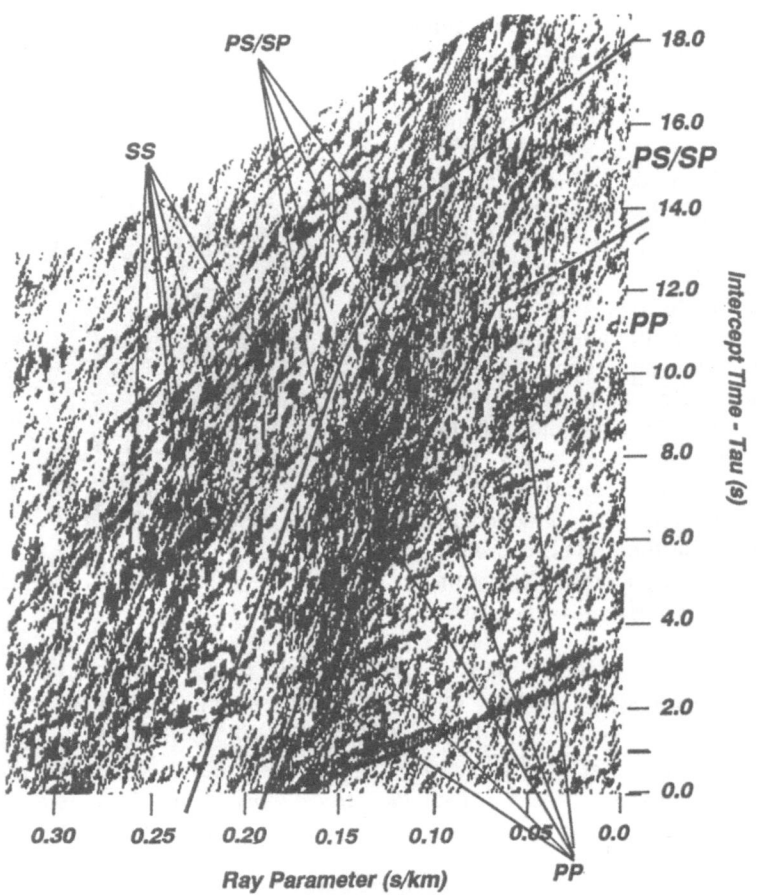

Fig. 10 a. Coherence envelope of the slant stack of the transverse component recorded at station KA (Fig. 1). The high amplitude contributions of the different wavefields are indicated by: PP P-wavefield, PS/SP converted wavefield, SS S-wavefield.

In order to unravel the different arrivals and to perform a quantitative interpretation we use a very well known mathematical transformation which is the Radon transform [McMechan et al., 1980; Phinney et a., 1981; Stofa et al., 1981]. This mathematical coordinate transformation changes the data from (t, x) time-offset space to (τ, p) or intercept time-ray parameter space. This transform adds the amplitudes along dip lines in the gather; each dip or slope corresponds to a different ray parameter and constitutes a trace in the new space. Under this transformation a straight arrival in the (t, x) gather maps onto a point in the (τ, p) space, a reflection, which has a hyperbolic moveout in the (t, x) space, maps onto an ellipse in the (τ, p) space. The P-wave energy is concentrated in the lower ray parameters because of the high apparent velocities, whereas the S-wave energy is

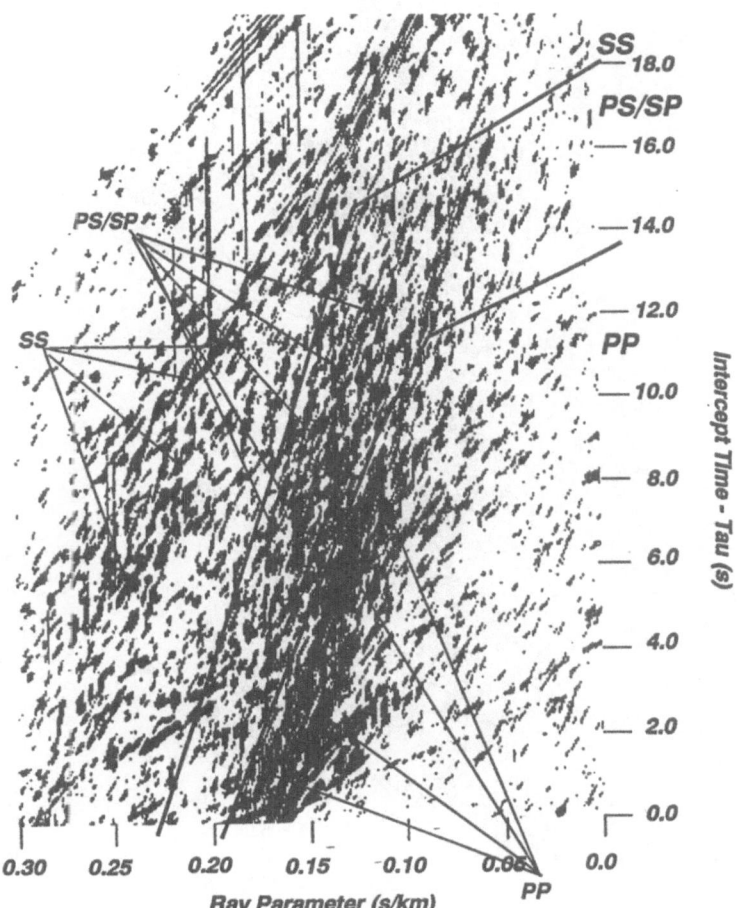

Fig. 10 b. *Coherence envelope of the slant stack of the radial component recorded at station KA (Fig. 1). The high amplitude contributions of the different wavefields are indicated by: PP P-wavefield, PS/SP converted wavefield, SS S-wavefield.*

characterized by higher ray parameters. The intermediate branches will correspond to P-to-S, or S-to-P converted energy. This specialized processing separates and enhances all the arrivals (Fig. 10a, b).

Both sections (the radial and the transverse) present well defined PP, SS, and PS/SP converted branches. All these branches present a finite width and appear as elongated clouds of points (several p, ray-parameters, for the same τ). The pre-critical branches do not reach the τ axis. Some of the branches, especially SS and PS/SP, exhibit gaps; other events have non-elliptical moveouts. We suggest that these two features are the results of lateral heterogeneities and out-of-plane reflections, respectively. Levander and Gibson [1991] demonstrated that lateral heterogeneities generate short horizontal events at near vertical

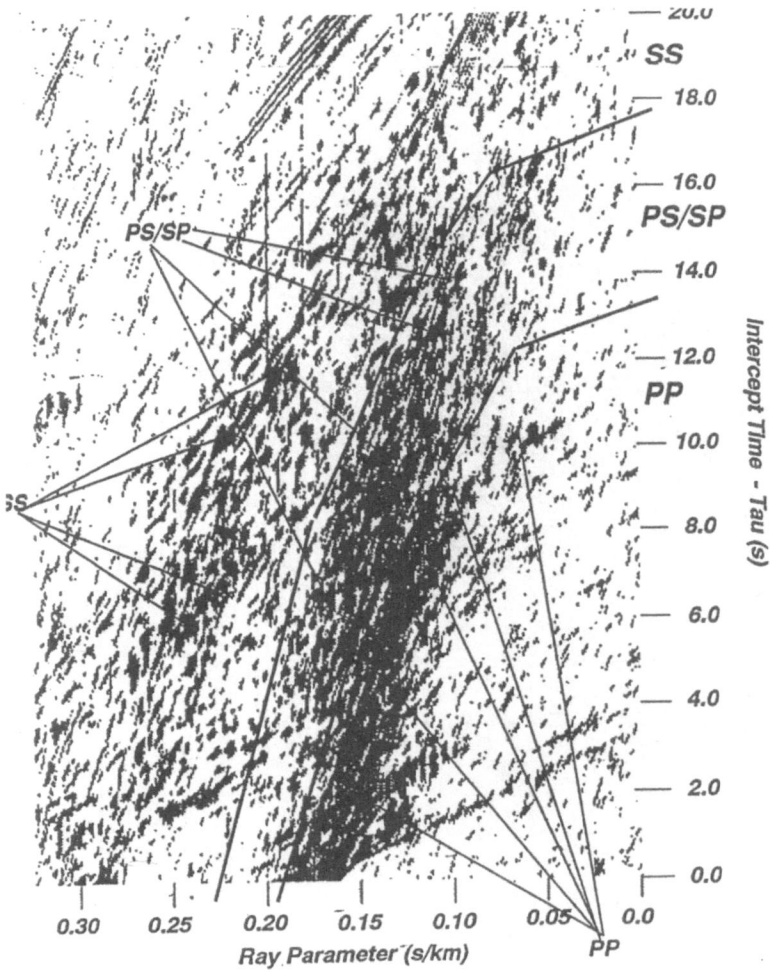

Fig. 11. Coherence envelope of the stack of the (τ, p) sections corresponding to the radial and transverse components recorded at station KA (Fig. 1). The high amplitude contributions of the different wavefields are indicated by: PP P-wavefield, PS/SP converted wavefield, SS S-wavefield.

incidence and coherent events that parallel principal arrivals at far offsets. This can produce the thickening of the branches in the (τ, p) space. Out-of-plane reflections can appear in the (t, x) time-offset space as dipping events that cross other principal arrivals. This can transform into non-elliptical events in the (t, p) space.

Large amplitude S-wave events are observed in the transverse component, indicating that more S-wave energy is recorded by this component (compare Fig. 10a and b). Stacking the radial and the transverse (τ, p) sections does not increase the continuity of the S-wave field, but deteriorates it (Fig. 11). This decrease in S-energy due to the stacking procedure suggests that the S-wave fields of the radial and transverse components are not in phase because they do not interfere constructively. One possible explanation is that a delay

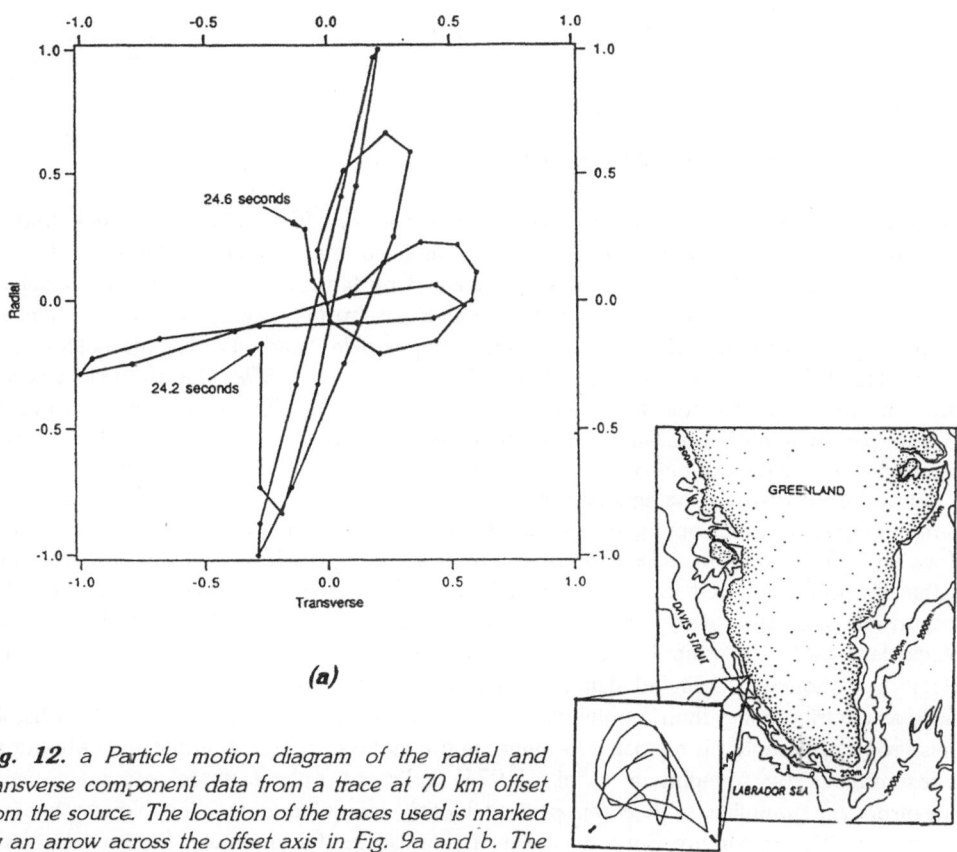

Fig. 12. a Particle motion diagram of the radial and transverse component data from a trace at 70 km offset from the source. The location of the traces used is marked by an arrow across the offset axis in Fig. 9a and b. The diagram starts a 24.2 s and ends at 24.6 s encompassing the $S\beta$ phase in Fig, 8b and c. An almost pure radial particle motion starts at 24.2, and at approximately at 24.4 s the transverse motion starts. The axes are the normalized amplitudes of the radial and transverse components. b. Map view showing the orientation of the particle motion diagram with respect to the coast line of Southwestern Greenland.

between the two wave fields may occur. Such characteristics would indicate that the S-wave events reach the receivers at slightly different times, suggesting S-wave velocity differences between the radial and transverse components and hence indicative of anisotropy. A delay of approximately 0.25s can be observed at an offset of 70 Km between the radial and the transverse components of the S$_j$S phase. The particle motion diagram drawn by using the two horizontal components indicates two distinct orientations for the particle motion. The maximum radial displacement is sensibly parallel to the coast line.

To determine well resolved P- and S-wave velocity-depth functions, we used an extension of an inversion scheme which uses the digitized (τ, p) [Bessonova et al., 1974; Bates and Kanasewich, 1976]. This inversion scheme couples the estimation of the P- and S-wave velocity-depth functions using the P-to-P and the S-to-P reflected and P-to-S and/or S-to-P converted energy. The inversion of synthetic data demonstrated the increase in resolution [Carbonell et al., 1994].

5. Discussion

5.1 ANISOTROPY AND POISSON'S RATIO

Seismic anisotropy has been measured in the upper crust [Crampin et al., 1984] and in the upper mantle. Laboratory measurements indicate that velocity differences due to anisotropy increase as a function of pressure [Christensen, 1989]. P-wave velocities measured in the laboratory show that several lower-crustal rock types have significant anisotropy, quartz-mica schist (10.7%), felsic amphibolite gneiss (7.3%), granulite-facies metapelite (5.5%), and amphibolite (5.4%) [Holbrook et al, 1992]. Horizontally layered structures also produce, anisotropic effects (transverse isotropy) [Crampin, 1989]. Common anisotropic minerals of lower crustal rocks are plagioclase, biotite, amphiboles, and pyroxenes. In the case of anisotropic rocks Poisson's ratios are azimuth dependent.

Laboratory measurements indicate that most metamorphic rocks are characterized by a certain degree of anisotropy, generally due to small scale layering and metamorphic fabrics (foliation). There are numerous examples of anisotropic rocks whose physical properties are within the range inferred for lower crustal rocks. The mineralogical composition can be responsible for anisotropy, the alignment of common anisotropic minerals such as plagioclase, biotite, amphiboles and pyroxenes. Siegesmund and Kruhl [1991] reported that plagioclase show weak mineral alignment in amphibolites from the Ivrea zone. Plagioclase is likely to contribute less than amphiboles and biotite to seismic anisotropy. Biotite is highly anisotropic, although it is a small constituent in ampholites and most granulites. Laboratory measurements by Manghnani et al., [1974] indicated only a three percent S-wave anisotropy in granulites with a ten percent biotite. Hornblende has the highest S-wave anisotropy. The maximum S-wave velocity is parallel to the lineations [Siegesmund and Vollbrecht, 1991; Ji and Salisbury, 1993]. The S-wave anisotropy in pyroxenes is the greatest oblique to the lineation, and the amount of anisotropy is approximately half that of hornblende in rocks with the same fabric [Siegesmund and Vollbrecht, 1991]. These laboratory measurements and the statistical distributions of density, P-wave velocity and anisotropy measurements displayed in Fig. 2, 3, 4 suggest that convining information on the anisotropy with knowledge of density and P-wave velocity can, in some cases, reduce the uncertainty in the correlation between physical properties and rock mineralogy. The presence of anisotropy narrows the possible composition to those rocks that are strongly anisotropic and suggests a foliated and possibly layered metamorphic rock assemblage. The general employment of single component (vertical) seismic instruments and the low resolution of refraction/wide-angle reflection experiments (mostly due to spatial under sampling) have prevented the use of anisotropy as a constraint for the determination of lower crustal composition in favor of estimates on Poisson's ratio [Holbrook et al., 1993].

In the presence of anisotropy, even for low anisotropy percentages (approximately between 4 and 6 %) the S-wave velocity derived from refraction and/or wide-angle reflection measurements can be characterized by uncertainties that can reach ±0.2km/s on average (Fig. 5). Assuming only this standard deviation in the S-wave velocity and no uncertainty for the P-wave velocity produces an error estimate for the Poisson's ratio of 0.03 (Fig. 6). By including the P-wave velocity uncertainties we obtain a Poisson's ratio error estimate that can approach 0.04 (Fig. 7). In general, careful geophysical studies involving the analysis, inversion and modeling of seismic wide-angle reflection/refraction report uncertainties for P- and S- wave velocities that approach 0.1-0.3 km/s The Poisson's ratio profiles derived from this measurements is characterized by 0.05-0.20 error estimates. For example: errors in Poisson's ratio ranging from 0.05-0.12 and from 0.05-0.20 have been obtained in the Basin and Range [Hawman et al., 1990], and for the Great valley and Allegheny plateau of eastern Pennsylvania [Hawman and Phinney, 1992], respectively. This large uncertainty reduces considerably the physical significance of Poisson's ratio profiles.

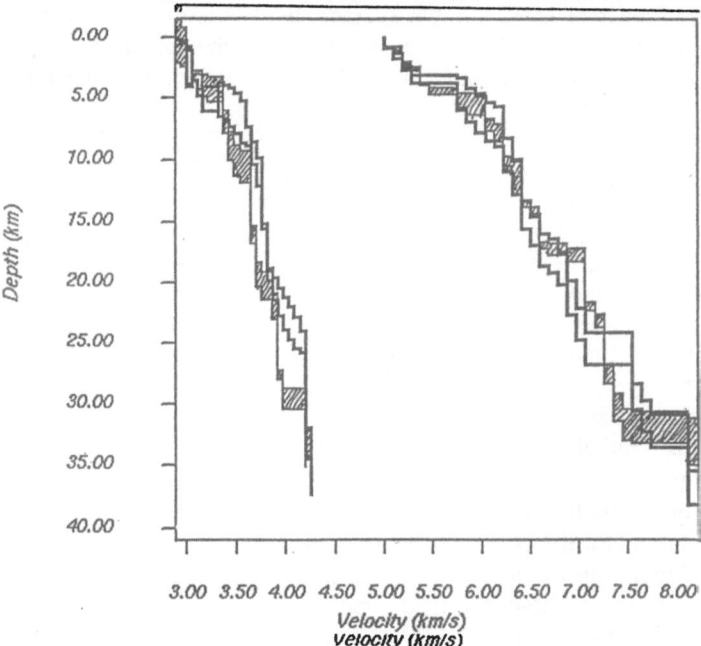

Fig. 13. *P- and S-wave velocity-depth functions obtained by the joint inversion scheme for the radial and transverse components. Bold lines (--) P-and S-wave velocity-depth functions obtained from the radial component receiver gather. Thin line (/////); P- and S-wave velocity-depth functions obtained from the transverse component receiver gather. For both components, only the velocity function plus the error and the velocity function minus the error are displayed for clarity.*

138

5.2 THE PASSIVE MARGIN OF SOUTHWESTERN GREENLAND

The radial and transverse horizontal recordings of station KA present a few differences. One qualitative differences is that the transverse component is more energetic than the radial component. This can be attributed to different crustal reflectivity for the two components, implying a certain degree of anisotropy. Nevertheless, as the S-wave energy is entirely generated by conversions at the marine-basement interface it is probably a source directivity effect at this surface. We used the horizontal component receiver gathers from station KA from the large aperture closely spaced wide-angle reflection data set acquired in southwestern Greenland (Fig. 1). Because this data set consists of three-component, densely sampled recordings (Fig. 8a, b, c), it is especially suitable for the inversion analysis. A number of S-wave arrivals and converted phases were observed. We digitized the intercept-time ray-parameter functions $\tau(p_{pp})$, $\tau(p_{ps})$ and $\tau(p_{ss})$ (Fig. 10a, b). These (τ, p) picks were used to invert the radial and transverse wave fields independently for velocity-depth functions.

The P-wave velocity-depth function indicates a crustal thickness under station KA of 35-38 km. The P-wave velocity-depth functions obtained from the radial and transverse components overlap each other up to a depth greater than 20 km. Between, 20-28 km, the velocity obtained from the transverse component is higher than the radial (Fig. 13); between 28-32 km depth, the velocity function obtained from the radial component is slightly higher. Both functions, however, display a high velocity layer at the bottom of the crust between approximately 32-37 km. On average, the P-wave crustal velocity function is characterized by two gradients, a high gradient where velocity increases from 5.0 km/s to values higher than 6.0 km/s at depths of 4-6 km and lower gradient extending to lower

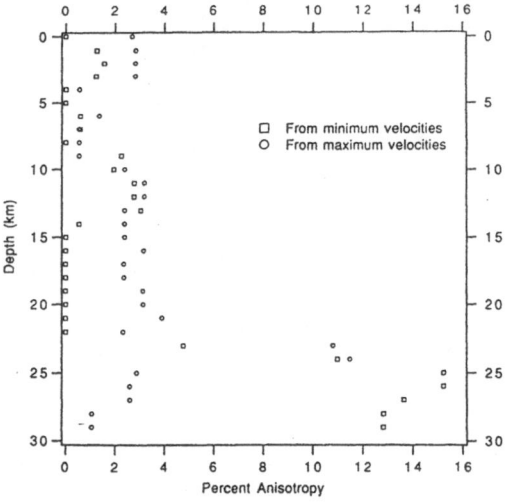

crustal depths. These P-wave velocity-depth functions are consistent with the P-wave velocity profiles derived from the vertical component recordings and obtained by an extremal inversion algorithm and 2-dimensional inversion schemes [Gohl, 1991; Gohl and Smithson, 1993].

The S-wave velocity-depth functions for the radial and transverse components show distinctive differences. Starting at approximately 7 km depth, the S-wave velocity-depth function obtained from the radial component appears to have higher values than the velocity-depth function obtained from the transverse component. A marked

Fig. 14. A plot of the percent anisotropy with the results of the inversion scheme, The squares and circles are the percents from the estimated minimum and maximum velocities.

difference between the two S-wave velocity functions is observed at 23-30 km depth; the S-wave velocity-depth function obtained from the radial component appears to have higher values than the S-wave velocity-depth function obtained from the transverse component. These differences suggest faster velocities for the radial component of the wave field at two different levels. This difference in velocity suggests an anisotropic crust with a significantly anisotropic zone near its base (Fig. 14) and just above the high velocity layer shown in the P-wave velocity-depth function. Such anisotropy could also explain the deterioration of the stack of the radial and transverse (τ, p) sections The average S-wave velocity differences between the radial and the transverse components can explain the S-wave energy decrease when the two horizontal components are stacked together. Suggesting that the wavefields differ by a small phase shift which prevents the waveforms to add constructively. For a 8.5 Hz signal (the frequency band of the data was 4-13 Hz) an average velocity difference like the one observed ($\approx$ 0.15) produces a time delay between two S-wave phases reflected at 20 km depth of 0.28 s which corresponds approximately to 2.5 λ (wavelength).

The different slope of phase S_lS is shown in Fig. 9a, b, the decrease in reflected energy observed in the stack of the radial and transverse (τ, p) sections (Fig. 11), the particle motion diagram (Fig. 12a, b), the results of the inversion scheme (Fig. 13) support a 10 % S-wave anisotropy at 25 km depth (Fig. 14). The P-wave velocities determined for the lower crust in southwestern Greenland 7.0-7.4 [Gohl and Smithson, 1993] correspond to a broad variety of rock types: hornblende-pyroxene gneiss. stronilite, mafic granulite, anortosite [Birch, 1960; Christensen, 1989; Fountain and Christensen, 1989; Bruke and Fountain, 1990; Holbrook et al., 1992] Anorthosites have an S-wave seismic anisotropy of approximately 4% (Fig. 4). These can be found in exposed rocks in the study area. Granulites with P-wave velocities of 6.5-7.2 km/s (Fig. 3) can have an average S-wave anisotropy of 2-3 %. This is mostly due to its mineralogy (granulite facies rocks are characterized by minerals with low anisotropy). The 10 % average anisotropy which characterizes amphibolites makes them the most appropriate rock type for the P-wave velocity. However, the 7.2 km/s velocity might be too high for rocks with a predominantly amphibolite composition. We must keep in mind that the velocity measurements correspond to average velocities, and small scale layering of lower velocity rocks, e. g., gneiss (with mean velocities of 6.4 km/s can decrease the P-wave velocity estimates. Additionally, tonalitic gneiss appears in outcrops in southwestern Greenland. Tonalitic gneiss contains significant amounts of plagioclase. Although plagioclase shows a weak mineral alignment, a layered structure with fine layers of amphibolite and tonalitic gneiss will give the appropriate P-wave velocities, and anisotropy values observed by the seismics [Birch, 1960; Babuska, 981; Fountain and Christensen, 1989; Fountain et al., 1990; Siegesmund and Vollbrecht, 1991].

Layering can be determined from the analysis of the spectral characteristics of the seismic phases [Berzon, 1965]. For P- and S-waves with the same frequency content, the S-waves have higher resolution because of their slower velocity resulting in shorter wave lengths. In our study the both wave types have a dominant frequency of 8 Hz. This frequency, however, is too low to provide any information on small scale layering. The long duration of the S-to-S reflections (S_lS for example), approximately of 0.6-0.8 s (Fig. 9a, b) which is about twice the duration of the P_lP provides some evidence for layering on a scale too fine for the P-wave phases to resolve. In summary, the high anisotropy percentage at 25 km depth (≈ 10%, Fig., 15) places an important constrain on the possible mineralogical compositions at that depth. At this depth velocities of 7.0-7.4 km/s are reported by Gohl and Smithson, [1993]. Laboratory measurements indicate that these velocity values correspond to a large variety of rock types for example: amphibolites, hornblende-pyroxene granulite, stronolite, diabase, and anortosite. However, the high S-wave seismic anisotropy favors an amphibolite (hornblende) composition at 25 km depth.

The analysis and modeling of gravity measurements by Speece [1992] across the southwestern Greenland margin, a profile west-east through Godthäbsfjord region (Fig. 15) is able to resolve a relatively high density (3.1 kg/m^3) wedge at the base of the crust just above the crust mantle transition zone. This density alone indicated mafic material. On the basis of densities alone, this value can be attributed to serpentinite, although it would correspond to the high end of the spectrum (Fig. 3). Mantle rocks with a peridotitic composition would also fall within the range of possible density and velocity values.

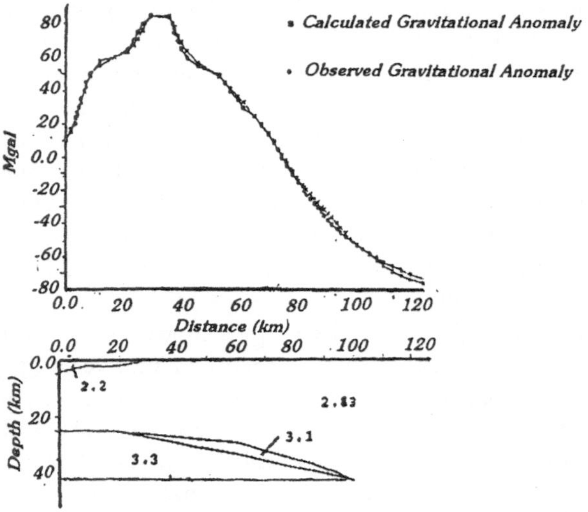

Fig. 15. Gravity profile west-east through Godhäbsfjord region. Model extends west from the Godthäbsfjord region into the Labrador Sea. Free-air anomaly west of 70 km mark along profile and complete Bouguer anomaly east of 70 km mark. Coast parallel, seismic wide angle profile crosses this gravity profile at 60 km mark. Calculated anomaly after the method of Talwani et al., [1959].

For a long time serpentinite has been considered to be a probable component of oceanic lower crust. More recently, it has been considered an active contributor to the building of lower continental crust beneath rifted areas and, especially, passive margins [Boillot et al., 1989; Boillot et al., 1991]. ODP drill sites and sea bottom sampling have detected serpentinites, which have been followed by multichannel seismic data close to the margin in Galicia, northwest Spain. Serpentinites are most probable beneath rifted areas where the crust has been intensively stretched and faulted. These conditions favor the hydrothermal circulation up to depths of 7-10 km within the uppermost, brittle lithosphere [Boillot et al., 1991]. The apparent crustal thickening by basal accretion of a serpentinite layer is known as undercrusting [Boillot et al., 1989]. In some cases undercrusting can explain lower crustal layering and also the young age and mobility of the Moho beneath a stretched continental crust. These characteristics can be explained by migration of the hydrothermal front adding new serpentinized layers to the crust, or the transformation of serpentinized mantle back into peridotite due to an uplift of the isotherms [Boillot et al., 1991]. This emplacement process is fundamentally different from underplating, where hot mantle material is added to the crust (upward magma crystallization). The small amount of exposed volcanics suggest that the Labrador Sea is a result of a nonvolcanic rifted basin. Thus, the over thickened crust observed in areas of the margins, and the high velocity layer imaged by the wide-angle seismic measurements can be a result of undercrusting processes. The gravity and seismics indicate a highly faulted basement, with an almost non-existing sedimentary cover [Speece, 1991; Keen et al., 1994] characteristics which favor hydrothermal circulation of water. Moreover, Serpentinites have a broad spectra in density and seismic wave velocities values (Fig. 2, 3), which makes it very difficult to differentiate them from other lower crustal rocks.

The S-wave seismic anisotropy is probably a result of alignment of anisotropic mineral (lattice- and shape-preferred orientation) [Mainprice and Nicolas, 1989] The Archean block of southwestern Greenland formed by the juxtaposition of several subterranes with different tectonic histories [Nutman et al., 1989] as indicated by the differences in intensity, orientation, and ages of the folding. The crust resulting from the amalgamation of blocks with different strain histories would not necessarily have a common mineral alignment. Nevertheless, a lower crustal plastic deformation caused by an increase in temperature from, for example, magmatic underplating, and the strain during a rifting episode would provide the appropriate physical environment for resetting the mineral alignment of the different subterranes originating seismic anisotropy. Although, the Labrador Sea basin is probably the result of a nonvolcanic rifting event, the over thickened crust in the Davis Strait and the exposed volcanics in the Davis Strait, Baffin Bay and Disko Island suggest an episode of active volcanism that could be associated with the development of a hot spot under the Davis strait [Hyndman, 1975]. White and McKenzie, [1989] suggested that the basaltic volcanics were the result of the intersection of a rift with an asthenospheric plume that caused rifting in the North Atlantic and migrated underneath Greenland. This volcanic episode could have been responsible for the emplacement of hot mantle material at the base of the crust. This hot material was a new heat source which notably increased the heat flow of the overlying rocks producing an increase in the plasticity and making them deform. Zoback, [1992] reported from borehole breakout measurements that the maximum horizontal compressive stress is approximately perpendicular to the southwest coast of Greenland. Plastic deformation of rocks under a stress field develops lattice and shape-preferred orientations within the lower crust [Tullis and Yund. 1987; Mainprice and Nicolas, 1989]. Feldspar. deformed under compressive stress. develops a strong preferred orientation of elongated grains perpendicular to the applied stress [Tullis and Yund, 1987] producing a maximum anisotropy parallel to the coast. The particle motion diagrams (Fig. 12) are consistent with this hypothesis.

The 7.0-7.4 km/s and 3.0-3.1 kg/m^3 lowermost-crustal wedge, presumably, resulted from mafic magmatic underplating. Viscous or even solid state flow in the magma probably aligned the minerals to develop shape-preferred orientation in this wedge [Mainprice and Nicolas, 1989]. Hence, this wedge is probably, seismically anisotropic. However, the small amount of S-wave energy recorded from these depths does not provide enough resolution to confirm this hypothesis. Finally, the seismic wide-angle reflection recordings and the gravity data(Fig. 15), support a model characterized by: 1) In the upper crust: a strongly faulted basement, with a very thin or even nonexistent sedimentary cover; 2) in the lower crust: a strongly anisotropic layer possibly consisting of interlayered amphibolites and tonalitic gneiss, and 3) at the bottom, just above the crust mantle transition, a high velocity wedge, probably of anisotropic, mafic mantle rocks which resulted from underplating processes during a volcanic rifting episode (Fig 16).

142

6. Conclusions

Resolution analysis studies reveal that P- and S- wave velocity measurements derived from wide-angle seismic reflection/refraction data are usually characterized by very low resolution. These large error estimates which are increased in the presence of seismic anisotropy render Poisson's ratio profiles physically meaningless. Nevertheless, seismic anisotropy estimates can constitute an additional constraint in the determination of the mineralogical composition of the deep crust from indirect seismic measurements. Hence, the non-uniqueness of the laboratory measured physical properties (P-wave velocities and densities) can be partly overcome by the inclusion of anisotropy.

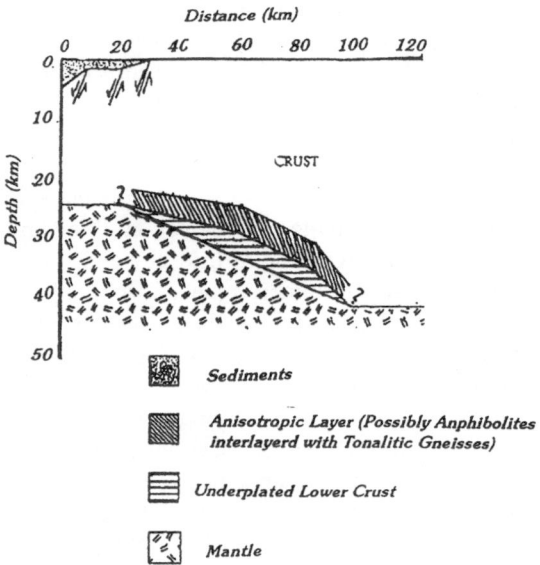

Fig. 16. *Interpretative cross-section along the gravity profile shown in Fig. 15 and based on the gravity and seismic reflection results.*

Anisotropy estimates can place new and more relevant constraints on the different rock types possibly present at the deep crust. We address indirect seismic anisotropy measurements along the south west coast of Greenland using three component, densely space large aperture (0-250 km offset, at 100-150 m spacing) seismic recordings. We use a mathematical inversion scheme that uses P-to-P, S-to-S reflections and P-to-S and/or S-to-P conversions for the determination of well resolve S- and P-wave velocity depth functions for the passive margin of southwestern Greenland. The inversion suggests a P-wave velocity structure characterized by two gradient zones: a relatively high gradient from the surface to approximately 5 km depth where velocities exceed 6.0 km/s, followed by a low gradient to the base of the crust where velocities reach 7.0 km/s. A high P-wave velocity layer (7.2-7.4 km/s) can be identified between 6-8 km above the Moho. Gravity modeling suggests relatively high densities 3.0-3.1 kg/m^3 for this layer. Independent analysis of the radially and transversely polarized horizontal components revealed average velocities of 4.9 ± 0.1 km/s and 4.5 ± 0.1 km/s respectively suggesting a seismically anisotropic crust. A time delay of 0.25 s between the radial and the transverse horizontal components is observed at offsets of 70 km. The particle motion diagrams indicate that the radially polarized S-wave is parallel to the southwest coast of Greenland. From the S-wave analysis, the ocean-continent transitional crust is clearly seismically anisotropic above a high velocity layer in the lower continental crust. The density and velocity values alone can not clearly differentiate between lower continental crust and serpentinized upper mantel. The anisotropy measured just above this high velocity layer, however, seems to favor an accretion of hot, mafic mantle material (underplating) at the base of the crust during a rifting episode. Possibly, magmatic underplating during Late Cretaceous Labrador Sea rifting heated the preexisting lower crust promoting plastic flow and enabling alignment of anisotropic minerals to produce the seismic anisotropy.

Acknowledgments.
This study was partially supported by NSF grants EAR-87098597, DPP-8821974, DPP-9023847 and EAR-8916706.

References

Aki, K., Richards, P.G., 1980. Quantitative Seismology: Theory and Methods, Freeman and Co.

Babuska, V., 1981. Anisotropy of Vp and Vs in rock-forming minerals. J. Geophys., 50: 1-6.

Babuska, V., Cara, M., 1991. Seismic Anisotropy in the Earth. Kluwer A., Dordrecht, pp 116, 146.

Bamford, D., 1977. Pn velocity anisotropy in a continental upper mantle. Geophys. J. R. Astro. Soc., 49:29-48.

Bates, B.C., Kanasewich, E.R., 1976, Inversion of seismic travel times using the tau method, Geophys. J. R. astr. Soc. 47, 59-72.

Berzon, I.S., 1965. The determination of a model of thinly layered medium by the simultaneous use of amplitude and phase spectrum characteristics of the layer. B. (Izv) Ac. Sc. USSR, 6: 363-367.

Bessonova, E.N., Fishman, V.M., Ryaboyi, V.Z., Sitnikova, G.A., 1974. The tau method for inversion of travel times-I. Deep seismic sounding data, Geophys. J. R. astr. Soc. 36, 377-398.

Birch, F., 1960. The velocity of compressional waves in rocks to 10 kbar. J. Geophys I. 36: 377-398.

Boillot, G., Féreud, G., Recq, M., Girardeau, J., 1989. Undercrusting by serpentinite beneath rifted margins: the example of the west Galicia margin (Spain), Nature. 341: 523-525.

Boillot, G., Besir, M.O., Comas, M., 1993, Seismic image of undercrusted serpentinite beneath a rifted margin, Terra Nova, 4: 25-33.

Braile, L.W., Chiang., C.S., 1986. The Continental Mohorovicic Discontinuity: Results from near-vertical and wide-angle seismic reflection studies. In: Barazangi, M., Brown. L., (Editors), Reflexion Seismology: Global Perspective. Am. Geophys. Union Geodyn Ser., 13: 257-272.

Bridgewater, D., Keto, L., McGregor, V.R., Myers, J.S., 1976. Archean gneiss complex of Greenland. In Escher, A., Watt, W.S., G. Greenland, pp 18-76, G. S. Greenland Copenhagen.

Brown, M., Friend, C.R.L., McGregor, V.R., Perkins, W.T., 1981. The late Archean Qorqut granite complex of southern West Greenland. J. Geophys. Res., 86: 10617-10632.

Bruke, M.M., Fountain, D.M., 1990. Seismic properties of rocks from an exposure of extended continental crust new lab measurements from the Ivrea zone. Tectonophys., 182: 119-146.

Carbonell, R., Smithson, S.B., 1991a. Large scale anisotropy within the crust in the Basin and Range province. Geology, 19, 698-701.

Carbonell, R., Clement, W.P., Smithson, S.B., 1994. Joint P- and S- wave velocity determination from reflected PP, SS and converted PS/Sp phases from large aperture seismic reflection measurements. Tectonophysics, 232: 379-389.

Chalmers, J.A., 1991. New evidence on the structure of the Labrador Sea/Greenland continental margin. J. Geol. Soc. London, 148: 899-908.

Chian, D., Louden, K., 1992, The structure of Archean-Ketilidian crust along the continental shelf of southwestern Greenland from a seismic refraction profile, Can. J. Earth Sci., 29: 301-313.

Christensen, N.I., 1966. Shear wave velocities in metamorphic rocks at pressures to 10 kilobars. J. Geophys. Res., 71: 3549-3556.

Christensen, N.I., 1989. Seismic velocities. In: R.S. Carmichael (Editor), Physical Properties of Rocks and Mnerals. CRC Press, Boca Raton, Fl, pp 431-546.

Christensen, N.I., Fountain, D.M., 1975. Constitution of the lower continental crust based on experimental studies of seismic velocities in granulite. Geol. Soc. Am. Bull., 86: 227-236.

Christensen, N.I., Zymanski, D.L., 1988. Origin of reflections from the Brevard fault zone. J. Geophys. Res., 93: 1087-1102.

Clarke, D.B., Pederson, A.K., 1976. Tertiary volcanic province of west Greenland. In: Escher, A., Watt, W.S., pp. 365-385, Geological Survey of Greenland, Copenhagen.

Clarke, D.B., Upton, G.J., 1971. Tertiary basalts of Baffin Island: Field relations and tectonic setting, Can. J. Earth Sci., 8: 248-258.

Crampin, S., 1984. Effective elastic constants for wave propagation through craked solids. Geophys. J.R. Astron. Soc., 76: 135-145.

Crampin, S., 1989. Suggestions for a consistent terminology for seismic anisotropy. Geophys. Prospect., 37: 735-770.

Crampin, S., Chesnokov, E.M., Hipkin, R.G., 1984, Seismic anisotropy -the state of the art II. Geophy. J. R. astro. Soc., 70: 1-16.

Denham, L.R., 1974. Offshore geology of northen West Greenland (69° to 75° N), Rep. Geol. Surv. Greenland, 63.

Díaz, J., Hirn, A., Gallart, J., Senos, L., 1993. Evidence for azimuthal anisotropy in southwest Iberia from deep seismic sounding data. Physics of the Earth and Planetary Interiors, 78: 193-206.

Fountain, D.M., Christensen, N.I., 1989. Composition of the continental crust and upper mantle; a review. In: Pakiser, L.C., Mooney, W.D., (Eds), Geophysical Framework of the Continental United States. Geol. Soc. Am. Mem., 172: 711-742.

Fountain, D.M., 1989. Growth and modification of lower continental crust in extended terrains: The role of extension and magmatic underplating. In: Mereu, R.F., Mueller, S., Fountain, D.M., (Ed) Properties and Processes of Earth's Lower Crust, Geophysical Monograph, Ser. 51; 6: 287-299.

Friend, C.R.L., Nutman, A.P., 1991. Refolded napppes formed during late Archean terrane assembly, Godthaabsfjord, southern west Greenland, J.Geol. Soc. Lon. 148, 507-519.

Fuchs, K., 1983. Recently formed elastic anisotropy and petrological models for the continental subcrustal litosphere in southern Germany. Phys. Earth Planet. Int., 31: 93-118.

Furlong, K.P., Fountain, D.M., 1986. Continental crustal underplating; Thermal considerations and seismic-petrologic consequences. J. Geophys. Res., 91: 8285-8294.

Geological Survey of Greenland, 1982. Geological map of Greenland, sheet 2, Frederikshaab Isblink- Soendre Stroemfjord, Geol. Surv. of Greenland, scale 1:500,000.

Gohl, K., 1991. Seismic Wide-Angle Studies of Early Archean and Proterozoic Crust in Greenland, Minnesota, and Wyoming. Ph.D. theisis, Univ. Wyoming, Laramie, WY, 189 pp.

Gohl, K., Smithson, S.B., 1993. Structure of the Archean crust and passive margin of southwestern Greenland from seismic wide-angle data J. Geophys. Res. 98: 6623-6638.

Gohl, K., Hawman, R.B., Smithson,S.B., Kristoffersen, Y., 1991. The structure of the Archean crust in sw Greenland from seismic wide-angle data: a preliminary analysis. In: Meissner, R., Brown, L., Durbaum, H.-J., Franke, W., Fuchs, K., Seifert, F., (Eds), Continental Lithosphere: Deep Crustal Reflections. Am. Geophys. Union Geodyn. Ser., 22: 53-57.

Hawman, R.B., Colburn, R.H., Walker, D.A., Smithson, S.B., 1990. Processing and inversion of refraction and wide-angle reflection data from the 1986 Nevada PASSCAL experiment. J. Geophys. Res., 95: 4657-4691.

Hawman, R.B., Phinney, R.A., 1991. Analysis of sparse wide-angle reflection data in the tau-p domain, Bull. Seism. Soc. Am. 81, 202-221.

Hawman, R.B., (1988). Wide-angle reflection studies of the crust and upper mantle beneath eastern Pennsylvania, Ph.D. thesis, Princeton University, Princeton, N.J.

Hawman, R.B., Phinney, R.A., (1992). Structure of the crust and upper mantle beneath the Great Valley and Allegheny Plateau of eastern Pennsylvania. 1. Comparison of linear inversion methods from sparse wide-angle reflection data, J. Geophys. Res. 97, 371-391.

Hinz, K., Schlueter, H.-U., Grant, A.C., Srivastava, P.S., Umpleby, D., 1979. Geophysical transects of the Labrador Sea: Labrador to southwest Greenland. Tectonophysics, 59: 151-183.

Hirn, A., 1977. Anisotropy in the continental upper mantle possible evidence from explosion seismology. Geophys. J. R. Soc., 49: 49-58.

Hoffman, P., 1989. Precambrian geology and tectonic history of Nort America. In: Bally, A.W., Palmer, A.R., (Eds), The Geology of North America. Geol. Soc. Am., Boulder, CO. 447-551.

Holbrook, W.S., Gajeski, D., Prodehl, C., 1987. Shear-wave velocity and Poisson's ratio structure of the upper lithosphere in Southwestern Germany. Geopys. Res. Lett., 14: 231-234.

Holbrook, W.S., Gajeski, D., Krammer, A., Prodehl, C., 1988. An interpretation of wide-angle compressional and shear wave data in Southwestern Germany: Poisson's ratio and petrological implications. J. Geophys. Res., 93: 12081-12106.

Holbrook, W.S., Mooney, W.D., Cristensen, N.I., 1992. The seismic velocity structure of the deep continnental crust. In: Fountain, D.M., Arculus R.J., Kay, R.M., (Editors). The Lower Continental Crust. Elsevier, Amsterdam, pp. 1-43.

Hyndman, R.D., 1973. Evolution of the Labrador Sea. Can. J. Earth Sci., 10: 637-664.

Hyndman, R.D., 1975. Marginal basins of the Labrador Sea and the Davis Strait hot spot. Can. J. Earth Sci., 12: 1041-1045.

Hyndman, R.D., Kelmperer, S.L., 1989. Lower-crustal porosity from electrical measurements and inferences about composition from seismic velocities. Geophys. Res. Lett., 16: 255-258.

Jackson, D.D., Matsu'ura, M., (1985). A bayesian approach to nonlinear inversion, J. Geophys. Res. 90, 581-591.

Ji, S., Mainprice, D., 1988. Natural deformation fabrics of plagioclase: Implication for slip systems and seismic anisotropy. Tectonophysics, 147: 145-163.

Ji, S., Salisbury, M.S., 1992, Sear-wave velocities, anisotropy and splitting in high grade mylonites. Tectonophysics, 221: 453-473.

Johnson, G.L., Sirivastava, S.P., Campsie, J., Rasmussen, M., 1982. Volcanic rocks in the Labrador Sea and environs and their relationship to the evolution of the Labrador Sea. In: Current research, Part B., Pap. Geol. Sur. Canada. 82-IB 7-20 Geological Survey of Canada.

Keen, C.E., Keen, M.J., Barret, D.L., Heffler, D.E., 1975. Some aspects of the ocean-continent transition at the continental margin of eastern North America In: van den Linden, W.J.M., Wade, J.A , Offshore Geology of Eastern Canada. Geol. Surv. Can. Pap. 74-30: 189-197.

Keen, C.E., Potter, P., Srivastava, S.P., 1994, Deep seismic reflection data across the conjugate margins of the Labrador Sea, Can. J. Earth Sci., (in press).

LASE Study Greoup, 1986. Deep structure of the U.S. East Coast passive margin from large aperture seismic experiments (LASE). Mar. Petrol. Geol., 3: 234-242.

Mainprice, D. Nicolas, A., 1989. Development of shape and lattice preffered orientations: application to the seismic anisotropy of the lower crust. J. Struct. Geol., 11: 175-189.

Manghnani, M.H., Ramananantoandro, R., Clark, Jr., S.P., 1974. Compressional and shear wave velocities in granulite facies rocks and eclogites to 10 kbar. J. Geophys. Res., 79: 5427-5446.

McGregor, V.R., 1973. The early Precambrian gneisses of the Godthaab distric. West Greenland. Philos. Trans. R. Soc. London A, 273: 343-358.

McGregor, V.R., 1979. Archean gray gneisses and the origin of the continental crust: Evidence from the Godhaab region, West Greenland. In: Barker, F., (Editor). Trondhjemites. Dacites and Related Rocks. Elsevier, Amsterdam, pp. 169-205.

McGregor,V.R., Nutman, A.P., Friend, C.R.L., 1986. The Archean geology of the Godthaabsfjord region, southern West Greenland. Lun. Planet. Inst. Tech. Rep., 86-04, pp. 113-169.

McKenzie, D.P., Bickle, M.J., 1988. The volume and composition of melt generated by extension of the lithosphere. J. Petrology, 29: 625-679.

McMechan, G.A, Ottolini, R., 1980. Direct observation of p-t curve in a slant-stacked wave-field, Bull. Seism. Soc. Am., 70, 775-790.

Menke, W., 1984. Geophysical data analysis: Discrete inverse theory, Academic Press, Inc.

Nicolas, A., Christensen, N.I., 1987. Formation of anisotropy in upper mantle peridotites. A review. In: Fuchs, K., Froidevaux, C., (Editors), Composition, Structure and Dynamics of the Lithosphere-Asthenosphere System. Am. Geophys. U. Geophys. Ser. 16, A.G.U., Washington, DC, pp 137-154.

146

Nur, A.M., Simmons, G., 1968. Stress-induced velocity anisotropy in rock: an experimental study. J. Geophys. Res., 74: 6667-6674.

Nutman, A.P., Friend, C.R.L., Baadsgaard, H., McGregor, V.R., 1989. Evolution and assembly of Archean gneiss terranes in the Godthaabfjord region southern west Greenland: structural, metamorphic and isotopic evidence. Tectonics, 8: 573-589.

Phinney, R.A., Chowdhury, K.R, Frazer, L.N., (1981).Transformation and analysis of record sections, J. Geophys. Res., 86, 359-377.

Roest, W.R., Srivastava, S.P., 1989. Sea-floor spreading in the Labrador Sea: A new reconstruction, Geology 17: 1000-1003.

Savage, M.K., Silver, P.G., Meyers, R.P., 1990. Observations of teleseismic shear-wave splitting in the Basin and Range from portable and permanent stations. Geophys. Res. Lett., 17: 21-24.

Sheridan, R.E., Musser, D.L., Glover, L.III., Talwani, M., Ewing, J.I., Holbrook, W.A., Purdy, G.M., Hawman, R.B., Smithson, S.B., 1993. Deep seismic reflection data of EDGE U.S. Mid-Atlantic continental margin experiment; implications for Appalachian sutures and Mezozoic rifting and magmatic underplating. Geology, 21: 563-567.

Siegesmund, S., Kruhl, J.H., 1991. The effect of plagioclase textures on velocity anisotropy and shear wave splitting at deeper crustal levels. Tectonophysics, 191: 147-154.

Siegesmund, S., Vollbrecht, A., 1991. Complete seismic properties obtained from microcrak fabrics and textures in an amphibolite from the Ivrea zone, Western Alps. Tectonophys. 199: 13-24.

Silver, P.G., Chan, W.W., 1988. Implications for continental structure and evolution from seismic anisotropy. Nature, 355: 34-39.

Speece, M.A., 1992. Geophysical Studies of Precambrian Regions: The Laramie Mountains, Wyoming and Godthaabsfjord Area, South West Greenland. Ph.D. thesis, Univ. of Wyoming, Laramie, WY. 110 pp.

Srivastava, P.S., 1983. Davis Strait: Structures, origin and evolution, In: Bott, M.H.P., Saxov, S., Talwani, M., Thiede, J., (Editors), Structure and Development of the Greenland-Scotland Ridge. Plenum, New York, NY, pp., 159-189.

Srivastava, P.S., Tapscott, C.R., 1986. Plate kinematics of the North Atlantic. In: Vogt, P.R., Tucholke, B.E., (Editors), The Western North Atlantic Region. The Geology of North America, M. Geol. Soc. Am., Boulder, CO, pp. 379-404.

Stergiopoulos, A.B., 1984. Geophysical crustal studies of the southwest Greenland margin, M.S. thesis, pp, 158.

Stoffa, P.L., Buhl, P., Diebold, J.B., Wenzel, F., (1981). Direct mapping of seismic data to the domain of intercept time and ray parameter: a plane wave decomposition. Geoph. 46, 255-267.

Suetnova, I.E, Carbonell, R., Smithson, S.B., 1994. Bright seismic reflections and fluid movement by porous flow in the lower crust, Earth Planet. Sci. Lett., 126: 161-169.

Talwani, M., Nutter, J.C., Houtz, R., Konig, M., 1979. The crustal structure underlying the magnetic quite zone on the margin of South Australia. In: Watkins, L., Montandert, L., Dickerson, P.W., Geological and Geophysical Investigations of Continental Margins, Am. Assoc. Petrol. Geolo. Mem., 29: 151-176.

Tullis, J., Yund, R.A:, 1987. Transition from cataclastic flow to dislocation creep of feldspar: Mechanics and microstructures. Geology, 15: 606-609.

Walden, J., Nur, A., 1984. Porosity reductionand crustal pore pressure development. J. Geophys. Res., 89: 11539-11548.

White, R., McKenzie, D., 1989. Magmatism at rift zones: The generation of volcanic continental margins and flood basalts. J. Geophys. Res., 94: 7685-7729.

Woodside, J.M., Verhoef, 1989. Geological and tectonic framework of eastern Canada as interpreted from potential field imagery. Pap. Geol. Sur. Canada. 88-26.

Zoback, M.L., 1992. First- and second-order patterns of stress in the lithosphere: The world stress map project. J. Geophys. Res., 97: 11703-11728.

RIFTED CONTINENTAL MARGIN OFF MID-NORWAY

JAKOB SKOGSEID & OLAV ELDHOLM
Department of Geology, University of Oslo
Pb. 1047 Blindern, N-0316 Oslo
Norway

ABSTRACT. New deep seismic profiles and velocity measurements on the mid-Norway margin image the crustal structure and provide data for estimating the tectono-magmatic dimensions of the Late Cretaceous-Paleocene rift episode which terminated with complete lithospheric separation and massive igneous activity during breakup and initial sea floor spreading. The breakup was preceded by a 15-20 m.y syn-rift phase developing a more than 2600 km long and about 300 km wide rift zone. Uplift of the central rift led to massive subaerial volcanism during breakup. Modeling of the margin history of vertical motion must consider large melt volumes emplaced during breakup, presently constituting lower crustal high-velocity bodies.

Introduction

The rifted Møre and Vøring continental margins off Norway (Fig. 1) have, for the past 20-25 years, been studied intensively by geophysical surveying and drilling. The understanding of the formation and development of the margin is closely linked to improvements in seismic acquisition and processing techniques. The initial investigations, which were based on single-channel, analog seismic profiles and analog refraction profiles, mostly recorded by sonobuoys, provided a first-order geological margin framework and suggestions about location of the continent-ocean boundary (Talwani & Eldholm, 1972). Later, multi-channel profiling (MCS), commercial drilling on the continental shelf and scientific drilling on the outer margin provided a regional, relatively detailed image of the upper crust (e.g., Myhre et al., 1992), and led to models for the tectono-magmatic events during continental breakup and early margin development, i.e. the volcanic margin formation (Eldholm et al., 1989). Recently, deep MCS profiling, wide-angle expanded spread (ESP) and ocean bottom seismometer experiments (Mjelde et al., 1993) by research institutions and industry have contributed new information about the structure and velocity distribution of the middle and lower crust which place important boundary conditions on quantitative models of margin evolution (Skogseid, 1994; Skogseid & Planke, in press).

Igneous activity during breakup

Off Norway, the Early Tertiary breakup and initiation of sea floor spreading were accompanied by transient volcanism. In fact, evidence of voluminous igneous activity during continental breakup is found along the entire North Atlantic rifted plate boundary (Eldholm et al., 1989; White & McKenzie, 1989; Leg 152 Shipboard Party, 1994). The extrusive complexes along the continent-ocean transition constitute conjugate volcanic margin segments extending for a distance of more

147

E. Banda et al. (eds.), Rifted Ocean-Continent Boundaries, 147–153.
© 1995 *Kluwer Academic Publishers. Printed in the Netherlands.*

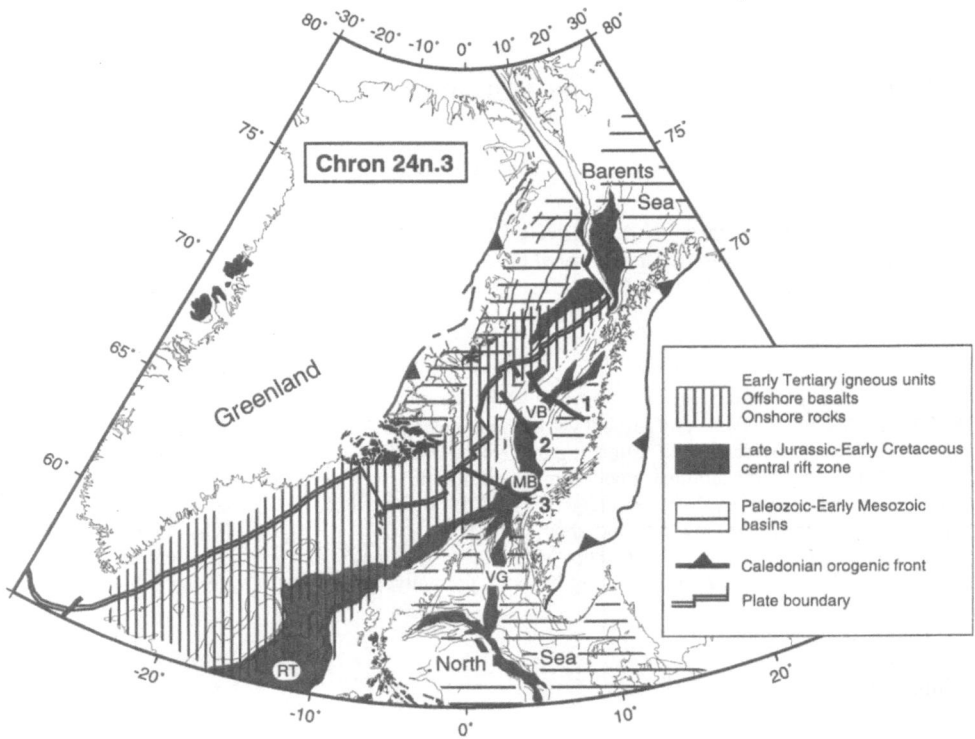

Fig. 1. Plate reconstruction to 53 Ma, about 2 m.y. after breakup between Eurasia and Greenland. Based on Skogseid & Planke (in press). MB, Møre Basin; VB, Vøring Basin; RT, Rockall Trough; VG, Viking Graben.

- Continental flood basalts
  - associated intrusives
- Extrusive complexes along continent-ocean transition
  - seaward dipping wedges and sub-horizontal units
  - associated intrusives
- Sills and low-angle dikes
- Volcanic vents
- Regional tephra horizons
- High-velocity lower crustal bodies
- Thick initial oceanic crust

Table 1. Geological features ascribed to North Atlantic igneous breakup activity.

than 2600 km. The volcanic margins reflect voluminous, transient extrusive activity over about a 3 m.y. period near the Paleocene-Eocene transition. The main phase of volcanism, leaving the extensive onshore and offshore flood basalts, took place during chron 24r.

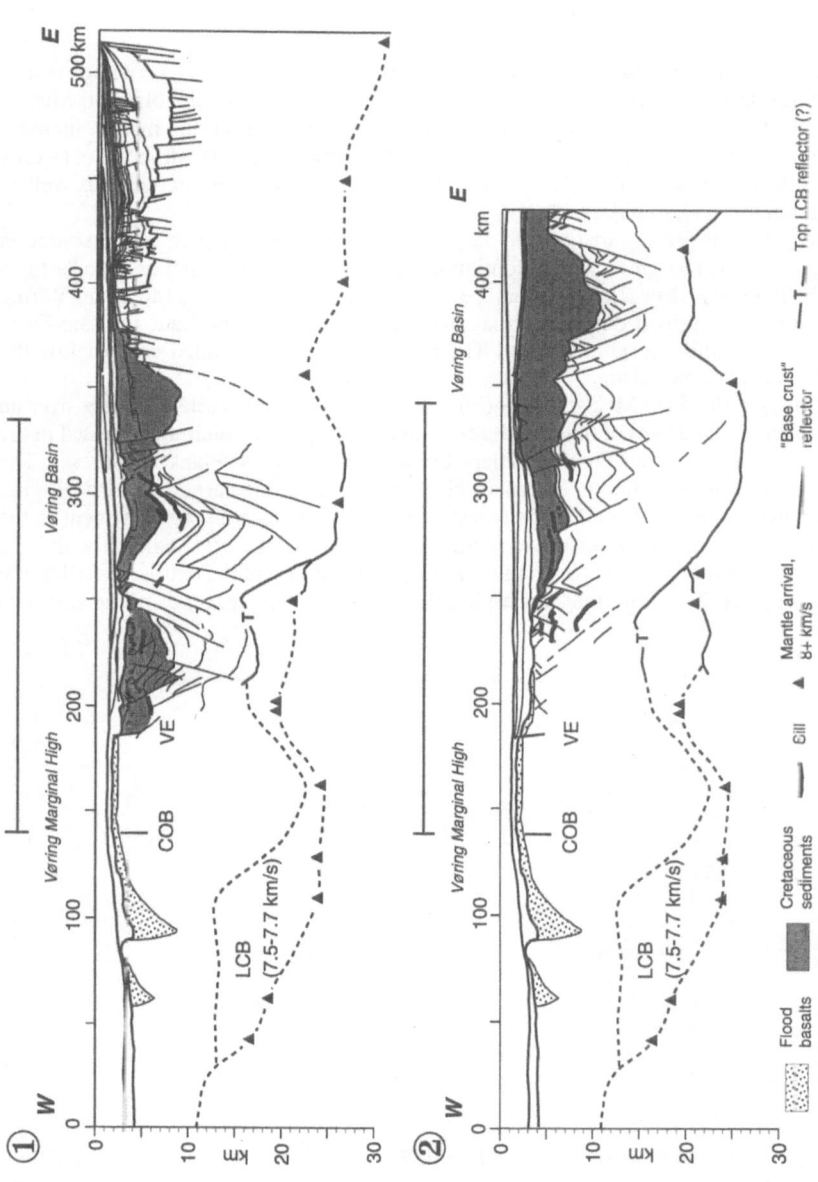

Fig. 2. Vøring margin crustal transects (Fig. 1) from ESP and deep MCS profiles (Eldholm & Grue, 1994; Skogseid & Planke, in press). Note that only one ESP transect exists west of the Vøring Escarpment. Therefore, both the northern (1) and southern (2) Vøring Basin transects have been linked to the Vøring marginal high ESP transect in the vicinity of the Vøring Escarpment (VE). Horizontal bar indicates region of Late Cretaceous-Paleozoic syn-rift tectono-magmatic activity. COB, continent-ocean boundary; LCB, lower crustal high-velocity body.

150

The massive magmatic activity during breakup is documented by several geological features (Skogseid & Eldholm, 1989; Skogseid et al., 1992b) (Table 1). Figure 1 shows the distribution of flood basalts emplaced during breakup, and the tectono-magmatic style of the mid-Norwegian margin is illustrated by selected crustal transects in Figures 2-3.

Crustal structure

The mid-Norwegian margin includes the deep Møre and Vøring basins (Fig. 1), centered west of the shelf edge, with maximum sediment thicknesses of at least 10-12 km (Eldholm and Mutter, 1986, Planke et al., 1991). The commercial exploration on the Møre and Vøring margin includes MCS-profiles recorded to 15 s two-way time. By performing relatively simple post-stack reprocessing of selected lines we are able to image deep sedimentary basin interfaces as well as intra-basement reflectors and, in places, Moho.

On the inner part of the Vøring margin the deep MCS-profiles image the entire crust and, in many profile segments, we recognize a deep, continuous reflector which corresponds to an 8+ km/s refractor in the ESP profiles (Fig. 2). This interface is interpreted as Moho. The Møre and Vøring sedimentary basins were formed during thermal subsidence following the Late Jurassic-Early Cretaceous episode of lithospheric extension. The crustal data show thinned crust below the deepest parts of the Cretaceous basins.

On the outer margin the deep MCS-profiles (Fig. 2) record a gradual crustal thinning over an 150-200 km wide region landward of the continent-ocean boundary. This thinning is related to the lithospheric extension prior to the Eearly Tertiary breakup (Skogseid & Planke, in press). The MCS-profiles also image voluminous extrusive complexes, including large seaward dipping wedges. Velocity measurements from ESPs suggest relatively thick oceanic crust adjacent to the extended continental crust, and high seismic velocities in the lowermost crust. Regionally, there is a first-order spatial correspondence between the lower crustal bodies and the extrusives (Eldholm & Grue, 1994). Figures 2-3 show that the high-velocity lower crust extends some distance

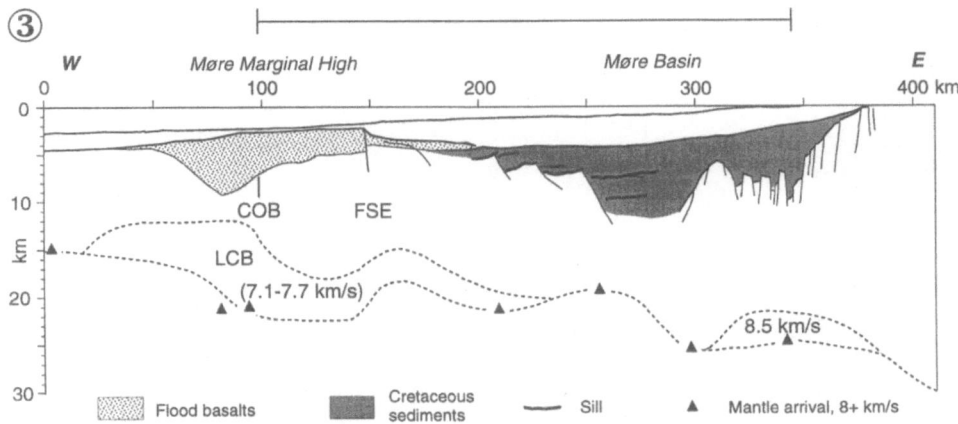

Fig. 3. Møre margin crustal transect (Fig. 1) based on ESP and MCS profiles (Olafsson et al., 1992; Skogseid, 1994). Vertical bar indicates region of Late Cretaceous-Paleozoic syn-rift tectono-magmatic activity. COB, continent-ocean boundary; LCB, lower crustal high-velocity body; FSE, Faeroe-Shetland Escarpment. The Cretaceous sediment thickness on the western flank of the Møre Basin is uncertain.

landward beneath the thinned continental crust in the Vøring and Møre basins. In this region, Moho is only recorded by the ESPs. However, we recognize a continuous intra-crustal reflector appearently corresponding to the top of the easternmost part of the high-velocity lower crustal body.

From seismic continuity, magnetic lineations and character, crustal structure and Ocean Drilling Program Site 642 which drilled through the basalts constituting the inner part of a seaward dipping wedge, Skogseid & Eldholm (1989) proposed a continent-ocean boundary located beneath the inner seaward dipping wedge (Fig. 2). The boundary separates an eastern area where the basalt flows rest on a characteristic basal horizon from a western area where there is no base to the flow sequences in the seismic record. This location is some tens of kilometers west of the Vøring Escarpment location previously proposed by Talwani & Eldholm (1972). This definition assumes that the term continent-ocean boundary relates to changing crustal properties at the top of the crystalline crust. If the entire crust is considered, the term "continent-ocean transition" will encompass the region between the continent-ocean boundary of Skogseid & Eldholm (1989) and the landward termination of the flow units east of the Vøring Escarpment (Fig. 2), a crustal region considered to be strongly intruded (Eldholm et al., 1989).

Tectono-magmatic dimensions

If we assume that the lower crustal body beneath the outer margin represents melts which were emplaced during the transient igneous breakup event (Table 1), the seismic data off mid-Norway allow us to make first-order estimates of volumes and emplacement rates for the extrusives as well as for the total crustal production. If this procedure is applied to the conjugate margins between the Charlie-Gibbs and Greenland-Senja fracture zones Eldholm & Grue (1994) have shown that the North Atlantic igneous breakup event produced at least $6.6 \times 10^6 \text{ km}^3$ of new crust including about $1.8 \times 10^6 \text{ km}^3$ of extrusive rock. Hence, the North Atlantic volcanic margins constitute a major, transient large igneous province.

Off Norway, the seismic data and the drilling results provide tectono-stratigraphic relations which we have interpreted in terms of a phase of lithospheric extension leading to breakup. The syn-rift phase lasted 15-20 m.y. On the Vøring margin in particular, we also observe spatial correlation of syn-rift normal faults, sills and low-angle dikes within Cretaceous sediments, volcanic vents at the top-Paleocene level and post-opening margin subsidence. All these features are limited to a region 200-250 km landward of the continent-ocean boundary. The syn-rift lithospheric extension is also corroborated by subsidence modeling (Skogseid et al., 1992a), and crustal thickness variations (Skogseid, 1994).

Both the data off Norway and its conjugate margin segments, and data from other North Atlantic conjugate margins, allow us to estimate the relative amounts of crustal extension for the Late Cretaceous-Paleocene syn-rift phase. Crustal structure and subsidence modeling including pre-rift extension suggest that this intra-continental rift phase, initiated near the Campanian-Maastrichtian boundary, resulted in about 140 km of lithospheric extension yielding average stretching factors in the range of 1.5-1.8. Furthermore, a rift zone more than 2600 km long and 300 km wide was affected by structural deformation and widespread igneous activity. These results allow plate reconstructions between Eurasia and Greenland back to Late Cretaceous times, and the continuity of reconstructed geological features provide an independent parameter to evaluate the magnitude of the relative movements (Skogseid & Planke, in press).

These structural dimensions indicate that the North Atlantic volcanic margins may be compared to other non-volcanic margins and that the main difference between the two margin "end-members" does not relate to the mode of lithospheric deformation. In fact, the extensive extrusive cover commonly hides the primary rift structures. Furthermore, the reduced rate of volcanic margin subsidence is caused by the melts being emplaced at the base of the crust. This concept implies that

152

the melt volume emplaced during breakup is primarily governed by: 1) the melt potential of the asthenosphere; and 2) the process of lithospheric deformation, i.e. the magnitude and duration of the syn-rift phase (Pedersen & Ro, 1992).

Margin evolution

The region between Eurasia and Greenland has experienced several episodes of extension between the end of the Caledonian Orogeny and the early Cenozoic onset of sea floor spreading. The structural data and subsidence modeling also yield estimates of the amount of post-Late Jurassic lithospheric deformation. We interpret that the intra-continental rifting in Late Jurassic-Earliest Cretaceous times caused ~100 km of lithospheric extension and subsequent Cretaceous basin subsidence from the Rockall Trough/North Sea areas in the south to the southwestern Barents Sea in the north (Fig. 1). There is also evidence of renewed rifting in the Rockall Trough and Labrador Sea in Early to Middle Cretaceous times probably associated with northward propagation of North Atlantic sea floor spreading. When seafloor spreading started in the Labrador Sea, in early Late Cretaceous times, the Rockall rift may have become extinct (Skogseid, 1994).

The seismic data and the drilling results provide a framework for developing a history of vertical motion and basin development on the Vøring margin since mid-Cretaceous times. This history is characterized by: 1) Paleocene uplift along the incipient plate boundary, 2) excess igneous activity and subaerial volcanism during and 2-3 m.y. after breakup, and 3) post-early Eocene margin subsidence. Superimposed on the post-opening margin subsidence, induced by lithospheric cooling and sediment loading, there is evidence of local doming and late Neogene regional uplift of the inner shelf and the adjacent landmass (Stuevold et al., 1992).

Because the melts constituting the lower crustal body extend landward of the continent-ocean boundary (Figs. 2-3), the post-opening outer margin subsidence will be reduced. Therefore, this body must be considered when calculating the history of vertical motion across the margin. Moreover, the spatial correlation of magmatism and extension may imply an increased asthenospheric melt potential over the entire rift zone. Skogseid (1994) estimates that a thermal anomaly of 50-80°C may produce the melts imaged by the seismic data.

The increased North Atlantic melt potential is related to the interaction of the Iceland plume and the lithosphere. Skogseid et al. (1992b) have suggested a history of tectono-magmatic events which is compatible with the observational data in the the North Atlantic and Baffin Bay. A main premise of this model is that the Eurasia-Greenland pre-Maastrichtian lithosphere, which was thinned by previous extension phases, had again become subjected to active extension when the plumehead impinged on the base of the lithosphere. Melts generated by pressure release moved preferentially towards the areas of maximum lithospheric relief, causing rapid weakening of overlying upper mantle and crust, which in turn resulted in a rapid release of large melt volumes.

Acknowledgements. The efforts of Tadeusz P. Gladczenko in data preparation are gratefully acknowledged. We thank the Norwegian Petroleum Directorate for providing key deep seismic profiles. This work was supported by the Research Council of Norway (NFR), as part of the IBS (Integrated Basin Studies) project under the JOULE II research programme funded by the Commission of European Communities (contract No. JOU2-CT 92-0110), and by NFR and Statoil as part of the German-Norwegian Geoscientific Cooperation. IBS contribution No. 20.

REFERENCES

Eldholm O. & Grue K., 1994. North Atlantic volcanic margins: Dimensions and Production Rates. J. Geophys. Res., 99, 2955-2968.

Eldholm O. & Mutter J.C., 1986. Basin structure of the Norwegian Margin from analysis of digitally recorded sonobuoys. J. Geophys. Res., 91, 3763-3783.

Eldholm O., Thiede J. & Taylor E., 1989. Evolution of the Vøring Volcanic Margin. In Eldholm O., Thiede, J. et al., Proc. ODP, Sci. Results, 104: College Station (Ocean Drilling Program), 1033-1065.

Leg 152 Shipboard Party, 1994. Drilling unearths "fire and ice" at southern Greenland margin. Eos, 75, 401-406.

Mjelde R., Sellevoll M.A., Shimamura H., Iwasaki T. & Kanazawa T., 1993. Ocean bottom seismographs used in a crustal study of an area covered wirh flood-basalt off Lofoten, N.Norway. Terra Nova, 5, 76-84.

Myhre A.M., Eldholm O., Faleide J.I., Skogseid J., Gudlaugsson S.T., Planke S., Stuevold L.M. & Vågnes E., 1992. Norway-Svalbard Margin: Structural and Stratigraphical styles. In Poag C.W. & de Graciansky P.C. (eds.), "Geologic evolution of Atlantic Continertal Rises", Van Nostrand Reinhold, New York, pp157-185.

Olafsson I., Sundvor E., Eldholm O. & Grue K., 1992. Møre Margin: Crustal structure from analysis of Expanded Spread Profiles. Mar. Geophys. Res., 14, 137-162.

Pedersen T. & Ro H.E., 1992. Finite duration extension and decompression melting. Earth Planet. Sci, Lett., 113, 15-22.

Planke S., Skogseid J. & Eldholm O., 1991. Crustal structure off Norway 62° to 70° North. Tectonophysics, 189, 91-107.

Skogseid J., 1994. Dimensions of the Late Cretaceous-Paleocene Northeast Atlantic rift derived from Cenozoic subsidence, Tectonophysics, 240, 225-247.

Skogseid J. & Eldholm O., 1989. Vøring Plateau continental margin: Seismic interpretation, stratigraphy and vertical movements. In: Eldholm O., Thiede J., Taylor E. et al., Proc. ODP, Sci. Results, 104: College Station (Ocean Drilling Program), 993-1030.

Skogseid J. & Planke S., in press. Seismic reflection and refraction imaging of crustal structure on NE Atlantic margins: Mesozoic continental rifting and Cenozoic volcanic margin formation. Basin Res.

Skogseid J., Pedersen T. & Larsen V.B., 1992a. Vøring Basin: subsidence and tectonic evolution. Norw. Petrol. Soc. Spec. Publ., 1, 55-82.

Skogseid J., Pedersen T., Eldholm O. & Larsen B.T., 1992b. Tectonism and magmatism during NE Atlantic continental break-up: the Vøring Margin. J. Geol. Soc. London, Spec. Publ., 68, 305-320.

Stuevold L., Skogseid J. & Eldholm O., 1992. Post-Cretaceous uplift events on the Vøring continental margin. Geology, 20, 919-922.

Talwani M. & Eldholm O., 1972. The continental margin off Norway: A geophysical study, Geol. Soc. Am. Bull., 83, 3573-3606, 1972.

White R.S. & McKenzie D., 1989. Magmatism at rift zones: The generation of volcanic continental margins and flood basalts, J. Geophys. Res., 94, 7685-7729, 1989.

THE EDGE EXPERIMENT AND THE U.S. EAST COAST MAGNETIC ANOMALY

Manik Talwani (Rice University and Houston Advanced Research Center)
John Ewing (Woods Hole Oceanographic Institution and Houston Advanced Research Center)
Robert E. Sheridan (Rutgers University)
W. Steven Holbrook (Woods Hole Oceanographic Institution)
Lynn Glover, III (Virginia Polytechnic Institute)

Abstract

The EDGE experiment offshore the U.S. east coast obtained near vertical incidence seismic reflection data, as well as wide angle seismic reflection and seismic refraction data. These results are combined with magnetic total intensity data in order to establish the origin for the East Coast Magnetic Anomaly (ECMA) and to hypothesize about geological events taking place at the time that opening was initiated in the North Atlantic.

Prominent seaward dipping reflectors (SDR's) which reach a depth of almost 25 km represent layers of volcanics which give rise to the ECMA. Because these reflectors extend below the depth where the Curie temperature is reached, and because the volcanic layers lie directly over material believed to be derived from asthenospheric melt, it is believed that no continental crust lies below the seaward dipping reflectors (except at their feather edge). The seaward dipping reflectors alone are responsible for the ECMA. The crustal section here represents the Initial Oceanic Crust which is thicker and possesses higher seismic velocities than normal oceanic crust. It should not be called "transitional crust," "modified continental crust," or "rift stage crust." The 7.2 - 7.5 km/sec layer is the lower part of the Initial Oceanic Crust rather than an underplating to a thinned continental crust. For the sub-horizontal volcanic layers (which give rise to the SDR's) to produce the ECMA it is necessary that they largely possess a single magnetic polarity. This can only take place if the subaerial sea floor spreading which produced these volcanic layers was initiated at a very high rate. As the spreading subsequently slows down, normal and reversely magnetized volcanic layers are presumably juxtaposed below each other, resulting in an effective cancellation of their magnetic effects and giving rise to a magnetic quiet zone. As the axis of sea floor spreading subsides to submarine depths the flow length of volcanic layers becomes small and typical sea floor spreading type magnetic anomalies can be produced. This model is also used to explain the coincidence of salt diapirs with the ECMA and also with the disappearance of typically hyperbolic echoes from oceanic basement as the margin is approached from the seaward side.

E. Banda et al. (eds.), Rifted Ocean-Continent Boundaries, 155–181.
© 1995 *Kluwer Academic Publishers. Printed in the Netherlands.*

Introduction

The EDGE experiment off the U.S. east coast involving near vertical incidence seismic reflection and seismic wide angle reflection and refraction experiments was carried out in 1990 (Sheridan et al., 1993; Holbrook et al., 1994; Glover et al.., in press) In this paper we use the seismic data, together with existing magnetic data, to determine the origin of the prominent East Coast Magnetic Anomaly (ECMA) and draw inferences about processes going on during the initiation of opening of the North Atlantic.

East Coast Magnetic Anomaly

The East Coast Magnetic Anomaly (ECMA) was discovered by Keller et al. (1954). Subsequent marine magnetic and aeromagnetic surveys have delineated it over large parts of the U.S. East Coast margin. A large number of studies have described the ECMA in great detail and attempted to explain it in terms of geological structures (Drake et al., 1963; Taylor et al., 1968; Keen 1969; Emery et al., 1970; Behrendt and Klitgord, 1980; Hutchinson et al., 1983; Alsop and Talwani, 1984; Nelson et al., 1985; McBride and Nelson, 1988; Austin et al., 1990). The ECMA, although varying in amplitude, shape, and the number of peaks and troughs within it, is clearly definable as a single geophysical entity along a large part of the U.S. and Canadian Atlantic margin (Figure 1, modified from Rabinowitz et al., 1983). In this study we focus on the area from offshore New Jersey to offshore Georgia. Aeromagnetic studies (Klitgord and Behrendt, 1977) clearly show that the character of the magnetic field changes on land. Over continental basement, which is generally at a shallow depth in the coastal area, the magnetic field is characterized by much shorter wavelength anomalies. A well known magnetic quiet zone lies eastward of the ECMA (Heirtzler and Hayes, 1967). Still farther east a well defined lineation called the Blake Spur anomaly lies at the eastern margin of the magnetic quiet zone (Figure 2).

The observed magnetic profiles along the LASE, EDGE and the USGS-32 lines are shown (Figures 4, 5, and 6). Since we did not want to be concerned with the very long wavelength regional anomaly we have added constant values to each of the observed total intensity magnetic anomaly profiles (Figures 4, 5, and 6) in such a way that the average total intensity value along each profile is zero.

Seismic reflection and refraction results

Five major wide angle reflection and refraction experiments have been carried out off the U.S. East Coast during the last 20 years. Three of these experiments, the LASE experiment (LASE Study Group, 1986); the EDGE experiment (Sheridan et al., 1993; Holbrook et al., 1994; Glover et al., in press), and the southern Carolina trough experiment (Austin et al., 1990; Oh et al., 1991) combined near vertical incidence seismic reflection with wide angle reflection and refraction experiments. LASE employed Expanding Spread Profiles (ESP's) for this purpose. The other two experiments used ocean bottom hydrophones and seismometers. In an earlier Carolina trough experiment, (Trehu et al., 1989), ocean bottom seismometer experiments were carried out separately and in a direction perpendicular to a seismic reflection profile (USGS-32, Hutchinson et al., 1983). In this paper we will discuss results obtained by the LASE, EDGE and the first Carolina trough experiment (Trehu et al., 1989). The results obtained in the second (southern) Carolina trough experiment (Austin et al., 1990, which includes line BA-6 shown in Figure 2) do not conflict with the results presented in this paper, but we will not discuss them in detail.

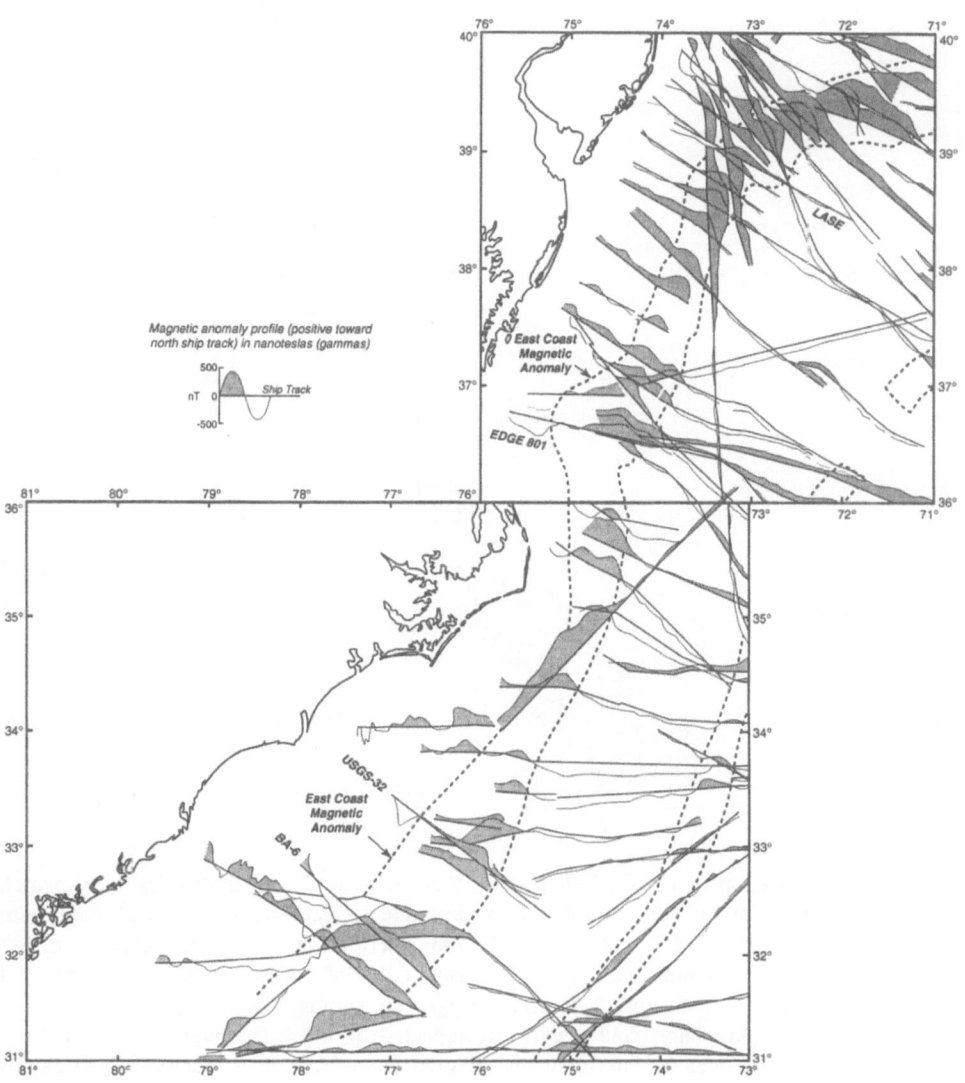

Figure 1   Total intensity magnetic values plotted along ship tracks off the U.S. east coast (modified from Rabinowitz et al., 1983).

158

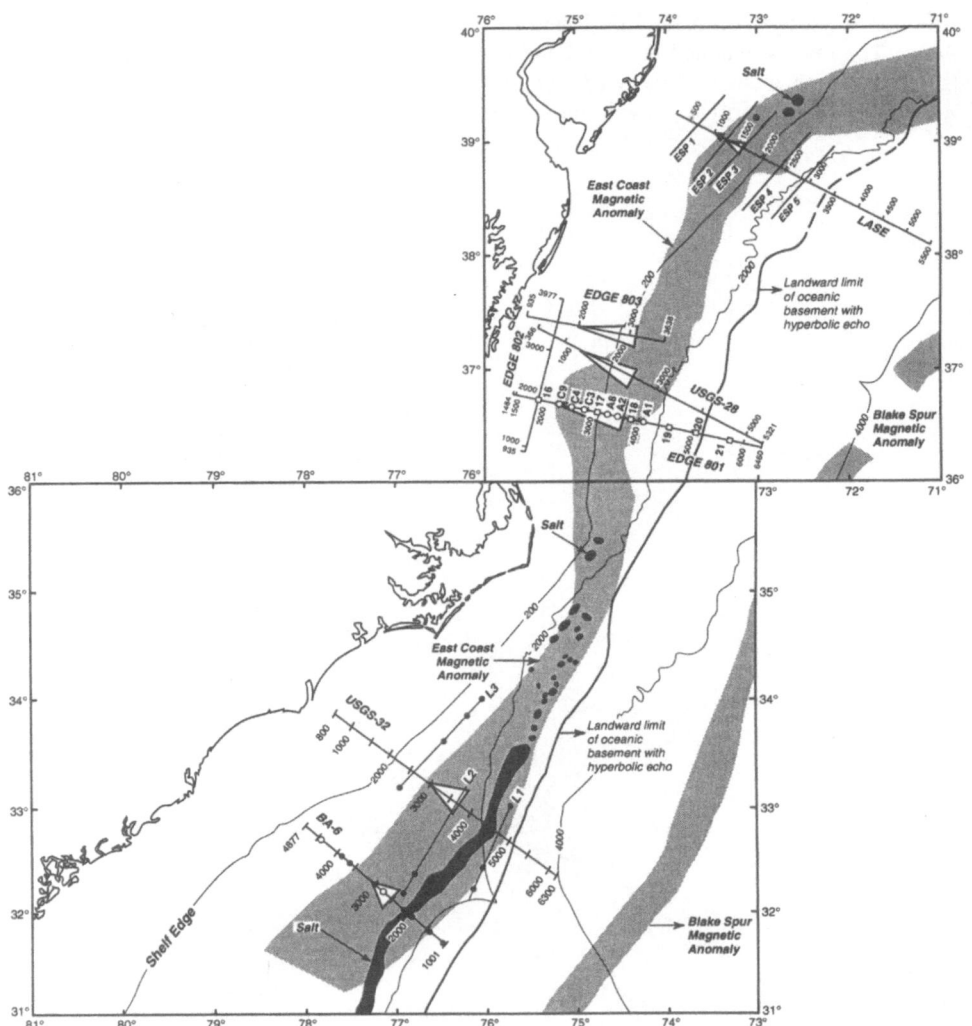

Figure 2   Location of LASE, EDGE, USGS-28, USGS-32, and BA6 lines along which
vertical incidence reflection, wide angle reflection and refraction data were
gathered [LASE Study Group, 1986; Sheridan et al., 1993; Holbrook et al.,
1994; Hutchinson et al., 1989; Trehu et al, 1989; Austin et al., 1990.] Shaded
areas indicate the East Coast Magnetic Anomaly (ECMA), and the Blake
Spur Anomaly. Salt diapirs are indicated by black shading; "Landward limit
of the oceanic basement" is adapted from Uchupi et al., 1983. Triangles
indicate the extent of seaward dipping reflectors, obtained from the reflection
profiles (Figures 4, 5, and 6). Also shown are the shelf edge, which is at
200m, and the 2,000 and 4,000 meter bathymetric contours. Shot points are
indicated along the profiles.

159

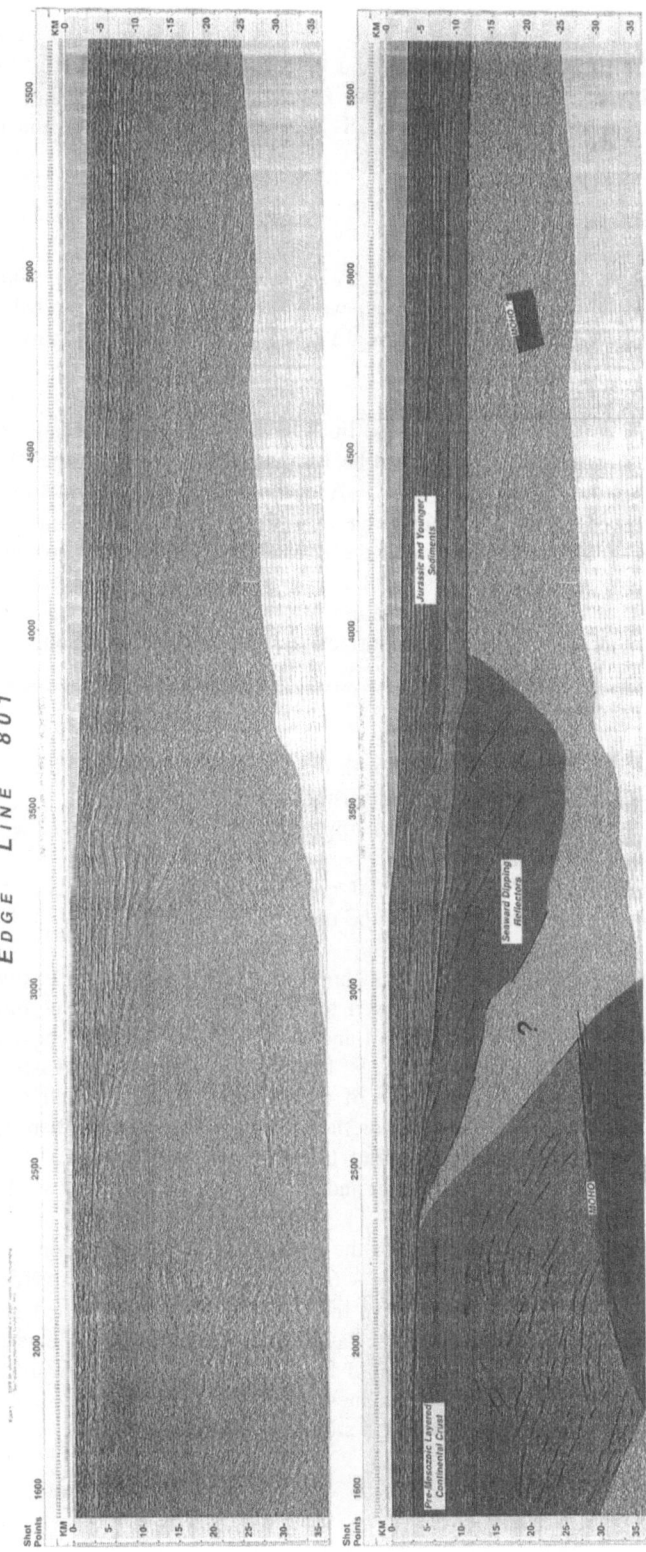

Figure 3   EDGE 801 reflection section plotted as a "depth" section. The corresponding "time" section was published by Sheridan et al., 1993.

(Large version in full-color at the back of the book)

The profiles that we will discuss were shot across the margin crossing two important sedimentary basins - the Baltimore Canyon trough and the Carolina trough. The LASE line crosses the Northern Baltimore Canyon trough, the EDGE lines 801 and 803 cross the Southern Baltimore Canyon trough, and the USGS-32 line crosses the Carolina trough (Figure 2).

SEISMIC REFLECTION RESULTS

Extensive earlier USGS surveys across the margin have delineated the shallower sedimentary part of the crustal section. In this paper we will concentrate on the results from the latest studies (LASE, EDGE and Carolina trough) to focus on the deeper parts of the crustal section.

The reflection results (Figures 3 - 8) along these lines have some common features. The sedimentary section, which is generally believed to be Jurassic and younger in age, thickens from the coast line towards the edge of the shelf. A carbonate bank is characteristically located at what is believed to be a paleo-shelf edge. A large thickness of Jurassic and younger sediments is present under the slope and the rise, and it gradually thins seawards toward the ocean basin. Typical oceanic basement is characterized by a reflection pattern consisting of hyperbolic echoes. The hyperbolic nature of the basement reflection under the deep ocean changes to a relatively flat reflection horizon towards the margin. Some authors

---

Figure 4   From top to bottom.
Traced seismic reflection record.
Crustal velocities from Holbrook et al. (1994). Solid and open circles locate the ocean bottom hydrophones and seismometers.
Assumed magnetic body chosen to approximately coincide with the SDR's and the ~6.5 km/sec layer.
The contribution of the continental crust to the total intensity magnetic anomaly has been subtracted from the observed anomaly.

A zero value for magnetization was chosen for the oceanic crust eastward of the ECMA. This value was chosen to reflect the fact that the magnetic anomaly in this region is very small and also for computational convenience. Since the magnetic anomaly arises from magnetization contrasts, small changes can be made in the value of this magnetization, as well as in the assumed value of magnetization for the continental crust, and then suitably alter the model for the magnetized body responsible for the ECMA by changing its value of magnetization and/or changing its geometry slightly.

The magnetic body (responsible for the ECMA) was chosen to coincide more or less with the SDR's and with the ~ 6.5 km/sec layer. The eastward end of this magnetized body is not seen in the SDR's. We have thinned it to the east to coincide with the 6.5 km/sec layer obtained by wide angle reflection. Alternatively, its magnetization could have been decreased eastward to shot point 4500 without reducing its thickness. If the body is limited to about shot point 3800, the observed anomaly is still nearly matched.

Southern Baltimore Canyon Trough - EDGE 801

Reflection

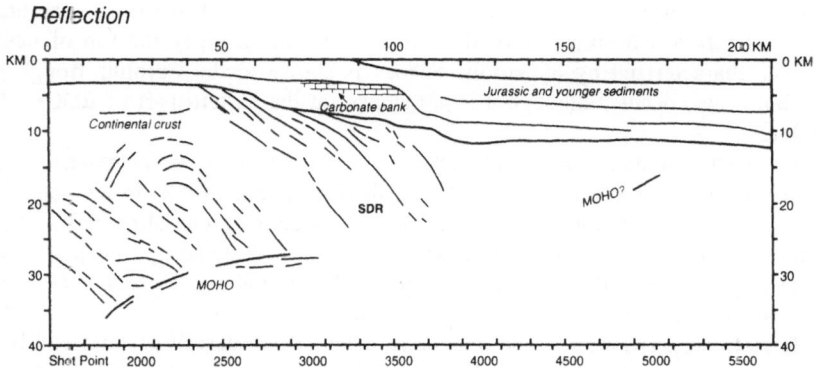

Wide Angle Reflection and Refraction

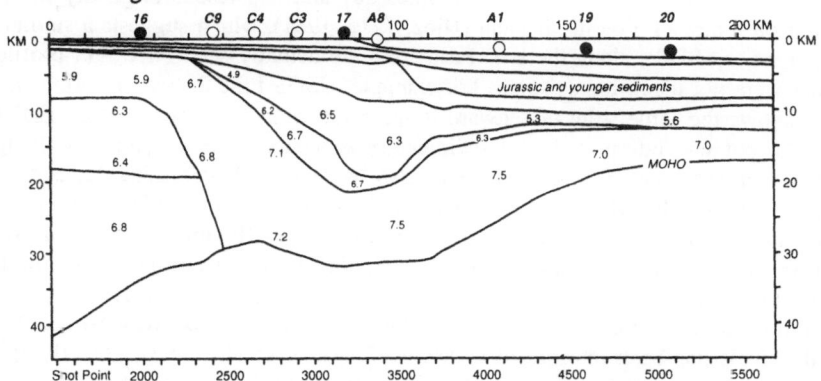

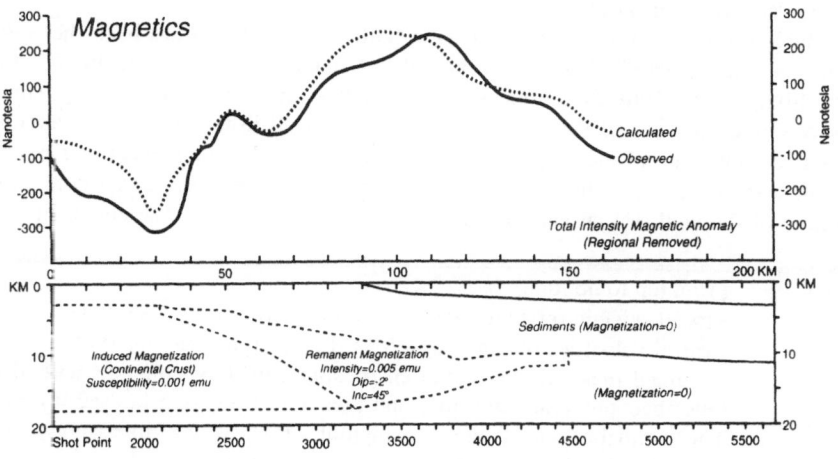

(for example Uchupi et al., 1983, Trehu et al., 1989) have designated the place where the change in echo character occurs as the landward limit of oceanic basement (Figure 2). As the basement reflection west of this limit gets smoother and begins to resemble a sedimentary reflection it becomes difficult to unambiguously identify the top of basement. The basement characterized by hyperbolic echoes is shown in the Carolina trough profile (Figure 6), but it also occurs, although it is not shown, at the eastern part of EDGE 801 and the LASE lines.

One of the most important findings of EDGE Lines 801 and 803 was an extensive suite of seaward dipping reflectors (Talwani et al., 1992; Sheridan et al., 1993). In the depth section (Figures 3, 4, and 7) the seaward dipping reflectors are seen to extend along the profile for a length of more than 70 km. They occur at depths less than 5 km and up to depths of almost 25 km. They appear to be thickest between 80 and 100 km along the EDGE 801 line and range from 10 to 15 km in thickness. This is perhaps the most extensive suite of seaward dipping reflectors seen anywhere. (As magnetic calculations described later imply, these SDR's may actually extend to 150 km along the EDGE 801 line, making the total width of this suite more than 100 km.) EDGE line 803 (not shown here) shows a similar pattern of SDR's. The suite of SDR's seen on EDGE lines 801 and 803 resemble greatly the suite of SDR's seen on the outer Voering plateau (Hinz, 1972, 1981), which suggests a similar origin for the two suites. SDR's along the U.S. East Coast are actually seen on USGS profiles shot earlier, but were not identified as such. For example, USGS Line 28 (see location in Figure 2) clearly shows the same suite of seaward dipping reflectors as is apparent on EDGE lines 801 and 803, but was interpreted as a sedimentary sequence by earlier authors (Klitgord et al., 1988). Similarly, an SDR sequence can also be seen on USGS Line 32, and we have indicated it in Figure 6. SDR's are also present on the LASE Line, although they were not interpreted as such and are present only over short distances (Figure 5). Hinz (1981), from some of the earliest deep-seismic reflection studies in the Baltimore Canyon trough, identified SDR's, but his results appear to have been generally disregarded.

A deep reflection horizon is clearly seen in the EDGE 801 lines between 10 and 70 km on the profile. This has been interpreted as Moho (Sheridan et al., 1993). Overlying it is a section with extensive layering, which Sheridan et al. (1993) interpreted as continental lower crust. A reflector which may be Moho is also seen near the seaward end of the EDGE 801 line over a short distance.

A notable discovery in the Carolina trough (Dillon et al., 1983) has been the presence of diapiric material, generally believed to be salt. Seismic reflection data in the Baltimore Canyon trough (from Gulf Oil Company archives) has also indicated, although much less extensively, the presence of salt diapirs. A discovery well in the Baltimore Canyon trough was located above one of the diapirs (Figure 2). Although the diapir itself was not reached, pore waters recovered from the well showed an appreciable increase in salt content towards the bottom of the well (Ed Driver, personal communication, 1982). The coincidence of the

Figure 5   From top to bottom.
            Traced seismic reflection record (modified from LASE Study Group, 1986).
            Crustal velocities obtained by ESP's (LASE Study Group, 1986).
            Assumed magnetic body chosen to approximately coincide with the SDR's (and their possible extension) and (generally) the ~6.5 km/sec layers.
            The contribution of the continental crust to the intensively magnetic anomalies has been subtracted from the observed anomaly.

Northern Baltimore Canyon Trough - LASE

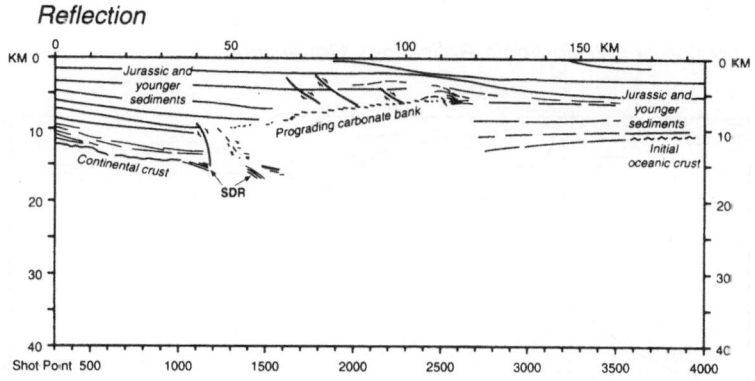

Reflection

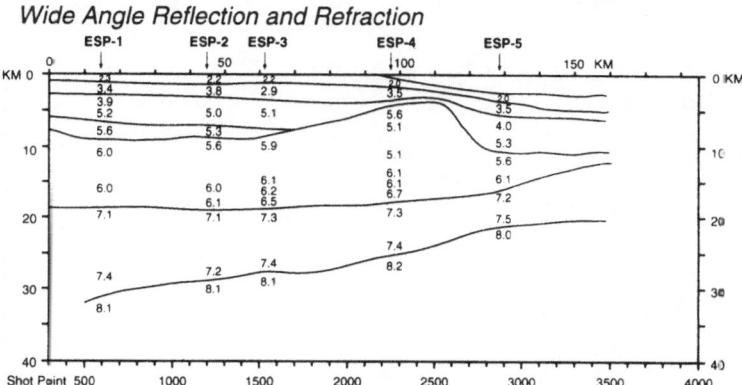

Wide Angle Reflection and Refraction

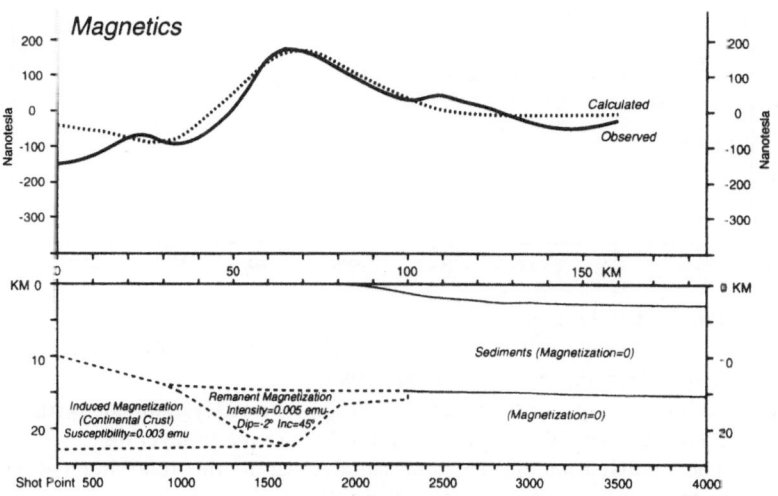

Magnetics

Carolina Trough - USGS 32

Reflection and Wide Angle Reflection - Refraction

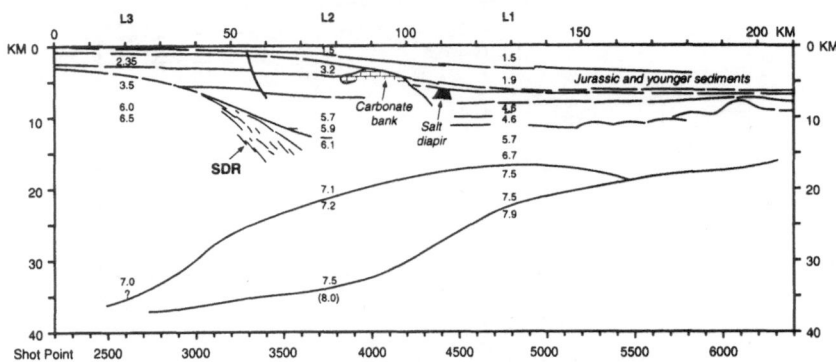

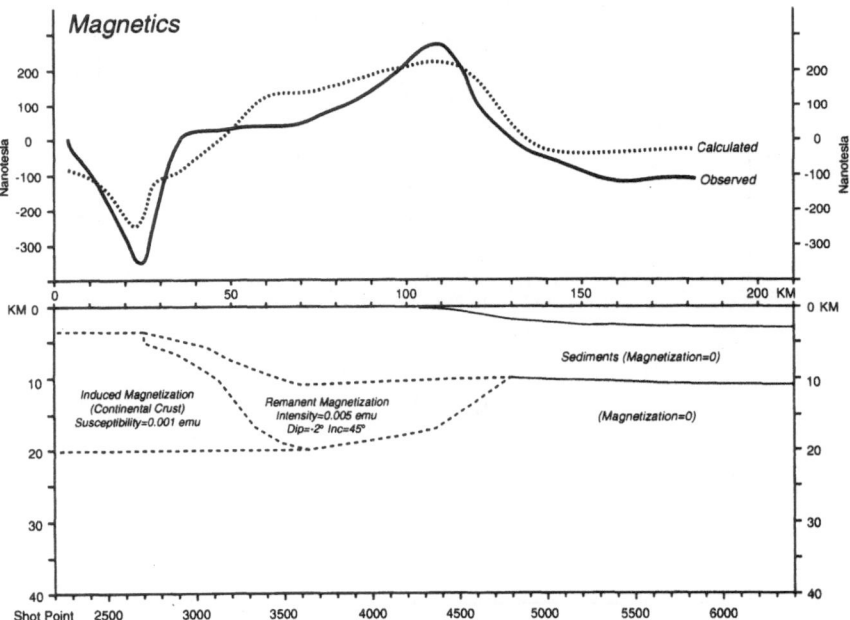

Figure 6   From top to bottom.

Traced seismic reflection section and crustal velocities obtained by ocean
bottom instruments (Trehu et al., 1989).

Assumed magnetic body chosen to approximately coincide with the SDR's
(and their possible extension) and the ~6.5 km/sec layering.

The contribution of the continental crust to the total intensity magnetic
anomalies has been subtracted from the observed anomaly.

salt diapirs to the ECMA is so striking that this coincidence begs for an explanation.

The reflection results from EDGE 801 line give more information about crustal layering than any other lines across the U.S. East Coast margin. The better definition and deeper penetration may, in part, be a consequence of the very powerful reflection equipment (airgun capacity greater than 10,000 cu.in.) used for this experiment. It is quite likely that the seaward dipping reflectors were not seen in the earlier surveys simply because the air gun sources were not powerful enough.

SEISMIC WIDE-ANGLE REFLECTION AND REFRACTION RESULTS

In the LASE, EDGE, and Carolina trough lines, continental crust is present on the landward side of the profiles, and oceanic crust (although with somewhat higher than normal velocities) is present on the seaward side (Figures 4, 5, and 6). Moho is seen to rise generally from between 35 and 45 km on the landward side to 15 to 20 km on the seaward side. The deeper than normal Moho on the seaward side is principally due to large thickness of overlying sediments, although the igneous crust also appears to be slightly thicker than normal, especially for the LASE Line. Data not shown in this paper clearly indicate that seaward of these lines the oceanic crust attains normal thickness as the ocean basin is approached.

Material with velocities between 7.2 km/sec and 7.5 km/sec at the base of oceanic crust was first reported by Ewing and Ewing (1959). It was shown to exist just above Moho in the LASE experiment. Similar material was discovered in the Carolina trough (Trehu et al., 1989) and in other parts of the world under continental margins (e.g. Norwegian and Greenland margins, Eldholm and Grue, 1994; Hatton margin, White et al., 1987). This material is considered to have been caused by igneous activity and is termed "underplating." White and McKenzie (1989) explain the igneous activity as a consequence of relatively small increases above normal in the temperature of the underlying asthenospheric mantle. The igneous rocks are assumed to be produced by partial melting of the asthenosphere due to decompression as it wells up passively to fill the space generated by the stretching and thinning lithosphere. Underplating has been considered an essential component of thinned crust underlying continental margins. Similar material is found in other instances of thinned continental crust where no subsequent margin has developed (e.g. Rhine graben, African rift valleys). Thus, material with velocity lying between about 7.2 to 7.5 km/sec is generally considered to be underplated material often lying below thin continental crust.

Relationship of ECMA to other geophysical features

The derivation of the bodies that give rise to magnetic anomalies is essentially ambiguous because many different bodies can give rise to the same magnetic anomaly. This ambiguity has resulted in suggestions of vastly different geologic solutions for the ECMA.

Two kinds of edge effect solutions have been proposed. One, proposed by (Keen, 1969), juxtaposed magnetized continental crust against non-magnetic oceanic mantle, and the other, proposed by a number of authors (e.g. Hutchinson et al., 1983 and 1990; Klitgord et al., 1988), juxtaposed weakly magnetized continental crust against strongly and reversely magnetized oceanic crust. The latest seismic results do not show a sharp crustal boundary

Southern Baltimore Canyon Trough - EDGE 801

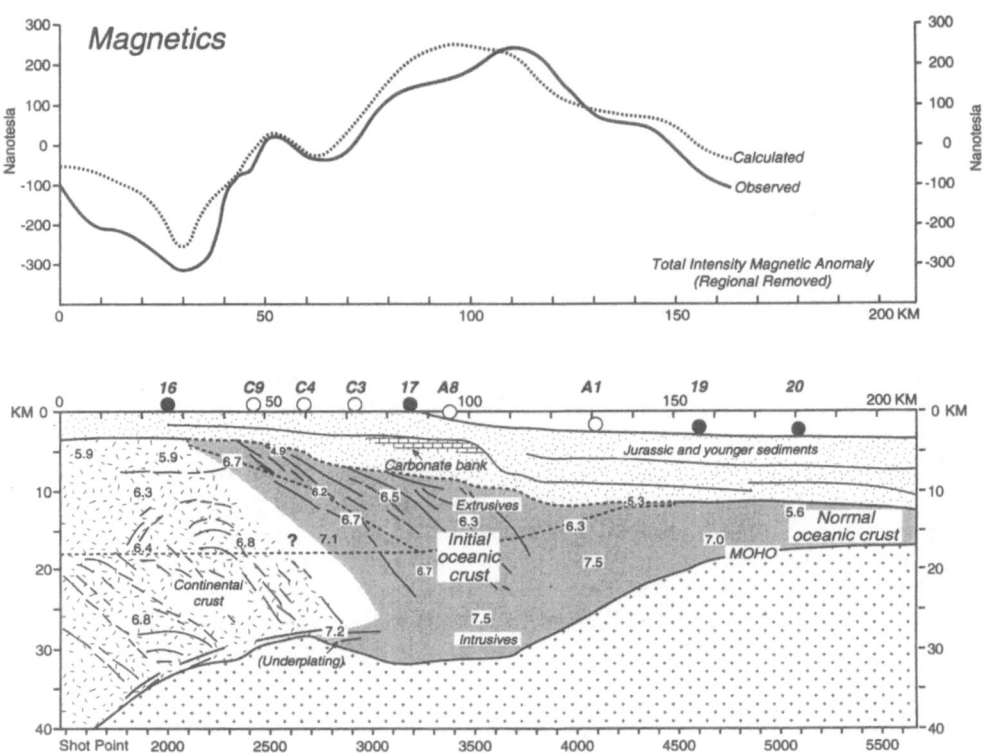

Figure 7   Composited crustal section based on seismic reflection, wide angle, refraction and magnetic data. The SDR's, except at their extreme landward edge, lie over high velocity oceanic crust. Together they constitute a thick Initial Oceanic Crust with higher than usual velocities.

There is only a small zone less than 20 kilometers wide separating the continental crust (indicated by lower crustal layering) and the initial oceanic crust (characterized by seaward dipping reflectors). This is seen as a zone of no reflectors and may represent a zone where new crustal material is intruded within blocks of continental crust. This narrow zone could be called a transitional zone.

The Initial Oceanic crust has greater thickness and higher velocities than normal oceanic crust, but it appears to grade eastwards into normal oceanic crust.

below the ECMA. In view of this seismic evidence it is not possible to sustain the hypothesis that an edge effect produces the ECMA.

A second hypothesis, first advanced by Drake et al., (1963) and later supported by a number of authors (e.g. Grow, 1980), has been that of a basement ridge giving rise to the ECMA. This hypothesis has proved untenable simply because the high velocity material below the shelf edge has turned out to be a carbonate bank (LASE study group, 1986) which is essentially non-magnetic, rather than a basement ridge.

The remaining hypothesis is the presence of a highly magnetic body below the ECMA that gives rise to it. This hypothesis has to be essentially true, but the geological nature of the body could not be determined unambiguously from earlier studies. Supporting evidence from seismology and gravity, which would have identified the nature of this body and supported a specific geologic solution, had been lacking. Seismic refraction or reflection measurements did not indicate any structural discontinuities which could have been associated with such a body. No evidence of any layering within basement, which could have been correlated with magnetization contrasts within basement, was available in earlier offshore seismic reflection profiles. No clearly identifiable gravity anomaly (which in this location would be complicated by the presence of the edge effect) has yet been correlated with the ECMA.

However, seismic reflection results from the latest experiments have removed the ambiguity. The first indication of the correlation of a seismically determined feature that appeared to have a relationship with the East Coast Magnetic Anomaly was the discovery by Austin et al. (1990) in the South Carolina trough of reflectors which they interpreted as Jurassic volcanics. They used this discovery to model the ECMA by postulating the magnetic body to be composed in part of volcanics associated with SDR's, and in part by normal oceanic crust.

EDGE 801 (and EDGE 803) results, which were used to identify the SDR's to a depth exceeding 20 km, demonstrate quite convincingly the relationship between these reflectors and the ECMA. Results obtained from measurements made on cored material recovered from drilling on the Voering Plateau (Schoenharting and Abrahamsen, 1989), show that the SDR's there (which appear to be very similar to those observed on the EDGE lines) are composed of volcanic material which is strongly magnetic, with high remanence. Although their feather edge lies on crust that contains continental material, it is generally believed that the SDR's represent material produced by subaerial sea floor spreading They extend down to a depth of about 25 km, and if the material below them is at a temperature above the Curie point, then they alone can give rise to the magnetic anomalies. It is difficult to determine the depth below the SDR's where the Curie temperature is reached, since we

---

As explained in the text, this model requires rapid rates of sea floor spreading as the ocean first opens. The magnetic anomaly s attributed to volcanic layers, normal and reversely magnetized volcanic layers being separated because of rapid rates of spreading. The subsequent slowing down gives rise to a magnetic quiet zone as long as spreading is subaerial, but typical marine magnetic anomalies are produced when the spreading center becomes submarine. Volcanic layers produced by subaerial spreading give rise to a smooth basement floor; those produced by submarine spreading give rise to typically hyperbolic echoes. Salt is produced in a basin lying between a subaerial spreading ridge and the continent.

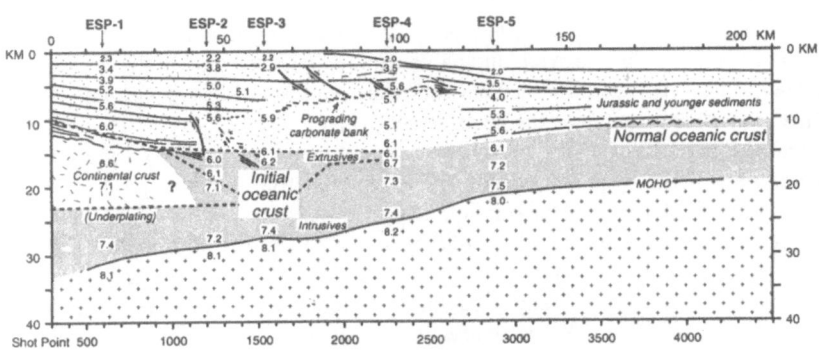

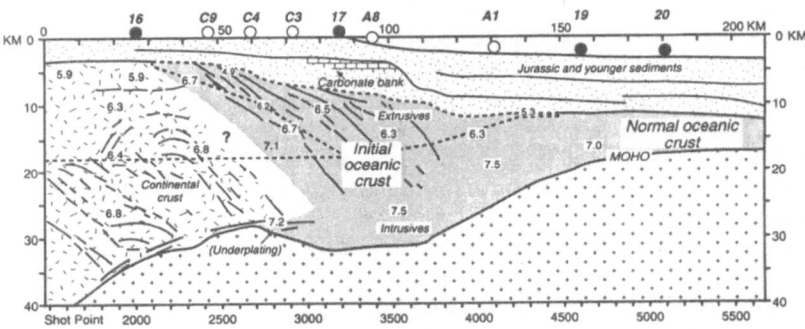

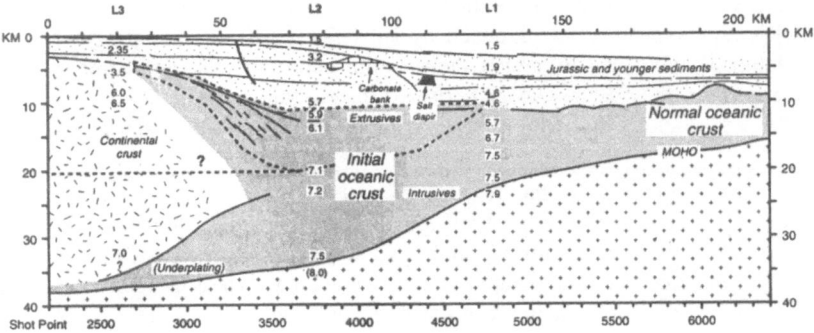

Figure 8   Results from EDGE 801 (Figure 7) are shown to be similar to results on the LASE and USGS-32 lines.

know neither the temperatures deep in the crust nor do we know precisely what magnetic minerals are involved. A global average curve given by Stacey (1977), gives a temperature of about 500°C at a depth of 25 km and O'Reilly and Bannerjee (1967), have obtained a maximum value of 425°C for the Curie point of basalt samples from the Deep Sea Drilling Project. It therefore seems reasonable to suggest that the material lying below the SDR's does not contribute to the ECMA and the volcanic flows (which appear as SDR's in the reflection records) are the sole contributors to the East Coast Magnetic Anomaly.

The SDR's do not appear to underlie the entire width of the ECMA. The volcanic layers may continue seaward of where they are last seen (115 km on the EDGE 801 line), but because of change of burial condition or other causes, do not appear as seismic reflectors. (In fact, there is a faint indication that SDR's exist near 150 km on this line.) The SDR's seen in the LASE line, as well as the USGS-32 line, are far less extensive as seen in the existing reflection records. This may be due to the smaller seismic sources that were used but we see no reason to believe that the ECMA in these locations is due to causes other than those demonstrated above for EDGE Line 801. The continuity of the ECMA along the U.S. East Coast strongly indicates a similar origin throughout its entire length.

The SDR's are associated with volcanic layers that have seismic velocities of upto about 6.5 km/sec, which is somewhat higher than that of SDR's observed elsewhere (e.g. in the Norwegian Sea, Eldholm and Grue, 1994), and they extend down to the 7.2 - 7.5 km/sec layer. Since both the volcanics and the underlying layer are attributed to material rising from the mantle, there is obviously no room for a thin continental crust lying in-between. In the absence of recovered material from drilling we cannot rule out the possibility that the 7.2 - 7.5 km/sec layer does not consist of pure asthenospheric melt, but contains some continental fragments. However, we consider this unlikely. In this location, as pointed out earlier, (Talwani et al., 1992; Holbrook et al., 1994) underplating is largely a misnomer. There is no older crust being underplated, (except at the extreme landward margin of this layer); there is only new igneous material with velocity 7.2 - 7.5 km/sec above Moho underlying SDR's with velocities around 6.5 km/sec. We term the two together as Initial Oceanic Crust. Seawards, the thickness of this crust, as well as the seismic velocity, decreases, ultimately reaching normal ocean basin characteristics seaward of the margin.

Magnetic calculations

The coincidence of the magnetic anomaly with the SDR's leads us to suggest that the volcanic layers (giving rise to the SDR's) and their seaward extensions (Figure 4, bottom) are the cause of the magnetic anomaly. If we had terminated the magnetic body at about 115 km along the profile, the calculated magnetic profile would not have been substantially different. East of this distance, where SDR's were not observed (or were very weakly imaged), we have generally allowed the magnetic body to correspond to the seismic layer with velocity of about 6.5 km/sec. We have done so because of the coincidence of this layer with the SDR's where they are present. Implicitly we have assumed that the volcanic layer giving rise to the SDR's exists some distance seaward of where it is clearly imaged as SDR's, but that the layer gets thinner. Alternatively, as far as magnetic calculations are concerned, we could have maintained a constant thickness of the volcanic layer but reduced the degree of magnetization, making it zero beyond 150 km instead of thinning the layer from about 110 km to 150 km along the profile. In these calculations we have assumed that the continental crust lying landward of the SDR's has induced magnetization. In fact, we have subtracted the magnetic contribution of the continental crust from the observed anomalies; the calculated magnetic anomaly thus represents the anomaly corresponding to the SDR's with assumed remanent magnetization. Eastward of this body the oceanic crust is assumed to have zero effective magnetization. We have made this assumption to reflect the observation that magnetic anomalies are absent here. Instead of a zero value, the magnetization could be slightly larger; also the magnetization of the continental crust could be changed. These changes can be easily accommodated by varying the shape and magnetization of the highly magnetic body (representing the SDR's) without altering the basic nature of the model presented here. In making our calculations we have assumed a magnetization of .005 cgs units for the highly magnetic body, and an inclination of 45 degrees and a declination of -2 degrees to roughly correspond to an average position of the early Jurassic pole. Even if the position of the Jurassic pole were varied (there is a controversy about the exact position of the pole during early Jurassic; Hagstrom, 1993), or a component of induced magnetization is allowed to change the value of the total magnetization, it will not significantly alter the nature of the body that would be required to match the observed magnetic anomaly.

Similar magnetic models were assumed for the LASE profile and for the USGS-32 trough profile. In each case, the magnetic body coinciding with the presence of the SDR's has been extended seaward, as in the case of the EDGE 801 line, to match the magnetic anomaly. The susceptibility of the continental material for the LASE line was chosen as .003 cgs units, which was the value chosen for an earlier study (Alsop and Talwani, 1984). If it is reduced to .001 cgs units, as for the other lines, the model can be shown to be only slightly altered to match the observed values.

Our model differs from that of Austin et al. (1990) in that they used blocks with vertical boundaries and varying magnetization There model is incompatible with the non-vertical geometry of the SDR's, which we believe are associated with the entire width of the ECMA. We have been able to model the ECMA along their line BA6 by a model similar to the other models off the U.S. east coast in this paper without invoking vertical boundaries or reversely magnetized bodies.

Other parts of the world

OUTER VOERING PLATEAU

We examine the relationship between SDR's and magnetic anomalies on the Outer Voering Plateau (Figure 9, modified from Talwani et al., 1981). As for the case of the ECMA offshore the U.S. East Coast, the SDR's coincide very nicely with the earliest marine magnetic anomaly, which is anomaly 24 on the Voering Plateau. An important difference between the magnetic field here and off the U.S. East Coast is the absence of a magnetic quiet zone, seaward from anomaly 24. Also, anomaly 24, which is broad and somewhat irregular in overall shape on the Outer Voering Plateau, becomes narrower and more typical of a marine magnetic anomaly as we proceed northwards off-shore Lofoten Islands. The well-developed marine magnetic anomalies 23, 22. etc., lie seaward of anomaly 24. The Voering Plateau, as well as the remainder of the Norwegian Sea, is at an abnormally shallow depth. There appears to be no association of salt diapiric activity with anomaly 24.

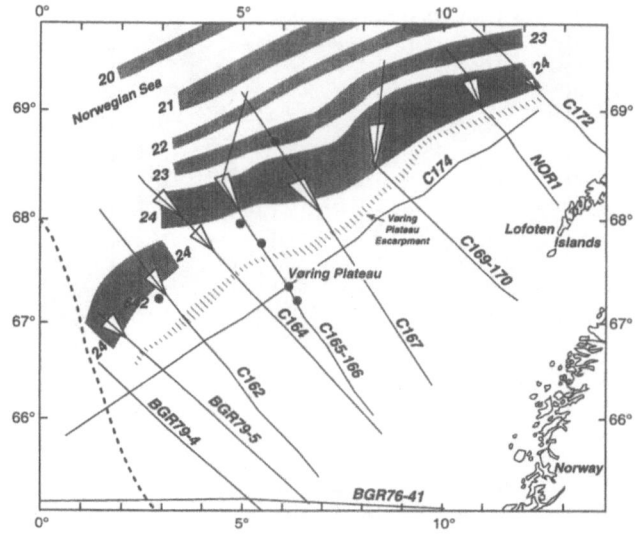

Figure 9   The magnetic anomalies on the Outer Voering Plateau (Talwani et al., 1981)
                are associated with a seaward dipping reflector (shown by triangles) similar
                to the way SDR's are associated with the ECMA. Anomaly 24 on the Outer
                Voering Plateau was produced by subaerial spreading; subsequent to
                anomaly 24 the spreading center became submarine and typical marine
                magnetic anomalies were produced. Drill holes are shown by solid dots.
                ODP site 642 drilled through seaward dipping reflectors at their landward
                edge and reveal that this edge lies on crust containing continental material.
                To the north-east (not shown in this figure) along anomaly 24 the width of
                the anomaly is reduced and it becomes very regular in shape, typical of
                marine magnetic anomalies. More recent data obtained by Olav Eldholm
                and his colleagues (e.g. Eldholm and Grue, 1994) give a more complete
                picture of seaward dipping reflector locations, but do not alter the
                relationships shown in this figure.

ARGENTINE MARGIN

A broad magnetic anomaly termed anomaly G (Rabinowitz and LaBrecque, 1979) exists on
the continental slope off Argentina (Figure 10). Immediately seaward of anomaly G are
Lower Cretaceous marine magnetic anomalies. Anomaly M4 is the oldest anomaly
identified here. Seaward dipping reflectors coinciding with anomaly G were observed by
Hinz and Popovici (1988). The situation here is unlike the ECMA in the absence of the
magnetic quiet zone and is more like the Outer Voering Plateau in that a succession of
identified marine magnetic anomalies lie immediately seaward of the anomaly associated
with the SDR's.

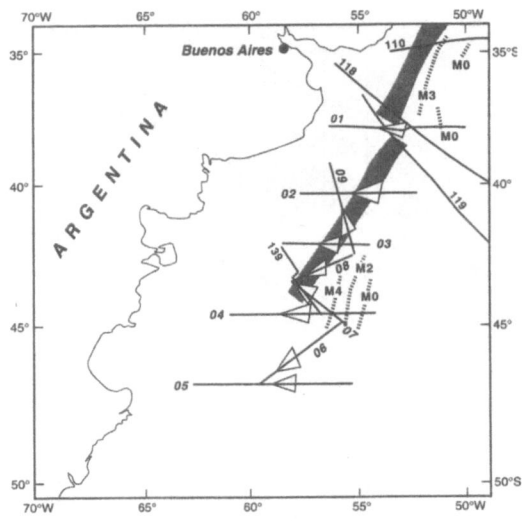

Figure 10   Seaward dipping reflectors indicated by triangles (Hinz and Popovici, 1988)
            are associated with prominent magnetic anomaly (anomaly G discovered by
            Rabinowitz and LaBrecque (1979). Immediately seaward of this anomaly
            the spreading center apparently becomes submarine with the generation of M
            series of typical marine magnetic anomalies.

OTHER AREAS

The SDR's and an associated magnetic anomaly have been followed northward from the
Argentine margin to the Brazilian margin (Schreckenberger and Roeser, 1994).

Zitellini and LaBrecque, 1994 reported a marine magnetic anomaly associated with
SDR's off South West Africa in a position that is conjugate to the Argentine margin SDR's.

Schreckenberger and Roeser (1994) examined and confirmed the seaward dipping
reflector sequence as a source for higher amplitude marine magnetic anomalies on a number
of continental margins and they have noted the presence of SDR's associated with the J
anomaly in the western North Atlantic.

At some locations a major part of the margin magnetic anomaly is satisfactorily
explained by the juxtaposition of a magnetic continental crust against a relatively non-
magnetic quiet zone crust. The magnetic anomaly off South Australia (Koenig and Talwani,
1977) and its conjugate Antarctic margin, as well as the magnetic anomaly off
Newfoundland (Fenwick et al., 1968), are such examples.

Some implications of the model

1. Seaward dipping reflectors are associated with the East Coast Magnetic Anomaly
(ECMA). It is likely that the volcanic layers which give rise to the SDR's are the major
cause, if not the sole cause of the ECMA. Even where SDR's are not seen (or are only
poorly imaged) in the reflection records, ECMA is attributed to volcanic material.

2. The association of volcanic layers and the marginal magnetic anomaly allows us to predict the presence of such layers on margins where a marginal anomaly is present, where no reflection measurements have been carried out, or where they have been carried out but SDR's have not been seen on existing reflection records.

3. The section along EDGE Line 801 also shows that the principal part of the SDR's lie on, and are a part of, the oceanic crust (defined as the crust produced by sea floor spreading, whether subaerial or submarine). Since continental crust is demonstrably absent, except below their feather edge (the SDR's lie almost entirely on material with velocity 7.2 - 7.5 km/sec, presumably produced from asthenospheric melt) it is erroneous to suggest that the crust in this region (say eastward of 70 km on EDGE Line 801 and Figure 7) may contain large blocks of continental crust. The situation appears to be clearly one of an entire oceanic crustal sequence with extrusives (indicated by SDR's) overlying intrusives. The latter have velocities lying between 7.2 and 7.5 km/sec. Several authors (e.g. White et al., 1987; Larsen and Jakobsdottir,1988,and Eldholm and Grue, 1994) have pointed out in earlier studies that material with velocity around 6.5 km/sec lies between the SDR's and the material with velocity between 7.2 and 7.5 km/sec. There has been a controversy whether this material represents oceanic crust or thinned continental crust. The EDGE results, by demonstrating that the SDR's extend down to the 7.2 to 7.5 km/sec layer and in their lower part represent material with velocity close to 6.5 km/sec, appear to have resolved this controversy.

4. If the magnetic anomaly is associated with volcanic layering produced by subaerial sea floor spreading, we run into a possible problem. Even after tilting, these layers are nearly horizontal. If sea floor spreading includes material of both normal and reverse polarity, then in the sub-horizontally layered pile, the normally and reversely magnetized layers should cancel their magnetic effects and the resulting anomaly should be zero or very small (Figure 11). To produce a magnetic anomaly with subaerial sea floor spreading and consequent production of lava flows of large lateral extent the rate of spreading must be high, or the polarity of the field must be constant for a long interval. In this case the volcanic layers of large horizontal extent are separated from each other and normal and reversed polarity material has enough separation so that magnetic anomalies can result. More precisely, the product of the rate of spreading and the time interval between polarity change should not be smaller than the linear extent of the volcanic flows. This leads us to the suggestion that marginal anomalies are present only because the North Atlantic Ocean began to open (between North America and Africa) with a fast spreading rate. otherwise the initial subaerial spreading would have produced a magnetic quiet zone.

Let us consider two alternative suggestions that would still allow sub-horizontal layers to give rise to a margin magnetic anomaly. One is that the magnetic anomaly is due to induced magnetization (Zitellini and LaBrecque, 1993). If this is so. then regardless of the initial polarity of the lava flows, the pile as a whole will have the same magnetization and can give rise to the magnetic anomaly. There are two objections to this. One, that volcanic layers (e.g. those examined in the cores of ODP Hole 642) have very high remanent magnetization. There is no suggestion that the major contribution to the magnetic anomaly from the volcanic flows is by material other than that which has strong remanent magnetization. Secondly, for rocks which have induced magnetization only, the value of susceptibility has to be unusually (and probably unacceptably) large in order that such rocks can produce the required anomaly (Emery et al., 1970).

174

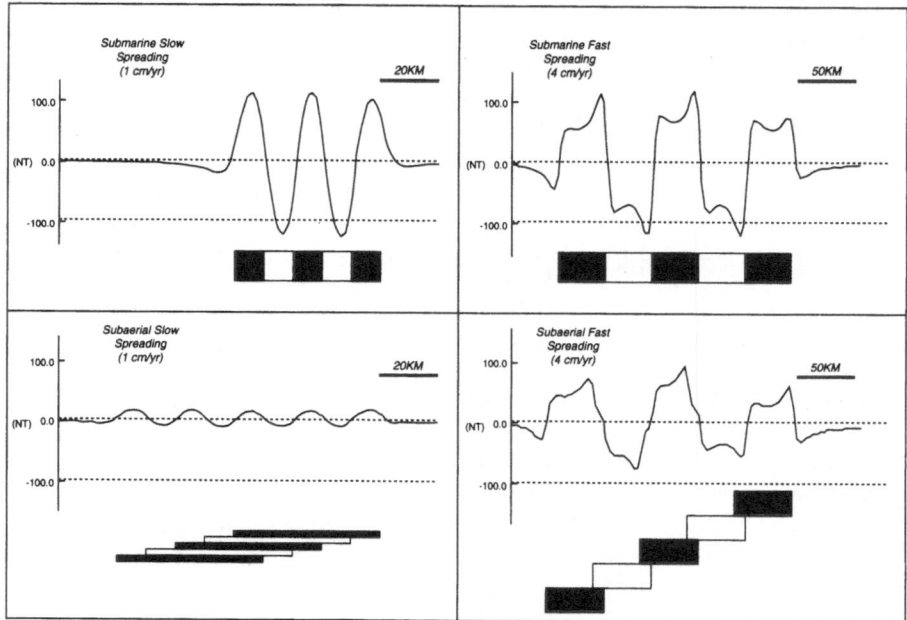

Figure 11   This cartoon shows that slow subaerial sea floor spread may not produce
large magnetic anomalies. Note that the horizontal scales are different for
the slow spreading ridges on the left and fast spreading ridges on the right.
The thickness of the anomaly producing layer is adjusted so that the same
amount of material is produced whether the spreading is subaerial with
volcanic layers of large lateral extent, or when the new crust being produced
underwater has volcanic layers with small flow lengths.

A second possibility is that the entire crust below the SDR's is continental crust. Over a
very short period of time (during an epoch of single polarity) all the SDR's were emplaced
and the original continental crust is somehow consumed during the process. If continental
crust were observed underneath the entire extent of the SDR's, this process could have some
validity. But if the entire crustal section is basically new oceanic crust and there is no trace
of continental crust, this hypothesis is untenable.

Support for initial fast spreading after opening comes from the Voering Plateau
anomalies, which have been dated. The rate of spreading associated with anomaly 24 is
much higher (even considering the anomalies on the conjugate margin, shifts in the injection
center, etc.) than the rate of spreading subsequent to that anomaly (Figure 12). The
abnormally large width of anomaly 24 also testifies to a high rate of initial spreading.
Support also comes from estimates of the rate of growth of the volcanics from studies on
cores recovered from ODP hole 642. Schoenharting and Abrahamsen (1989) state that the
volcanics represented by the SDR's grow at a rate that is an order of magnitude larger than
the growth of late Tertiary Icelandic plateau basalts.

5. When opening is due to subaerial sea floor spreading, the magnetic anomalies will be
very small unless the spreading is very fast, or the polarity interval is long. Submarine sea
floor spreading, on the other hand, will normally produce typical marine magnetic

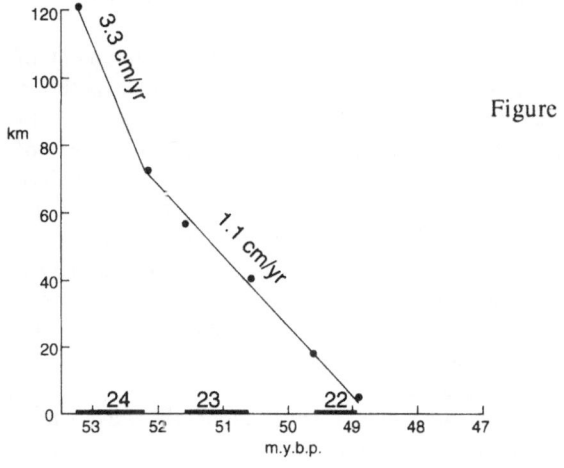

Figure 12   The rates of spreading calculated from the position of anomalies on the Outer Voering Plateau (Figure 10) constructed by determining anomaly widths roughly along the spreading direction show that the rate of spreading was three times as high during anomaly 24 time than during subsequent times.

anomalies, even at slow spreading rates: the assumption, of course, being that the very short length of submarine lava flows and dikes approximate a very narrow zone in which new magnetic material is introduced as a result of sea floor spreading. (We avoid here the further complication arising from a wide zone of injection.) It is likely, then, that as the sea floor spreading slows down after its initial fast rate which gave rise to the ECMA, the sub-horizontal layering of piles of subaerially produced volcanic flows effectively possess a very small resultant magnetization, giving rise to a magnetic quiet zone. (The existence of a single polarity during part of the Jurassic complicates the situation in the North Atlantic.) As the sea floor spreading changes from subaerial to submarine, marine magnetic anomalies will appear again even if the spreading is relatively slow. We propose that this explanation might hold for magnetic quiet zones in other oceans of the world that exist near the margins. It follows from what we have stated above that if the initial spreading is subaerial and slow there will be no margin magnetic anomaly, but instead a margin quiet zone will result. On the other hand, if fast subaerial spreading is immediately followed by slow submarine spreading then there will be no quiet zone but the margin magnetic anomaly will be followed by normal marine magnetic anomalies. The latter situation appears to hold for the outer Voering Plateau and Argentine margins.

We note that along some marine magnetic lines SDR's are present that are not associated with margin magnetic anomalies (Oh et al., 1995) These would also be associated with subaerial spreading.

6. If we accept the presence of subaerial sea floor spreading throughout the extent of the ECMA then it may be possible to explain the presence of salt diapirism in the Carolina and Baltimore Canyon troughs. As the axis of subaerial sea floor spreading retreats from the continent, the ocean lying between the spreading axis and the continent may be relatively isolated if barriers exist to the north and to the south. A relatively small ocean basin with limited circulation could lead to deposition of salt. When the axis of sea floor spreading becomes submarine the size of the ocean suddenly doubles and its depth also increases. This leads to more vigorous water circulation and decreases the possibility of desiccation and salt production.

We note that no salt has been discovered on the Outer Voering Plateau nor on the Argentine Margin where subaerial sea floor spreading changed very quickly after its initiation

initiation to submarine sea floor spreading. (Higher paleo-latitudes may also have contributed to the non-evaporitic environment.)

7. It is believed that submarine sea floor spreading with consequent development of pillow lavas is responsible for a rough oceanic basement which gives rise to hyperbolic echoes in reflection records. It follows that when the spreading is subaerial, long, smooth volcanic flows will give rise to basement which is much less rough and therefore will not give rise to the typical hyperbolic echoes associated with oceanic basement. The boundary between these two types of basement may have been interpreted by earlier authors as the landward limit of oceanic basement. This boundary should coincide with the boundary between the magnetic quiet zone and the landward limit of the marine magnetic anomalies. We have not tried to determine if this correlation is exact, because the transition between subaerial and submarine sea floor spreading may not have been sudden, and because of the possible complications of a part of the quiet zone in the North Atlantic arising from the presence of a single magnetic polarity during the Jurassic.

8. If the width of the ECMA indicates the width of extrusives along the East Coast margin we can make a rough calculation of the total volume of extrusives lying off the U.S. East Coast in the area of the study. If the extrusives extend about 1000 km along the length of the margin and are about 100 km wide and have an average thickness of 10 km, then the total volume of extrusives is approximately one million cubic kilometers. Presumably the conjugate margin also contains a million cubic kilometers of extrusives, thus adding up to a total of two million cubic kilometers. The comparison between this situation and the situation in the Norwegian Sea is interesting. In the latter, Eldholm and Grue (1994), estimate a total volume of 1.8 million cubic kilometers for the extrusives. However, this subaerial sea floor spreading quickly changed to submarine sea floor spreading, although the entire Norwegian Sea continues to have an elevated ocean floor. In the Central North Atlantic, that is, off the U.S. East Coast, and presumably off the conjugate North African margin, subaerial sea floor spreading lasted for a longer time but the ocean floor assumed normal oceanic depths very shortly thereafter. As discussed by Holbrook and Kelemen (1993), a satisfactory explanation for this large volume of igneous material is lacking.

Opening of the North Atlantic

We can begin to surmise the events at the time of opening of the Atlantic Ocean off the U.S. East Coast using the interpretation of the magnetic and seismic data. The existence of a thick but wholly oceanic initial crust that starts on the shelf itself changes our ideas about the continuity between the initial rifting and the subsequent opening process. Following Sawyer and Harry (1992), it may seem that there is a time gap between the initial thinning process, which is focused on a point of crustal weakness, and a subsequent opening process, which is associated with weakness in the mantle. The opening process, then, is not so much a thinning of the crust but may be more appropriately thought of as a rupture with new crustal material being added very quickly. A small part of this new material underlies the continental crust and is termed underplating, and some of it overlies the continental crust and forms the feather edge of the seaward dipping reflectors. The zone lying between the SDR's and the layered continental crust in EDGE Line 801 is a remarkably narrow zone and has no reflectors. Thus, this may be the only location (no more than about 20 km wide), where the notion of a transitional crust consisting of intrusives mixed with continental blocks may apply. This is in contrast to the much wider zone lying between confirmed continental crust and the "landward limit of oceanic basement" which has traditionally been identified as "transitional' crust.

The Initial Oceanic Crust is thick (ranging off the U.S. East Coast from more than 20 km along the EDGE 801 and USGS-32 lines and slightly under 15 km along the LASE profile). The seismic velocities are higher than those associated with normal oceanic crust. The top of the basement forms a smooth reflector which is difficult to distinguish from a sedimentary horizon.

At the opening of the Atlantic the rate of the subaerial sea floor spreading was high and the volcanic layers (indicated by the SDR's) were emplaced within a single epoch of normal polarity, giving rise to the East Coast Magnetic Anomaly. Subsequent slowing down of the spreading caused a juxtaposition of the normally and reversed sub horizontal layers, reducing the effect magnetization of the total volcanic pile and giving rise to a magnetic quiet zone.

While the spreading axis was still subaerial, a relatively shallow ocean with limited water circulation lay between it and the continent. Salt was deposited during a period of restricted circulation. Subsequent salt diapiric activity is seen in the reflection records.

As the axis of spreading subsided to submarine depths, the ocean became wider and deeper with consequent vigorous water circulation and the possibility of any salt production ceased. Subsidence of the spreading axis to submarine depth also changed the nature of the basement reflector - from a smooth horizon to one that is rough and characterized by typical hyperbolic reflectors. The boundary between the two kinds of top of basement reflector has been called the "landward limit of oceanic basement." The short length of the submarine volcanic flows also made it possible to produce typical marine magnetic anomalies even when the spreading continued at low rates. The oceanic crust thus formed attained normal oceanic crustal seismic velocity and thickness.

Acknowledgements

This work was supported by the National Science Foundation (Grant #EAR-8721194) and a grant from Texaco. We thank Dale Sawyer, Olav Eldholm, and John McBride for critical reviews of the manuscript.

References

Alsop, L.E., and M. Talwani, The east coast magnetic anomaly, *Science* 226, 1189-1191, 1984.

Austin, J.A. Jr., P.L. Stoffa, J.D. Phillips, J. Oh, D.S. Sawyer, G.M. Purdy, E. Reiter, and J. Makris, Crustal structure of the Southeast Georgia embayment-Carolina Trough: Preliminary results of a composite seismic image of a continental suture (?) and a volcanic passive margin, *Geology, 18,* 1023-1027, 1990.

Behrendt, J.C, and K.D. Klitgord, High-sensitivity aeromagnetic survey of the U.S. Atlantic continental margin, *Geophysics* 45, 1813-1846, 1980.

Chowns, T.M., and C.T. Williams, Pre-Cretaceous rocks beneath the Georgia Coastal Plain - regional implications, in: G.S. Gohn, ed., pp.L1-L42, *Studies Related to the Charleston, South Carolina, Earthquake of 1886 - Tectonics and Seismicity*, USGS Prof. Paper 1313, 1983.

Dillon, W.P., P. Popenoe, J.A. Grow, K.D. Klitgord, B.A. Swift, C.K. Paull, and K.V. Cashman, Growth faulting and salt diapirism; Their relationship and control in the Carolina Trough, eastern North America, in *Studies in Continental Margin Geology, Amer. Assoc. Pet. Geol..*, J.S. Watkins and C.L. Drake, eds, Memoir 34, 21-46, 1983.

Drake, C.L., J. Heirtzler, and J. Hirshman, Magnetic anomalies off eastern North America, *J. Geophys. Res.*, 68, 5259-5275, 1963.

Ed Driver, personal communication, 1982.

Eldholm, O., and K. Grue, North Atlantic Volcanic Margins: Dimensions and production rates, *J. Geophys. Res.*, 99, 2955-2968, 1994.

Emery, K.O., E. Uchupi, J.D. Phillips, C.O. Bowin, E.T. Bunce, and S.T. Knott, Continental rise off eastern North America, *Am. Assoc. Petrol. Geol. Bull.*, 54, 44-108, 1970.

Ewing, J., and Ewing, M., Seismic-Refraction Measurements in the Atlantic Ocean Basins, in the Mediterranean Sea, on the Mid-Atlantic Ridge, and in the Norwegian Sea, *Bull. Geol. Soc. Amer.*, 70, 291-318, 1959.

Fenwick, D.K.B., M.J. Keen, C.E. Keen, and A. Lambert, Geophysical studies of the continental margin northeast of Newfoundland, *Canad. J. Earth Sci. 5*, 483-500, 1968.

Glover III, L.G., R.E. Sheridan, W.S. Holbrook, J. Ewing, M. Talwani, R. Hawman, and S. Smithson, The "Taconic" suture problem - Piedmont to Offshore - North Carolina to Maine, *GSA* (in press).

Grow, J.A., Deep structure and evolution of the Baltimore Canyon Trough in the vicinity of the C.O.S.T. no. B-3 well, in *Geological Studies of the C.O.S.T. No. B-3 Well, United States Mid-Atlantic Continental-Slope Area*, edited by P.A. Scholle, USGS Circ. 833, 117-125, 1980.

Hagstrom, J.T., North American Jurassic APW; The current dilemma, *EOS, 74*, 65, 1993.

Heirtzler, J.R., and D.E. Hayes, Magnetic boundaries in the North Atlantic, *Science*, 157, 185-187, 1967.

Hinz, K., The seismic crustal structure of the Norwegian continental margin in the Voering Plateau, in the Norwegian deep sea, and on the eastern flank of the Jan Mayen Ridge between 66° and 68° N, *24th Int. Geol. Congr., Sec. 8*, pp.28-36, 1972.

Hinz, K., A hypothesis on terrestrial catastrophes: Wedges of very thick oceanward-dipping layers beneath passive continental margins - their origin and paleoenvironmental significance, *Geol. Jahrbuch, ser. E, 22*, 3-28, 1981.

Hinz, K., and A. Popovici, On a multichannel seismic reconnaissance survey of the Argentine Eastern Continental Margin by S.V. EXPLORA, Bericht ueber eine digitalseismische uebersichtsvermessung des Atlantischen Kontinentalpandes vor Argentinien mit S.V. EXPLORA, 22 Dezember 1987 - 15 Januar 1988, Bundesanstalt fuer Geowissenschaften und Rohstoffe, Hanover, Germany, 1988.

Holbrook, W.S., and P.B. Keleman, Large Igneous province on the U.S. Atlantic margin and implications for magmatism during continental break up, *Nature*, 364, 433-436, 1993.

Holbrook, W.S., G.M. Purdy, J.A. Collins, R.E. Sheridan, D.L. Musser, L. Glover, M. Talwani, J.I. Ewing, R. Hawman, and S. Smithson, Deep velocity structure of rifted continental crust, U.S. Mid-Atlantic margin, from wide-angle reflection/refraction data: *Geophysical Research Letters, .19*, 1699-1702.

Holbrook, W.S., G.M. Purdy, R.E. Sheridan, L. Glover III, M. Talwani, J. Ewing, and D. Hutchinson, Seismic structure of the U.S. Mid-Atlantic Continental Margin, *J. Geophys. Res.*, 99, 17871-17891, 1994.

Hutchinson, D.R., J.A. Grow, K.D. Klitgord, and B.A. Swift, Deep structure and evolution of the Carolina trough, in: J.S. Watkins and C.L. Drake, eds., pp.129-152, *Studies in Continental-Margin Geology*, Am. Assoc. Petrol. Geol. Mem. 34, 1983.

Hutchinson, D.R., K.D. Klitgord, and A.M. Trehu, Alternative Interpretation: "Integration of COCORP deep reflection and magnetic anomaly analysis in the southeastern United States: Implications for the origin of the Brunswick and East Coast magnetic anomalies," *Geol. Soc. Am. Bull.* 102, 271-274, 1990.

Keen, M.J., Possible edge effect to explain magnetic anomalies off the eastern seaboard of the U.S., *Nature 222*, 72-74, 1969.

Keller, R. Jr., J.L. Meuschke, and L.R. Alldredge, Aeromagnetic surveys in the Aleutian, Marshall, and Bermuda Islands, *Trans., Am. Geophys. Un. 35*, 558-572, 1954.

Klitgord, K.D., and J.C. Behrendt, Aeromagnetic-anomaly map of the United States Atlantic continental margin, *USGS Misc. Field Studies Map MF-913*, 1977.

Klitgord, K.D., D.R. Hutchinson, and H. Schouten, U.S. Continental margin: structural and tectonic framework, in: *The Geology of North America, vol 1-2: The Atlantic Continental Margin, U.S., The Geological Society of America*, 19-55, 1988.

Koenig, M., and M. Talwani, A geophysical study of the southern continental margin of Australia: Great Australian Bight and western sections, *Geol. Soc. Am. Bull. 88*, 1000-1014, 1977.

Larsen.H.C., and S.Jakobsdottir, Distribution,crustal properties and significance of seawards-dipping sub-basement reflectors off E. Greenland, Geol. Soc. Spec. Publ.London, 39, 95-114, 1988.

LASE Study Group, Deep structure of the US East Coast passive margin from large aperture seismic experiments (LASE), *Mar. Petrol. Geol. 3*, 234-242, 1986.

McBride, J.H., and K.D. Nelson, Integration of COCORP deep reflection and magnetic anomaly analysis in the southeastern United States: Implications for the origin of the Brunswick and East Coast magnetic anomalies: *Geol. Soc. Am. Bull. 100*, 436-445, 1988.

Nelson, K.D., J.H. McBride, J.A. Arnow, J.E. Oliver, L.D. Brown, and S. Kaufman, New COCORP profiling in the southeastern United States, Part II: Brunswick and East Coast magnetic anomalies, opening of the north-central Atlantic Ocean, *Geology 13*, 718-721, 1985.

Oh, J., J.A. Austin Jr., J.D. Phillips, M.F.Coffin, and P.L. Stoffa, Presence of Seaward Dipping Reflectors on the Blake Plateau: Implications for Seafloor Spreading Events, *Geology*, *23*, 9-12, 1995.

Oh, J., J.D. Phillips, J.A. Austin Jr., and P.L. Stoffa, Deep-penetration seismic reflection images across the United States continental margin, in *Continental lithosphere: Deep seismic reflections, Geodynamics Series Volume 22*, edited by R. Meissner, L. Brown, H.-J. Durbaum, W. Franke, K. Fuchs and F. Seifert, 225-240, American Geophysical Union, Washington, D.C., 1991.

O'Reilly, W., and S.K. Banerjee, The mechanism of oxidation in titanomagnetites: a magnetic study, *Mineralog. Mag.*, 36, 29-37, 1967.

Parson, L.M., and the ODP Leg 104 Scientific Party, Dipping reflector styles in the N.E. Atlantic Ocean, in Eds: Morton, A.C., and Parsons, L.M., Early Tertiary opening of the N.E. Atlantic, *Geol. Soc. Special Publication No. 39*, 57-68, 1988.

Rabinowitz, P.D., and J. LaBrecque, The Mesozoic South Atlantic Ocean and evolution of its continental margin, *J. Geophys. Res.*, 84, 5973-6002, 1979.

Rabinowitz, R.D., Heirtzler, J.R., and Cande, S.C., Magnetic anomaly profiles along ship tracks, in *Eastern North American Continental Margin and Adjacent Ocean Floor 28° to 36°N and 70° to 82°W (Atlas 5)*, Editors G.M. Bryan and J.R. Heirtzler, and 34° to 41°N and 68° to 78°W (Atlas 4), Editors J.J. Ewing and R.D. Rabinowitz, in *Ocean Margin Drilling Program, Regional Atlas Series*, Marine Science International, Woods Hole, 1983.

Sawyer, D.S., and D.L. Harry, A dynamic model of extension in the Baltimore Canyon trough, *Tectonica*, 11, 420-436, 1992.

Schoenharting. G., and N Abrahamsen, Paleomagnetism of the volcanic sequence in Hole 642E, ODP Leg 104, Voering Plateau, and correlation with Early Tertiary basalts in the North Atlantic: *Proc. ODP, Sci Results,* 104: College Station, TX (Ocean Drilling Program), 911-920, 1989.

Schreckenberger, B., and H.A. Roeser, Seaward Dipping Reflector Sequences as Sources for High Amplitude Marine Magnetic Anomalies, *EOS 75,* 16, 131, 1994.

Sheridan, R.E., D.L. Musser, L. Glover, III, M. Talwani, J.I. Ewing. W.S. Holbrook. G.M> Purdy, R. Hawman, and S. Smithson, Deep Seismic reflection data of EDGE U.S. mid-Atlantic continental margin experiment: Implications for Appalachian sutures and Mesozoic rifting and magmatic underplating, *Geology.* 21. 563-567, 1993.

Stacey, F.D., Physics of the Earth, 2nd ed., New York, Wiley, 414 p., 1977.

Talwani, M., J.C. Mutter, and O. Eldholm, The initiation of opening of the Norwegian Sea, *Oceanologia Acta,* no. SP., 23-30, 1981.

Talwani, M., J. Ewing, R.E. Sheridan, D.L. Musser, L. Glover III. S. Holbrook and M. Purdy, EDGE lines off the U.S. Mid-Atlantic margin and the East Coast Magnetic Anomaly, *EOS,* 73, , 490, 1992.

Taylor, P.T., I. Zietz, and L.S. Dennis, Geologic implications of aeromagnetic data for the eastern continental margin of the United States, *Geophysics 33,* 755-780, 1968.

Trehu, A.M., A. Ballard, L.M. Dorman, J.F. Gettrust, K.D. Klitgord. and A. Schreider, Structure of the lower crust beneath the Carolina trough. U.S. Atlantic continental margin, *J. Geophys. Res. 94B,* 10585-10600, 1989.

Uchupi, E., J.T. Crosby, S.T. Bolmer Jr., J. D. Eusden Jr., J.I. Ewing, J.K. Costain, R.J. Cleason and L. Glover III, in *Eastern North American Continental Margin and Adjacent Ocean Floor 28° to 36°N and 70° to 82°W (Atlas 5),* Editors G.M. Bryan and J.R. Heirtzler, and 34° to 41°N and 68° to 78°W (Atlas 4), Editors J.J. Ewing and R.D. Rabinowitz, in *Ocean Margin Drilling Program, Regional Atlas Series.* Marine Science International, Woods Hole, 1983.

White, R.S., G.K. Westbrook, S.R. Fowler, G.D. Spence, P.J. Barton, M. Joper., J. Morgan, A. Bowen, C. Prescott, and M.H.P. Bott, Hatton Bank (northwest U.K.) continental margin structure, *Geophys. J.R. Astron. Soc.,* 89, 265-272, 1987.

White, R., and D. McKenzie, Magmatism of Rift Zones: the generation of volcanic continental margins and flood basalts, *J. Geophys. Res.,* 94, 7685-7229, 1989.

Zitellini, N., and J.L. LaBrecque, Stochastic approach to modeling the ocean continent transition, IUGG 7th Scientific Assembly, Buenos Aires, Argentine, *IAGA Bulletin No. 55, Part C,* p.415.

STRUCTURE OF ATLANTIC OCEANIC CRUST AROUND CHRON M16 FROM DEEP SEISMIC REFLECTION PROFILES

J. J. DAÑOBEITIA*, R.W. HOBBS**, J.H. McBRIDE**, T. A. MINSHULL*** and J. GALLART*.
* Institute of Earth Sciences, CSIC, 08028 Barcelona, Spain.
** BIRPS, Bullard Laboratories, Madingley Road, Cambridge CB3 0EZ, England.
*** Department of Earth Sciences, Bullard Laboratories, Madingley Road, Cambridge, CB3 0EZ, England.

ABSTRACT. Deep seismic images around isochron M16 (142 Ma) are presented from three different sites on both flanks of the North Atlantic ocean. The similarity of the acquisition parameters and of the reflective patterns, allows us to perform an identical post-stack processing sequence in order to interpret jointly the three data-sets. Despite the scattered energy produced at the contact between the sedimentary cover and the top of the igneous basement, a reasonable generalised velocity-depth function has been obtained with which to perform seismic migration. The final results provide clearer images of structures in the igneous crust than previously obtained.

The migrated sections show in general rather similar images on the two plates (North America and West Africa), some distinctions can also be deduced from the individual profiles. A common feature on the three profiles is the blocky character of the top of the igneous basement, which is cut by a series of offsets ranging from 200 to 600 ms. Most of them can be extended down to the base of the upper crust as dipping reflections, and can be interpreted as normal faults. On the American profile, two steeply dipping events show continuity down to the lower crust, dying out as low-angle normal faults (detachment). On the American plate profile, the geometry of the faults suggests a domino-style faulting with most of the steeply dipping events striking ridgewards. On the African plate profiles, the majority of the dipping events also show a ridgeward dip. The upper crust is made up of short and gently dipping events mostly located beneath the basement ridges. An interesting feature is seen on the American profile, where a pair of almost parallel reflections located beneath a basement ridge show an intermittent continuity with a lateral extension of about 5 km. This event could be interpreted as magmatic in origin possibly produced during a high melt production period. The lower crust shows scarce reflectivity on the African profiles, whereas on the American ones it shows a great variety of events. In general the reflectivity is higher on the older side of the lines. From the available data-set it seems that the deep seismic images obtained on both sides of the Atlantic are fairly symmetrical. Some exceptions could be ascribed to small differential changes, between ridge segments, in the spreading rate.

1. Introduction

Over the past ten years much work has been undertaken on both sides of the North Atlantic in order to understand better the internal structure of oceanic crust by means of near normal incidence reflection technique (Mutter et al., 1984; NAT Study Group, 1985; White et al., 1990; Banda et al., 1992; Morris et al., 1993; White et al., 1994; McBride et al., 1994b). Old multichannel seismic data showed an absence of crustal reflectivity that was consistent with a continuous variation of velocity with depth, and with a lack of impedance contrasts.

183

E. Banda et al. (eas.), Rifted Ocean-Continent Boundaries, 183–196.

However, more recent research has shown high quality images of deep crustal structure and widespread intracrustal reflectivity patterns including steeply dipping events whose significance is still a matter of debate (e.g., White et al., 1990; Banda et al., 1992).

The production of accurate images of seismic reflection data is highly dependent on the acquisition and processing procedures, careful migration being necessary. In this paper we apply a similar post-stack processing flow to all data-sets with a special emphasis on the strategy used to define a velocity-depth function in order to analyse the tectonic significance of selected intracrustal reflectors imaged on both flanks of the North Atlantic. We will focus our interest on the area corresponding to a smooth change of spreading velocity that took place near anomaly M16 (140 Ma) (Klitgord and Schouten, 1986; Roest et al., 1992). We use three data sets that cross-cut this anomaly; on the African plate, three reflection lines shot obliquely to the spreading fabric west of the Canaries (Banda et al., 1992) and, a series of parallel lines spaced at 4 km over the Cape Verde abyssal plain (OCEAN) (McBride et al., 1994b); on the North-American plate, we have a widely spaced (10-20 km) grid survey with lines parallel and orthogonal to the spreading direction around the Blake Spur fracture zone (Morris et al., 1993).

2. Geophysical and Tectonic Setting

Detailed magnetic anomaly maps are available for the three areas studied (Hayes and Rabinowitz, 1975; Klitgord and Schouten, 1986; Verhoef et al., 1991). Recent North Atlantic reconstruction based on seafloor spreading anomalies together with fracture zones traces provides the appropriate framework for dating the Mesozoic series (Roest et al., 1992). The negative anomaly M16 intersects the three lines (Figure 1). This anomaly

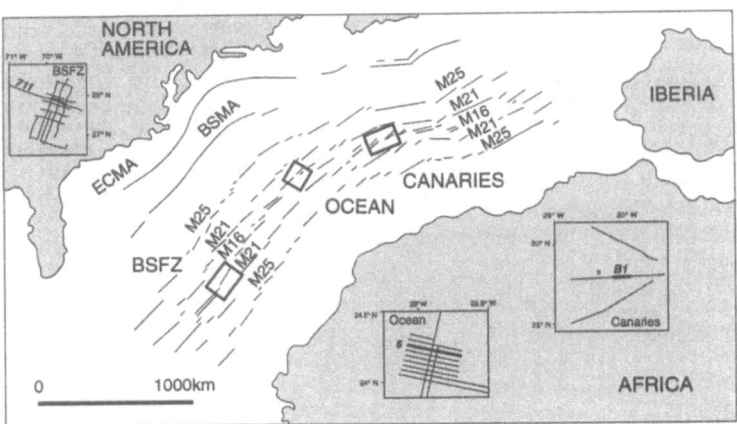

Figure 1.- *Reconstruction of the Central Atlantic at chron M16 (142 Ma), with identified spreading magnetic anomalies together with the East Coast Magnetic anomaly (ECMA) and the Blake Spur Magnetic anomaly (BSMA). The boxes show the three studied areas; Blake Spur fracture zone (BSFZ); Cape Verde abyssal plain (OCEAN), West Canaries survey (CANARIES). Modified from* **White et al. (1994).**

probably marks a smooth change in the spreading velocity during the transition between the Late Jurassic to Early Cretaceous time in the North Atlantic. On the African plate, Roest et al. (1992), have reported a gradual decrease in the half-spreading rate based on stage poles derived from total separation poles. These half-spreading velocities range from 10-17 mm/yr between M16-M21 (Jurassic) to 7-8 mm/yr between M0-M16 (Cretaceous). On the North American plate similar half-spreading rates ranging from 9 mm/yr to 13 mm/yr have been obtained (Klitgord and Schouten, 1986; Minshull et al., 1991). Whether this change has consequences in the internal crustal structure is still not known, although some authors have correlated the change in spreading velocity with a change from smooth to rough basement topography (Sundvik et al., 1984; Sundvik & Larson, 1988).

The average half-spreading rate calculated for the study area is about 10 mm/yr. The area is bounded by small-offset fracture zones that may distort the deep crustal configuration. The velocity-depth structure is well constrained in the Blake Spur and Cape Verde abyssal plain areas by wide-angle data (Minshull et al., 1991; White et al., 1994). We made use of this velocity information during the seismic migration sequence.

3. Data Acquisition and Pre-stack Processing

The three data-sets used in this study have similar acquisition parameters (Table I). For the Blake Spur and Canaries data a cable length of 3 km with 240 and 120 channels respectively was used. The source was slightly larger at the Canaries with 6276 cu. in (105 litres) as opposed to the 5821 cu.in (96 litres) used for the Blake Spur data. For the Cape Verde profiles the parameters were improved with a cable length of 4.5 km, 180 channels and an array of 7118 cu in (117 l). In order to avoid out-of-the-plane energy coming from the rough buried igneous topography, a wide (16 m for BSFZ and 50-60 m for Ocean and Canaries) tuned air-gun-array was designed. Increasing the width of the source array from 16 m to 60 m enhances the cross line directivity of the source over the bandwidth 10-50 Hz focussing more the energy into the plane of the profile, hence reducing the out-of-the plane energy (Parkes et al., 1984; Hobbs & Snyder, 1992). A substantial difference between the Cape Verde survey and the other surveys is the application of a pseudo-random time delay to suppress the wrap around multiples, which typically appear at these water depths. The method consits in decreasing the coherency of the multiples by destructive interference as they are stacked (McBride et al., 1994a).

TABLE I
ACQUISITION PARAMETERS

	CANARIES	CAPE VERDE	BLAKE SPUR
Recording System	DFS V/DSS	DFS V/GDR1000	DFS V
Record length	16 s	18 s	12 s,15 s
Sample rate	4 ms	4 ms	4 ms
High cut filter/slope	64 Hz/72 dB	64 Hz/72 dB	-----
Low cut filter/slope	3.5 Hz/18 dB	5.3 Hz/18 dB	-----
Shot-point Interval	50 m	50 m	22 s
Primary Navigation	GPS	GPS	GPS
Total airgun volume	6278 in3 (105 l)	7118 in3 (117 l)	5821 in3(96 l)
Number of guns	34	36	10
Nominal pressure	2000 psi	2000 psi	2000 psi
Array width	50 m	60 m	-----
Airgun depth	7.5 m	7.5 m	10 m
Number of groups	120	180	240
Group length	25 m	25 m	12.5 m
Cable depth	12 m	15 m	10 m
Streamer active length	3000 m	4500 m	3000 m

186

The processing sequence up to the final stacks are similar for the three surveys with resampling, common depth point (CDP) sorting, normal moveout (NMO) corrections, mutes and stacking. For the Canaries and Cape Verde data, an additional dip filter and an amplitude recovery using average decay curves were applied. The final stack sections from the three lines studied are displayed in Figure 2.

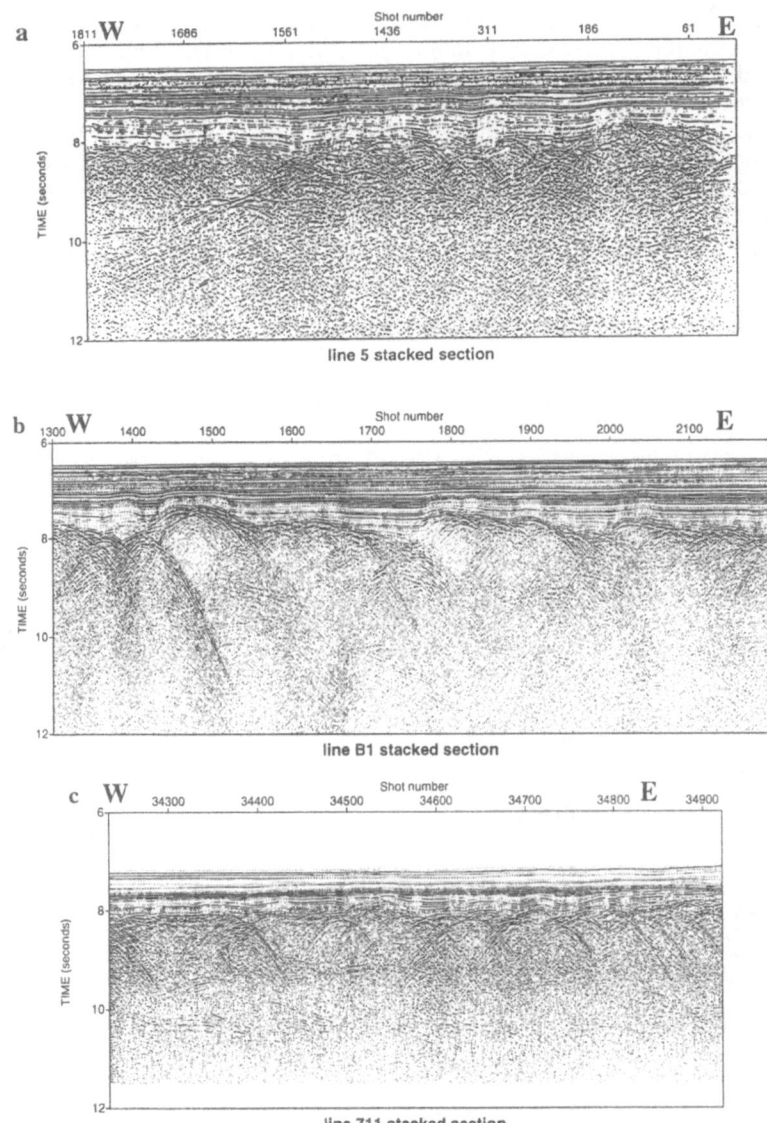

Figure 2.- *Final stacked sections; a) line 5 (OCEAN) with f-k filter to reduce the diffractions, b) Line B1 (CANARIES), and c) line 711 (BSFZ).*

4. Post-stack Processing

Post-stack processing was performed in order to enhance deep structures mainly at mid and lower crustal levels. The stacked data were first resampled to 8 ms, and adjacent traces were added to have 25 m common mid point (CMP) spacing. The section was then filtered in the frequency-wavenumber (f-k) domain to reduce noise with low apparent velocity and to increase reflection coherency with a filter of ±1.78 km/s, and ±1.56 km/s for the Canaries and Cape Verde survey respectively.

Migration collapses or focuses the recorded wavefield into its correct subsurface position. A fundamental requirement for the success of the migration operator is a good knowledge of the velocity field. The strategy to define this velocity field consisted of integrating all the known data, i.e. stacking velocities, refraction velocities and Stolt migrated time sections (Stolt, 1978), to finally derive a generalised velocity function for 2-D finite difference migration. Stacking velocities can be considered to be equivalent to root-mean-square velocities, which are related to interval velocities for a simple 1-D model. In the three selected profiles, the pelagic sediments are flat lying and consist of several bright marker reflectors. These can be easily identified on stacking velocity semblance plots. Hence, reliable velocity estimates can be determined for the sediments. This procedure fails at the basement interface, as it is not 1-D and the semblance becomes confused with scattered energy from nearby diffractions. It is beneath the basement surface that velocities determined from refraction become important. However, there still remains the problem of a good definition of the basement surface. This was achieved by a f-k time migration using the velocities determined from the sediments. The

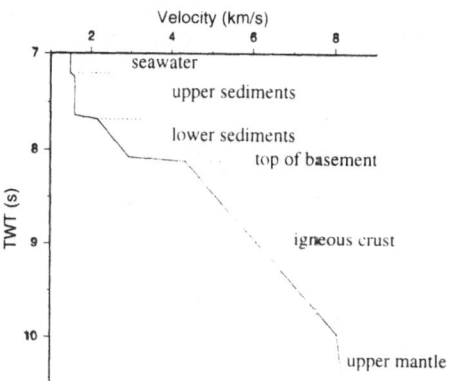

Velocity-depth function

Figure 3.- *Mean interval velocity-depth function used for migration and depth conversion. The velocity varies vertically and horizontally depending on the basement topography. Sediment velocities are derived from stacks, and subbasement velocity is determined from available wide-angle data.*

two velocity datasets were merged at the basement interface. The average final velocity-depth function is shown in Figure 3. This velocity varies vertically and laterally depending on the basement topography. With this velocity field a fast 2-D finite differences migration algorithm was applied to derive an improved image (Claerbout, 1989). The top of the igneous basement was poorly defined because of the inherent filtering effect due to the truncation of the expansion of the wave equation. Subsequently a finite difference migration was performed imaging up to 50 °, which resulted in a much better definition of the steeply dipping events. The seismic section is a 2-D representation of the 3-D seismic wave field, and therefore it also includes out-of-the-plane energy. The profiles analysed in this study consisted mainly in lines in the dip direction, so that the 2-D assumption can be considered adequate. Of course a correct imaging of the subsurface, in the presence of dipping events, can only be achieved by a 3-D migration. A complete discussion of 3-D versus 2-D is out of the scope of this paper. To improve the seismic image at the basement a pre-stack depth migration is under preparation.

5. Description of Profiles

5.1. *Cape Verde (Line 5, OCEAN)*

The seafloor is around 6.4-6.5 s (4800 m) (Fig. 4a). A series of continuous, subhorizontal and well-layered reflections can be followed all along this line. At about 600 ms below the seabed a strong reflection could mark a main lithological boundary (s-reflector, Fig. 4b) as proposed by McBride et al., 1994b, who interpreted this layer as the contact Middle/Lower Miocene identified at the nearby DSDP-139 (Hayes et al., 1979). Another strong reflector made up of a series of sub-parallel reflections is situated 1000 ms beneath the seabed and shows a remarkable continuity. Faulting within the sediments is imaged in some places associated with the edge of the basement blocks (f, Fig. 4b). In general the sediment thickness can be considered fairly constant but increasing slightly westwards.

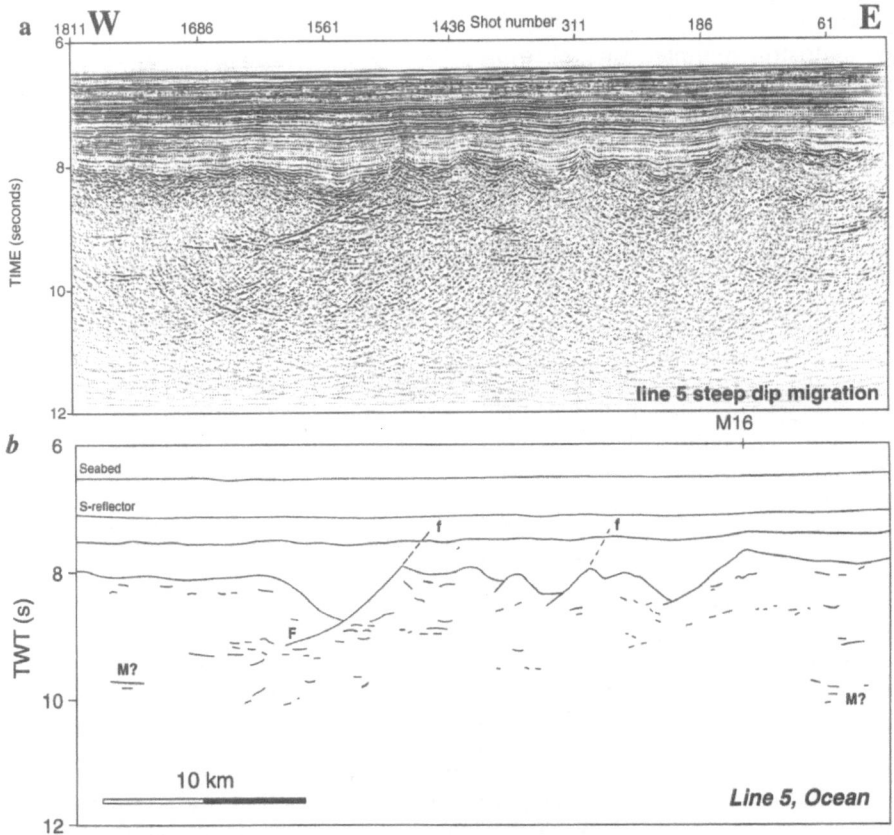

Figure 4.- *(a) Steep dip migrated of line 5, performed with a finite-difference algorithm using a 2-D velocity field. Shot point interval is 50 m. (b) Line-drawing interpretation of section line 5. The main seismic characteristics are labelled as follows: "F" is major normal fault; "f" faulting within the sediments; "M" tentative Moho reflection.*

The basement interface appears on the migrated section (Fig. 4a) as progressively increasing its blocky character eastward, mainly between the two prominent basement offsets which show a kind of en echelon structure with a 300 ms step (Fig. 4a). These two basement ridges have slightly different structural patterns. The westernmost ridge (sp 1450-1560, Figs. 4a and b) is clearly associated with a dipping event that shows continuity downward to the mid-crust where it seems to terminate at about 9.2 s. This has been interpreted by McBride et al. (1994b) as a normal fault (F, Fig. 4b) based on the basement offset and the truncation of horizontal reflections within the igneous crust. The faulting geometry in this migrated section seems to be listric with dips of 25-30° (Fig. 4a), and in the depth converted section the reflector dies out at about 11 km depth (Figure 5).

The second basement offset (sp 120-180, Figs. 4a and 4b) shows a similar dip, but this trend is restricted to the upper crust, its downward continuation being uncertain. The upper crust is marked on the eastern side by a series of short, westwards gently dipping events at about 700 ms beneath the basement. The western side (young) of the profile shows more coherent reflectivity than on the eastern side. This reflectivity is mainly made up of short and discontinuous subhorizontal reflections, some of which show a gentle westward dip, mostly located beneath the small basement ridges (500-1000 ms) and restricted to the upper and mid-crust. The scarce lower crust reflectivity appears to stop suddenly at the contact with the normal fault (Figs. 4a and 4b). These regions of diffuse reflectivity, imaged in the lower crust, have been described as the expression of ductile deformation (Mutter & Karson ,1992). The Moho can be tentatively identified at the edges of the profile (M, Fig. 4b).

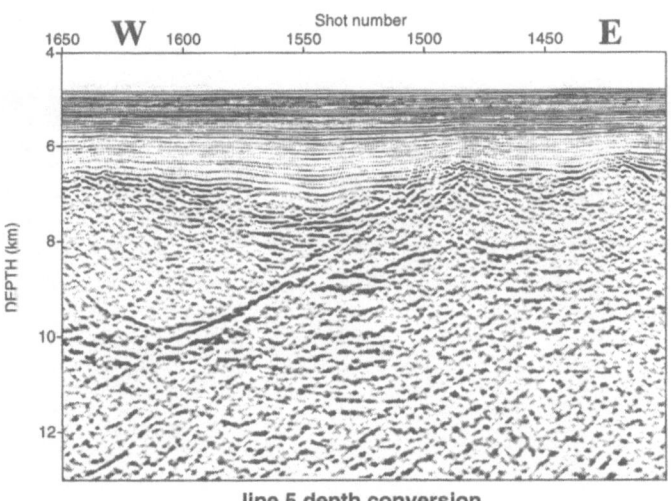

line 5 depth conversion

Figure 5.- *Depth converted section of a part of line 5, achieved with the same velocity function as for the migration of the line. The fault offsetting the top of the igneous crust (1 km) can be extended down to 11 km. The highest reflectivity is concentrated at the footwall.*

5.2. Canaries (Line B1)

The water depth for the Canaries data-set is comparable to the one for line 5 (Fig. 6a). A strong continuous sub-horizontal reflection at 600 ms beneath the seafloor (s-reflector, Fig. 6a and 6b), probably represents the same lithological boundary as in the OCEAN area. The hummocky nature of the bottom sediments suggest compaction. The character of the top of the basement differs from the OCEAN area in the offset size of the basement blocks, but the average horizontal distances between major ridges remain the same (~16 km). The two major basement ridges offsetting 600 ms are observed on the younger side of the profile (i.e., west of M16).

The reflections in the upper crust appear as short (1-2 km) discontinuous events. At mid-crustal levels, a weak and discontinuous reflection is observed (10 s) dipping gently westward on the older side of the profile (**w**, Figs. 6a and 6b). On the younger side, between the two prominent basement ridges, the crust shows a complicated reflection pattern probably disturbed by out-of-the-plane energy.

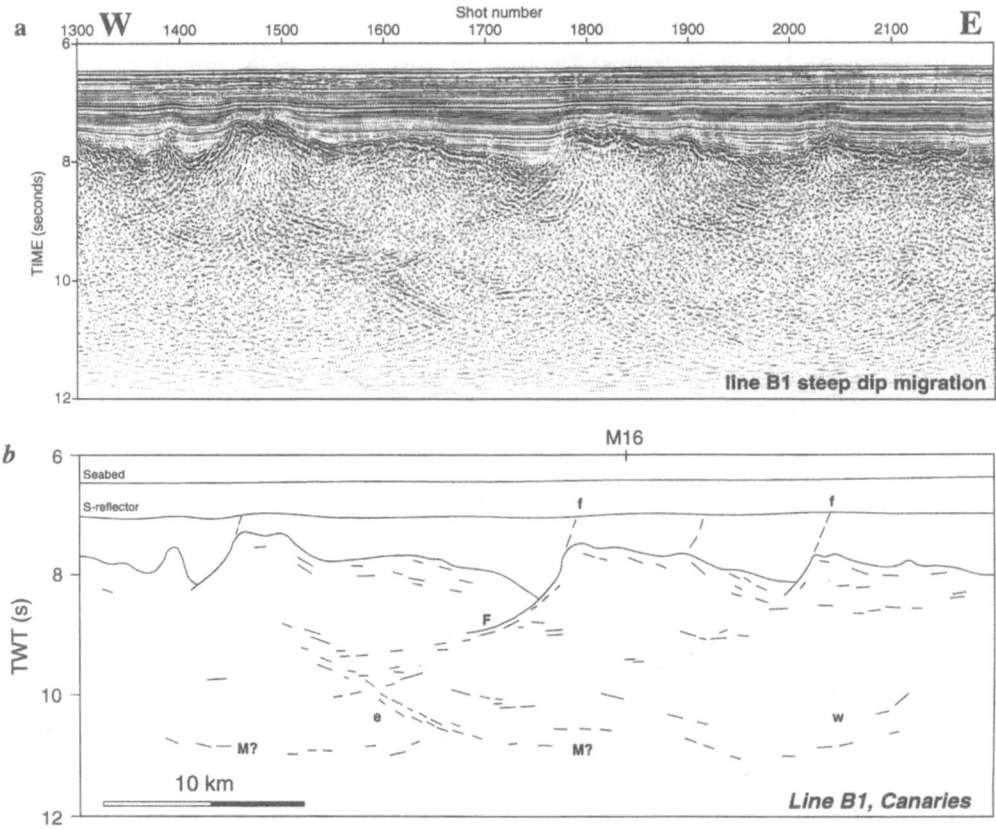

Figure 6.- (a) *Steep dip migrated section of line B1, performed with a finite-difference algorithm using a 2-D velocity field. Shot point interval is 50 m.* (b) *Line-drawing interpretation of line B1. Labelling as in figure 4a. In addition, "e" and "w" show eastward and westward banded packages of dipping reflections.*

The central basement ridge (sp 1750-1850, Fig. 6a) is clearly associated with a steeply dipping event (**F**, Figs. 6a and 6b) which occurs in two parallel reflections facing west and continuing downward into the crust as a steep linear reflection terminating in a bow tie at the midcrust (9.0-9.2 s). In order to determine the cause of this event we performed several migrations using a range of possible velocities. The result of this test suggests that this event could be related to a fault plane, as shown in the depth converted section, which flatten-outs at about 10 km depth (Figure 7). The lower crust displays a scarce reflectivity on the older side with a more reflective character observed at the younger side of the profile. The other main dipping event observed on this profile shows an opposite strike (i.e eastwards), and is characterised by packages of banded dipping events located below the two basement ridges (**e**, Figs. 6a and 6b). It can be traced from mid-crustal levels down to the upper mantle (11 s), just cross-cutting a series of short and discontinuous horizontal reflections that are tentatively associated with the Moho.

The absence of wide-angle data in the area makes it difficult to constrain the location of the Moho boundary. The similarity of the Cape Verde and Canary regions, in terms of tectonics and age of basement structures, facilitates the extrapolation of smooth velocity information from the former region. Under this assumption we tentatively correlate the Moho discontinuity with a series of short and discontinuous reflections, mostly observed on the older side of the profile, at a depth of 10.2-10.4 s, where the basement topography is smoother (**M**, Fig. 6b).

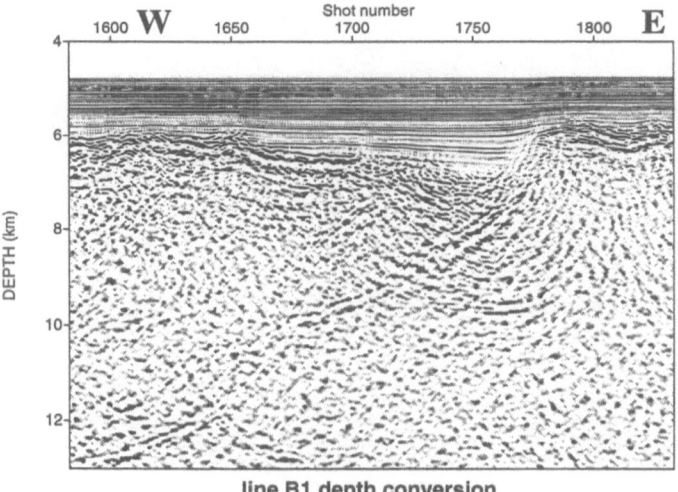

line B1 depth conversion

Figure 7.- *Depth converted section of a part of line B1, achieved with the same velocity function as for the migration. The fault offsetting the top of the igneous crust (> 1 km) can be extended down 10 km.*

5.3. Blake Spur (Line 711)

The seafloor in this profile is at about 7.0 s increasing slightly westward (Figure 8a). The sedimentary cover is characterised by a series of high amplitude sub-horizontal reflections. From these reflections two main groups can be distinguished. The shallower, at a depth of 400-500 ms beneath the seafloor, is made up of multicyclic sub-parallel reflections with a remarkable continuity and no noticeable disruptions. The second one is found at 700-750 ms beneath the seafloor and shows a similar pattern except that it is disrupted by faulting within the sediments (**f**, Fig. 8b) and at a basement scarp by onlap truncation. This probably indicates tectonic activity post-ridge formation (Fig. 9). The basement is blocky in character mainly on the younger side of the profile where a series of small basement offsets are observed. In the central part of the profile a basement ridge of 3 km length and 200 ms offset is observed (**R**, Figs. 8a and 8b). At the eastern edge of the basement steps a structure like a half-graben can be observed in the migrated sections (Figs. 8a and 8b), which is filled with westward tilted sediments (Fig. 9). This observation is associated with some eastward dipping events that can be tentatively traced down into the lower crust. This

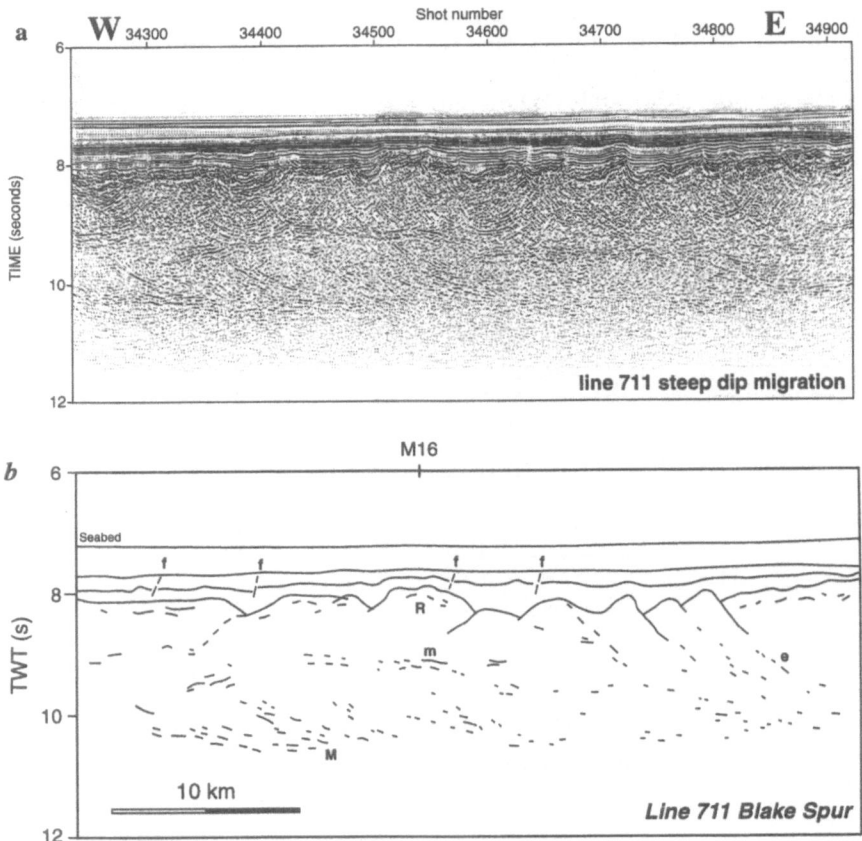

Figure 8 .- *(a) Steep dip migrated section of line 711, performed with a finite-difference algorithm using a 2-D velocity field. Shot point interval is 50 m. (b) Line-drawing interpretation of section line 711. Labels as in previous figures. "R" refers to the central ridge and, "m" is interpreted as magmatic reflectivity beneath it.*

configuration suggests that these basement highs could be associated with extensional faults. The process could be related to the rotation of fault blocks away from the paleo-ridge axis. This tilting is indicated by a failure of the crust which attain a depth of 8 km, as

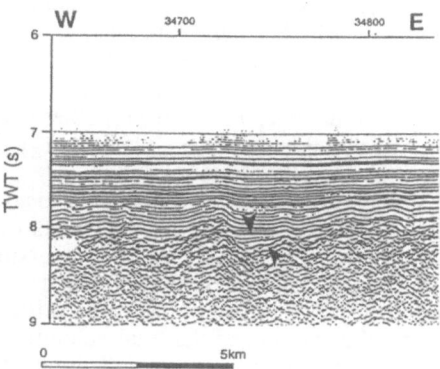

Figure 9.- *Detailed image of part of line 711, showing the syn-sedimentation activity. The arrows indicate the change of thickness between these two sedimentary facies.*

reported in earthquake studies at the ridge crest (Toomey et al., 1988). Moreover, syn-rift sedimentation can be inferred from the shape of the sedimentary wedges (Figure 9). It seems that on the American plate profile the basement blocks are more closely spaced (3 km) than on the African profiles (5 km). If these basement offsets correspond to volcanic ridges (Banda et al., 1992), partitioning within a spreading segment could be accommodated by differential asymmetric spreading between segments, producing a distinct tectonic style away from the rift valley.

In the upper crust sub-horizontal or gently dipping events (5°) with small lateral coherence are observed. At the base of the upper crust low coherent energy is visible, except below a basement ridge in the central part of the profile (**R**, Fig. 8a). At about 9.1-9.3 s (**m**, Figs 8a and 8b) a pair of parallel banded reflections 1-2 km long can be traced for about 5 km, with an approximate depth between 2-3 km, calculated from velocities obtained nearby (Minshull et al., 1991 and 1993). This event could be interpreted as primary magmatic in origin produced during a possible high melt production period. A hydrothermal origin for this event cannot be completely ruled out, a robust hydrothermal front having been reported to occur at the spreading axis down to 3 km (White et al., 1990). It is also possible that these short events are caused by some scattered energy from the near dipping reflections. From the depth and the attitude of these short reflections, which are located at the boundary between upper-crust and mid-crust, we prefer a magmatic origin consistent with episodic igneous intrusion (Henstock et al., 1993).

The lower crust on the Blake Spur profiles in contrast to the African profiles shows a considerable variety of events. Most of them are eastward dipping and end on contact with a relatively pronounced sub-horizontal package of short and discontinuous reflections at a mean travel-time depth of 10.5 s (**M**, Figs. 8a and 8b). Some of these reflections appear to be cut by banded packages of dipping reflections, mostly beneath the ridge. These later reflections have been correlated by Minshull et al (1991) with the Moho discontinuity based on wide-angle data.

6. Discussion and Conclusions

Accurate imaging from seismic reflection data is highly dependent on acquisition and processing. This is especially true for oceanic abyssal plain data where the targets are remote, typically more than 4500 m deep, and where the velocity contrasts between pelagic sediments and oceanic basement rocks are high. Changes in the p-wave velocity from about 3 km/s to more than 4.5 km/s rapidly rising to over 6 km/s have been reported in many oceanic areas. The three data-sets used in this study satisfy the requirements of large sources rich in low frequencies for imaging deep structures together with a relatively wide tuned seismic array to avoid as far as possible out-of-the plane energy. In addition, a long receiver cable, was used to enhance the signal-to-noise ratio.

Migration is sensitive to spatial smoothing of the data which leads to an underestimation of the correct velocity field. Hence, to minimise the spurious effects of the processing, a similar post-stack sequence was applied, followed by depth conversion. The strategy to define a reliable velocity function consisted in merging the stacking velocities determined from the sediments with a sub-basement velocity determined from refraction data (Minshull et al., 1991; White et al., 1994). We then smoothed the total velocity field to remove any abrupt discontinuities, and then depth-converted the images. The resulting images show a good resolution, particularly for the upper igneous crust, where we can better resolve fault structures.

The main goal of this study was to analyse whether or not any regular feature within the igneous crust could be deduced beneath the North Atlantic ocean during the Mesozoic when a reported change of spreading velocity occurred. In general we can say that from the three data-sets, no clear systematic change could be deduced, but some specific characteristics can be noted from the individual profiles. In the first place, the blocky character of the top of the igneous basement as defined by offsets appears to be associated with steeply dipping events which flatten out at upper, mid and lower-crustal levels on all the profiles. Most of these dipping reflections can be ascribed to normal faults on the basis of the clear basement offsets, which appear on a variety of different migration trials.

Another interesting observation is the sedimentary wedges located at the edges of the basements offsets. On the American plate profile, it is possible to interpret some syn-tectonic sedimentation, due to the tilted sedimentary facies, unlike the African plate, where we do not observe such features, although the hummocky character of the bottom sedimentary reflections persists on all the profiles.

The upper crust is characterised mostly by flat or gently dipping reflectivity and is generally concentrated 1.5-2.0 km beneath the top of the igneous basement. The lateral continuation of these features is typically restricted to 1-2 km, although some bright sub-horizontal reflections have been imaged in the Blake Spur and Canary areas up to distances of 15-20 km (Banda et al, 1992; Morris et al., 1993). Two conspicuous sub-horizontal reflections located at about 2-2.5 km beneath a basement ridge on the American profile, have been tentatively associated with the layering produced at the contact between the upper and middle crust. This interpretation poses many problems. It is not easy to fit a slow spreading ratio with an increase of the melt supply, which is suggested by the length of this reflection. An alternative explanation is that deep hydrothermal circulation could account for the continuation and depth of such a reflection (White et al., 1990).

On the eastern edge of the American plate profile, a pair of dipping reflections can be tentatively continued down to the lower crust, where they seem to die at low angles. On both plates the majority of the dipping reflections dip towards the Cretaceous paleo-ridge, but some opposite dips are also observed, indicating that faulting plays an important role in the crustal accretion process at or near the mid-ocean ridges (Alexander, 1993). In places where some tie-lines are available the uppermost reflectivity agrees with isochron and flow lines. This is not always the case for mid-crustal and ridgeward dipping events, as shown by the closely spaced lines collected in the Cape Verde area (McBride et al., 1994b).

The mid and lower crusts are in general sparse in the African profiles. Only one weak coherent reflection, which strikes opposite to the trend of the normal faults, is imaged from the mid-crust down to the Moho. In contrast, the lower crust, on the American profile, displays a series of subhorizontal layered reflection packages, which can be followed along the whole profile. Finally, differences in the basement blocks between the two plates could be ascribed to differential assymetrical spreading between ridge segments, as reported in the Mid-Atlantic ridge (Grindlay et al., 1991)

Acknowledgements

This research was partially supported by the BBV Foundation which granted a three month stay of JJD at BIRPS (Cambridge), by a EC Science programme under the contract SC1CT92-0782 and, by a British-Spanish collaboration programme number 127 B.

References

Alexander, R.J., Harper, G.D. and Bowman, J.R., Oceanic faulting and fault-controlled subseafloor hydrothermal alteration in the sheeted dicke complex of Josephine ophiolite, J. Geophys. Res., 98, 9731-9759, 1993.

Banda, E., Ranero, C.R., Dañobeitia, J.J., and Rivero, A., Seismic boundaries of the estern Central Atlantic Mesozoic crust from multichannel seismic data, Geol. Soc. Am. Bull., 104, 1340-1349, 1992.

Claerbout, J.F., Imaging the Earth's Interior, Blackwell Scientific Publications, 1985.

Grindlay, N.R., Fox, P.J., and Vogt, P.R., Morphology and tectonics of the Mid-Atlantic ridge (25°-27°30'S) from seabeam and magnetic data, 97, 6983-7010, 1992.

Hayes, D.E., Pimm, A.C., Beckmann, J.P., Benson, W.E., Berger, W.H., Roth, P.H., Supko, P.R. and von Rad, U., Initial Reports of Deep Sea Drilling Project, vol XIV, Washington (U.S. Government Printing Office), pp. 157-215, 1972.

Hayes, D.E., and Rabinowitz, P.D., Mesozoic magnetic lineations and the magnetic quiet zone off northwest Africa, Earth Planet. Sci. Lett., 28, 105-115, 1975.

Henstock, T.J., Woods, A.W., and White, R.S., The accretion of oceanic crust by episodic sill intrusion, J. Geophys. Res., 98, 4143-4161, 1993.

Hobbs, R., and Snyder, D., Marine seismic sources used for deep seismic reflection profiling, First Break, 10, 417-426, 1992.

Klitgord, K.D., and Schouten, H., Plate kinematics of the Central Atlantic, in Vogt, P.R., and Tucholke, B.E., eds., The geology of North America, the western North Atlantic region, Volume M: Boulder, Colorado, Geophysical Society of America, pp 351-377, 1986.

McBride, J.H., Henstock, T.J., White, R.S., and Hobbs, R.W., Seismic reflection profiling in deepwater: avoiding spurious reflectivity at lower-crustal and upper-mantle traveltimes, Tectonophysics, 232, 425-453, 1994a.

McBride, J.H., White, R.S., Henstock, T.J., and Hobbs, R.W., Complex structure along a Mesozoic seafloor spreading ridge: BIRPS deep seismic reflection. Cape Verde Abyssal Plain, 119, 453-478, 1994b.

Minshull, T.A., White, R.S., Mutter, J.C., Buhl, P., Detrick, R.S., Williams, C.A., and Morris, E., Crustal structure at the Blake Spur Fracture Zone from expanding spread profiles, J. Geophys. Res., 96, 9955-9984, 1991.

Minshull, T.A., and Singh S.C., Shallow structure of oceanic crust in the Western North tlantic from seismic waveform inversion and modeling, J. Geophys. Res., 98, 1777-1792, 1993.

Morris, E., Detrick, R.S., Minshull, T.A., Mutter, J.C., White, R.S., Su, W., and Buhl P., Seismic Structure of Oceanic Crust in the Western North Atlantic, 98, J. Geophys. Res., 98, 13879-13903, 1993.

Mutter, J.C., Detrick, R.S., and North Atlantic Transect Study Group, Multichannel seismic evidence for anomaously thin crust at Blake Spur fracture zone, Geology, 12, 534-537, 1984.

Mutter, J.C., and Karson, J.A., Structural processes at slow-spreading ridges, Science, 257, 627-634, 1992.

NAT Study Group, North Atlantic Transect: A wide-aperture, two-ship multichannel seismic investigations of the ocean crust, J. Geophys. Res., 90, 10321-10341, 1985.

Parkes, G.E., Hatton, L., and Haugland, T., Marine source array directivity: a new wide airgun array system, First Break, 2, 9-15., 1984.

Roest, W.R., Dañobeitia, J.J., Verhoef, J., and Collette, B.J., Magnetic anomalies in the Canary Basin and the Mesozoic evolution of the Central North Atlantic, Mar. Geophys. Res., 14, 1-24, 1992.

Stolt, R.H., Migration by Fourier transform, Geophysics, 43, 23-48, 1978.

Sundvik, M., Larson, R.L., and Detrick, R.S., Rough-smooth basement boundary in the western North Atlantic basin: Evidence for seafloor-spreading origin, Geology, 12, 31-34, 1984.

Sundvik, M., and Larson, R.L., Seafloor spreading history of the western North Atlantic basin derived from Keathly Sequence and computer graphics, Tectonophysics., 155, 49-71, 1988.

Toomey, D.R., Soloman, S.C. Purdy, G.M., Microearthquakes beneath the median valley of the Mid-Atlantic Ridge near 23°N: Tomography and tectonics, J. Geophys. Res., 93, 9093-9112, 1988.

Verhoef, J., Collette, B.J., Dañobeitia, J.J., Roeser, H.A., and Roest, W.R., Magnetic anomalies off West Africa (20-38°N), Mar. Geophys. Res., 13, 81-103, 1991.

White, R.S., McBride, J.H., Henstock, T.J., and Hobbs, R.W., Internal structure of a spreading segment of Mesozoic oceanic crust, Geology, 22, 597-600, 1994.

White, R.S., Detrick, R.S., Mutter, J.C., Buhl, P., Minshull, T.A., and Morris, E., New seismic images of the crustal structure, Geology, 18, 462-465, 1990.

CONSTRAINTS ON RIFTING PROCESSES FROM REFRACTION AND DEEP-TOW MAGNETIC DATA: THE EXAMPLE OF THE GALICIA CONTINENTAL MARGIN (WEST IBERIA)

JEAN-CLAUDE SIBUET [1], VÉRONIQUE LOUVEL [1,2],
ROBERT B. WHITMARSH [3], ROBERT S. WHITE [4],
SUSAN J. HORSEFIELD [4,5], BERTRAND SICHLER [1],
PIERRE LÉON [6] and MAURICE RECQ [7]

[1] *Ifremer Centre de Brest, B.P. 70, 29280 Plouzané, France, e-mail jcsibuet@ifremer.fr*

[2] *EOPGS, 5 rue R.-Descartes, 67084 Strasbourg Cedex, France*

[3] *IOSDL, Wormley, Surrey GU8 5UB, U.K.*

[4] *University of Cambridge, Bullard Laboratories, Madingley Road, Cambridge, CB3 0EZ, U.K.*

[5] *now at Esso Exploration and Production UK Ltd, Leatherhead, KT22 8UY, U.K.*

[6] *Ifremer Centre de Toulon, Zone portuaire de Brégaillon, B.P. 330, 83507 La Seyne/Mer Cedex, France*

[7] *U.B.O., B.P. 809, 29285 Brest Cedex, France.*

ABSTRACT. The Galicia margin, which is characterised by the presence of a series of tilted fault blocks and underlying seismic reflectors well imaged by multichannel seismic (MCS) reflection profiles, is a classic example of a rifted continental margin. Since the early 1980's, several models have been proposed for the formation of this margin. All these models were based mainly on seismic imagery obtained from MCS data with age control provided by data collected during DSDP leg 47B and ODP leg 103. To understand fully the evolution of this margin, pertinent geophysical data are needed, especially to constrain seismic reflection interpretations. We discuss a new preliminary seismic refraction model and deep-tow magnetic data collected across the Galicia continental margin and show how they constrain the deep structure in this region and the precise location of the ocean-continent transition.

West of the peridotite ridge, the oceanic crust is reduced to a thickness of 5 km which increases to normal oceanic thickness some 10 - 15 km further west. The base of the crust consists of 7.4 - 7.5 km/s material interpreted as serpentinized peridotite. East of the peridotite ridge, the thinned continental crust overlies a 7.4 km/s transitional layer with a limited landward extent of 25 km. The 7.4 km/s material is interpreted as upper mantle peridotite serpentinized by sea-water percolating through the extremely thin crust disrupted by normal faults created during rifting. At the ocean-continent boundary a peridotite ridge has been drilled. Near-bottom magnetic anomalies recorded just above the sea-bottom show that oceanic crust abuts the peridotite ridge on its west side and that the magnetization of the peridotite ridge is similar to that of the adjacent thinned continental crust, that is about ten times smaller than the oceanic crust. Seismic velocities

E. Banda et al. (eds.), Rifted Ocean-Continent Boundaries, 197–217.

are consistent with the whole ridge being a serpentinized peridotite body. Drilling data, as well as the presence of a normal fault along its eastern flank, suggest that the ridge has been uplifted at the end of the rifting phase. As the extensional rate during rifting and the half spreading rate during the emplacement of early oceanic crust are both lower than 10 mm/yr, theoretical modelling supports the presence of serpentinized peridotite at the base of both the thinned continental and the oldest oceanic crusts, with only a limited generation of melt.

The S reflector is a detachment fault located east of the peridotite ridge which follows the top of the peridotite body over 20 km (which explains its highly reflective character on seismic reflection profiles), cuts through much of the lower continental crust and then merges with the top of the lower continental crust, which may represent the brittle-plastic interface at the time of rifting.

We favour a composite pure and simple shear model of formation of the Galicia margin and its conjugate margin in which a detachment surface (the S reflector), active since the beginning of rifting, crops out on the Canadian side and merges beneath the Galicia margin with the brittle-plastic interface. The peridotite ridge is emplaced during the late stages of rifting by mantle upwelling through the highly fractured, thin veil of continental crust. After the onset of spreading, peridotite ridges are left on both sides of the ocean. The lower portion of the thin oceanic crust immediately adjacent to the peridotite ridges consist of serpentinized peridotite and is the result of poor magma supply for several M.y. immediately after continental breakup.

1. Introduction

The Galicia margin is characterised by the presence of a series of tilted fault blocks and underlying seismic reflectors well imaged by multichannel seismic reflection (MCS) profiles. Since the early 1980's, several models have been proposed for the formation of this margin. First, Le Pichon and Sibuet (1981) suggested that the margin was formed by pure shear. Then the simple shear model (Wernicke, 1985) was pushed to an extreme by Boillot et al. (1987a), who proposed that a peridotite ridge (PR) which appears to mark the landward edge of oceanic crust, was emplaced by tectonic denudation of the upper mantle. Finally, Sibuet (1992b) suggested that the Galicia margin and its conjugate margin were formed by pure shear except for the upper portion of the crust where a detachment fault, coincident with a strong laterally continuous intracrustal reflector (the S reflector), was active and merged eastward beneath the upper Galicia margin with the brittle-plastic limit. All these models were based mainly on the interpretation of MCS data, with age control provided by data collected during Deep Sea Drilling Project (DSDP) leg 47B (Sibuet and Ryan, 1979) and Ocean Drilling Program (ODP) leg 103 (Boillot et al., 1988).

In this paper, we reinterpret MCS data in the light of preliminary seismic refraction results of the Réframarge cruise (1987) and of the modelling of sea-surface and near-bottom magnetic data collected during the Fluigal cruise (1993).

2. Seismic Reflection and Refraction data

The geophysical dataset used here was collected in April 1987 aboard the R/V Le Suroît as part of the Réframarge cruise (Figure 1). Line 6 was positioned to coincide with MCS profile GP 101, acquired and processed by the Institut Français du Pétrole (Mauffret and Montadert, 1987) (Figure 2, courtesy L. Montadert). Three digital ocean bottom seismometers (DOBS), operated by the

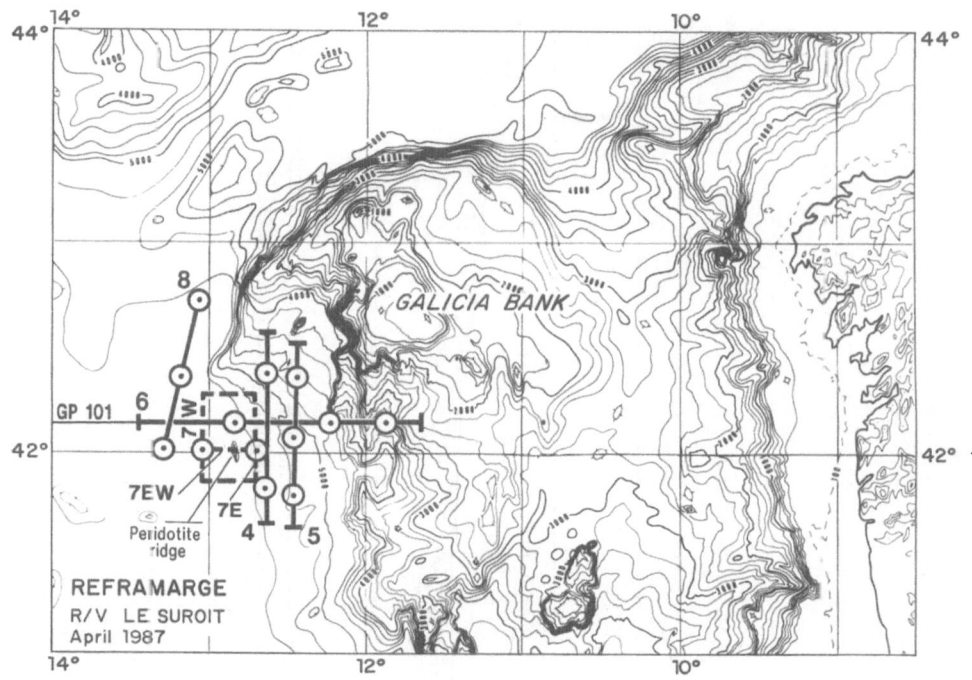

Figure 1. The east-west seismic refraction Line 6 was shot during the Réframarge cruise along the MCS profile GP 101. Circles with dots indicate locations of DOBS along profiles. Bathymetric map from Lallemand and Sibuet (1986).

Institute of Oceanographic Sciences Deacon Laboratory (IOSDL) (Kirk et al., 1982) and each with an hydrophone and an internal three-component geophone package were deployed on this line. The source was an array of eight 16 litre (1000 cubic inch) airguns fired every two minutes, giving a shot spacing of about 300 m. The MCS profile GP 101 was used to provide control on the sediment structure and the depth-to-basement along Line 6.

The DOBS data were travel-time and amplitude modelled at Bullard Laboratories (Cambridge University), IOSDL and Université de Bretagne Occidentale (UBO) using a synthetic seismogram computer program based on Maslov asymptotic ray theory initially developed by Chapman and Drummond (1982). North-south Lines 4, 5 and 8 (Figure 1) were modelled first to provide controls at the intersections with Line 6 (Horsefield, 1992). Additional constraints are provided by short north-south Lines 7E and 7W (Figure 1) located on each side of the peridotite ridge (Whitmarsh et al., 1993). Preliminary modelling was also performed on east-west Line 7EW which cuts across the peridotite ridge (Recq et al., 1991). Modelling of Lines 6 and 8 is presently under revision. The model of Line 6 presented here is still preliminary (Figure 3). It has been

200

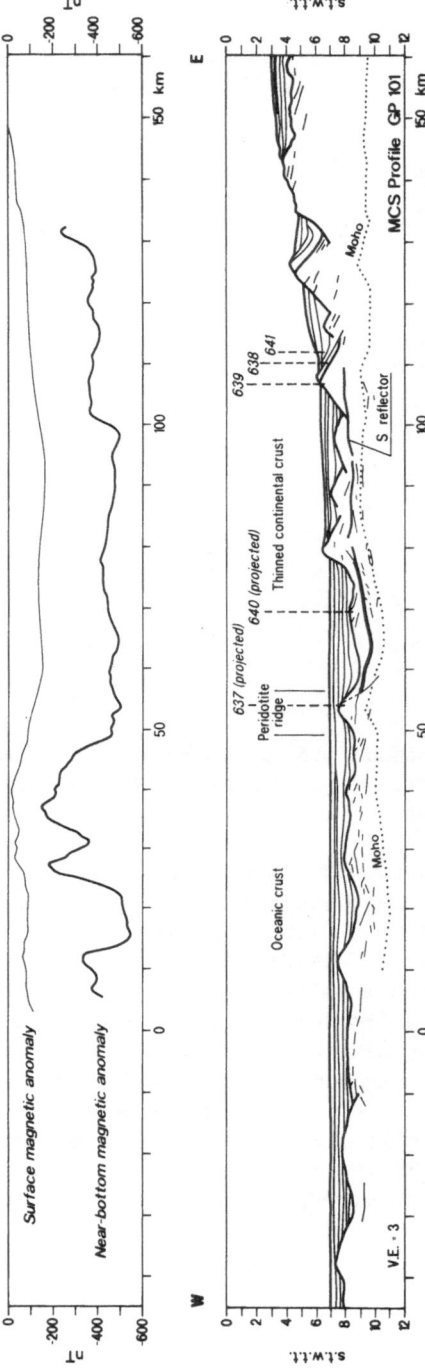

Figure 2. Line drawing of the MCS reflection profile GP 101 along which seismic refraction Line 6 was shot. Location of Moho is deduced from Réframarge seismic refraction data. Sea-surface and near-bottom magnetic anomalies from Fluigal (1993) and Seagal (1981) cruises.

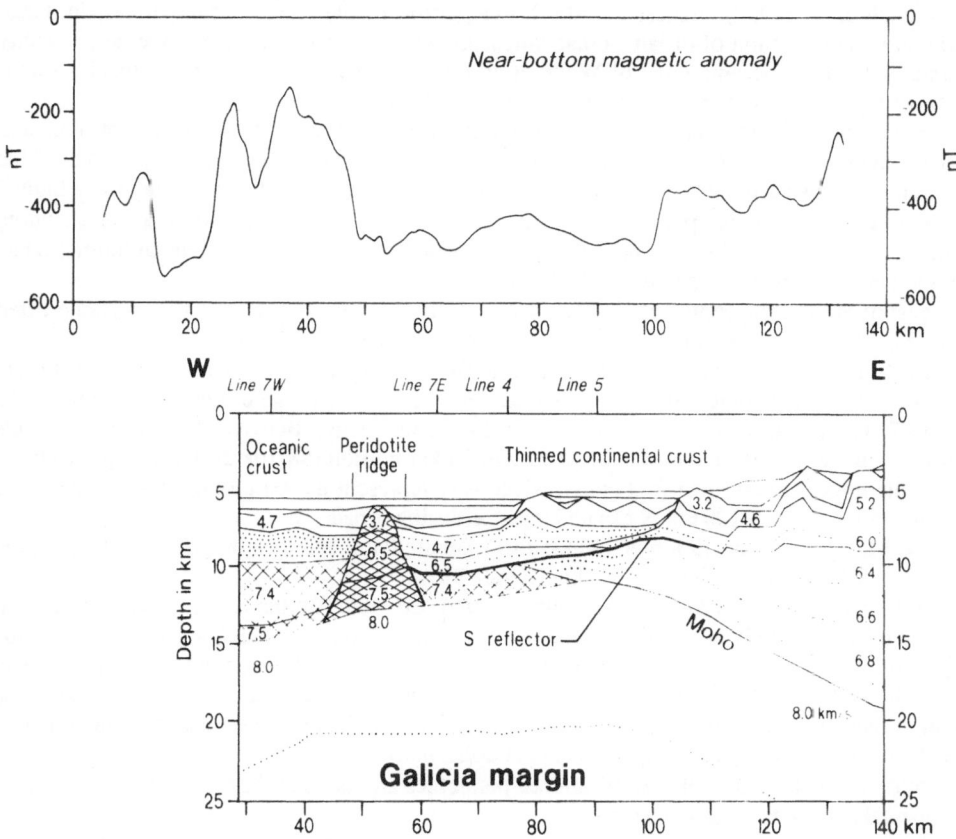

Figure 3. Preliminary crustal section of the Galicia margin obtained from seismic refraction data. Lines 7W, 7E, 4 and 5 shot parallel to the margin intersect Line 6 (see locations in Figure 1). The S reflector identified on the MCS profile GP 101 has been depth converted using refraction velocities. Dotted areas are serpentinized peridotite. Near-bottom magnetic anomalies from the Fluigal cruise (1993).

constructed from Horsefield (1992) east of the peridotite ridge, Whitmarsh et al. (1995b) west of the peridotite ridge, from crossing Lines 4, 5 (Horsefield, 1992), 7E, 7W (Whitmarsh et al., 1993) and extrapolated from Line 7EW (Recq et al., 1991). The main features of the model are:

- west of the peridotite ridge, the oceanic crust is reduced to a thickness of 5 km which increases to normal oceanic thickness at 34 km, some 10-15 km further west of the PR. The base of the crust consists of 7.4 - 7.5 km/s material underlain by 8.0 km/s. Although the oceanic crust is thinner than normal oceanic crust just west of the peridotite ridge, between 34 and 43 km,

Whitmarsh et al. (1993) suggest that the lower portion of the crust consists of serpentinized peridotite. This portion of oceanic crust, whose thickness does not lie within the new velocity bounds for normal oceanic crust of White et al. (1992), would be primarily the result of poor magma supply for several M.y. immediately after continental breakup.

- east of the peridotite ridge, Lines 7E and 4 show a thin crustal structure which is attributed to thinned continental crust overlying a transitional layer with velocities of 7.4 km/s and larger. Line 5 does not display such lower crustal velocities which implies that this 7.4 km/s body has a limited landward extent. It is interpreted as upper mantle peridotite serpentinized by sea-water coming through the extremely thinned continental crust disrupted by several generations of normal faults (Le Pichon et al., 1983; Whitmarsh et al., 1993).

- east of 90 km, the continental crust has a velocity structure typical of a non-volcanic rifted margin in which the upper and lower crusts are thinned by the same extent.

- the S reflector identified on the MCS profile GP 101 (Figure 2) was depth converted (Figure 3) using computed seismic refraction velocities. From the eastern edge of the peridotite ridge (58 km) to 77 km, S follows the top of the presumed peridotite body. Between 77 and 88 km, S cuts across the lower continental crust. In its eastern 20 km portion (88-108 km), S merges with the top of the lower crustal layer (6.0 km/s), which may represent the brittle-plastic interface at the time of rifting. East of 108 km, S can no longer be distinguished.

- the Moho rises from 19 km at 140 km to 11 km at 90 km and then increases its depth in the direction of the peridotite ridge to 12.5 km (Figure 3).

- drilling at ODP Site 637 recovered serpentinized peridotite (Boillot et al., 1987b). A detailed seismic refraction study across the peridotite ridge (Line 7EW) shows that velocities vary in the upper part of the peridotite ridge from 3.7 to 6.5 km/s. The 6.5 km/s velocity found 3 km beneath the seafloor gradually increases to 7.5 km/s at 5 km depth (Recq et al., 1991). Therefore, the steadily increasing velocity is consistent with decreasing serpentinization and it seems possible that the whole ridge may be a serpentinized peridotite body.

- refraction data suggest that serpentinized peridotites are found at the base of both the thinned continental and the thin oceanic crusts.

The velocity-depth model of Line 6 was converted to two-way travel-time (twtt) to enable a direct comparison to be made with the MCS profile GP 101 (Figure 2). Lower crustal reflections are identified below S. The deep reflections correspond to the Moho whose position is shown in twtt from MCS and refraction data west of 108 km and from refraction data alone east of 108 km (Figure 2). The Moho below the thinned continental crust is at 9-10 sec twtt with a slight increase from 9 sec twtt between 125 and 90 km to 10 sec twtt at 60 km.

On MCS profile GP101, S is represented by a very strong continuous reflection between 58 and 77 km. Elsewhere, S reflections are generally faint and discontinuous. From west to east we interpret S in the following way:

- east of the peridotite ridge (58-77 km) S is the contact between the thinned continental crust and the lower crustal serpentinized peridotite body. S corresponds to an abrupt refraction velocity contrast of 0.9 km/s (Figure 3). This is confirmed by comparisons of the S reflector waveform with that of the seafloor. Reston et al. (1994), using a detailed velocity model obtained through depth-focusing error analysis, have demonstrated that the western portion of the S reflector coincident with the 58-77 km strong seismic reflection is a continuous feature which corresponds to a single reflector with a positive polarity. They show that this brightest portion of the S reflector represents a sharp increase in acoustic impedance (reflection coefficient of ca. 0.2) though a velocity contrast of 0.9 km/s will give only a 0.07 reflection coefficient. Therefore, from MCS data analysis, S is probably there a single interface with a high velocity contrast.

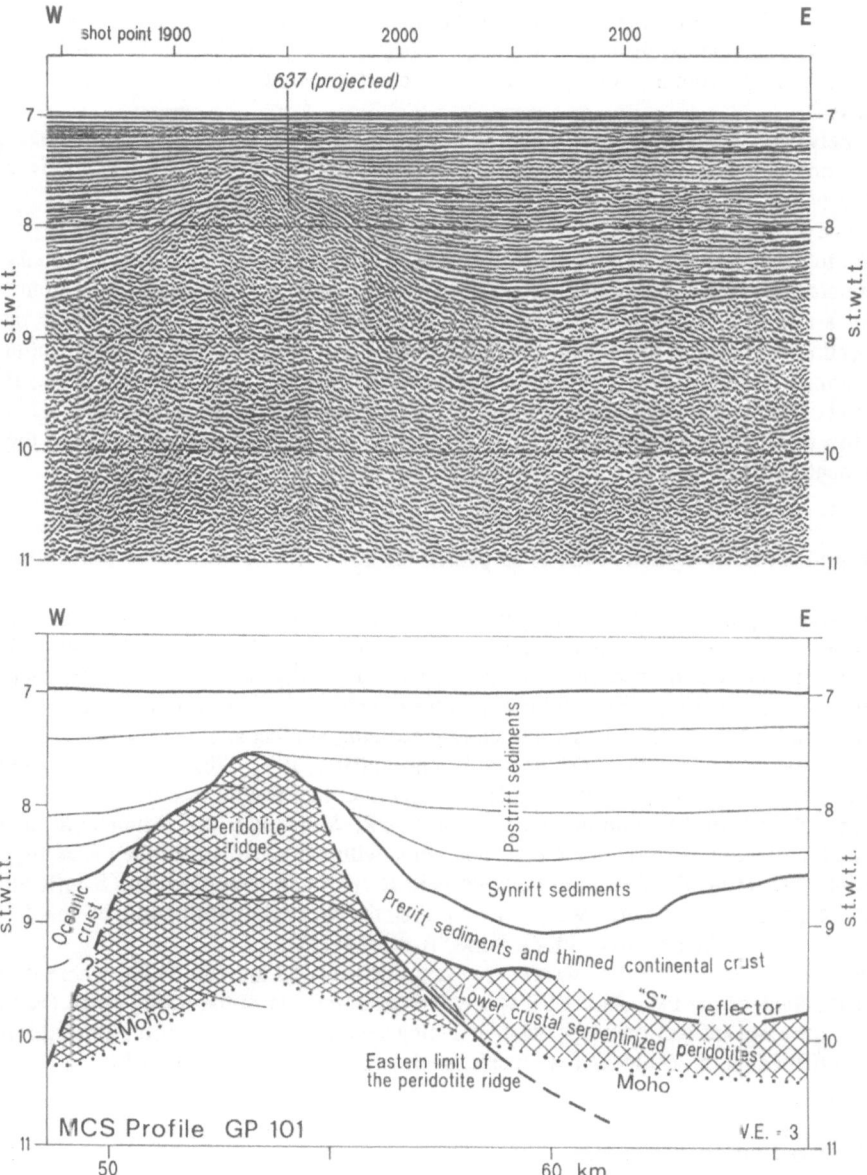

Figure 4. Detail of the GP 101 migrated section (courtesy L. Montadert) in the area of the peridotite ridge and corresponding interpretation. The S reflector intersects the eastern edge of the peridotite ridge which can be considered as a normal fault down to 11 sec twtt, active at the end of rifting. The western limit of the peridotite ridge is deduced from modelling of the near-bottom magnetic anomalies. Location of Moho from refraction results.

- between 77 and 88 km, S is represented by a series of discontinuous and less energetic reflections. These reflections cut across part of the lower portion of the thinned continental crust. Whatever the horizontal shear motion along S, since S is sub-parallel to the isovelocity lines in the seismic model (Figure 3), only a weak velocity contrast, and therefore little reflected energy, is expected.

- between 88 and 108 km, S is characterised by even fainter and more discontinuous reflections. S is here coincident with the top of the lower crust (Figure 3) which, following Sibuet (1992a and b), we interpret as the brittle-plastic interface at the time of rifting.

- east of 108 km, S is absent.

An interesting observation raised by the geometry of the S detachment fault within the thinned continental crust is that S cuts through much of the lower crust from top to bottom and then follows the roof of the peridotite body. One explanation could be linked to the process of serpentinization, which favours the concentration of shear where the degree of serpentinization is a maximum because this is where most "lubricant" minerals are present (Sibuet et al., 1995). At the end of rifting, the S detachment surface could have propagated from an intra-crustal location to the top of the serpentinized peridotite body when it was created at the base of the thinned continental crust.

3. Detailed interpretation of the MCS profile GP 101 in the vicinity of the peridotite ridge

The detailed portion of the MCS profile GP 101 located in the vicinity of the PR (Figure 4) is used here to discuss the lateral extent of the serpentinized peridotite body. The western limit of the PR does not clearly appear on reflection data but will be discussed in a later section. The eastern limit of the ridge can be followed on the migrated section down to 11 sec twtt, within the upper mantle. This feature is not a seismic processing artefact as it has also been identified on several other seismic sections (Shipboard Scientific Party, 1994; Sibuet et al., 1995) and by Tim Reston (personal communication, 1994). We interpret this feature as a normal fault which was active at the end of rifting during the uplift of the PR. S intersects this feature at an angle of 30° on the time section (Figure 4) and does not merge with the upper part of the PR as suggested by Boillot et al. (1987a). Attention must be paid to the interpretation of the S reflector and its western prolongation; here, we point out that the western termination of the S reflector corresponds to its intersection with the buried flank of the PR.

Figure 3 shows that S dips westwards except between 58 km and 64 km where it climbs up over the flank of the PR. As the PR was emplaced at the end of the rifting phase (Féraud et al., 1988), the eastward portion of the crust, including the S reflector, the prerift sediments and part of the synrift sediments, could have been uplifted there during the ascent of the PR.

4. Near-bottom magnetic data

We have installed a 3-component magnetometer on the Ifremer deep-tow sidescan sonar named SAR. The sensor is located at the end of a 1.3-metre long glass-epoxy container fixed behind the SAR in order to reduce its magnetic effects (Figure 5). It is composed of three double axis fluxgates provided by the Sinomag company (Grenoble) and mounted at 90° in a permaglass block. The precision of the mechanical positioning is ± 0.3°. With the electronic angular compensation device, this value decreases to ± 0.001°. An analog computer gives the amplitude

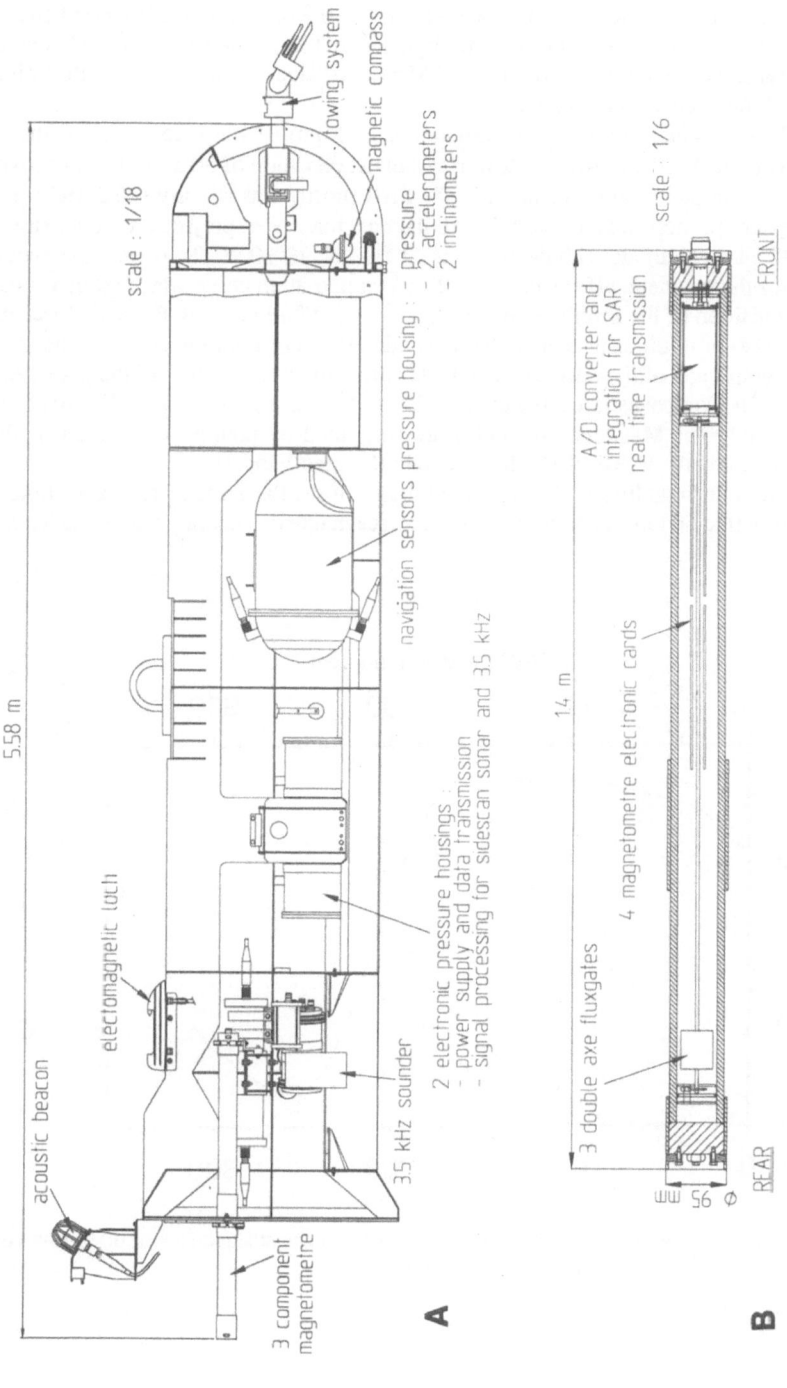

Figure 5. Scheme of the SAR (A) and of the pressure housing which contains the 3-component magnetometer. This device (B) is installed behind the fish with sensors ouside the fish in order to reduce the magnetic effects of the SAR.

of the total magnetic field from the three recorded orthogonal components of the magnetic field with an accuracy of 5.10^{-5} oersted (5 nT). The noise is $0.3.10^{-5}$ oersted in the 0-1 kHz band. The three analog components and the total field are digitised and integrated into the SAR real time system of data transmission to the ship (Figure 5). Magnetic values are recorded together with the SAR navigation parameters every 1.5 second.

During the Fluigal cruise (July 1993), a 133-km long SAR profile was recorded along the MCS profile GP 101 (Figure 1). The SAR was towed at a mean elevation of 80 ± 30 metres above the sea-bottom. Owing to bad weather conditions, the SAR profile was slightly offset from profile GP101 in its western portion, west of the PR. All the data have been projected onto an east-west line located at 42°08.5'N latitude with the origin at 42°08.5'N; 13°30'W. To correct the measured magnetic field for the magnetic effects of the SAR, a compensation curve was established before and after the acquisition of the profile by describing a navigation loop with the SAR located at a water depth of 1000 m in order to reduce the magnetic effects of both the ship and the sea bed (Figure 6). The amplitude of the deep-tow magnetic correction as a function of the SAR heading is about 950 nT. The compensation curve differs slightly from a sinusoid, especially for headings between 020° and 160°. Magnetic anomalies are computed by taking into account both the compensation curve and the IGRF 1990 reference field (Figures 2 and 3).

Due to bad weather conditions during the Fluigal cruise, the surface magnetic field was recorded only from 0 to 50 km. Therefore, the sea-surface magnetic anomaly was extracted from

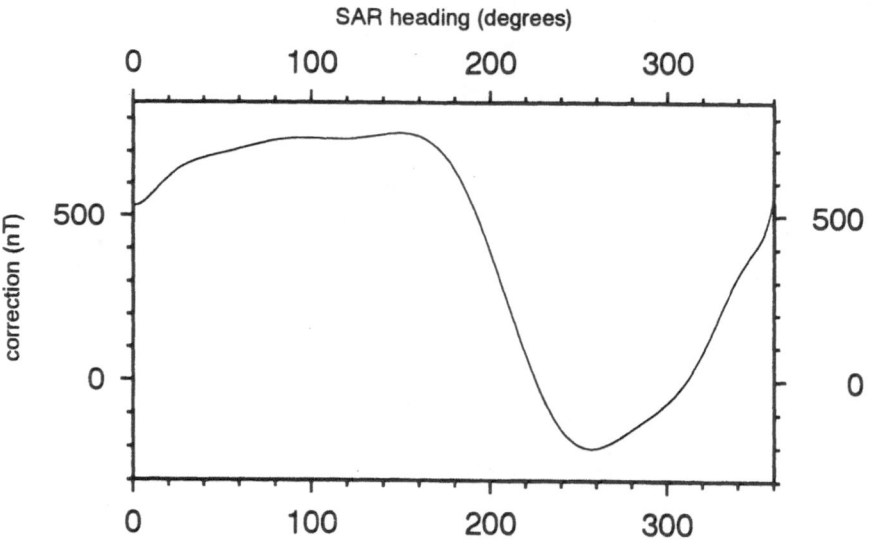

Figure 6. Compensation curve as a function of the azimuth of the sidescan sonar deep-tow vehicle on which the 3-component magnetometer is mounted.

Fluigal and Seagal (January 1981) cruise data during which a closely spaced series of geophysical profiles was acquired. The maximum amplitude of the near-bottom magnetic anomaly along the reference profile is about 400 nT compared to 150 nT for the sea-surface magnetic anomaly (Figure 2). Above the thinned continental crust and up to the PR, the near-bottom magnetic anomaly is generally low except for perturbations which can be correlated with the tilted fault blocks identified on the seismic reflection profile (Figure 2), which means that the magnetic source layer follows or mimics the basement topography. The large magnetic anomaly, located between 22 km and 50 km is offset oceanward with respect to the location of the PR and is not associated with the PR. The PR itself seems to be associated with the small 50 nT near-bottom magnetic anomaly located between 50 km and 54 km.

Natural remanant magnetization (nrm) values of serpentinized peridotite samples are generally high but are very scattered. For mid-Atlantic ridge serpentinized peridotites, the nrm values range from 0 to 10 A/m with an average value of 3.5 A/m (Dunlop and Prévot, 1982; Nazarova, 1994). In the Mariana and Izu-Bonin regions, the nrm intensity of the serpentinized peridotites ranges from 0 to 0.6 A/m (Stokking et al., 1992). Though no magnetization measurements were performed on samples collected on the PR (site 637, ODP leg 103, (Boillot et al., 1988)), average nrm measurements on serpentinized peridotites drilled at sites 897 and 899 (ODP Leg 149 Shipboard Scientific Party, 1993) are 0.3 and 1.5 A/m, respectively (Whitmarsh et al., 1995a; Zhao, 1995). During the process of serpentinization, magnetite is a secondary product of serpentinization and peridotites acquire a crystallisation remanence (crm), presumably directly along the prevailing geomagnetic field at the time serpentinization occurs (Dunlop and Prévot, 1982). For serpentinized peridotites, this crm is the only nrm, the primary chrome spinels being paramagnetic. The production of secondary magnetite is not correlated solely with the degree of serpentinization. It could also be linked to the abundance and composition of the parent minerals (Shive et al., 1988). Therefore, simple linear relationships between density (empirically correlated with seismic velocities and degree of serpentinization) and magnetic properties should not be expected.

Seismic velocities vary substantially within the peridotite body. This suggests that the peridotite rocks are serpentinized, although no clear indications of the variation of the degree of serpentinization with depth can be established. Regardless of the magnetization of the peridotite body, in a first trial using a 2-D model of the source layer we have assumed that the PR was not magnetized. A 1-km thick body of magnetized oceanic crust abutting the western PR flank and following the basement topography with a normal magnetization of 5 A/m roughly explains the near-bottom and surface magnetic anomalies (Figure 7). Anomalies were calculated at the points of magnetic measurements, i.e. at the sea-surface for the surface ship magnetic anomaly and at the measuring depth for the near-bottom magnetic anomaly (about 80 m above the sea-bottom). Since the oceanic domain located west of the Galicia margin was formed during the Cretaceous constant polarity interval, no magnetic reversals are expected. Figure 7 shows that near-bottom magnetic anomalies are mostly explained by depth variations in the oceanic basement topography. A better fit is obtained by changing the thickness of the magnetized layer (Figure 7). Such variations in thickness could correspond to alternating magmatic and amagmatic cycles, but this remains to be confirmed by the final refraction modelling. The 4-km lateral offset between the computed (thin continuous line) and observed (thick line) anomalies is probably due to the difference in location of the seismic profile GP 101 and the end of the SAR profile, west of the PR.

In the thinned continental domain, the regional near-bottom magnetic anomaly is mostly explained by the edge effect of the oceanic magnetic body (Figure 7). However, the shape of the small near-bottom magnetic wavelengths (10 to 20 km) seems to be correlated with the shape of

208

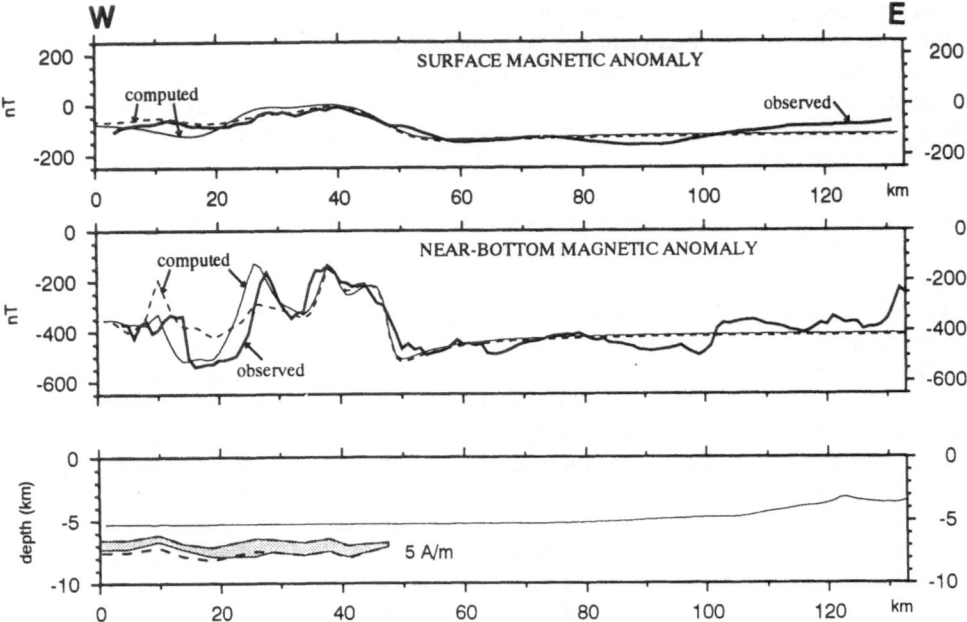

Figure 7. Modelling of the near-bottom and sea-surface magnetic anomalies with only a body of oceanic crust abutting against the peridotite ridge, a) with a constant 1-km thick uniformly magnetized oceanic crust in the direction of today's field (grey body extended to the dashed line, dashed line for the computed anomaly), b) with a variable thickness of oceanic crust (grey body and thin continuous line for the computed anomaly).

the tilted fault blocks (Figure 2) and the computed sea-surface anomaly progressively departs from the regional anomaly in a landward direction. We therefore introduced a body of magnetized continental crust including the PR (0.5 A/m, ten times less than the oceanic crust), which thickens landwards. A similar thickening was established by the inversion of Goban Spur magnetic data, assuming a constant magnetization (Louvel et al., 1995). Sediments were assumed to be effectively non-magnetized. Figure 8 shows a better regional correlation between observed and computed surface magnetic anomalies. The amplitudes of the small near-bottom magnetic anomalies roughly correspond to the computed anomalies. Discrepancies could be attributed to the fact that recording is too close to the magnetic sources or that the geometry adopted for the surface of the tilted blocks is not correctly estimated for the lower part of the margin, as early Cretaceous synrift sediments of very low magnetization were deformed at a late stage during the most recent episode of rifting.

As no significant near-bottom magnetic anomaly corresponds to the PR, its magnetization is probably weak (close to the magnetization of the thinned continental crust) which was unexpected. Thus, the magnetic modelling demonstrates that the oceanic crust abuts the western limit of the

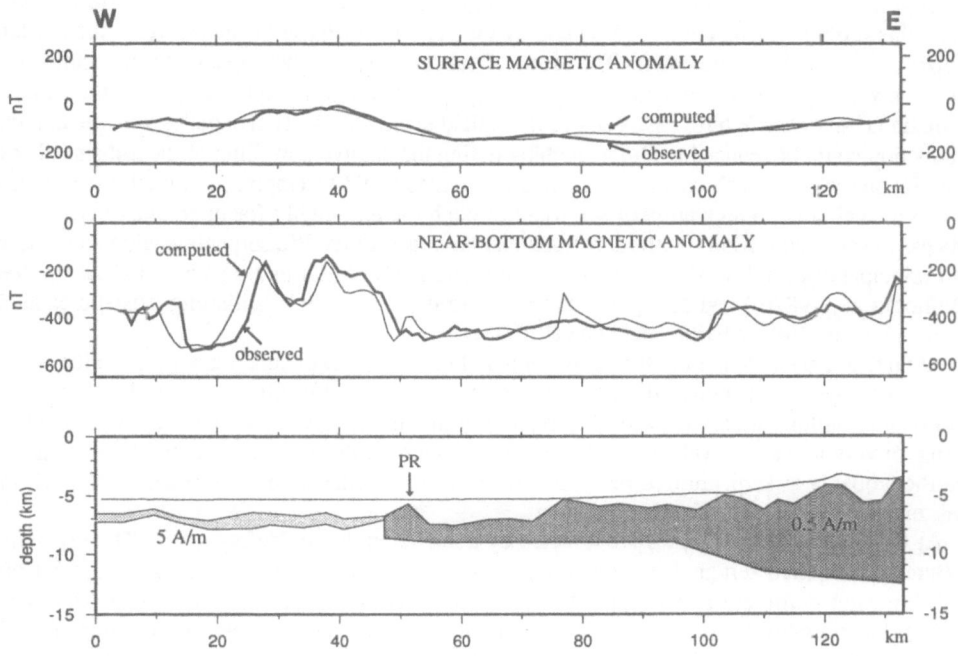

Figure 8. Modelling of the near-bottom and sea-surface magnetic anomalies with a 5 A/m body of oceanic crust of variable thickness abutting against the peridotite ridge, a 0.5 A/m body of thinned continental crust and non-magnetized sediments. Magnetization is uniform in the direction of today's field.

PR. No significant magnetic contrast exists between the PR and the adjacent thinned continental crust. Whitmarsh et al. (1995a) came to the same conclusion from modelling deep-tow profiles in the Iberian Abyssal Plain.

5. Discussion

The seismic structure across the Galicia margin shows that the continental crust thins progressively oceanwards, reaching a minimum thickness of 3 to 4 km above the 7.4 km/s serpentinite body (58-77 km, Figure 3). Continental crust as thin as 3 to 4 km implies stretching factors of 7.5 to 10. For a low asthenospheric potential temperature at the time of rifting and a long duration of the rifting episode (25 M.y., from Berriasian (140 M.y.) to latest Aptian (115 M.y.) (Boillot et al., 1988)), the expected volume of melt generated by adiabatic decompression (Bown and White, 1994, 1995b) is negligible regardless of the amount of crustal thinning. As the adjacent oceanic crust is thinner (5 km) than the normal 7-km thickness (White et al., 1992), mantle potential temperatures were 50-100°C cooler than the temperature reached during normal seafloor spreading (Horsefield et al., 1994; Whitmarsh et al., 1993). Such an indication of cold underlying mantle at the time of break-up is suggested by the low measured heat flow values along this profile (Louden et al., 1991) and by the very deep adjacent oceanic crust (about 1 km deeper than normal oceanic crust). The ascent of subcontinental asthenosphere during continental break-up was so slow that enough heat was lost by conduction to inhibit or largely diminish melt formation (Bown and White, 1995a and b; Horsefield et al., 1994). This explains why the magma production could be reduced to almost nothing during the passive upwelling of the asthenospheric material and why, in such conditions, mantle peridotites could be emplaced through the thin veil of continental crust. Percolation of sea-water through the thin highly fractured crust induced the process of serpentinization. In the constructional processes of the PR, serpentinization could have two principal effects: lowering the density of ultramafic rocks (2.4 to 2.9 g/cm^3 at ODP site 637 (Boillot et al., 1987b)) and acting as a lubricant (formation of talc) facilitating isostatic balance (Bougault et al., 1993; Charlou and Donval, 1993).

At ODP site 637, the peridotite underwent only limited melt extraction (about 7%, which is less than other abyssal peridotites (Evans and Girardeau, 1988)). For Girardeau et al. (1988), this observation outlines a very cold thermal regime of the lithosphere for which the hot mantle was rising, or very low ascent velocities. Therefore, we interpret the 7.4 km/s material located east of the PR probably as serpentinized peridotite mixed with a limited amount of frozen melt (<10%). This excludes a true underplating process.

At ODP Site 637, peridotites are affected by a ductile mylonitic deformation. The mylonitic texture was acquired at high, but decreasing, temperatures (1000 to 860°C) at rather shallow depth (< 7 km), not definitely in a normal ductile simple shear zone, as suggested by Boillot et al. (1994), but possibly from the edge of a rising mantle dome as suggested by Girardeau et al. (1988). Furthermore, the observed shear directions trend east or east-northeast, i.e. perpendicular to the direction of the PR at the location of ODP site 637, and the shear sense is normal. These two observations are consistent with a rising body as previously suggested.

6. New constraints on models of formation of the Galicia margin

On the basis of the plate kinematic reconstructions of Sibuet and Collette (1991) and Srivastava et al. (1990), Sibuet (1992b) published a composite two-way travel time section across the Galicia-southeastern Flemish Cap continental margins. A master fault associated with the steep continental slope, was identified on the available seismic profiles (Parson et al., 1985), confirmed by data acquired during the Erable cruise (Flemish Cap margin, 1992) and interpreted as the outcropping portion of a large detachment zone affecting the continental crust. The base of the southeastern Flemish Cap detachment zone is assumed to merge with the S reflector (Sibuet, 1992a and b).

We favour a composite pure and simple shear model for the formation of these two conjugate margins initially proposed in a schematic manner by Sibuet (1992a and b): pure shear for the whole lithosphere of the continentward portion of the Galicia margin and simple shear for the brittle crust of the southeastern Flemish Cap margin and the oceanward portion of the Galicia margin. In this model, a detachment surface (Figure 9A) has been active since the beginning of rifting and has propagated eastwards from the Canadian side to the Iberian side as extension increased. This detachment surface, which crops out on the Canadian side, affects the upper portion of the continental crust and probably merges at depth with the brittle-plastic interface, as suggested by our refraction data and by theoretical modelling (Radel and Melosh, 1987). Below

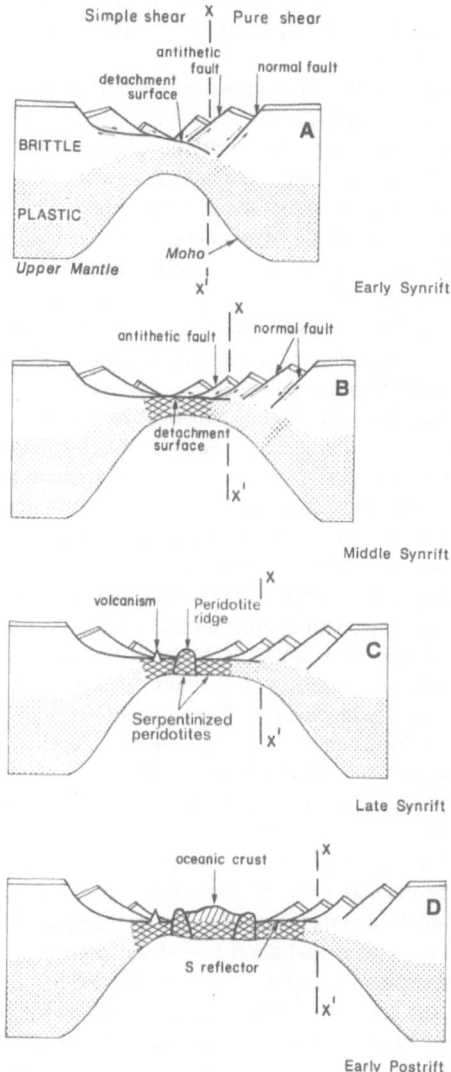

Figure 9. Proposed model of formation of the western Galicia margin: the xx' dashed line divides the system of conjugate margins into two asymmetric parts. To the right of xx'. the pure shear mechanism involves the whole lithosphere. To the left of xx', the simple shear mechanism involves the upper brittle part of the crust and pure shear the remaining lower part of the lithosphere. As extensional rates during rifting and the early stage of spreading are rather low (half rates <10 mm/yr) serpentinized peridotites are emplaced at the base of both the thinned continental and oceanic crusts. At the end of the rifting phase, the final break-up occurs where the lithosphere is extremely thinned and is associated with intrusion of upper mantle material (serpentinized peridotites). After the onset of spreading these peridotite ridges are left on both sides of the ocean.

the Galicia margin, this surface corresponds to the S reflector. In this hypothesis, the lateral offset along the detachment surface decreases eastwards from the Canadian side to zero where the S reflector disappears (Figures 9A and B).

Below the upper part of the Galicia margin, where the detachment surface dies out, there is no lateral motion along the brittle-plastic interface. The xx' vertical line, which intersects the detachment surface at its point of disappearance, divides the system of conjugate margins into two parts: (1) the left-hand side for which the basic mechanism within the brittle crust is simple shear; (2) the right-hand side where we suggest that the pure shear model is not only applied to the brittle portion of the crust but also to the whole lithosphere. More generally, we suggest that the main mechanism which affects the whole lithosphere could be pure shear (Etheridge et al., 1989; Keen et al., 1989; Kuznir and Egan, 1989); it is only within the upper part of the crust that the fundamental mechanism of rifting could be either simple or pure shear. Therefore, we consider that pure shear is the basic lithospheric mechanism and that simple shear is an upper crustal phenomenon, whatever its local importance.

During the rifting, and especially for large extensional factors, the continental crust thins and is highly fractured. Sea-water can easily reach the upper mantle, generating serpentinized peridotite just below the continental crust and then incorporating this material at the base of the crust. The rate of extension is very low during rifting (100 km of extension during at least 25 M.y., that is less than 5 mm/yr half rate) and only a small amount of melt (<10%) is expected to rise in the serpentinized peridotite (Bown and White, 1995a and b).

At the end of the rifting phase, upper mantle material composed of highly serpentinized peridotites less dense than the surrounding crustal material, can be uplifted and "fossilised" at the ocean-continent boundary (Figure 9C), at the place where the crust is the most greatly thinned and fractured and where sea-water circulation is maximum. Normal faults are thus expected on the outer sides of the uplifted peridotite ridge(s). They put into contact the serpentinized peridotite bodies located beneath the thinned continental crust with the uplifted PR. The break-up did not occur where the Moho was shallowest but within the PRs where the lubricant effect could favour the final continental disruption, leaving normal faults only on the landward side of the PRs.

The formation of early oceanic crust occurred there at slow spreading rates (8 mm/yr half rate from Sibuet and Collette (1991)). Theoretical modelling (Bown and White, 1995a and b) shows that at slow spreading rates oceanic crust of normal thickness cannot be created because of the reduced amount of melt, which favours the emplacement of serpentinized peridotite at the base of the oceanic crust. An alternative explanation is that the poor melt supply is due to abnormal cooling of the upper asthenosphere (Whitmarsh and Miles, 1995). Subsequent to the creation of oceanic crust (Figure 9D), the continental lithosphere cooled and the brittle-plastic interface sank, leaving the former brittle-plastic interface within the present-day brittle upper crust. In this scheme PRs should be observed on both sides of the Atlantic, as suggested earlier (Sibuet, 1987).

7. Conclusions

Travel-time and amplitude modelling of seismic refraction data collected across the Galicia continental margin show that:

- west of the peridotite ridge, the oceanic crust is reduced to a thickness of about 5 km which increases to normal oceanic thickness some 10-15 km further west. The base of the crust consists of 7.4 - 7.5 km/s material interpreted as serpentinized peridotite.

- east of the peridotite ridge, the thinned continental crust overlies a 7.4 km/s layer also interpreted as serpentinized peridotite. This body has a limited landward extent of 25 km.

- from the peridotite ridge to the east, the S reflector, which is a major detachment surface, follows the contact between the thinned continental crust and the 7.4 km/s body, then cuts across the lower portion of the thinned continental crust and merges the top of the lower crust interpreted as the brittle-plastic interface at the time of rifting.

- seismic velocities within the peridotite ridge increase greatly from the top to the base, indicating that the ridge is a highly serpentinized peridotite body.

- as the extensional rate during rifting and the half spreading rate during the emplacement of early oceanic crust are both lower than 10 mm/yr, theoretical modelling (Bown and White, 1995a and b) supports the emplacement of serpentinized peridotite at the base of both the thinned continental and the oldest oceanic crusts mixed with only a limited amount of melt (<10%). However, the poor melt supply below the oceanic crust could be due also to abnormal cooling of the upper asthenosphere (Whitmarsh and Miles, 1995).

Near-bottom magnetic anomalies show that:

- the oceanic crust abuts the west side of the peridotite ridge. Forward modelling with a constant normally magnetized oceanic crust fits remarkably well the regional observed near-bottom magnetic anomalies.

- the magnetization of the peridotite ridge is similar to that of the adjacent thinned continental crust, that is about ten times smaller than the oceanic crust.

Seismic refraction, reflection and near-bottom magnetic data confirm that the peridotite ridge is uplifted along a normal fault located on its eastern flank which cannot have been emplaced by mantle tectonic denudation. Serpentinized peridotite is found at the base of both the thinned continental and oceanic crusts but the peridotite ridge definitely marks the ocean-continent boundary. We favour a composite pure and simple shear model of formation of the Galicia margin and its conjugate margin in which a detachment surface (S reflector), active since the beginning of rifting, crops out on the Canadian side and merges beneath the Galicia margin with the brittle-plastic interface. The peridotite ridge is emplaced during the late stages of rifting by mantle upwelling through the highly fractured, thin veil of continental crust. After creation of oceanic crust, peridotite ridges are left on both sides of the ocean.

Acknowledgements

We thank J.-F. Rolin, N. Lantéri, P. Jegou (Ifremer Engineering Department) and P. Plasseraud (Ifremer technical staff in charge of the SAR) for their help in the development and integration of the 3-component magnetometer in the SAR system. Thanks are also due to the SAR operational team. The instrument was tested and became operational during its first cruise. We are indebted to the Masters and crew of the R/V Le Suroît and the Fluigal and Réframarge scientific parties for their help at sea under difficult conditions. We thank L. Montadert for providing migrated sections of the MCS profile GP 101 and D. Carré for drafting the figures.

214

References

Boillot, G., Beslier, M.-O., and Girardeau, J., 1994. Nature, structure and evolution of the ocean-continent boundary: the lesson of the West Iberia margin. *In* Banda, E., Talwani, M. and Torne, M., eds., Rifted ocean-continent boundaries, Kluwer, this volume.

Boillot, G., et al., 1987a. Tectonic denudation of the upper mantle along passive margins: a model based on drilling results (ODP Leg 103, western Galicia margin, Spain). Tectonophysics, 132, p. 335-342.

Boillot, G., Winterer, E.L., Meyer, A.W., and Shipboard Scientific Party, 1987b. Introduction, objectives and principal results: Ocean Drilling Program Leg 103, West Galicia margin. *In* Boillot, G., Winterer, E.L., Meyer, A.W. and Shipboard Scientific Party, eds., Initial Reports of Ocean Drilling Project, U.S. Government Printing Office, Washington, Leg 103, (Part A), p. 3-17.

Boillot, G., Winterer, E.L., Meyer, A.W., and Shipboard Scientific party, 1988. Proceedings ODP, Leg 103 Scientific Results. Ocean Drilling Program, Government Printing Office, Washington D.C., College Station, Texas.

Bougault, H., et al., 1993. Fast and slow spreading ridges: structure and hydrothermal activity, ultramafic topographic highs, and CH4 output. Journal of Geophysical Research, 98, p. 9643-9651.

Bown, J.W., and White, R.S., 1994. Variation with spreading rate of oceanic crustal thickness and geochemistry. Earth and Planetary Science Letters, 121, p. 435-449.

Bown, J.W., and White, R.S., 1995a. The effect of finite extension rate on melt generation at continental rifts. Journal of Geophysical Research, in press.

Bown, J.W., and White, R.S., 1995b. Finite duration rifting, melting and subsidence at continental margins. Proceedings of the NATO-ARW workshop on rifted ocean-continent boundaries, 11-14 May 1994, Mallorca, Spain, this volume.

Chapman, C.H., and Drummond, R., 1982. Body wave seismograms in inhomogeneous media using Maslov asymptotic theory. Bulletin of the seismological Society of America, 72, p. 277-317.

Charlou, J.-L., and Donval, J.-P., 1993. Hydrothermal methane venting between 12°N and 26°N along the mid-Atlantic ridge. Journal of Geophysical Research, 98, p. 9625-9642.

Dunlop, D.J., and Prévot, M., 1982. Magnetic properties and opaque mineralogy of drilled submarine intrusive rocks. Geophysical Journal of the Royal Astronomical Society of London, 69, p. 763-802.

Etheridge, M.A., Symonds, P.A., and Lister, G.S., 1989. Application of the detachment model to reconstruction of conjugate passive margins. *In* Tankard, A.J. and Balkwill, H.R., eds., Extensional tectonic and stratigraphy of the north Atlantic margins, Tulsa, AAPG, AAPG memoir 46, p. 23-40.

Evans, C.A., and Girardeau, J., 1988. Galicia margin peridotites: undepleted abyssal peridotites from the north Atlantic. *In* Boillot, G., Winterer, E.L., Meyer, A.W. and Shipboard Scientific party, eds., Ocean Drilling Program, College Station, Texas, Government Printing Office, Washington D.C., Proceedings ODP, Leg 103 Scientific Results, p. 195-207.

Féraud, D., Girardeau, J., Beslier, M.-O., and Boillot, G., 1988. Datation $^{39}Ar/^{40}Ar$ de la mise en place des péridotites bordant la marge de Galice (Espagne). Comptes Rendus de l'Académie des Sciences de Paris, 307, p. 49-55.

Girardeau, J., Evans, C.A., and Beslier, M.-O., 1988. Structural analysis of plagioclase-bearing peridotites emplaced at the end of continental rifting: hole 637, ODP leg 103 on the Galicia margin. *In* Boillot, G., Winterer, E.L., Meyer, A.W. and Shipboard Scientific party, eds., Ocean

Drilling Program, College Station, Texas, Government Printing Office, Washington D.C., Proceedings ODP, Leg 103 Scientific Results, p. 209-223.

Horsefield, S.J., 1992. Crustal structure across the continent-ocean boundary. University of Cambridge, 215 p.

Horsefield, S.J., Whitmarsh, R.B., White, R.S., and Sibuet, J.-C., 1994. Crustal structure of the Goban Spur rifted continental margin, NE Atlantic. Geophysical Journal International, 119, p. 1-19.

Keen, C.E., Peddy, C., de Voogd, B., and Matthews, D., 1989. Conjugate margins of Canada and Europe: Results from deep reflection profiling. Geology, 17, p. 173-176.

Kirk, R.E., Langford, J.J., and Whitmarsh, R.B., 1982. A three component ocean bottom seismogram for controlled source and earthquake seismology. Marine Geophysical Researches, 5, p. 327-341.

Kuznir, N.J., and Egan, N.N., 1989. Simple-shear and pure-shear models of extensional sedimentary basin formation: application to the Jeanne d'Arc basin, Grand Banks of Newfoundland. In Tankard, A.J. and Balkwill, H.R., eds., Extensional tectonic and stratigraphy of the north Atlantic margins, Tulsa, AAPG, AAPG memoir 46, p. 305-322.

Lallemand, S., and Sibuet, J.-C., 1986. Tectonic implications of canyon directions over the Northeast Atlantic continental margin. Tectonics, 5, p. 1125-1143.

Le Pichon, X., Angelier, J., and Sibuet, J.-C., 1983. Subsidence and stretching. In Watkins, J.S. and Clarke, C.L., eds., Studies in continental margin geology, Tulsa, AAPG Memoir 34, p. 731-741.

Le Pichon, X., and Sibuet, J.-C., 1981. Passive margins: a model of formation. Journal of Geophysical Research, 86, p. 3708-3710.

Louden, K.E., Sibuet, J.-C., and Foucher, J.-P., 1991. Variations in heat flow across the Goban Spur and Galicia Bank continental margins. Journal of Geophysical Research, 96, p. 16131-16150.

Louvel, V., Dyment, J., and Sibuet, J.-C., 1995. Thinning of the Goban Spur continental margin and formation of the early oceanic crust: constraints from forward modelling and inversion of marine magnetic anomalies. Geophysical Journal International, in press.

Mauffret, A., and Montadert, L., 1987. Rift tectonics on the passive continental margin off Galicia (Spain). Marine and Petroleum Geology, 40, p. 49-70.

Nazarova, K.A., 1994. Serpentinized peridotites as a possible source for oceanic magnetic anomalies. Marine Geophysical Researches, 16, p. 455-462.

ODP Leg 149 Shipboard Scientific Party, 1993. ODP drills the west Iberia rifted margin. EOS, 74, p. 454-455.

Parson, L.M., Masson, D.G., Pelton, C.D., and Grant, A.C., 1985. Seismic stratigraphy and structure of the east Canadian continental margin between 41 and 52°N. Canadian Journal of Earth Sciences, 22, p. 686-703.

Radel, G., and Melosh, H.J., 1987. A mechanical basis for low-angle normal faulting in the Basin Range. EOS, Trans. AGU, 68, p. 1449.

Recq, M., Whitmarsh, R.B., and Sibuet, J.-C., 1991. Anatomy of a lherzolite ridge, Galicia margin. Terra abstracts, 3, p. 122.

Reston, T.I., Krawczyk, C.M., and Wilkens, D., 1994. Detachment faulting and continental breakup: the S reflector offshore Galicia. Annales Geophysicae, Supplement 1 to vol. 1, p. C43.

Shipboard Scientific Party, 1994. Site 897. *In* Sawyer, D.S., Whitmarsh, R.B., Klaus, R.B. and Shipboard Scientific Party, eds., Proc. ODP, Init Repts., College Station, TX (Ocean Drilling Program), 149, p. 41-113.

Shive, P.N., Frost, B.R., and Peretti, A., 1988. The magnetic properties of metaperidotitic rocks as a function of metamorphic grade: implications for crustal magnetic anomalies. Journal of Geophysical Research, 93, p. 12187-12195.

Sibuet, J.-C., 1987. Contribution à l'étude des mécanismes de formation des marges continentales passives. Thèse Doctorat d'Etat, Université de Bretagne Occidentale, Brest, 351 p.

Sibuet, J.-C., 1992a. Formation of passive margins: a composite model applies to the conjugate Galicia and southeastern Flemish Cap margins. Geophysical Research Letters, 19, p. 769-772.

Sibuet, J.-C., 1992b. New constraints on the formation of non-volcanic continental Galicia-Flemish Cap conjugate margins. Journal of the Geological Society of London, 149, p. 829-840.

Sibuet, J.-C., and Collette, B., 1991. Triple junctions of Bay of Biscay and North Atlantic: new constraints on the kinematic evolution. Geology, 19, p. 522-525.

Sibuet, J.-C., and Ryan, W.B.F., 1979. Initial Reports of the Deep Sea Drilling Project. U.S. Government Printing Office, Washington, 47B, part 2.

Sibuet, J.-C., Thomas, Y., Marsset, B., Louvel, V., Savoye, B., and Le Formal, J.-P., 1995. Detailed relationship between tectonics and sedimentation from Pasisar deep-tow seismic data acquired at ODP Site 897 in the Iberian Abyssal Plain. *In* Sawyer, D.S., Whitmarsh, R.B., Klaus, R.B. and Shipboard Scientific Party, eds., Proc. ODP, Scientific Results., College Station, TX (Ocean Drilling Program), 149, in press.

Srivastava, S.P., Roest, W.R., Kovacs, L.C., Schouten, H., and Klitgord, K., 1990. Iberian plate kinematics: a jumping plate boundary between Eurasia and Africa. Nature, 344, p. 756-759.

Stokking, L.B., Merill, D.L., Haston, R.B., Ali, J.R., and Saboda, K.L., 1992. Rock magnetic studies of serpentinite seamounts in the Mariana and Izu-Bonin regions. *In* Fryer, P., Pearce, J.A. and Stokking, L.B., eds., Proceedings of the Ocean drilling Program, Scientific Results, College Station, Texas, Government Printing Office, Washington D.C., 125, p. 561-579.

Wernicke, B., 1985. Uniform-sense normal simple shear of the continental lithosphere. Canadian Journal of Earth Sciences, 22, p. 108-125.

White, R., and McKenzie, D., 1989. Magmatism at rifted zones: the generation of volcanic continental margins and flood basalts. Journal of Geophysical Research, 94, p. 7685-7729.

White, R.S., McKenzie, D., and O'Nions, K., 1992. Oceanic crustal thickness from seismic measurements and rare earth element inversions. Journal of Geophysical Research, 97, p. 19683-19715.

Whitmarsh, R.B., Miles, P., Sibuet, J.-C., and Louvel, V., 1995a. Geological and geophysical implications of deep-tow magnetometer observations near ODP Sites 897, 898, 899, 900 and 901 on the west Iberia continental margin. *In* Sawyer, D.S., Whitmarsh, R.B., Klaus, R.B. and Shipboard Scientific Party, eds., Proc. ODP, Scientific Results, College Station, TX (Ocean Drilling Program), 149, in press.

Whitmarsh, R.B., and Miles, P.R., 1995. Models of the development of the West Iberia continental margin at 40°30'N deduced from surface and deep-tow magnetic anomalies. Journal of Geophysical Research, in press.

Whitmarsh, R.B., Pinheiro, L.M., Miles, P.R., Recq, M., and Sibuet, J.-C., 1993. Thin crust at the ocean-continent transition off Iberia and ophiolites. Tectonics, 12, p. 1230-1239.

Whitmarsh, R.B., White, R.S., Sibuet, J.-C., Horsefield, S.J., and Recq, M., 1995b. The ocean-continent boundary off the western continental margin of Iberia - III. Crustal structure across the Galicia Bank margin. Geophysical Journal International, in prep.

Zhao, X., 1995. Magnetic signatures of serpentinites and peridotites from ODP Leg 149 and their implications. *In* Sawyer, D.S., Whitmarsh, R.B., Klaus, R.B., and Shipboard Scientific Party, eds., Proc. ODP, Scientific Results., College Station, TX (Ocean Drilling Program), 149, in press.

NATURE, STRUCTURE AND EVOLUTION OF THE OCEAN-CONTINENT BOUNDARY : THE LESSON OF THE WEST GALICIA MARGIN (SPAIN)

G. BOILLOT
Laboratoire de Géodynamique Sous-Marine
B.P. 48
06230 Villefranche-sur-Mer
France

M. O. BESLIER
Laboratoire de Géodynamique Sous-Marine
B.P. 48
06230 Villefranche-sur-Mer
France

J. GIRARDEAU
Laboratoire de Pétrologie Structurale
Faculté des Sciences et des Techniques
2, rue de la Houssinière
44072 Nantes Cedex
France

ABSTRACT. The available geological data on the West Galicia Ocean-Continent boundary are briefly reviewed. Four parallel belts of basement terranes were sampled from the oceanic domain to the passive margin: basaltic seafloor, sheared and serpentinized mantle rocks cropping out on the "peridotite ridge", mylonitized underplated gabbro, and upper continental crust resting on a strong seismic reflector considered to be a detachment.

The geological and geophysical data support a simple shear model, according to which the lithosphere stretching and thinning were accommodated by two sets of conjugate shear zones. In conclusion we advocate a "basement strategy" in passive margin and Ocean-Continent Boundary studies. the basement rocks providing better evidence of the rifting processes than sediments.

1. Introduction

The west Iberia margin (Fig. 1) is one of the best places in the world for imaging and sampling a typical, non volcanic rifted Ocean-Continent boundary (OCB), because of the thin and discontinuous sedimentary cover resting on the basement. Geological and geophysical studies started more than 20 years ago, and the amount of data bring considerable constraints on models of continental break-up. In this short paper, we present a) a review of the geological data from the west of Galicia (Spain) collected by drilling (ODP program) or by diving with the French submersible Nautile, which made possible the identification of terranes imaged by seismic reflection near the OCB ; b) a simple shear model, based on analogic experiments, accounting for available geological data ; and c) a brief discussion on the "basement strategy" we adopted in the last years in order to further passive margin studies

E. Banda et al. (eds.), Rifted Ocean-Continent Boundaries, 219–229.
© 1995 *Kluwer Academic Publishers. Printed in the Netherlands.*

220

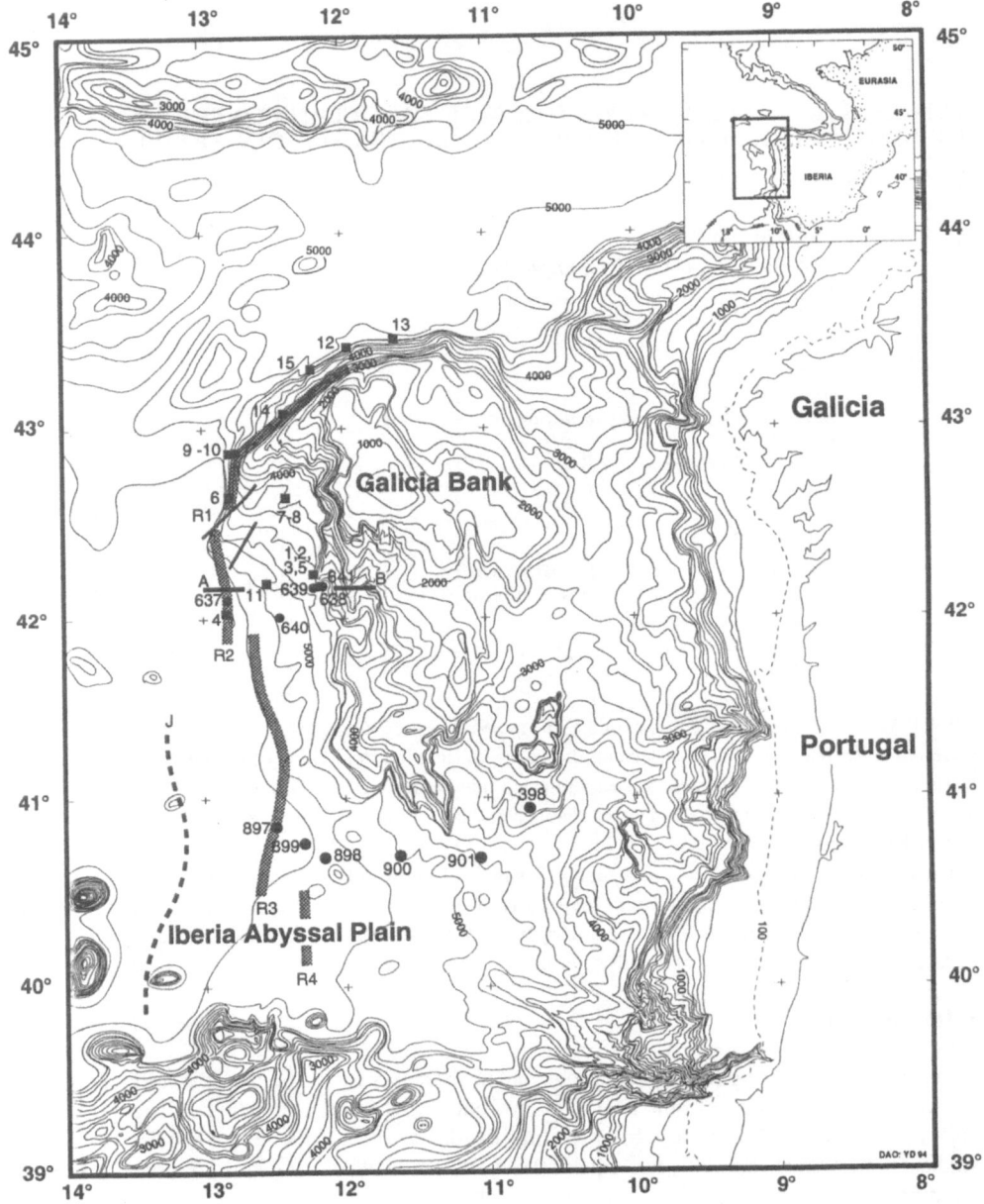

Fig. 1 - The west Iberia margin north of 39°N. Bathymetry after Lallemand et al., 1982. Sites of drilling (circles) and of diving (squares) are indicated. R1-4: segments of the peridotite ridge after Beslier et al., 1993. A-B: location of the cross-section shown on Fig. 2.

and, given the success of this strategy, we assign some new targets for drilling or diving cruises proposed for the future.

The west Iberia margin results from the break-up of Iberia and North America in late Jurassic and early Cretaceous times. The initiation of seafloor spreading propagated northward from the Tagus abyssal plain to the Iberia Abyssal Plain (IAP) and to the west Galicia margin (Mauffret et al., 1989 ; Mougenot, 1989 ; Whitmarsh et al., 1990, 1993 ; Pinheiro et al, 1992). In this paper, we limit the presentation and the discussion to the Galicia segment of the margin which is the best known at that time (see Mauffret and Montadert, 1987 and Boillot et al., 1989a for more exhaustive synthesis). A schematic and interpreted cross-section of the margin and adjacent OCB along 42°N (Fig. 1) is presented in Fig. 2.

2. Review of geological data and constraints

In this section, we summarize the results of geological studies on the Galicia OCB, starting from the ocean and moving landward. The following § are referred to 1-9 in Fig. 3, A and B.

2.1. Basaltic, oceanic seafloor was observed and sampled by diving at sites 12, 13 and 15 (Fig. 1) on the north-west slope of the Galicia Bank, a place where the Mesozoic OCB, including a piece of the bay of Biscay lithosphere, was uplifted and escaped the Cainozoic subduction of Eurasia beneath Iberia (Malod et al., 1993). Petrological and geochemical studies of the samples showed that the basalt is tholeiite, free of continental contamination. Rare earth element patterns as well as isotopic ratio (Nd and Sr) grade from gently enriched to moderately depleted composition. Geochronological $^{40}Ar/^{39}Ar$ dating of two samples indicate 100 +/- 5 Ma whereas the break-up occurred 114 Ma ago in that area. Clearly the sampled basaltic layer is part of the post-rift oceanic crust (Kornprobst et al., 1988 ; Malod et al., 1993). Moreover, thin oceanic crust is inferred from refraction and magnetic data to the west of the Ocean-Continent Boundary off Galicia and Portugal (Whitmarsh et al, 1993).

2.2. The Ocean-Continent Boundary is marked by a basement ridge made of serpentinized peridotite, which was sampled by dredging, drilling and diving in several places (Boillot et al., 1980, 1988; Boillot, Winterer et al., 1987, 1988; Sawyer, Whitmarsh, Klaus et al., 1994). From the southern part of the Iberia Abyssal Plain to the northern edge of the Galicia Bank, the ridge is made-up of four distinct segments (Beslier et al., 1993: Fig. 1).

In spite of its ultramafic composition, the peridotite ridge resembles a crustal tilted block bounded by normal faults (Mauffret and Montadert, 1987 ; Fig. 2). It is covered by syn-rift sediments on its eastern, landward slope, and can be considered to be a late extensional structure of the rift separating Iberia and Newfoundland before the initiation of seafloor spreading.

2.3. In the Galicia margin, the ultramafic body marking the OCB is made up of serpentinized plagioclase bearing harzburgite or lherzolite (Evans and Girardeau, 1988). The serpentinization of the initial peridotite occurred in static conditions as a result of hydrothermal activity (Agrinier et al., 1988). There is no evidence of serpentinite diapirism, the fabric of the initial peridotite being preserved in all the samples.

The mylonitic texture of the peridotite was acquired in a ductile normal simple shear zone at high and decreasing temperature (> 1000 - 850°) and relatively shallow depth, under large deviatoric stress (180 Mpa ; Fig. 4 ; Girardeau et al., 1988 ; Beslier et al., 1990). At the drill Site 637 (Fig. 1), the shear zone dips toward the continent (S2 on Fig. 3, B and C). 39

222

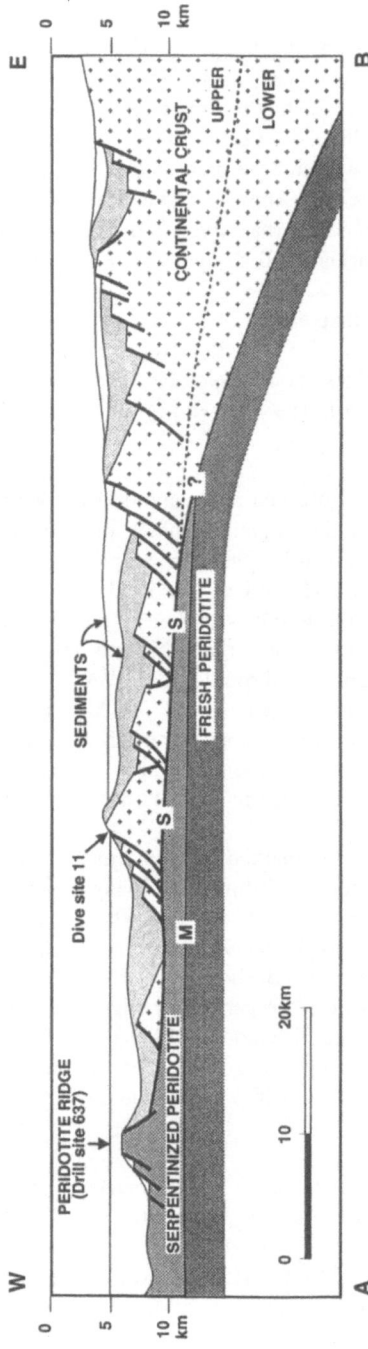

Fig. 2 - Schematic and interpreted cross section of the Galicia margin along 42°N, after Boillot et al., in press, slightly modified. Fig. 3, A is a cartoon of the western part of this cross-section. Location on Fig. 1 (A-B).

Ar/^{40}Ar dating of syn- and post-kinematic brown amphiboles indicate a syn-rift age for the shearing (122 Ma ; Féraud et al., 1988). The tectonic unroofing and serpentinization of mantle rocks occurred just before seafloor spreading started 114 Ma ago in this part of the north Atlantic (Boillot, Winterer et al., 1987, 1988).

2.4. At the top of the Galicia margin peridotite ridge, we observed a 100-150 m thick layer of chlorite-bearing schist derived from gabbro, retromorphosed and mylonitized in green-schist facies conditions. Relative to the ductile deformation of underlying peridotite, the sense of shear within the schist is contrary although the direction is parallel, and the deformation occurred at lower pressure and temperature (Beslier et al., 1990). For these reasons, we relate the sheared schist with the lower crust shear zone S_1 on Fig. 3, B and C. Crystals of zircon included in the rock are 122 Ma old, indicating a syn-rift age for the crystallisation and underplating of the gabbro (Schärer et al., submitted). It is also possible that the flaser gabbro recently drilled at site 900 during ODP Leg 149 is underplated material (Sawyer, Whitmarsh, Klaus et al., 1994).

2.5. The brittle crust of the margin was faulted and cut-up into tilted blocks, 15-20 km wide, by syn-rift extensional tectonics (Montadert et al., 1979; Mauffret and Montadert, 1987; Sibuet et al., 1987 ; Thommeret et al., 1988; Mougenot, 1989). Samples recovered by diving on the top of the last (oceanward), deepest tilted block of the margin, 20 km to the east of the Galicia peridotite ridge (site 11; Fig. 2), indicate Hercynian basement from the upper continental crust (Boillot et al., 1988 ; Beslier et al, 1990).

2.6. In the same area, in the part of the Galicia margin close to the OCB, the crustal basement is separated into two layers by a strong seismic reflector labelled S (de Charpal et al., 1978). The S reflector is now interpreted as the seismic signature of a syn-rift detachment (Wernicke and Burchfield, 1982; Boillot et al., 1989a, in press ; Hoffmann and Reston, 1992; Sibuet, 1992; Reston et al., in press). However, the actual nature of terranes resting between S and the Moho (the lower seismic crust) remains controversial : either it is continental material (Sibuet, 1992), or partly serpentinized peridotite (Boillot et al., 1989b, 1992 : Beslier and Brun, 1991), or underplated gabbro.

2.7. Beneath the peridotite ridge bounding the margin, the Moho is located several kilometres deeper than the seafloor (Whitmarsh et al., 1993). At that place, the Moho corresponds to the boundary between fresh and serpentinized peridotite, i.e. to the paleo-hydrothermal front, possibly corresponding to a paleo-isotherm. On both sides of the ridge, the question of the nature of the lower crust and of the Moho beneath the basaltic layer (oceanward ; § 1) or the thinned upper continental crust (landward; § 5) remains open. In figs. 2 and 3, we consider the lower crust to be mainly made of serpentinized peridotite, the Moho being the limit between fresh and serpentinized mantle rocks. However, underplating by gabbro can also contribute to the building up of the lower crust (see § 2.4).

2.8. Pre-rift lower continental crust was recovered in the form of exotic rounded blocks of granulite, Archean in age, sampled in the northern margin of Spain, and, locally, on the edge of the Galicia and Vigo seamounts (Guerot et al., 1989). The boulders are not Pleistocene ice drifts, as some of them were sampled by diving from an Albian conglomerate (Malod et al., 1980). However, the pre-rift lower continental crust was never recovered in situ in the Iberia margin at that time, and remains a target for further investigations.

224

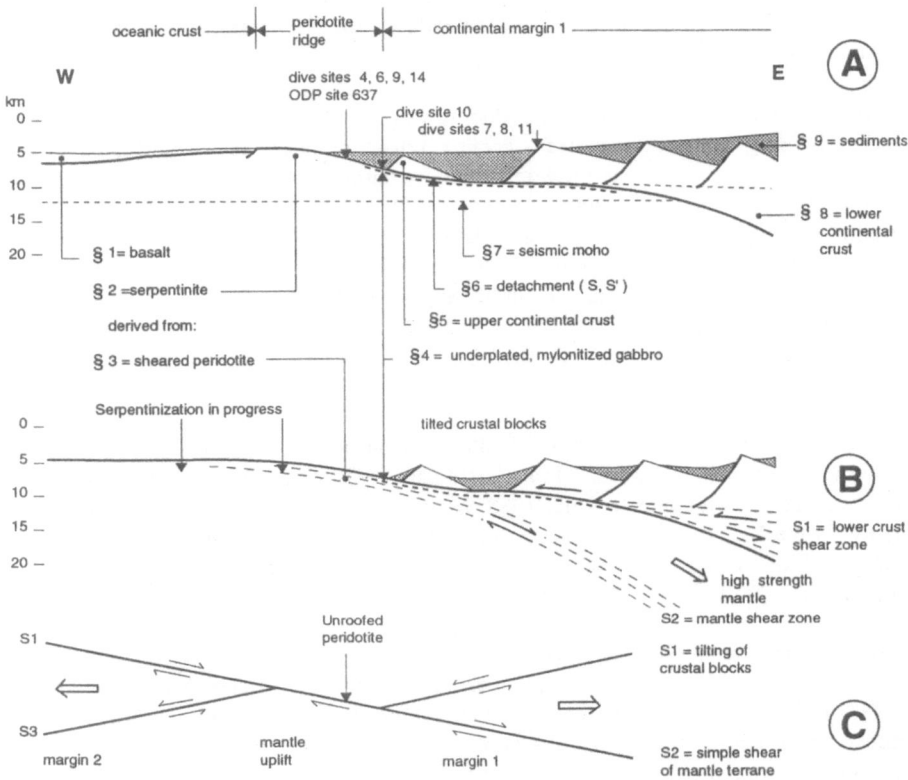

Fig. 3 - A : Cartoon of the crustal structure at the west Galicia ocean-continent boundary. Numbers relate to § of the section 2 of the paper. B : kinematic model accounting for the crustal structure depicted in A; C : diagram of the lithosphere thinning accommodated by conjugate simple shear zones S_1, S_2 and S_3 depicted in B.

2.9. Pre-, syn- and post-rift sediments of the West Iberia margin were drilled during IPOD and ODP legs 48b, 103 and 149 (Sibuet, Ryan et al., 1979 ; Boillot, Winterer et al., 1988 ; Sawyer, Whitmarsh, Klaus et al., 1994). From these data, we know that the last phase of the rifting lasted about 25 Ma before the continents broke up, the age of the break-up being 135, 125 and 114 ma in the Tagus abyssal plain, Iberia Abyssal Plain and Galicia margin, respectively (Whitmarsh et al., 1993).

3. The model : simple, conjugate shear zones

To account for the west Galicia continental break-up and OCB structure, we propose a simple shear model based on analog modelling by Beslier and Brun (1991), and Brun and Beslier (in press) (Fig. 3, B and C). According to the model, the thinning of the lithosphere was accommodated by two sets of conjugate shear zones S_1 and S_2 and not only by a single, lithospheric shear zone as postulated earlier (Boillot et al., 1987). S_1, located in the lower continental crust, accounts for the sense of tilting of the crustal blocks and for the sense of shearing of underplated mylonitized gabbro sampled at the top of the peridotite ridge (§ 2.4). S_2, located in the mantle, was partly drilled at ODP site 637, and accounts for the ductile shearing of peridotite at high and decreasing temperature (§ 2.3 and Fig. 4), and for the final unroofing of mantle terrain at the continental rift axis (now the OCB). Actually, Fig. 3 is a cartoon : it is fairly certain that several active shear zones of the S_1 and S_2 families coexisted within the lithosphere during its syn-rift thinning and extension.

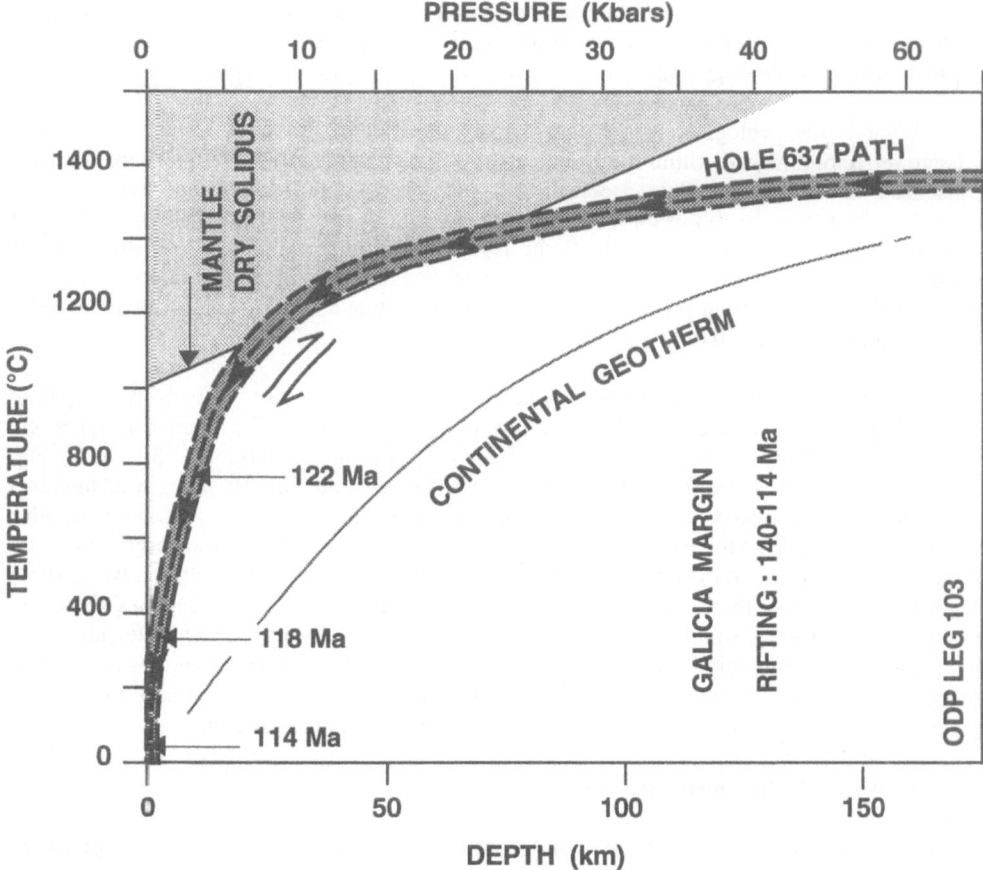

Fig. 4 - Pressure-temperature path of the peridotite drilled at ODP hole 637 (Leg 103) after Girardeau et al., 1988, slightly modified. Ages are from Féraud et al., 1988; Boillot et al.,1989b and Schärer et al., (in press).

4. Discussion

Three questions at least remain open for further research and discussions.

4.1. What were the respective parts of pre-rift, lower continental crust of Iberia (§ 2.8), syn-rift underplated gabbro (§ 2.4) and undercrusted serpentinite (§ 2.3) in the building up of the present lower seismic crust of the deep Galicia margin ? During their syn-rift uplift, the margin peridotites underwent about 9% of partial melting (Evans and Girardeau, 1988). Where are the products of the partial melting ?

The basaltic seafloor sampled oceanwards (2.1) cannot be an answer, as the basalt is younger than the emplacement of the poorly depleted peridotite on the seafloor. More probably, the melt products were either trapped in the upper part of the mantle, and/or underplated, as was probably the chlorite-bearing schist sampled at dive site 10 (Fig. 3, A). To make further, we need geophysical tools and methods to distinguish between serpentinized peridotite, gabbro and pre-rift lower continental crust, all terranes having close seismic velocities, though with very different geological significance. A related problem is the actual nature of the seismic Moho depending on places (phase change, perhaps related to an isotherm, between fresh and serpentinized peridotite, or contact between peridotite and crustal terranes, with or without gabbro).

4.2. What is the geological nature and tectonic significance of the S reflector and other seismic reflectors imaged within the Galicia margin basement near the OCB ? There is now a consensus for considering S as a detachment. However, according to some authors, it is related to shear zone S_1, S_2 or both (as suggested on Figs. 2 and 3). To discriminate, we need samples from the terranes overlying, at the level of, and underlying S. Then, the sense of shear, and the nature of the lower crust beneath S as well, will be determined, and the understanding of the rifting processes will certainly constitute a step forward. This is indeed a very exciting target for ODP.

4.3. Is the crustal structure of the Galicia margin typical of rifted passive margins ? A partial answer is given by the results of ODP Leg 149 in the IAP (Sawyer, Whitmarsh, Klaus et al., 1994) : here the ridge marking the OCB is made of serpentinized peridotite (Site 897 ; Fig. 1), as it is the case off Galicia . In fact, the ultramafic belt bounding the margin in this area seems to be wider than to the west of the Galicia Bank (mantle-derived terranes were also sampled at Site 899). Moreover some intra-crustal reflectors resembling the S reflector were imaged in the IAP (S" reflector of Beslier et al., 1993). These similarities between two distinct segments of the West Iberia margin suggest that the crustal structure at the Galicia OCB is not exceptional. However, we need discovery of other regional examples of peridotite ridges and intra-crustal reflectors within the thinned crust of passive margins to be sure that the Galicia margin features are actually representative of rifted margins, and thus that simple shear is the main mechanism for lithosphere stretching and thinning at continental rifts.

5. Conclusion : the basement strategy

One lesson from of the west Galicia margin studies is that the memory of rifting processes (stretching, thinning and break-up of continental lithosphere) is preserved in the margin basement probably better than in the overlying sediments: petrological, structural, geochronological and geochemical studies of oriented samples enable the reconstitution of the pressure- temperature paths of the basement rocks in space and time. Moreover, as the simple shear (detachment) played a major part in the continental rifting, deep structural levels of the lithosphere were uplifted and unroofed, and are now cropping out in tectonic windows in the deepest part of the passive margin. This fact allows us to approach the deep part of the lithosphere beneath the west Iberia passive margin, including the upper mantle. Those areas must be entirely explored, especially by drilling.

In fact, the basement strategy has already resulted in major progress in passive margin studies:

- discovery of tectonic windows open to the mantle (the peridotite ridge) and mantle evolution beneath a continental rift ;

- competitive parts of syn-rift hydrothermal activity and peridotite partial melting in the formation of new lower seismic crust (undercrusting by serpentinization of the uppermost mantle ; underplating by crystallisation of gabbro at the crust-mantle boundary);

- imaging and sampling of simple shear zones in the mantle (ODP Leg 103, site 637, for example), or in the lower, newly formed crust (probably the flaser gabbro drilled at ODP Leg 149 Site 901).

Other major problems can be addressed by the basement strategy applied to OCB research at starved margins, such as the actual nature of the seismic Moho and lower seismic crust on both sides of the OCB ; partial melting related to rifting or kinematics and dynamics of the deformation along syn-rift shear zones. Indeed the venture is only just beginning.

Acknowledgements - We thank T.J. Reston and the anonymous English reviewer for useful comments and suggestions.

Contribution N° 663 of the "Groupe d'Etude de la Marge Continentale et de l'Océan" (URA 718 of CNRS - INSU and University of Paris).

References

Agrinier, P., Mevel, C. and Girardeau, J., 1988. Hydrothermal alteration of the peridotite cored at the Ocean-Continent Boundary of the Iberian margin : petrologic and stable isotope evidence. *In* Boillot, G., Winterer, E.L. et al.. *Proc. of Ocean Drilling Program, Scientific results (College Station, TX)*, 103, 225-234.

Beslier, M.O., Girardeau, J., and Boillot, G., 1990. Kinematics of peridotite emplacement during North Atlantic Continental rifting, Galicia, NW Spain. *Tectonophysics*, 184, 321-343.

Beslier, M.O. and Brun, J.P., 1991. Boudinage de la lithosphère et formation des marges passives. *C.R. Acad. Sci. Paris*, 313, 951-958.

Beslier, M.O., Ask, M. and Boillot, G., 1993. Ocean-Continent boundary in the Iberia abyssal Plain from multichannel seismic data *Tectonophysics*, 218, 383-393.

Boillot, G., Grimaud, S., Mauffret, A., Mougenot, D., Kornprobst, J., Mergoil-Daniel, J. and Torrent, G., 1980. Ocean-Continent Boundary off the Iberian margin : a serpentinite diapir west of the Galicia Bank. *Earth Planet Sci. Lett.*, 48, 23-34.

Boillot, G., Winterer, E.L. et al., 1987, *Proceedings of the Ocean Drilling Program, Initial Reports : College Station, TX*, 103, 663p.

Boillot, G., Recq, M., Winterer, E.L., Meyer, A.W., Applegate, J., Baltuck, M., Bergen, J.A., Comas, M.C., Davies, T.A., Dunham, K., Evans, C.A., Girardeau, J., Goldberg, G., Haggerty, J., Jansa, L.F., Johnson, J.A., Kasahara, J., Loreau, J.P., Luna-Sierra, E., Moullade, M., Ogg,

J., Sarti, M., Thurow, J., and Williamson, M., 1987. Tectonic denudation of the upper mantle along passive margins : a model based on drilling results (ODP Leg 103, Western Galicia Margin, Spain). *Tectonophysics*, 132, 335-342.

Boillot, G., Winterer, E.L. et al., 1988 : *Proceedings of the Ocean Drilling Program, Sci. Results : College Station, TX*, 103, 858 p.

Boillot, G., Comas, M.C., Girardeau, J., Kornprobst, J., Loreau, J.P., Malod, J., Mougenot, D. and Murillas, M., 1988. Preliminary results of the Galinaute cruise. In Boillot, G., Winterer, E.L. et al., *Proceedings of the Ocean Drilling Program, Sci. Results : College Station, TX*, 103, 37-51.

Boillot, G., Mougenot, D., Girardeau, J. and Winterer, E.L., 1989a. Rifting processes on the West Galicia margin, Spain. In Extensional Tectonics and stratigraphy of the north Atlantic margins (Tankard, A.J. and Balwill, H.R., eds). *Am. Ass. Petr. Geol. Mem.*, 46, 363-377.

Boillot, G., Féraud, G., Recq M. and Girardeau, J., 1989b. Undercrusting by serpentinite beneath rifted margins. *Nature*, 341, 523-525.

Boillot, G., Beslier, M.O. and Comas, M, 1992. Seismic image of undercrusted serpentinite beneath a rifted margin. *Terra Nova*, 4 , 25-33.

Boillot, G., Beslier, M.O., Krawczyk, C.M., Rappin, D. and Reston, T.J., in press. The formation of passive margins : constraints from the crustal structure and segmentation of the deep Galicia margin (Spain). *Geol. Soc. Sp. Publ., London.*

Brun, J.P. and Beslier, M.O., in press. Mantle exhumation at passive margins. *Earth and planet. Sci. Letters.*

De Charpal, O., Guennoc, P., Montadert, L. and Roberts, D.G., 1978. Rifting, crustal attenuation and subsidence in the Bay of Biscay : *Nature*, 275, 706-711.

Evans, C.Y., and Girardeau, J., 1988, Galicia margin peridotites : undepleted abyssal peridotites from the North Atlantic. In Boillot, G., Winterer, E.L. et al.. *Proceedings of the Ocean Drilling Program, Sci. Results : College Station, TX,*. 103, 195-207.

Féraud, G., Girardeau, J., Beslier, M.O. & Boillot, G., 1988. Datation ^{39}Ar-^{40}Ar de la mise en place des péridotites bordant la marge de la Galice (Espagne) : *C.R. Acad. Sci. Paris*, 307, 49-55.

Girardeau, J., Evans, C.A. and Beslier, M.O., 1988. Structural analysis of plagioclase-bearing peridotites emplaced at the end of continental rifting : hole 637, ODP leg 103 on the Galicia margin. In Boillot, G., Winterer, E.L. et al. *Proceedings of the Ocean Drilling Program, Sci. Results : College Station, TX*, 103, 209-223.

Guerot, C., Peucat, J.J., Capdevila, R. and Dosso, L., 1989. Archean protoliths within Early Protorozoic granulitic crust of the west European Hercynian belt : possible relics of the west African craton. *Geology*, 17, 241-244.

Hoffmann, H.J. and Reston, T.J., 1992. The nature of the S reflector beneath the Galicia Bank rifted margin. Preliminary results from pre-stack depth migration. *Geology*, 20, 1091-1094.

Kornprobst, J., Vidal, Ph. and J.A. Malod, 1988. Les basaltes de la marge de Galice (NO de la Péninsule Ibérique) : hétérogénéité des spectres de terres rares au niveau de la transition continent/océan. Données géochimiques préliminaires : *C.R. Acad. Sci. Paris*, 306, 1359-1364.

Lallemand, S., Mazé, J.P., Monti, S. et Sibuet, J.C., 1985. Présentation d'une carte bathymétrique de l'Atlantique nord-est : *C.R. Acad. Sci. Paris*, 300, 145-149.

Malod, J.A., Boillot, G., Capdevila, R., Dupeuble, P.A., Lepvrier, C., Mascle, G., Müler, C. et Taugourdeau-Lantz, J., 1980. Plongées en submersible au sud du Golfe de Gascogne et structure de la pente du Banc Le Danois. *C.R. Somm. Soc. Geol. Fr.*, 3, 76-6.

Malod, J.A., Murillas, J., Kornprobst, J. and Boillot, G., 1993.Oceanic lithosphere at the edge of a Cenozoic active continental margin (northwestern slope of Galicia Bank, Spain). *Tectonophysics*, 221, 195-206.

Mauffret, A. and Montardet, L., 1987, Rift tectonics on the passive continental margins off Galicia (Spain) : *Marine and Petroleum Geology*, 4, 49-70.

Mauffret, A., Mougenot, D., Miles, P.R. and Malod, J., 1989. Cenozoic deformation and Mesozoic abandonned spreading centre in the Tagus Abyssal Plain (west of Portugal). Results of a multichannel seismic survey. *Can. J. Earth Sciences*, 26, 1101-1123.

Montadert, L., De Charpal, O., Roberts, D.G., Guennoc, P. and Sibuet, J.C., 1979. Northeast Atlantic passive continental margin : rifting and subsidence processes. In Talwani, M. and Ryan, W.B.F. (Eds.), Deep Drilling results in the Atlantic ocean : Continental margin and paleoenvironment, *M. Ewing Series 3*, 154-186.

Mougenot, D., 1989. Geologia da margem portuguesa. Documentos tecnicos, *Instituto Hidrografico*, Lisboa, Portugal, 259 pp.

Pinheiro, L.M., Whitmarsh, R.B. and Miles, P.R., 1992. The ocean-continent boundary off the western continental margin of Iberia-II. Crustal structure in the Tagus Abyssal Plain. *Geophys. J. Int.*, 109, 106-124.

Reston, T.J., Krawczyk, C.M. and Hoffmann, H.J., in press. Detachment tectonics during Atlantic rifting : analysis and interpretation of the S reflection, the west Galicia margin. *Geol. Soc. Sp. Publ., London*.

Sawyer, D.S., Whitmarsh, R.B., Klaus, A. et al., 1994. *Proceedings of Ocean Drilling Program, Initial Reports, College Station, TX*, 149, 719p.

Schärer, U., Kornprobst, J., Beslier, M.O., Boillot, G. and Girardeau, J, in press. Gabbro and related rock emplacement beneath rifting continental crust : U-Pb geochronological and geochemical constraints for the Galicia passive margin (Spain). *Earth Planet. Sci. Lett.*

Sibuet, J.C., Ryan, W.B.F. et al., 1979 : *Initial Reports of the Deep Sea Drilling Project*, Washington, D.C., U.S. Government Printing Office, 47, 2, 787 p.

Sibuet, J.C., Amortila, P. and Le Pichon, X., 1987, Physiography and structure of the western Iberian continental margin off Galicia from Sea-beam and seismic data. In Boillot, G., Winterer, E.L. et al. *Proceedings of the Ocean Drilling Program, Initial Reports : College Station, TX*, 103, 77-97.

Sibuet, J.C., 1992. New constraints on the formation of the non-volcanic continental Galicia-Flemish Cap conjugate margins. *Jour. of the Geol. Soc., London*, 149, 829-840.

Thommeret, M., Boillot, G. and Sibuet, J.C., 1988. Structural map of the Galicia margin. In Boillot, G., Winterer, E.L. et al. *Proceedings of the Ocean Drilling Program, Sci. Results : College Station, TX*, 103, 31-36.

Wernicke, B., and Burchfiel, B.C., 1982. Modes of extensional tectonics : *Journal of Structural Geology*, 4, 105-115.

Whitmarsh, R.B., Miles, P.R. and Mauffret, A., 1990. The Ocean-Continent boundary of the western continental margin of Iberia - I. Crustal structure at 40°30'N. *Geophys. J. Int.*, 103, 509-531.

Whitmarsh, R.B., Pinheiro, L.M., Miles, P.R., Recq, M. and Sibuet, J.C., 1993. Thin crust at the western Iberia ocean-continent transition and ophiolites. *Tectonics*, 12, 1230-1239.

DETACHMENT FAULTING AND CONTINENTAL BREAKUP: THE S REFLECTOR OFFSHORE GALICIA.

C. M. KRAWCZYK and T. J. RESTON
GEOMAR Research Centre,
Christian-Albrechts Universität zu Kiel,
Wischhofstr. 1-3,
D24148 Kiel,
Germany

ABSTRACT. The west Galicia rifted margin is a non-volcanic rifted margin, characterised by landward tilted fault blocks bound by oceanward dipping faults. The fault blocks are underlain on time sections by an undulating group of bright reflections, the so-called Galicia S reflector. We use a variety of advanced processing techniques to constrain the nature of S, including prestack depth migration to construct detailed velocity models of the section overlying S, and to provide optimum images of portions of the data in depth. Our results show that S passes continuously beneath the tilted fault blocks of the deep margin. We interpret S and associated reflectors as a brittle detachment fault system, similar to those observed in the western USA. We infer that the detachment was active during the last phase of extension immediately prior to continental breakup. To the west, S appears to be truncated by east-dipping reflections coming off a ridge of peridotite. This ridge, drilled during ODP Leg 103 is characterised by top-to-the-east shear structures. We relate these east-dipping reflections, not S, to this shear zone and propose that this east-dipping structure (rather than S) was responsible for the exhumation of the mantle rocks. Our results are discussed in the context of the evolution of the margin.

1. Introduction

The highly extended continental margin to the west of the Galicia Banks (offshore Spain) is an ideal place to study extreme continental extension as it is covered by only a thin and discontinuous sedimentary cover, meaning that deep crustal levels are present close to the seafloor. Furthermore, it is crossed by a series of old (1975-1980) but remarkably high quality seismic reflection profiles (Figure 1 - Mauffret and Montadert, 1987), which, supplemented by the results of drilling on ODP Leg 103 (Boillot et al., 1988a) and subsequent sampling from submersible (Boillot et al., 1988b), reveal the detailed stratigraphy and structure of this margin.

Rifting took place from 140 Ma until 114 Ma when continental breakup occurred (Boillot and Winterer, 1988 and papers therein; Mauffret and Montadert, 1987; Boillot et al., 1988a). The seismic profiles show that just east of a ridge of serpentinised peridotite, taken to represent the continent-ocean boundary (Boillot et al., 1988a), the margin is characterised by east-tilted blocks bound by west-dipping normal faults. Underneath these blocks, a bright, but locally complex and generally (at least on time sections) undulating reflection or group of reflections, the so-called S reflector (de Charpal et al., 1978), is observed at about 8-10 s TWT. As the westernmost blocks have been shown by samples obtained through drilling and submersible diving (Boillot et al., 1988b) to be cored by basement material, the S reflector is demonstrably overlain by crystalline basement and does not represent an intra-sedimentary boundary.

231

E. Banda et al. (eds.), Rifted Ocean-Continent Boundaries, 231–246.
© *1995 Kluwer Academic Publishers. Printed in the Netherlands.*

232

In addition to a core of crystalline basement, the tilted blocks have been shown by drilling to consist of prerift Tithonian carbonates and synrift (Hauterivian-Valanginian) clastic sedimentary units (described as Synrift I - Mauffret and Montadert, 1987; Boillot et al., 1988a). Thus at least part of the synrift sedimentary sequence predates tilting and the accompanying normal faulting. In contrast, later Aptian-Hauterivian units (interpreted as Synrift II - Mauffret and Montadert, 1987) were deposited within the halfgrabens between the fault blocks, in places thickening towards the faults, and postdate the onset of the second phase of faulting. Thus two phases of rifting can be identified (Mauffret and Montadert, 1987), associated with these two sedimentary sequences (Synrift I and Synrift II), one from 140 Ma to 125 Ma, the other from 125 Ma until eventual breakup at 114 Ma.

Although it is generally accepted that the S reflector is related to the extreme extension of the continental lithosphere, the precise nature of the S reflector is the subject of much current debate. It has been interpreted within different models for lithospheric extension as a brittle-ductile transition (de Charpal et al., 1978), as part of a lithosphere-penetrating normal fault, dipping either west (Winterer et al., 1988) or east (Boillot et al., 1988a), as an intracrustal detachment fault (Sibuet, 1992), as the crust-mantle boundary underlain by serpentinised peridotites and cut by late normal faulting immediately prior to continental breakup (Boillot et al., 1989), and as the top of a layer of mafic underplate (Horsefield, 1992). Thus a study of the S reflector is required to understand the evolution of this margin, and to further our understanding of the mechanics of extreme continental extension in general.

Previous interpretations of the S reflector, and of the structure of the margin, have been made on the basis of time sections, with only simple depth conversions based on general velocity models (Sibuet, 1992). On migrated time sections, S appears to undulate beneath the fault blocks (Figure 2): although it has long been recognised (Mauffret and Montadert, 1987) that some of this may represent the effects of velocity pull-up and push-down, without detailed velocity information it has not been possible to determine the true

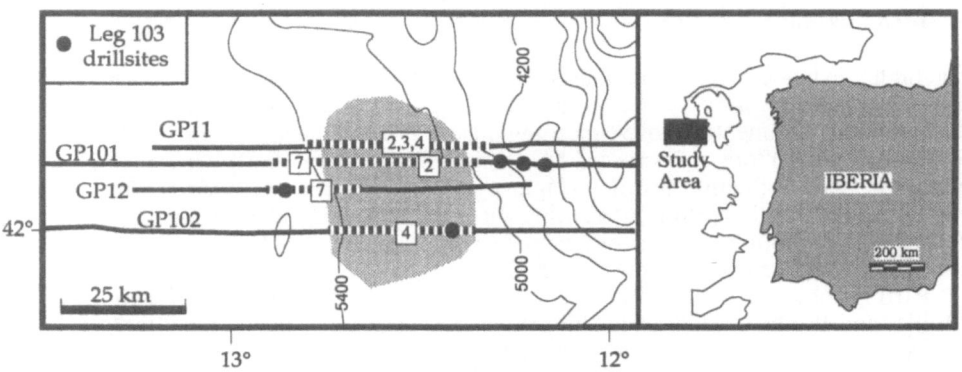

Figure 1: Map of the Galicia margin showing the four margin-normal profiles that image the S reflector. Approximate areal extent of the S reflector indicated by shading. Data shown indicated by striped line and numbered by figure.

geometry of S in two dimensions, let alone three. Furthermore, the image of the S reflector on some of the migrated time sections is discontinuous and complex: in places S appears as a reflection only a couple of cycles thick, but elsewhere S has been described (Sibuet, 1992) as a complex band of reflections up to a second thick: towards the eastern end of the profiles shown in Figure 2, no single S reflector can be identified with confidence, but rather a complex zone of reflectivity is observed. Finally, the relationship of S to the overlying block-bounding faults is far from clear on the time sections due to the complex, discontinuous image and the effects of velocity pull-up and push-down, leading to the suggestion that S could in places be offset by the block-bounding faults (Boillot et al., 1989).

This paper sets out to constrain the nature of S and its relationship to the overlying faults and other key structures of the margin by providing an improved image of critical portions of the data in depth. Furthermore, some key results from analysis of the waveform of the S reflector will be presented, as these demonstrate conclusively that S cannot be the brittle ductile transition. The results will then be placed in the context of the known geology of the margin, and a model for lithospheric extension leading to continental breakup advanced.

2. Method

Interpretations of the S reflector have until recently been hampered by its apparently variable character (Figure 2), the uncertainty regarding its relationship to the overlying block-bounding faults, and the lack of accurate velocity information to convert a seismic *time* section into a more geologically meaningful depth section. We have addressed these problems by applying the process of iterative prestack depth migration to derive a detailed velocity model in the section overlying S, and to provide an improved image of the structure in *depth* (Figure 3) rather than time. Depth migration also corrects for raypath bending due to refraction at velocity interfaces and gradients (e.g. Yilmaz, 1987). Such effects have been shown to be particularly important at rifted margins (Peddy et al., 1986), where normal faulting has juxtaposed tilted blocks of high velocity basement rocks against low velocity sediments, or even water.

By performing the depth migration simultaneously in the shot and in the receiver domains *before stack* and summing within the migration process, the smearing effects of conventional CMP stacking are avoided, thus providing a clearer image (Sherwood, 1989). Furthermore, by performing this process *iteratively* it is possible to construct a detailed velocity model through depth-focussing error analysis (Denelle et al., 1986; Reston, et al., 1995). During prestack depth migration simultaneously in both the shot and receiver domains, maximum focussing of reflected and diffracted energy occurs when the downward continuation operator reaches the actual depth of the reflector, if the correct velocity function is used. If an incorrect velocity function is used, then there is a discrepancy between the depth of maximum focussing and the depth to the reflector. The resulting depth-focussing error can then be used to correct the velocity field for the layer in question.

It is necessary to build up the velocity model one layer at a time, as the overburden velocities affect those determined for deeper levels. Thus we first check that the waterbottom reflection focusses with water velocity (and so check the line geometry), then incorporate that velocity into the next iteration when the velocity of the topmost sedimentary layer is determined, again through depth-focussing error analysis. The procedure is repeated until the entire section has been analysed for velocity, resulting in an optimised depth migration, and an interval velocity model that is both detailed and

234

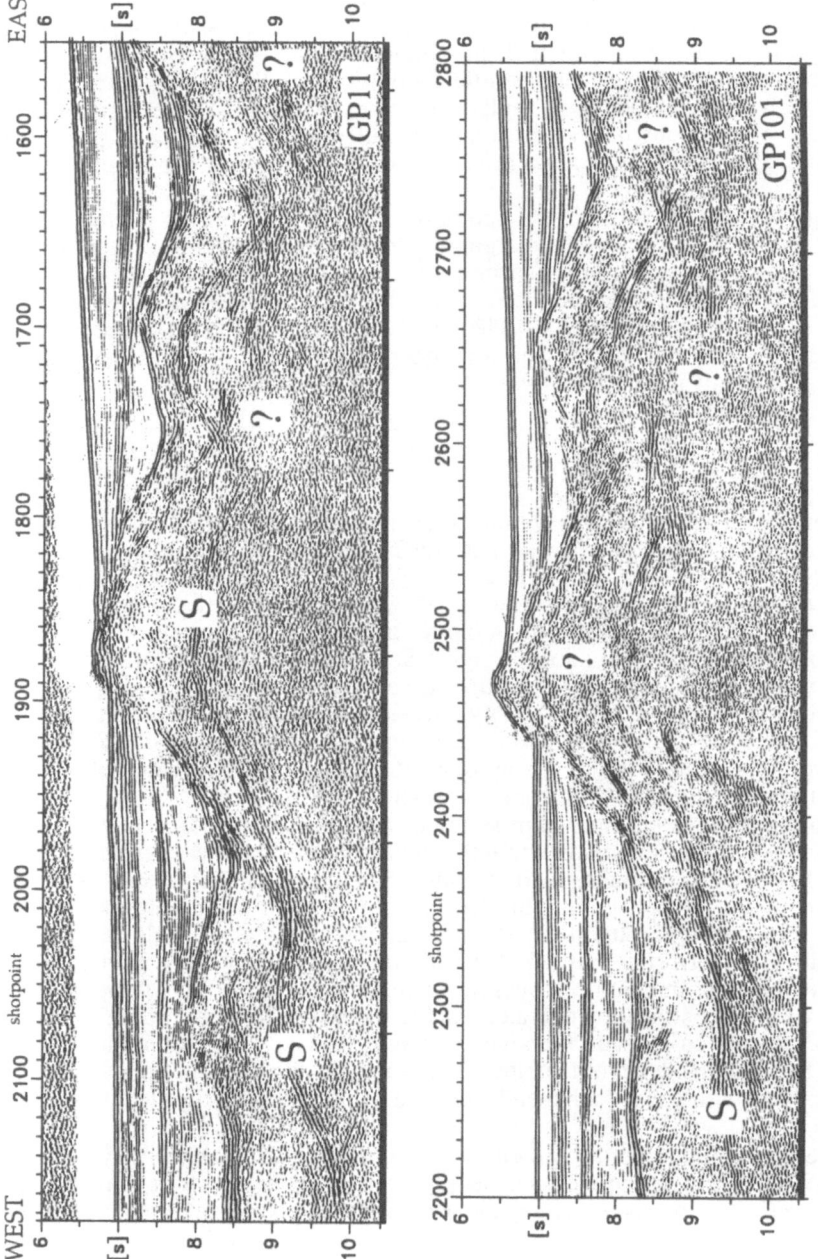

Figure 2: Time migrated sections of profiles GP11 and GP101 (for location see Figure 1). The S reflector is imaged as a locally complex, undulating group of bright reflections beneath the tilted fault block topography characteristic of the margin. Shot interval 50 m.

geologically meaningful. The layers are generally chosen to follow stratigraphic marker horizons and unconformities within the sedimentary sequence: in the basement the layers are defined by the presence of strong and continuous reflections as such reflections provide the only opportunity to pick velocity using depth-focussing error analysis. Velocity within the layers can vary both vertically and horizontally, allowing the construction of a detailed velocity model with relatively few layers. In practice we have found that little or no improvement is achieved by having many layers, and that the use of vertically varying velocity (defined as a compaction factor) within a few layers generally provides a better image than many constant velocity layers. For the data analysed here, the postrift sequence generally comprised three or four velocity layers, the synrift II another, the synrift I and prerift one layer, and the basement above S another layer. Generally no reliable picks were obtained below S, so the velocity there was arbitrarily assigned a value of 6000-6500 km/s.

Iterative prestack depth migration does have two main drawbacks. First, it is very computer intensive, so we apply it only to those portions of the data likely to benefit most (e.g. where tilted fault blocks are present). Second, it requires bright continuous reflections to provide the reliable foci necessary to construct an accurate velocity model. Without such a model, the method can produce substantially worse images than less velocity-sensitive processing. Consequently we also use standard time-migrated profiles in our interpretation.

3. Results

The benefit of applying iterative prestack depth migration is apparent in a comparison between Figure 2a and Figure 3. Figure 2a is the result of optimum standard processing through to time migration after DMO stack of a portion of profile GP11. Note that on this section, although some faults are clearly imaged, others are not and can only be inferred from the tilted block - half graben structure of the margin. Furthermore, the internal structure of the fault blocks is not clear. However most fundamentally, the S reflector appears to undulate dramatically beneath the fault blocks and that towards the east of the section, its image becomes very confused.

In contrast, Figure 3 is the product of eight iterations of prestack depth migration, resulting in the optimum section shown here. The most striking result is that S no longer undulates beneath the fault blocks but rather has a relatively simple domal shape. Furthermore, the clearer image reveals that rather than being complex and variable, S appears to have a remarkably consistent character, and hence presumably internal structure.

The depth section (Figure 3) plus the velocity model (Figure 4) derived from depth-focussing error analysis reveal the internal structure of the fault blocks. In particular the combination of the improved image and the detailed velocity model allows the contact between the base of the tilted sedimentary sequence and the top basement to be picked with more confidence than before. This allows the identification of a further two faults that had not previously been recognised from the time sections.

There remain patches of the section, perhaps regions of poor signal penetration where the image of S is discontinuous. However, the simple domal shape of S on the depth section means that S can be extrapolated through these patches with confidence. We emphasise that we do not consider that the S reflector itself is discontinuous but only that the reflection from it is. Rather we interpret the smooth overall shape of S as an indication that the reflector is a continuous feature, passing beneath the fault blocks without disruption.

236

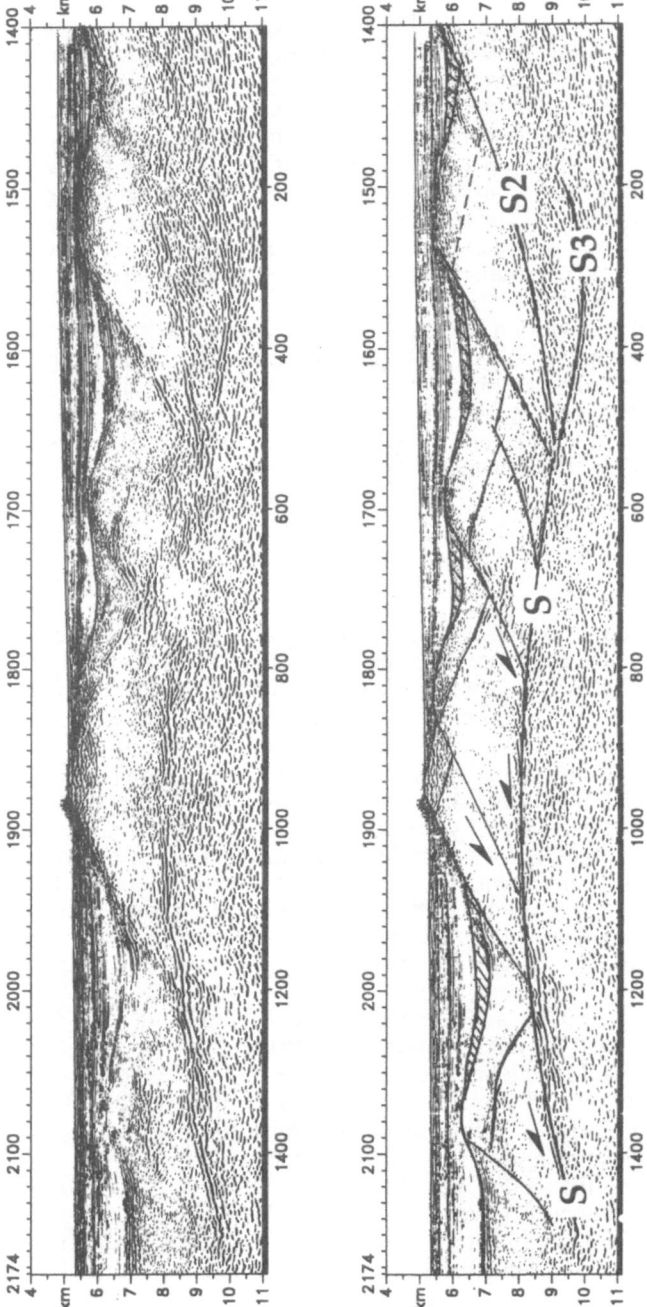

<u>Figure 3</u>: Prestack depth migration of profile GP11. The S reflector is imaged as a smooth, relatively continuous, simple domal structure, which splits into three at the eastern end of the profile. The structure of the fault blocks is more clearly imaged now, revealing previously unidentified faulting. The faults appear to detach onto S. Synrift II is shaded. No vertical exaggeration.

The only portion of the data where the identification of S remains ambiguous is at the eastern end of the profile. Here, where the time section showed a complex and discontinuous image of S, the improved imaging from prestack depth migration has resolved three separate structures, of similar continuity and character, all of which are connected to S at shotpoint 1670. One rises at c. 25° towards a half-graben (Figure 4) and is probably a normal fault, another (S2) rises more gently towards the east, but again to the boundary between two fault blocks and is thus also probably a normal fault, whereas the third (S3) dips to the east, flattens at about SP 1550 and then appears to turn up towards the surface, before disappearing at about SP1500.

We have previously applied the same technique to part of GP102 (Hoffmann and Reston, 1992). This profile had been previously interpreted as showing that the block-bounding faults cut and offset the S reflector (Boillot et al., 1989). Although on this profile the faults are not themselves unambiguously imaged even after depth migration, the detailed velocity structure derived from depth-focussing error analysis, coupled with the geometry of the tilted sedimentary units show that this profile is characterised by similar features to GP11: S passes as a continuous feature beneath the blocks, which contain both basement and tilted sedimentary units (Figure 4b). Critically, S does not appear offset by the block-bounding faults. It is now relevant to consider what sort of structure S might represent.

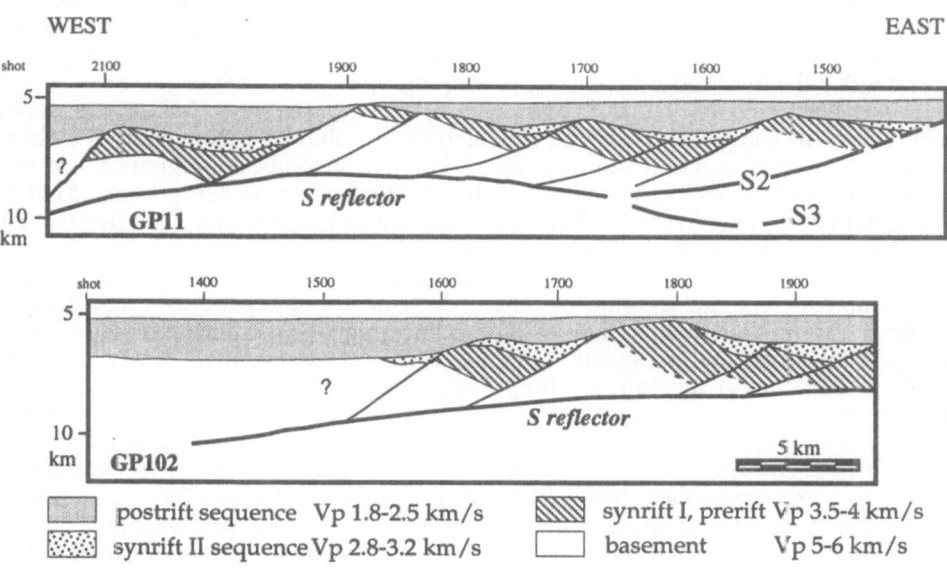

Figure 4   Simplified velocity models for profiles GP11 and GP102 produced by iterative prestack depth migration. The velocity information allows the identification of the various units, and hence aids interpretation. Interpreted faults and the S reflector are also shown. No vertical exaggeration.

4. Discussion: the Nature of the S Reflector

A number of bright intracrustal reflections from a variety of settings, including the western United States (Goodwin et al., 1989; Litak and Hauser, 1992) and the Southern Appalachians (Pratt et al., 1991; Barnes and Reston, 1992), have been shown to be reflections from thin high velocity layers, interpreted as igneous intrusions. Several such bright features have been drilled in the Siljan structure of Sweden (Juhlin et al., 1990), and shown to be igneous intrusions. Hence one possibility that must be considered is that the S reflector represents a thin igneous intrusion of some kind. If so, it must have intruded at or after the end of rifting, as it is not offset by the block-bounding faults.

A simple comparison (Figure 5) of the waveform of S with that of the seafloor demonstrates that this interpretation is rather unlikely. S has a waveform virtually identical to that of the seafloor reflection, which can be shown by a comparison with the seafloor multiple to represent a reflection from a single interface. Hence S can also be taken as a reflection from a single interface rather than from the thin layer expected for an igneous intrusion (Reston et al., 1995).

It is also unlikely that S represents the top of a large igneous intrusion, again intruded at the end of rifting as it is not offset by block-bounding faults. First, no reflection corresponding to the base of such an intrusion is observed (Reston et al. 1995), and second, it seems unlikely that the base of the fault blocks/top of the intrusion would be such a smooth surface in depth: instead one would expect that as the overlying fault blocks floated on top of the crystal mush of such an intrusion, the thicker portions of the blocks would have developed keels relative to the thinner portions of the blocks, according to the principle of isostasy. Instead, the remarkably smooth shape of S indicates that the lower plate to S must possess significant flexural strength.

It has been suggested that S represents a transition between brittle and ductile deformation (de Charpal et al., 1978), with deformation above S accommodated by block faulting, and that beneath S by ductile flow. However three lines of argument lead us to discount this hypothesis. First, analysis of the amplitude and waveform (Figure 5) of the S reflector (Reston et al., 1995) show that the brightest portions of the S reflector (e.g. profile GP12 shotpoints 450-750) represents a sharp increase in acoustic impedance (a reflection coefficient of c. 0.2 - Reston et al., 1995), in effect a single interface. This is consistent with a sharp tectonic boundary of some form, separating quite different lithologies above from below (the lower plate to S might be continental lower crust, or perhaps serpentinised upper mantle), but not with a transition zone. Second, the depth sections show that at the end of rifting S was in places less than 1 km beneath the seafloor, and is never more than 3 km beneath the top of the fault blocks. Even in regions of extreme extension, this is within the shallow seismogenic zone (e.g., Jackson, 1987) where brittle deformation should dominate. Finally, as for the igneous intrusion hypothesis, if S were a brittle-ductile transition it would seem unlikely that the fault block topography could be supported, as the ductile lower crust would presumably have little flexural strength. S is however a remarkably smooth feature, passing beneath lateral changes in fault-block thickness without distortion. Because of these objections, we consider it most unlikely that S represents the brittle-ductile transition. Instead, we propose that the relationship between S and the overlying faults is most consistent with an interpretation of S as a brittle detachment fault, probably similar to those mapped out in the western United States.

The relationship between the block-bounding faults and the S reflector is particularly important as it provides one of the strongest constraints on the role of S during rifting. On profile GP101 (Figure 6), it can be seen that one of the block-bounding faults (which detaches onto S) also bounds a wedge-shaped packet of sediments, which thickens

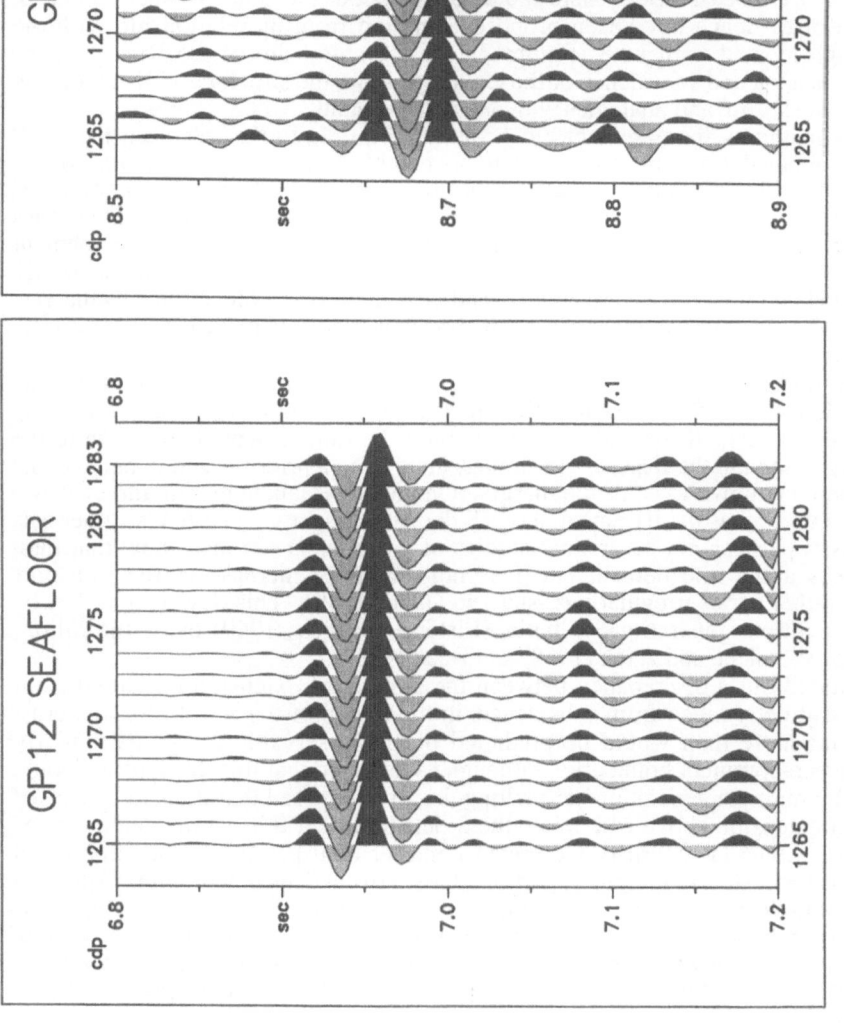

Figure 5: Waveform of the S reflector in comparison with that of the seafloor (minimum-phase filtered to the same bandwidth as S, simulating the effects of attenuation). As the waveforms are almost identical, it is not only clear that S is positive polarity, but also that it is probably a reflection from a simple interface. S may be the tectonic contact between an upper crustal upper plate and a high velocity and density lower plate.

towards the fault. These have been identified as Synrift II sediments (Mauffret and Montadert, 1987; Winterer et al., 1988) deposited immediately prior to breakup during extension along that fault. As it is precisely those faults active during the last phase of rifting (i.e. those that cut the Synrift I sequence and bound the Synrift II sequence) which appear to detach onto S, we suggest that at least S was active right up to continental breakup.

4.1. SENSE OF MOVEMENT ALONG S

If S is a detachment fault analogous to those observed in the western United States (e.g. Lister and Davis, 1989), then movement along S should have a consistent sense. Some have interpreted this to be top-to-the-west (Winterer et al., 1988) but most recent workers (Boillot et al., 1988; Sibuet, 1992) propose that S accommodated top-to-the-east movement. As this controversy is clearly important to an understanding of the evolution of this margin, we discuss briefly here the evidence about the sense of shear along S. Various lines of evidence have been proposed, ranging from its relationship with the peridotite ridge located west of S (Figure 7), the structural dip of the Biscay S, the measurable extension and subsidence of the Galicia margin, the structure of the conjugate margin off Flemish Cap, and analogies with detachment faults.

The faults bounding the basement blocks above S all dip to the west (e.g Figure 4) and all accommodated top-to-the-west normal movement. If S accommodated top-to-the-east shear, all the block-bounding faults would be antithetic to S; if S accommodated top-to-the-west shear, the faults would be synthetic to S. Mapping of detachment terranes has shown that faults overlying a detachment are generally synthetic to that detachment (e.g., Lister and Davis, 1989 - see also Figure 8a): by analogy the S detachment would have accommodated top-to-the-west motion and would cut up to the surface to the east (the breakaway). However, the location of the breakaway to S is unclear as reflector S3 cannot be traced to the surface.

Le Pichon and Barbier (1987) used the thickness of basement within fault blocks overlying the Biscay S reflector to infer that the Biscay S cut to depth under the margin, that is to the north-east. If the Galicia S were to do the same it would cut down to the east. However, although the features are similar, there is no a priori reason to believe that they should both cut down towards the margin. On the west Galicia margin the evidence is somewhat equivocal as the basement above S does not increase consistently either east or west. This is perhaps not surprising: studies of detachment terranes show that fault block thickness above the detachment does not generally increase in the transport direction, but rather varies irregularly (Lister and Davis, 1989). This observation may be explained by multiple generations of faults (Lister and Davis, 1989) or by the rolling hinge model of Wernicke and Axen (1988) and Buck (1988).

Sibuet pointed out a discrepancy between the amount of extension measured from faulting, and that determined from subsidence: in the region of the S reflector, the crust was thinned far more than would be predicted by the observable faults, particularly towards the ocean-continent boundary. Although as he noted that this discrepancy could be explained by multiple generations of faulting, he also proposed that it could be caused by differential extension above and below the S detachment if S accommodated top-to-the-east shear. Sibuet (1992) also interpreted a master fault parallel to the continental slope of the SE Flemish Cap margin as part of a large crustal detachment fault. Although he could not follow the eastward continuation of this structure towards the continent-ocean boundary, Sibuet proposed that this fault represented the westward continuation of the S reflector, and located the breakaway to S on the Newfoundland side. However, as discussed below, it is possible that such an east-cutting detachment can be present but not

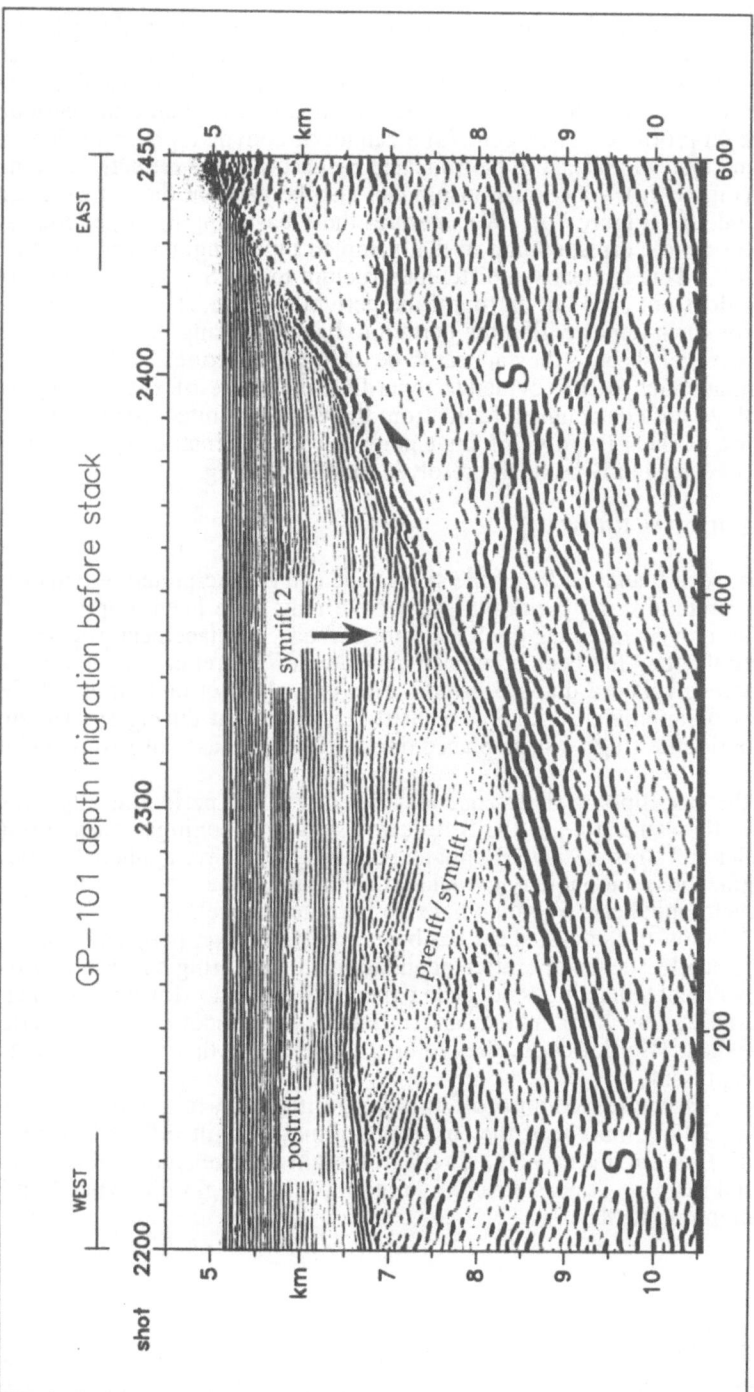

Figure 6: Detail of profile GP101, showing a block-bounding fault detaching onto the west-dipping S reflector. Synrift sediments located in the hangingwall to this fault were deposited during the latest phase of rifting and have not been disturbed since.

be a continuation of the S reflector.

Another argument that has been used to infer top-to-the-east shear along S is the discovery of top-to-the-east shear structures on the top of the peridotite ridge, interpreted as the westward continuation of an east-cutting S (Boillot and others, 1988a; 1989). However, the migrated sections show that S cannot be directly traced onto the peridotite ridge. For instance on profile GP12 (Figure 7a) S cannot be convincingly identified west of shotpoint 900 and instead appears to be truncated by east-dipping reflections coming off the peridotite (Figure 7a). This relationship can also be seen on the corresponding portion of profile GP101 (Figure 7b). We interpret these east-dipping reflections as a top-to-the-east shear zone, corresponding to that sampled by diving and drilling on the peridotite ridge itself. This is clearly a different structure to the S reflector, indicating that the S reflector does not crop out on top of the peridotite ridge. Hence the sense of shear determined for shear zones within the peridotite has no bearing on that of S: they can have opposite sense of shear as they appear to be different structures.

We conclude that much of the evidence regarding the sense of shear along S is equivocal. By analogy with detachment faults from the western United States we infer a top-to-the-west sense of shear (Figure 8b), but cannot rule out alternative interpretations. The sense of shear of S may only be finally resolved through drilling.

4.2. S AND THE PERIDOTITE RIDGE

The relationship between S and the peridotite shear zone can be interpreted in one of two ways. In one interpretation, the peridotite shear zone may have been emplaced after breakup, that is after movement along the S reflector ceased. Emplacement of peridotite after the end of crustal rifting has been previously proposed (Winterer et al., 1988; Sibuet, 1992), both of whom invoked a diapiric origin. However, Boillot and others (1988a) interpret sediments onlapping the ridge as Synrift II, deposited during final rifting, implying that the peridotite ridge was emplaced during the last phases of rifting and not after breakup.

Alternatively, the east-dipping shear zone associated with the peridotite ridge might have been active at the same time as the S reflector, and be a conjugate structure (cf Reston, 1990; Beslier and Brun, 1991). Boillot and others (1995) have applied this model to the Galicia margin, and point out that it would predict that the shear zone bounding the peridotite (and in their model also S) would have both top-to-the-west and top-to-the-east senses of shear. Only top-to-the-east shear has been observed there, perhaps because of the erosion of the mantle shear zone after denudation, highlighting the need to drill through such a detachment where it has not been exposed to erosion. However, further to the north, a similar shear zone marking the eastern limit of the peridotite is characterised by top-to-the-west shear (Boillot et al., this volume), suggesting that such a conjugate deformation zone is present.

The east-dipping structures coming off the peridotite ridge might represent the continuation on the Galicia margin of the east-dipping master fault inferred by Sibuet (1992) from Flemish Cap. In this case, the S reflector might be unrelated to the structure of the Newfoundland margin. A possible configuration of the margin immediately prior to breakup is shown in Figure 8b.

5. Conclusions

Our results show that the block-bounding faults appear to detach onto the underlying S reflector, which nowhere appears offset, but instead displays a smooth, simple geometry.

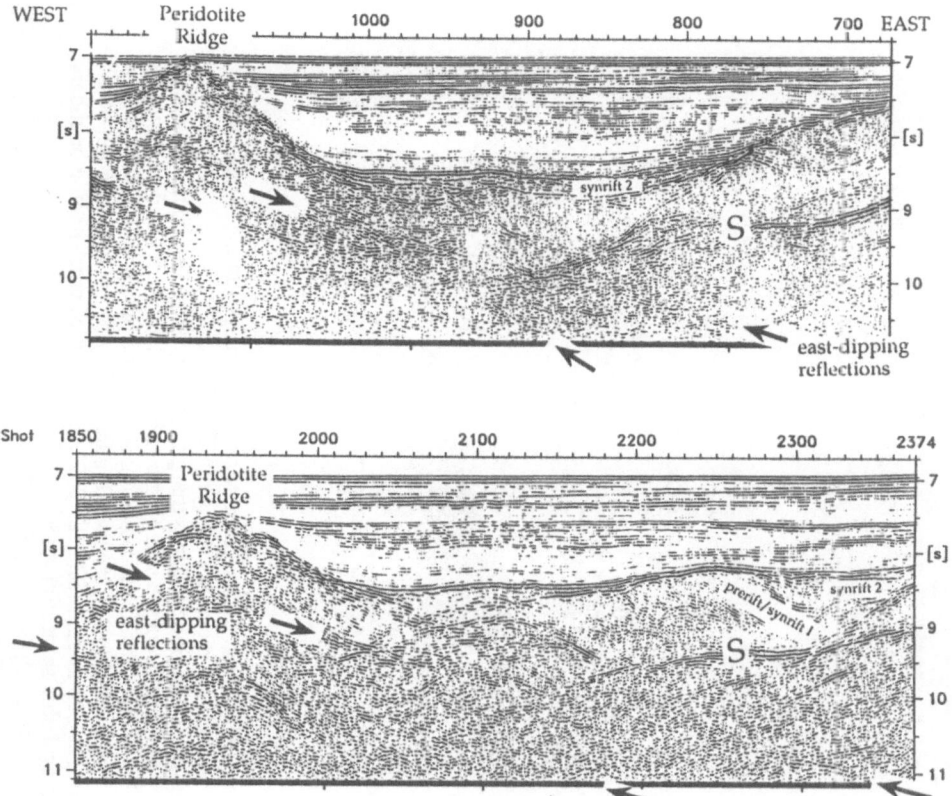

Figure 7:
Detail of the region between S and the peridotite ridge on two migrated profiles, GP12 and GP101.
West-dipping S appears to be truncated by east-dipping reflections coming from the peridotite. These
may be shear zones related to the exhumation of the peridotite. Shot interval 50 m.

244

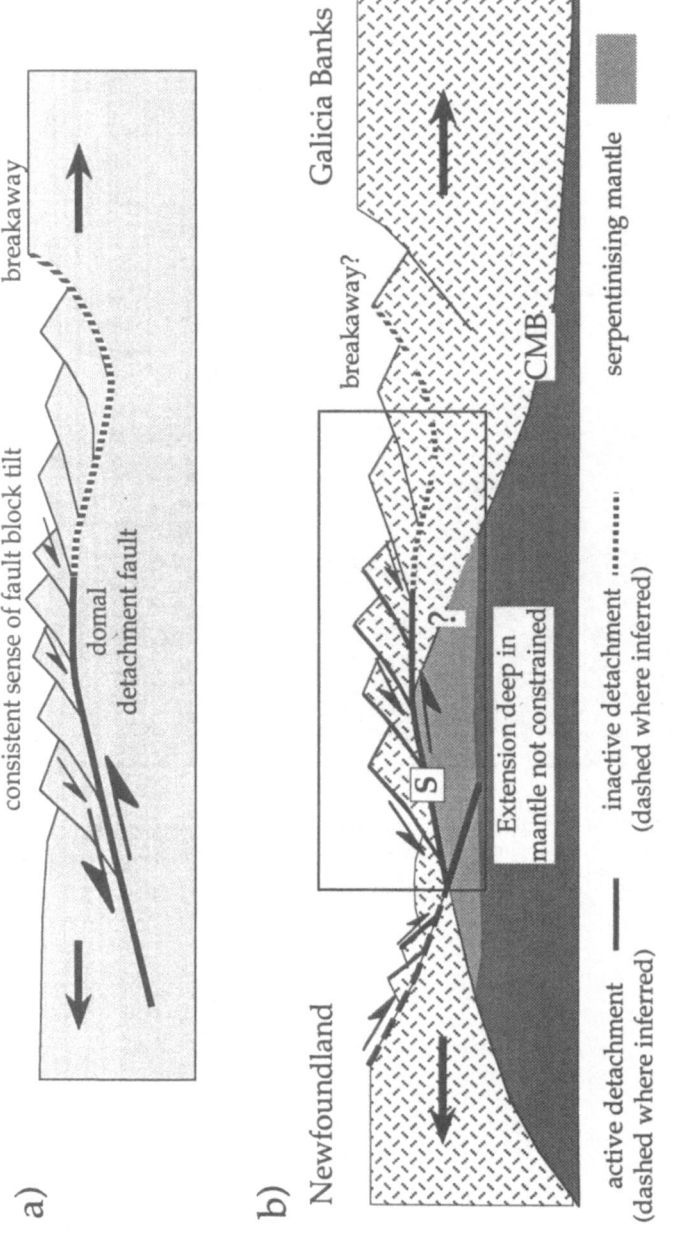

<u>Figure 8:</u> a) Cartoon (from Lister and Davis, 1989) summarising geometry of detachment fault terranes. Note that detachment fault is synthetic to overlying block-bounding faults.
b) Modified detachment model applied to the Galicia Banks margin, shortly before breakup. Box marks portion of system imaged by profiles discussed here (Figures 3, 6 and 7). S acts as detachment to overlying faults (Figures 3 and 6) but is truncated by east-dipping shear zone linked with emplacement of peridotites sampled at surface. This east-dipping structure (Figure 7) may be the continuation of a detachment interpreted on SW Flemish Cap margin (Sibuet, 1992). Location of crust-mantle boundary (CMB) beneath rift unclear, due to probable serpentinisation.

The block-bounding faults represent only the last phase of extension, as the sediments tilted within the blocks include earlier synrift units. From these observations we infer that S was a detachment fault, active just before breakup. Towards the east, S appears to split into a couple of separate features, which we interpret as part of the same low-angle normal fault/detachment fault system. Towards the west, S appears to be truncated by east-dipping reflections related to the peridotite ridge. These may have been responsible for the emplacement of the peridotite, and may be related to the master fault described by Sibuet from the conjugate Flemish Cap margin (Figure 8b).

The sense of shear of S is not fully constrained: the sense of tilt of the fault blocks suggests shear along S was top-to-the-west, whereas other workers have presented a variety of arguments that shear was top-to-the-east. Although we consider it most likely that S cuts down to the west and accommodated top-to-the-west shear, this issue can perhaps only be fully resolved by drilling S. Clearly this is a topic of considerable importance, as it has implications not only for the structure of this margin but also for that of the conjugate SW Flemish Cap margin.

6. Acknowledgements

This work was made possible by generous financial support from the Deutsche Forschungsgemeinschaft (DFG) under grant Re 873/1. The data were made available by the Institut Français du Pétrole through Dr. J. Wannesson, whose help we gratefully acknowledge. Dirk Klaeschen, D. Wilkens and H.-J. Hoffmann have at various times contributed to the processing of the data. We would like to thank Gilbert Boillot and Marie-Odile Beslier for much discussion about the nature of S. Finally, one of us (TJR) wishes to thank Enric Banda for an invitation to attend the Mallorca workshop, and the organisers and participants for making that meeting so enjoyable.

7. References

Barnes, A. and Reston, T., 1992, A study of two mid-crustal bright spots from southeast Georgia (USA), Geophysical Journal International, 108, 683-691.

Beslier, M-O. and J. P. Brun, 1991, Boudinage de la lithosphère et formation des marges passives, C. R. Acad. Sci. Paris, v. 313, p. 951-958.

Boillot, G. and Winterer, E.L., 1988. Drilling on the Galicia margin: retrospect and prospect. In Boillot, G., Winterer, E.L., et al., Proc. ODP, Sci. Results, 103: College Station, TX (Ocean Drilling Program), 809-828.

Boillot, G., Girardeau, J. and Kornprobst, J., 1988a. The rifting of the Galicia margin: crustal thinning and emplacement of mantle rocks on the seafloor. In Boillot, G., Winterer, E.L., et al., Proc. ODP, Sci. Results, 103: College Station, TX (Ocean Drilling Program), 741-756.

Boillot, G., Comas, M.C., Girardeau, J., Kornprobst, J., Loreau, J.P., Malod, J., Mougenot, D. and Moullade, M., 1988b. Preliminary results of the Galinaute cruise: dives of the submersible Nautile on the western Galicia margin, Spain. In Boillot, G., Winterer, E.L., et al., Proc. ODP, Sci. Results, 103: College Station, TX (Ocean Drilling Program), 37-51.

Boillot, G., Féraud, G., Recq, M. and Girardeau, J., 1989. "Undercrusting" by serpentinite beneath rifted margins: the example of the west Galicia margin (Spain). Nature, 341: 523-525.

Boillot, G., Beslier, M.-O., Krawczyk, C., Rappin, D., and Reston, T., The formation of a passive margin: constraints from the crustal structure and segmentation of the deep Galicia margin. in: Tectonics, Sedimentation and Paleoceanography of the North Atlantic Margins. (edited by R Scrutton) Geol. Soc. London Spec. Volume, in press 1995.

Boillot, G., Beslier, M.-O., and Girardeau, J., 1995. Nature and evolution of the ocean-continent boundary: the lesson of the west Iberia margin. This volume.

246

Buck, W. R., 1988, Flexural rotation of normal faults, Tectonics, v. 7, p. 959-973.

de Charpal, O., Guennoc, P., Montadert, L. and Roberts, D.G., 1978. Rifting, crustal attenuation and subsidence in the Bay of Biscay. Nature, 275: 706-711.

Denelle, E., Dezard, Y. and Raoult, J., 1986. 2-D prestack depth migration in the (S-G-W) domain. Extended abstract, 56 th SEG Meeting, Houston.

Goodwin, E., Thompson, G., and Okaya, D., 1989, Seismic identification of basement reflectors: the Bagdad Reflection Sequence in the Basin and Range Province-Colorado Transition Zone, Arizona, Tectonics, v. 8, p. 821-831.

Hoffmann, H.J. and Reston, T.J., 1992. The nature of the S reflector beneath the Galicia Bank rifted margin. Preliminary results from pre-stack depth migration. Geology, 20: 1091-1094.

Horsefield, S., 1992, Crustal Structure across the Continent-Ocean Boundary. Unpubl. PhD thesis, Cambridge University, 215 pp.

Jackson, J., 1987, Active normal faulting and crustal extension, in Coward et al., eds., Continental Extensional Tectonics, Geol., Soc. London Spec Publ., 28, p. 3-17.

Juhlin, C., 1990, Interpretation of the reflections in the Siljan Ring area based on results from Gravberg-1 borehole, Tectonophysics, v. 173, p. 345-360.

Le Pichon, X. and F. Barbier, 1987, Passive margin formation by low angle faulting within the upper crust: the northern Bay of Biscay margin, Tectonics, v. 6, p. 133-150.

Lister, G.S. and Davis, G.A., 1989. The origin of metamorphic core complexes and detachment faults formed during Tertiary continental extension in the northern Colorado River region, U.S.A. J. Struct. Geol., 11: 65-94.

Litak, R and Hauser, E., 1992, The Bagdad Reflection Sequence as tabular mafic intrusions: Evidence from seismic modelling of mapped exposures, GSA Bull, v. 104, p. 1315-1325.

Mauffret, A. and Montadert, L., 1987. Rift tectonics on the passive continental margin off Galicia (Spain). Mar. Pet. Geol., 40: 49-70.

Peddy, C., Brown, L., and Klemperer, S., 1986, Interpreting the deep structure of rifts with synthetic seismograms, AGU Geodynamics Series, v. 13, 301-310.

Pratt, T., Hauser, E., Hearn, T., and Reston, T., 1991, Reflection Polarity of the Midcrustal Surrency Bright Spot beneath southeastern Georgia: Testing the Fluid Hypothesis. JGR, 96B, 10145-10158.

Reston, T. J., 1990. The lower crust and the extension of the lithosphere: kinematic analysis of BIRPS deep seismic data, Tectonics, v. 9, 1235-1247.

Reston, T.J., Krawczyk, C.M. and Hoffmann, H.J., 1995. Detachment tectonics during Atlantic rifting: analysis and interpretation of the S reflector, west Galicia margin. in: Tectonics, Sedimentation and Paleoceanography of the North Atlantic Margins, (edited by R. Scrutton) Spec. Publ. Geol. Soc. London, in press.

Sibuet, J.C., 1992. New constraints on the formation of the non-volcanic continental Galicia - Flemish Cap conjugate margins. J. Geol. Soc. London, 149: 829-840.

Sherwood, J., 1989, Depth sections and interval velocities from surface seismic data, Leading Edge, v. 8, 9, p 44-49.

Wernicke, B., and Axen, G., 1988, On the role of isostasy in the evolution of normal fault systems, Geology, v. 16, p. 848-851.

Winterer, E., Gee, J. and van Waasbergen, R., 1988, The source area for lower Cretaceous clastic sediments of the Galicia margin: geology and tectonic and erosional history, n Boillot, G., Winterer, E.L., et al., Proc. ODP, Sci. Results, 103: College Station, TX (Ocean Drilling Program), 697-732.

Yilmaz, O., 1987, Seismic Data Processing, SEG publication.

LITHOSPHERIC TRANSITION FROM CONTINENTAL TO OCEANIC IN THE WEST IBERIA ATLANTIC MARGIN.

M. TORNE, M. FERNANDEZ, J. CARBONELL, and E. BANDA
Institute of Earth Sciences "Jaume Almera", CSIC.
Martí i Franqués s/n. 08028-Barcelona (Spain).

ABSTRACT. The lithospheric structure and degree of isostatic compensation of the Western Iberian Margin is studied along two profiles -A and B- which run perpendicular to the margin from the deep ocean to the continental platform. The profiles cut across the Iberian and Tagus Abyssal Plains, respectively. The lithospheric modelling, which is based on finite element technique, combines elevation, heat flow, and gravity data. This together with the computation of geoid anomalies allows us to better constrain how the structure of the crust and lithospheric mantle evolves across the margin. Our model results show that the crustal transition from continent to ocean is quite gentle and occurs over distances of about 115km, resembling the crustal attitude observed in other passive continental margins. The total lithospheric thickness remains constant along Profile A (110km) whereas along Profile B the lithosphere thickens from 110km below stable Iberia to 120 below the Tagus Abyssal Plain thinning towards the Madeira Tore Rise up to 95km.

1. Introduction

The western continental margin of Iberia is a non-volcanic rifted margin that extends from Cape Finisterre to the North (43°N) to Cape San Vicente to the South (37°N) (Fig. 1). The margin was formed by rifting when Iberia separated from the Grand Banks of Canada in Late Mesozoic times. From South to North it can be divided into three different segments: the Tagus Abyssal Plain (TAP), the Iberia Abyssal Plain (IAP) and the Galicia Bank (Fig. 1) which rifted successively as continental rifting propagated northward during the Early Cretaceous and was succeeded by seafloor spreading (Whitmarsh et al., 1993). In the southern Iberia Abyssal Plain, Whitmarsh et al. (1993) have proposed, from modelling surface and deep-tow magnetometer profiles, that seafloor spreading began at about 130 Ma at a rate of 10 mm/yr, while in the Galicia Bank region the break-up uncomformity is dated at about 114 Ma (Sibuet et al., 1979; Boillot et al., 1987). In contrast, the age of continental break-up in the Tagus Abyssal Plain is not well established. Following the work of Pinheiro et al. (1992) -based on seismic refraction and magnetic anomalies- sea-floor spreading would have begun at about 133 Ma, which is in accordance with the generally accepted hypothesis of northward propagation of sea-floor spreading in this region of the Atlantic.

The crustal structure of both plains is relatively well known from seismic wide-angle reflection and refraction data and from magnetic and gravity modelling (e.g., Purdy, 1975; Mauffret et al., 1989; Whitmarsh et al., 1990; Pinheiro et al., 1992). These workers have mainly focused on the location and crustal structure of the ocean-continent transition (OCT). A recent work by Whitmarsh et al. (1993) summarizes the main findings along the western

E. Banda et al. (eds.), Rifted Ocean-Continent Boundaries, 247–263.
© *1995 Kluwer Academic Publishers. Printed in the Netherlands.*

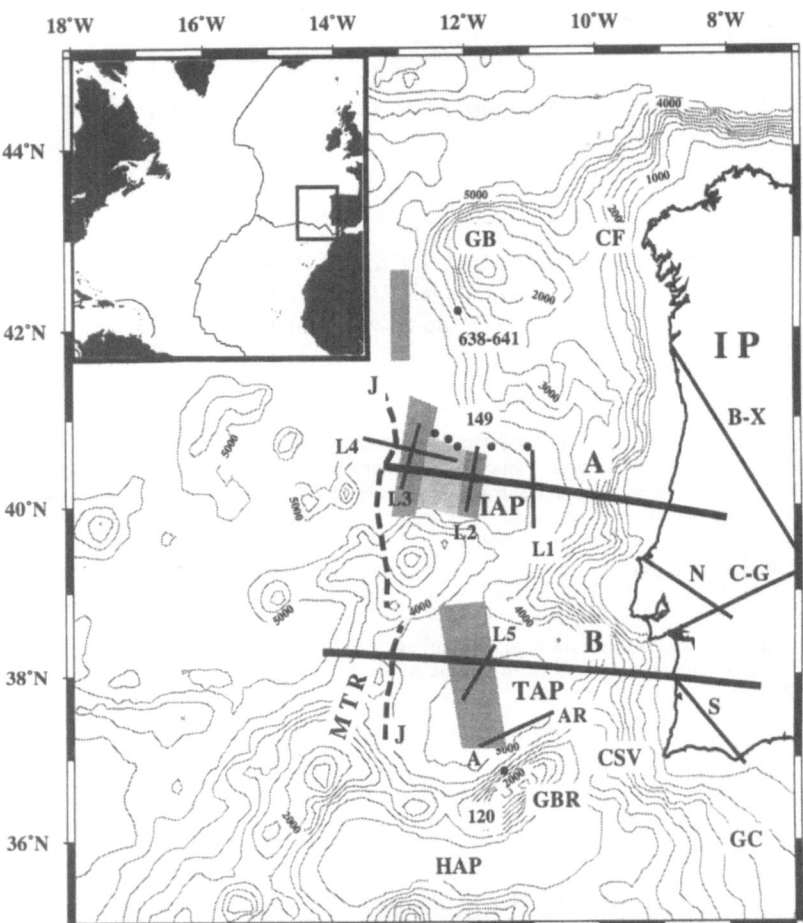

Fig. 1.- Location of the study area and modelled profiles (A and B). The thick bold lines show the location of the lithospheric sections presented in this work. The thin bold lines represent interpreted seismic refraction lines. At sea: lines L1-L4 (Whitmarsh et al., 1990), line L5 (Pinheiro et al., 1992), line A-AR (Purdy, 1975). On land: line N (Moreira et al., 1982), line S (Mueller et al., 1973), lines B-X and C-G (ILIHA DSS Group, 1993). Stippled areas indicate the possible extent of thin crust underlain by serpentinized peridotite; lighter stippled area indicates lack of seismic velocity data (as defined by Whitmarsh et al., 1993). GB: Galicia Bank; CF: Cape Finisterre; IAP: Iberian Abyssal Plain; IP: Iberian Peninsula; TAP: Tagus Abyssal Plain; MTR: Madeira Tore Rise; GBR: Gorringe Bank Region; CSV: Cape San Vicente; HAP: Horseshoe Abyssal Plain; GC: Gulf of Cadiz. Numbered solid circles denote Deep Sea Drilling Program and Ocean Drilling Program sites. (Modified from Whitmarsh et al., 1993).

continental margin of Iberia in terms of age of break-up and recognition, from geophysical observations, of the OCT. These authors conclude that the OCT lies between 12°10′ and about 12°45′ in the Iberia Abyssal Plain (Fig. 1) and is characterized by a very thin oceanic crust (2-4km) underlain by 7.3-7.6 km/s material which could correspond to a layer of serpentinized peridotite. To the East of the OCT the velocity structure and crustal thickness found by these authors resembles that of a thinned continental crust. More controversy concerns, however, the age and location of the oceanic crust in the Tagus Abyssal Plain. Mauffret et al. (1989), using multichannel seismic data, favour a two-stage spreading history model in which the eastern part of the abyssal plain would be underlain by a Late Jurassic-Earliest Cretaceous (M21-16) spreading center that was abandoned, whereas the western side would be underlain by a younger crust (from M10 to J) formed at a ridge jump. Recently Pinheiro et al. (1992), on the basis of seismic refraction and magnetic data, have proposed that the OCT would lie more to the West between 12°00' to 11°30' (Fig. 1). The results of Pinheiro et al. also show that this region is characterized by a very thin crust (2km), with velocities in the range of 4.4-6.3 km/s, which is underlain by low upper mantle velocities between 7.6 and 7.9 km/s. The eastern side of the Plain is underlain by a 6.5km thick continental crust (Purdy, 1975) whereas the western part of the Plain, adjacent to the Madeira-Tore rise, is underlain by oceanic crust (Pierce and Barton, 1991)

Available seismic refraction results along the continental shelf and Iberian Massif (e.g. Mueller et al., 1973; Mendes-Victor et al., 1980; Banda, 1988; ILIHA DSS Group, 1993) show that the crustal thickness is about 30km close to the shoreline. The average velocity of the crust varies from 6.1 km/s to 6.3 km/s, while the Pn velocity falls within the range of 8.1-8.2 km/s. The crust thickens slightly towards the central part of Iberia reaching average values of about 32km with an average velocity of the crystalline crust of about 6.2-6.3 km/s, and a poorly constrained Pn velocity of 8.0-8.1 km/s (Banda et al., 1981).

Although there is some knowledge of the crustal structure at the OCT and Iberian Peninsula, little is known about the mode of crustal and lithospheric mantle transition along the margin and about the thermal structure of the lithosphere in the study region. To address this problem, we have modelled two lithospheric profiles that run perpendicular to the margin from the deep ocean to the stable Iberian Massif (Fig. 1). The northern profile -hereinafter referred to as Profile A- runs perpendicular to the margin cutting across the Iberia Abyssal Plain at a latitude of about 40°N while the southern profile -hereinafter referred to as Profile B- extends, West to East, from the Madeira Tore Rise to the Iberian Massif crossing the Tagus Abyssal Plain. The lithospheric modelling is based on a finite element code that combines three coupled geophysical concepts: the thermal field, the gravity field, and local isostatic equilibrium (Zeyen and Fernàndez, 1994). Furthermore, to better constrain the deeper structure of the lithosphere across the margin, we also used the geoid since it is more sensitive to density inhomogeneities at subcrustal levels.

The main objectives of this work are twofold: firstly to study the mode of crustal and lithospheric transition from ocean to continent, and secondly to determine the thermal lithospheric structure across the central and southern parts of the Western Iberia Margin.

2. Observations

The data used in this study to constrain the lithospheric models were compiled from different sources. Bathymetry and free-air gravity anomalies were obtained from ETOPO5 (1986) and Haxby (1988) respectively. The scarce surface heat flow data offshore were taken from Louden et al. (1991) and Pollack et al. (1991). Onshore, the available surface heat flow values were taken from Fernàndez et al. (1995); Camelo (1987), and Duque and Mendes-Victor (1988, 1993). The geoid heights were derived from Topex-Poseidon altimeter missions (Rapp et al., 1991). To remove the regional trend from the geoid data, we removed harmonic coefficients from the OSU91A set, up to degree and order ten, leaving a geoid anomaly with wavelength components shorter than 4000km.

Maps of these input parameters are shown in Figs. 2 and 3. Figure 2a shows the bathymetry data and heat flow measurement distribution used in this study. As can be observed the IAP and TAP are bounded by the 4000m isobath. Whereas to the West the IAP opens to the deep ocean, the western boundary of the TAP is bounded by the northern end of the Madeira-Tore Rise. This bathymetric feature, which soals to up to 1000 m, is believed to have a hotspot origin (Pierce and Barton, 1991). The eastern boundary of both plains is the irregular continental margin of Portugal. The transition from deep waters to the shallow waters of the continental shelf is marked by a narrow zone with steep gradients, particularly along Profile A (Fig. 2a). On land the topography is quite flat reaching values of up to 200m on the easternmost side of both profiles (Fig. 2a).

In the study area there are some irregularly distributed heat flow measurements (Fig. 2a). Offshore, the only area which is quite well constrained is the Galicia Bank, where a recent work by Louden et al. (1991) shows that values are in the range of 30 to 45 mW/m^2 increasing seawards. According to these authors, the observed anomalous trend would be related to variations of the bottom water temperatures caused by deep eddies. These values increase slightly to 45-50 mW/m^2 on the western side of the IAP. Similar values are observed to the South of the TAP, although the data show more scatter. Onshore and along the continental shelf, the available values come from oil, mining and water exploration wells, which show average values between 70 and 80 mW/m^2 in Profile A, and from 65 to 85 mW/m^2 in Profile B (Fig. 2a).

Free-air gravity anomalies were taken from Haxby (1988). Offshore, the regional gravity field of the study area (Fig. 2b) is dominated by gravity lows (-50 mGal), which are clearly associated with bathymetry lows of the IAP, TAP (Figs. 1, 2a and 2b), bounded by gravity highs that reach maximum values of about 200 mGal in the southern border of the TAP, in the Gorringe Bank region (Figs. 1 and 2b). Of particular interest is the increase observed in the gravity anomaly along the outer continental shelf where values of up to 30 mGal are reached. This free-air gravity anomaly pattern, characterized by a negative trough over the continental slope/rise and an associated positive peak over the outer continental slope, has been observed in many passive continental margins and has been partially interpreted as the effect of the topography and its isostatic compensation (e.g. Rabinowitz, 1982). More recently, Watts and Mare (this volume) have modelled the so called "edge effect" in terms of the response of the lithosphere to sediment loading and it is used to characterize the long-term thermal and mechanical properties of the lithosphere. Further seaward, in deep waters, the value of the free-air gravity anomalies oscillates around 0 mGal probably indicating that the area is in isostatic equilibrium.

Owing to the strong correlation, observed in the study area, between bathymetry and free-air gravity anomalies, which is to be expected because of the large density contrast between the crust and sea water, we calculated the Bouguer gravity anomaly (Fig. 3a). The map was obtained by applying the Bouguer slab and terrain corrections with a reduction density of 2670 kg/m^3. Terrain corrections were applied from 4 to 40km. The Bouguer anomalies range from -50 mGal on land to 350 mGal in deep waters reflecting the subseafloor contributions to the gravity field.

The geoid anomaly map (Fig. 3b) shows that both plains are associated with local "lows" of about 4-5m, whereas the continental shelf is associated with geoid "highs" of up to 4m. The transition from the continental region to the oceanic domain is characterized by a decrease of 8-9m in the geoid anomalies, which are separated by a steep gradient that correlates with the increase in water depth from the continental shelf toward both plains.

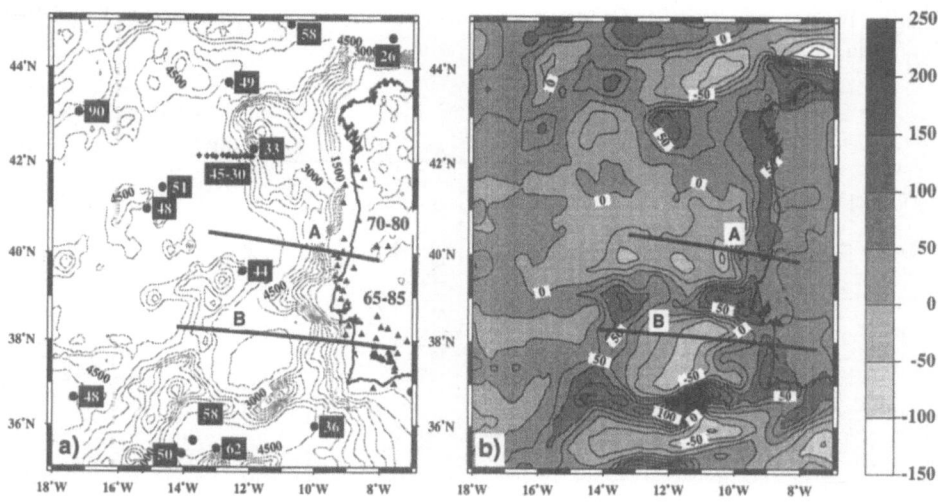

Fig. 2.- a) Shaded bathymetry relief and heat flow density map in mW/m² of the Western passive continental margin of Iberia. Bathymetry is based on ETOPO5 (1986). Contour interval 500m Thick solid lines -labelled A and B- show location of the interpreted lithospheric profiles. Heat flow values correspond to: numbered black solid diamonds (Louden, 1991,; numbered filled circles (Pollack et al., 1991); onshore values (Fernàndez et al., 1995; Camelo, 1987 and Duque and Mendes-Victor, 1988, 1993). b) Free-air gravity anomaly map of the Western Iberia Margin based on Haxby (1988). Contour interval 25 mGal.

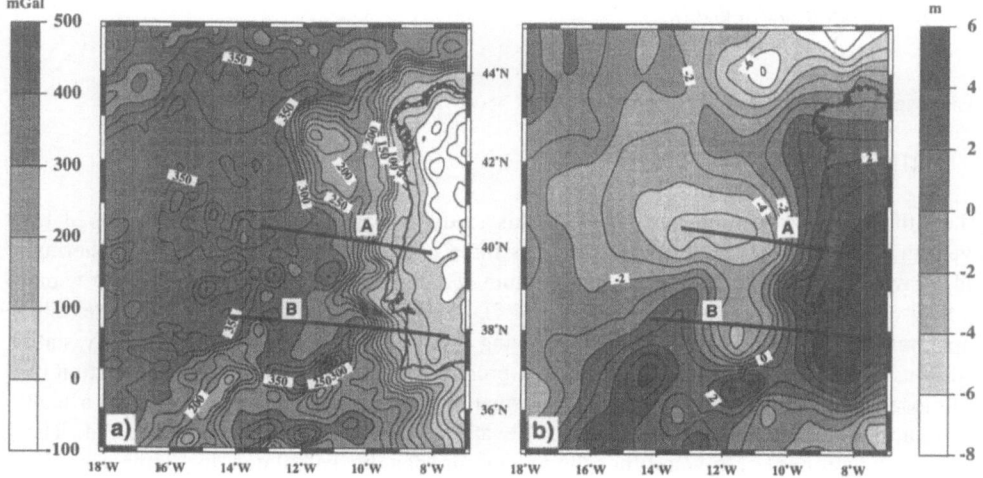

Fig. 3.- a) Bouguer gravity anomaly map of the study area. Contour interval 50 mGal. Density reduction is 2670 kg/m³. See text for more details. b) Geoid anomaly calculated from Topex-Poseidon altimeter missions (Rapp et al., 1991). Wavelenght components greater than 4000km have been removed. Contour interval 1m. Other symbols are identified in Fig. 2.

Finally, Fig. 4 summarizes the wide-angle and refraction results, along the OCT and Iberian Peninsula, considered in this work.

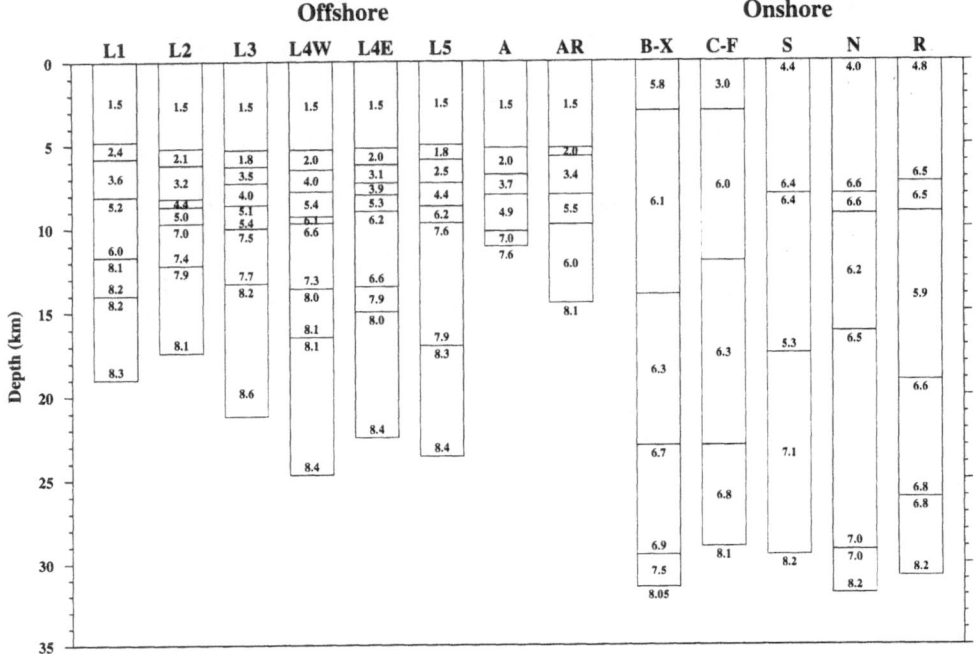

Fig. 4.- Velocity/depth structures from seismic refraction lines shot over the study region (See Fig. 1 for location and legend).

3. Lithospheric modelling

3.1. METHODOLOGY

The lithospheric modelling presented in this study is based on the work developed by Zeyen and Fernàndez (1994) which combines three coupled geophysical concepts: thermal field, gravity field and local isostatic equilibrium. This model, to which we added the geoid anomaly calculation, allows us to deduce the 2D thermal structure of the lithosphere. The temperature distribution is determined by solving the heat transport equation in steady state conditions and by considering radiogenic heat production. This provides a first estimation of the lithospheric thickness. However, since thermal data are usually scarce and present a high uncertainty, additional constraints such as elevation, gravity, and geoid are required. The absolute elevation of a given lithospheric column can be calculated by comparing its buoyancy with that corresponding to the mid-oceanic ridges when local isostasy is assumed (Lachenbruch and Morgan, 1990):

$$E = \frac{\rho_a - \rho_l}{\rho_a} L - H_0 \qquad\qquad E \geq 0$$

$$E = \frac{\rho_a}{\rho_a - \rho_w}\left[\frac{\rho_a - \rho_l}{\rho_a}L - H_0\right] \qquad E \leq 0$$

where E is elevation above sea level; ρ_a is the density of the asthenosphere; ρ_l is the mean density of the lithosphere; ρ_w is the density of sea water; L is the lithospheric thickness; and H_0 is the depth below sea level of an unloaded asthenospheric column.

Therefore, elevation depends on both the mean density of the lithosphere and its thickness. The density distribution in the crust and its geometry are obtained by combining gravity and seismic data, whereas for the lithospheric mantle, we consider a temperature dependent density referred to the density of the asthenosphere which is assumed constant. Therefore, the mantle density increases upward with decreasing temperature following (Parsons and Sclater, 1977; Lachenbruch and Morgan, 1990):

$$\rho_m(z) = \rho_a(1 + \alpha(T_a - T(z)))$$

where $\rho_m(z)$ and $T(z)$ are the density and temperature of the lithospheric mantle at a depth z, respectively; α is the thermal expansion coefficient; and T_a is the temperature at the lithosphere-asthenosphere boundary.

Finally, the geoid anomaly is computed assuming local isostatic equilibrium and assuming that the density anomalies are slowly varying horizontally. Under these conditions, the geoid anomaly is proportional to the density-moment function (Haxby and Turcotte, 1978), providing additional information about the density-depth distribution. In summary, surface heat flow, gravity, geoid and elevation are coupled by the density distribution and thickness of the lithosphere each of these data sets being sensitive to different depth ranges.

Calculations are performed by using a finite element code (Zeyen and Fernàndez, 1994). The lithospheric section is divided into a mesh of triangular elements, where the top of the model corresponds to the Earth's surface and the bottom to the boundary of the lithosphere-asthenosphere. The boundary conditions to solve the heat transport equation are: fixed temperature at the surface (15°C) and base of the lithosphere (1350°C), and no heat flow across the lateral boundaries. The lithospheric section is split into a number of bodies each of which is defined by: the thermal conductivity; heat production and its variation at depth; and density and its dependence on temperature and pressure. These material parameters are automatically assigned to each element of the mesh according to the body in which they are located. The Bouguer gravity anomaly is calculated using Talwani's 2D algorithm for each element of the mesh (Talwani et al., 1959) which allows for lateral and depth density variations. The absolute elevation and geoid anomaly are calculated in every column of the mesh. The depth of compensation to calculate the elevation and gravity and geoid anomalies is taken at the maximum depth reached by the lithospheric mantle, and the space between this depth and the base of the model is filled with asthenospheric material with a constant density. For a more detailed description of this modelling technique, the reader is referred to the work by Zeyen and Fernàndez (1994).

The lithospheric sections -Profiles A and B- were divided into a mesh of 98x84 and 116x79 nodes respectively. The horizontal grid size is 5km in the model area, whereas the vertical grid size varies at depth, being smaller at upper crustal levels where geophysical constraints allows for a more detailed geometry. To avoid border effects in gravity calculations both sections were extended up to 800km beyond the limits of the profiles. Figures 5 and 6 show the lithospheric geometries presented in this work.

3.2. PROFILE A

Profile A, with a total length of 450km, runs perpendicular to the margin cutting across the IAP at a latitude of about 40° (Fig. 1). To obtain the elevation and gravity profiles, one data point every 10km was calculated by projecting elevation and gravity values within a 40km wide strip onto the profile. Measured heat flow values were projected considering a 100km wide strip. We selected this transect because of the available geophysical data. The transect is coincident -at sea- with one of the profiles of the deep multichannel seismic reflection survey carried out in August 1993 within the IAM project (Banda et al., 1995). Therefore, the basic geometry of the sedimentary layer was adapted from the constraints provided by the seismic reflection data. At sea, the gross crustal structure was taken from four seismic wide-angle and refraction profiles (Whitmarsh et al., 1990) which run perpendicular and parallel to the modelled profile (Figs. 1 and 5). On land, the total crustal thickness and crustal structure were taken from previous seismic refraction results of Mendes-Victor et al. (1980), Mueller et al. (1973), Sousa Moreira et al. (1983) and ILIHA DSS Group (1993) (Fig. 4). Densities were obtained from the empirical relationship between P-wave velocity and density of Nafe and Drake (1961) for the sediments, and Christensen and Shaw (1970) for the crystalline crust. A constant density of 2200 kg/m^3 was assigned to the marine sediments. In the upper crust we considered a linear increment of density with pressure following the relation $\rho = 2670*(1 + 1.05*10^{-10}$ P) where P is given in Pa. In the lower continental crust we used a constant density of 2900 kg/m^3. In the transition zone we took a density of 2700 kg/m^3 for the low P-wave velocity upper crust (Profile A), and a density of 3000 kg/m^3 for the 2.1km thick layer of 7.6 km/s material (Whitmarsh et al., 1990). The density values and other rock-parameters used in the model are summarized in Table 1.

3.3. PROFILE B

Profile B, with a total length of 590km, extends from the northern end of the Madeira-Tore Rise to the southern end of the Iberian Massif running across the TAP at a latitude of about 38°N (Fig. 1). The elevation and gravity profiles were obtained in the same manner as in Profile A. This transect is also coincident with a deep multichannel seismic reflection profile of the IAM project (Banda et al., 1995) from which we constrained the basic geometry of the sedimentary layer. For the oceanic part of the transect, the gross crustal structure is mainly based on the work of Pinheiro et al. (1992) and Purdy (1975) for the TAP and on Pierce and Barton (1991) for the Madeira-Tore Rise (Fig. 1). On land, we took available results from the seismic refraction modelling of Mueller et al. (1973) and ILIHA DSS Group (1993) (Fig. 4). For consistency we used the same rock-parameters and densities as defined in Profile A (Table 1).

4. Modelling results

Figures 5 and 6 present the results of the lithospheric modelling along the two model Profiles A and B. Panels a), b), c), and d) show the observed and calculated elevation, heat flow, and Bouguer and geoid anomalies while panel e) shows the lithospheric structure along the profiles. As can be seen the model results are in fairly good agreement with the data observed with the exception of some local misfits which can be attributed to short wavelength features that are not included in the models.

	Material	Heat production [m W m^{-3}]	Thermal cond. [W m^{-1} K^{-1}]	Density [kg m^{-3}]
Upper Crust	1	1.9	3.0	2670*(1+1.05*10^{-10}P)*
Sediments	2	1.2	2.2	2200
Oceanic Crust	3	0.3	2.1	2850
Lower Crust	4	0.3	2.1	2900
Thinned Continental Crust	5a	0.3	2.1	2700
Anomalous Oceanic Crust	5b	0.3	2.1	2800
Anomalous uppermost mantle	6	0.02	3.2	3200
Lithospheric mantle	7	0.02	3.2	**
Madeira-Tore Rise	8	0.3	2.1	2650*(1+3.2*10^{-10}P)*
Asthenosphere	9	------	-------	3200

*Table 1.- Rock parameters used in Profiles A and B. *Linear increment with pressure where P is given in Pa. **Lithospheric mantle density is temperature dependent being: 3200 $(1+4.1*10^{-5}(T_a - T(z))$; $T_a = 1350°C$.*

Along Profile A, differences between the observed and calculated elevation of up to 300m are seen from km 150 to 200 (Fig. 5a). These short wavelength misfits are attributed to local topography of the basement and/or sediments or to flexural effects which are not considered in the model. Local misfits of up to 250m are also seen along Profile B (from km 175 to 225 and from km 325 to 375), which we attribute to the gross geometry adopted for the sediments and top of the basement (Fig. 6a). In general, however, the calculated regional trend fits the observed one, which suggests that the study area is basically in local isostatic equilibrium.

Figures 5b and 6b show the calculated and measured heat flow values with error bars which indicate a high degree of scatter, particularly on land measurements. At sea we considered a mean value of around 45 mW/m^2 along both profiles, while on land we adopted a mean value of about 75 mW/m^2. Since at sea there are no measured heat-flow values around Profile B, we considered the same mean value as for Profile A.

The discrepancies observed in both profiles between the observed and calculated Bouguer gravity anomalies have wavelengths of less than 25km with differences in the Bouguer gravity anomaly values of up to 15 mGal (Figs. 5c and 6c). These misfits are likely to be related to variations in the shallowest structures which are not considered in the adopted geometry.

The calculated isostatic geoid anomaly is in good agreement with the one observed (Figs. 5d and 6d). The most important differences are located at the easternmost end of Profile B (about 1.5m). This misfit is clearly related to the presence of a strong geoid anomaly located South of the Iberian Peninsula.

Figures 7a and b show the temperature distribution calculated in the lithosphere along Profiles A and B, respectively, where the base of the lithosphere corresponds to the 1350°C isotherm. The temperature distribution is mainly affected by heat production in the upper continental crust and the differences in thermal conductivity between the crust and the lithospheric mantle. A striking feature is the variation in the Moho temperature along the profiles. Below stable Iberia the temperature at the base of the crust would be about 550°C while in the oceanic part of both profiles this value decreases to 150-200°C.

As discussed above the gross crustal structure was taken from available data, in particular the total crustal thickness on land and in the OCT zone. However, to our knowledge, little is known of the deep crustal transition from stable Iberia to the thinned crust of both the Tagus and Iberia Abyssal Plains. Our results show that the crustal thinning, from 31 to 22-21km, is

Profile A

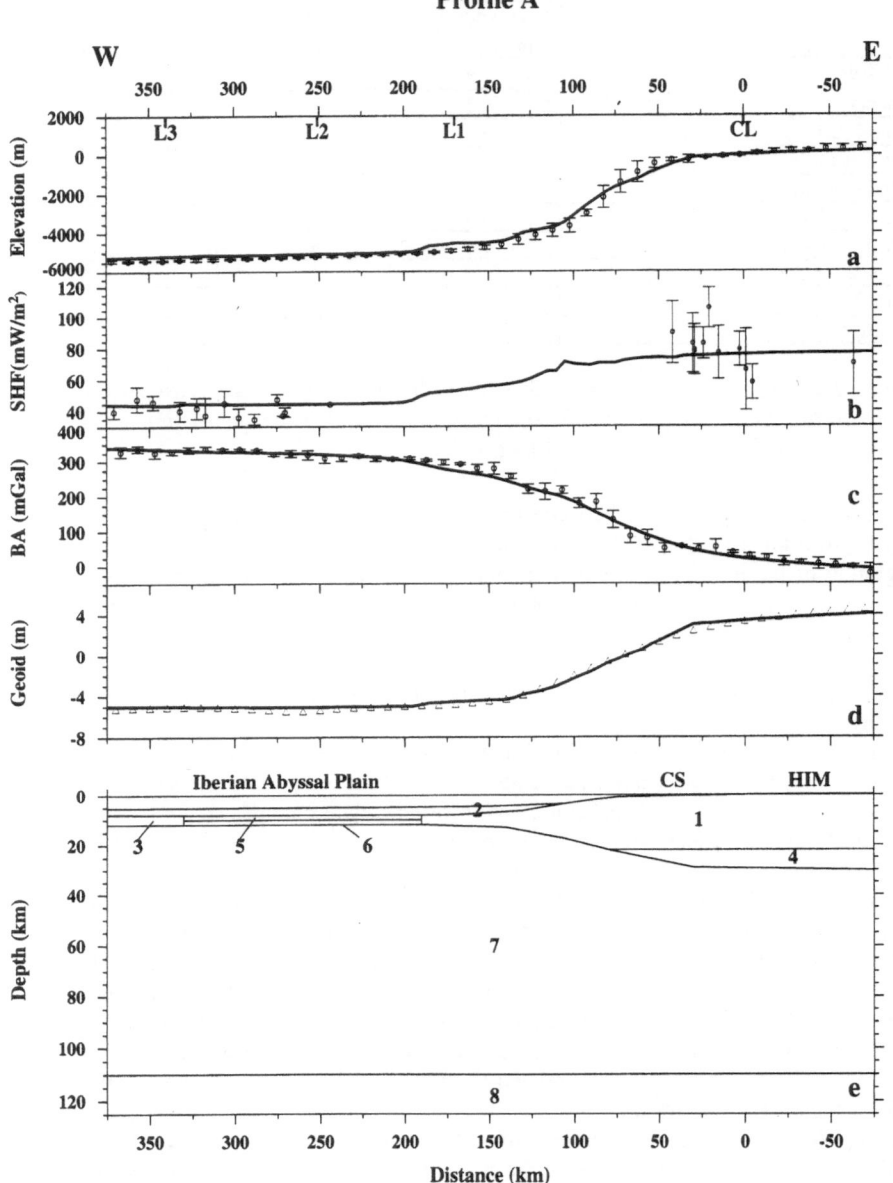

Fig. 5.- Results along model Profile A. a) Elevation. b) Surface heat flow. c) Bouguer gravity anomaly. Dots with bars indicate measured data and continuous lines calculated values. d) Geoid anomaly. Open triangles correspond to measured data and continuous line to calculated values. e) Model geometry. Description of bodies numbered from 1 to 8 is given in Table 1. CL: Coast line; L1, L2 and L3: location of seismic refraction profiles of Whitmarsh et al. (1990); CS: Continental Shelf; HIM: Hercynian Iberian Massif.

Profile B

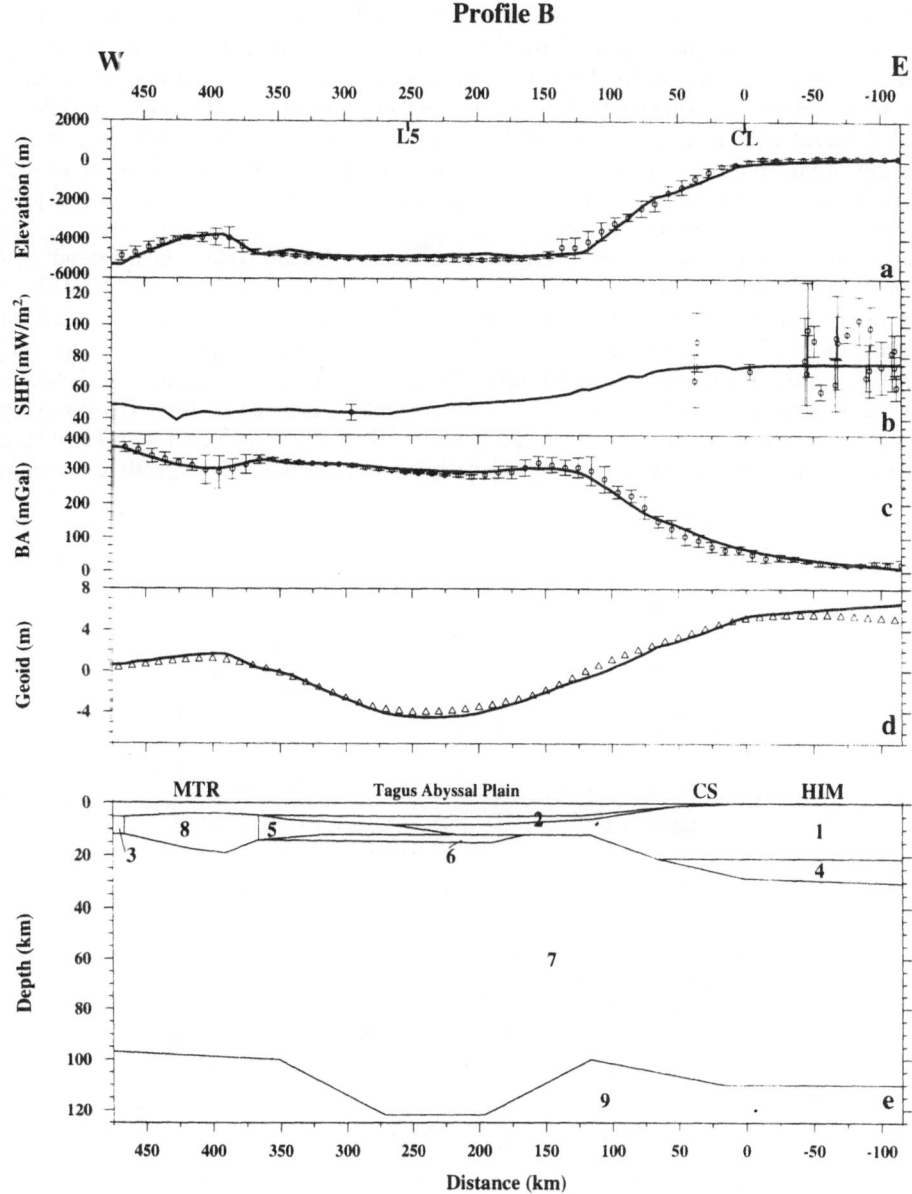

Fig. 6.- Results along model Profile B. Same legend as in Fig. 5. MTR: Madeira-Tore Rise.

fairly gentle and occurs across horizontal distances of about 125km. This crustal attitude is also observed in other rifted passive margins such as those of Nova Scotia (Fenwick et al., 1968), East Coast US (LASE Study Group, 1986), or Hatton Bank (White et al., 1987). Our results also show that an important part of the crustal thinning is accommodated by the lower crust. This relative thinning between the upper-middle crust and lower crust suggests that different stretching factors apply to these crustal levels.

258

A key question in the study area, which has been addressed by many authors over the last decades, is the exact location and structure of the zone of transition from continental to oceanic crust (OCT), particularly in the Tagus Abyssal Plain. As pointed out earlier, there is still some controversy surrounding the nature and location of this area in both Plains. We tried different end models in terms of location, extent and density distribution along the transition zone. With our modelling, however, we are unable to opt for any of these models since the mean density distribution between them is quite similar. Therefore, we adopted an intermediate solution in both profiles.

In Profile A, we assumed a two-layered transition zone, 140km wide, which extends a little bit more to the East than previously suggested (e.g. Whitmarsh et al., 1993). This is in agreement with the results obtained in ODP Leg 149, where drilling revealed a much wider expanse of peridotitic material than previously expected from geophysics (Sawyer et al., 1994). In Profile B the model adopted differs from that proposed by Pinheiro et al. (1992) in that the total thickness of the low velocity upper mantle is reduced to 3km. The geometry of the ocean-continent transition zone (OCT), however, should be considered with some caution since with our modelling we were unable to decipher whether this occurs gently or strongly.

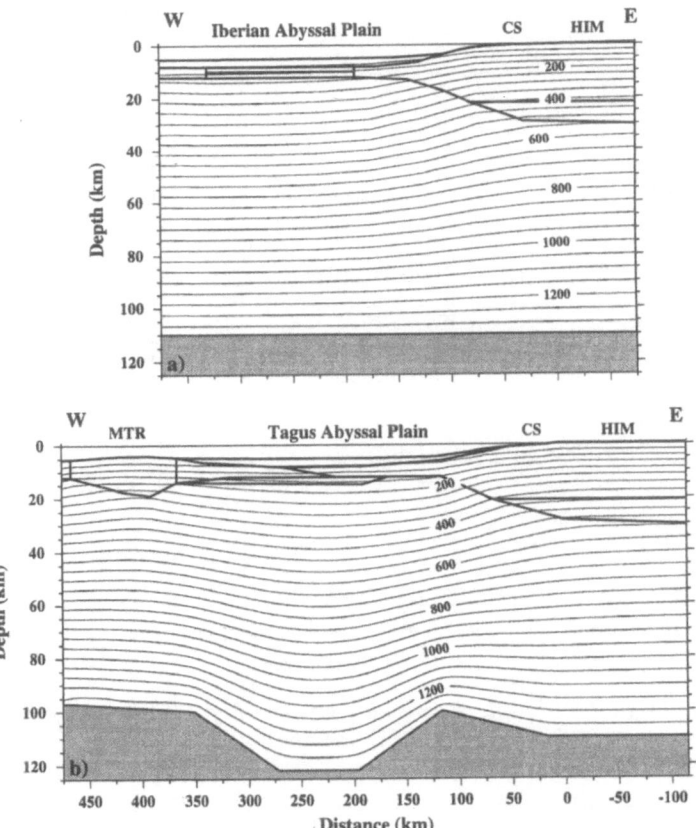

Fig. 7.- Lithospheric temperature distribution obtained along Profile A (panel a) and Profile B (panel b). Isolines every 50°C. The bottom of the model corresponds to the 1350°C isotherm. CS: Continental Shelf; HIM: Hercynian Iberian Massif; MTR: Madeira-Tore Rise.

The most outstanding result of this study is the geometry of the base of the thermal lithosphere. Major differences are observed between both profiles. Along Profile A, the base of the thermal lithosphere remains constant at about 110km depth. By contrast, in Profile B we could not keep the same flat geometry for the base of the lithosphere as for that of Profile A, since in this case we obtained a clear misfit between the elevation and geoid data observed and those calculated below the TAP. It was necessary to increase the lithospheric thickness up to 120km beneath the plain to fit both data sets.

5. Discussion and conclusions

The modelling presented in this study is based on a steady-state thermal regime, local isostasy and temperature dependent density for the lithospheric mantle. The steady-state for temperature distribution is a reasonable assumption since the age of the oceanic lithosphere in the study region is older than 100 My. Furthermore, on land there have been no major tectono-thermal events since Hercynian times, although some alpine and recent deformation have been recorded in central and southern Portugal. Nevertheless transient effects would only affect heat-flow results since elevation, gravity and geoid reflect the present-day lithospheric structure.

More debatable, however, is the assumption of local isostasy. The results of oceanic flexure studies show that the elastic thickness (T_e) increases with the age of the oceanic lithosphere at the time it was loaded and that there is good agreement between T_e and the depth to the 300-600°C oceanic isotherms (Watts, 1978). This result has been interpreted by Watts (1978) as indicating that when oceanic lithosphere increases in age it cools and becomes more rigid in its response to surface loads. Although there is general agreement on the mechanical response of the oceanic lithosphere, further controversy exists on the long-term mechanical properties of the lithosphere in passive continental margins. Watts and Marr (this volume) have used the free-air gravity anomaly "edge effect" to study the strength of the lithosphere along the East Coast US Atlantic and African margins. They conclude that while in the East Coast US margin, T_e is low (about 5km) the African margins appear to be very different as regards their strength. For a complete discussion of this topic the reader is referred to Watts and Marr (this volume).

The assumption that the density distribution in the lithospheric mantle depends only on temperature is perhaps the most debatable one. Although thermal models assume that the density of the lithospheric mantle depends only on temperature, taking as reference the density of the asthenosphere, (e.g., Parsons and Sclater, 1977, McKenzie, 1978), experimental results show that density varies also with pressure and phase changes. In fact, in the study region, seismic results indicate a constant P-wave velocity or even a slight increase in those velocities through the lithospheric mantle (ILIHA DSS Group, 1993). This result would imply that density remains constant or increases slightly at depth if a linear v_p-density relationship is assumed Zeyen and Fernàndez (1994) have calculated and compared the results of considering a linear decrease in density with increasing temperature and a constant density for the lithospheric mantle. These authors conclude that although the deduced thermal structure of the lithosphere does not vary significantly, there are important conceptual differences between both models. As pointed out by Zeyen and Fernàndez (1994) a model, in which the density of the mantle remains constant, implies that thermal subsidence is mainly driven by the cooling of the asthenospheric mantle. Moreover, a density model of the lithospheric mantle accounting for T, P and phase changes would imply that neither the density of the asthenosphere nor its temperature would remain constant everywhere as defined in thermal models. Obviously, such a model, which is beyond the scope of this work, would modify the location of the depth of compensation resulting in a different lithospheric structure.

Bearing in mind the above mentioned considerations, the integrated interpretation of elevation, heat-flow, gravity and geoid data allows us to present the gross regional crustal and lithospheric configurations across the central and southern parts of the West Iberia Margin.

The results obtained indicate that the continental crust of stable Iberia thins towards the margin in a similar way in both profiles across horizontal distances of about 125km. As mentioned above, our modelling is unable to decipher the exact location and extent of the ocean-continent transition zone. Despite this, both profiles show some differences in the crustal structure of the OCT, which are in agreement with the available seismic and gravity data.

At subcrustal levels, the transition from continental to oceanic along Profile A does not show noticeable variations in the lithospheric thickness. In contrast, Profile B shows a thickening of about 15km below the Tagus Abyssal Plain. This anomalous region, in terms of density distribution, could be related either to the proximity of the area to the Azores-Gibraltar fracture zone, which delineates the Africa-Eurasia plate boundary, or to the fact that the western side of the plain is bounded by the Madeira-Tore rise which is believed to have a hotspot origin (Pierce and Barton (1991). An alternative explanation is to consider a departure of local isostasy in that region. However, the modelling results suggest that the whole region is in local isostatic equilibrium.

The thickness observed of the oceanic lithosphere roughly coincides with that predicted by the plate model as defined by Parsons and Sclater (1977). Our results, however, do not compare with those predicted by the plate model of Stein and Stein (1992) since these authors use a higher temperature at the base of the lithosphere. Major differences are noted between the total thickness of the continental lithosphere observed in our modelling (about 110km) and that deduced from seismic data. Surface-wave analyses show that the lithospheric thickness below the western side of the Iberian Peninsula is about 80-100km (Badal et al., 1991). Recently, ILIHA DSS Group (1993), using refraction and wide-angle reflection profiling, have suggested that the total thickness of the lithosphere could reach minimum values of about 95km. Whether the seismic results are comparable with our model results is, at least, debatable since the "seismic" lithosphere might or might not be equivalent to the "thermal" lithosphere.

In summary, our model despite integrating several observables, can be described as "classic". Further research into the nature and structure of the subcrustal lithosphere and asthenosphere is needed to understand the processes of margin formation. We feel that although this kind of modelling can be tuned it will not conclusively identify the real lithosphere-asthenosphere boundary since it is based on a number of assumptions that, despite being generally accepted, are sometimes questionable.

Acknowledgements

Partial funding was from EC project JOU2-CT92-0177. Additional funding comes from CICYT project AMB93-1362-CE. We are indebted to C. Keen, H. Zeyen and an anonymous reviewer for their helpful comments.

References

Banda, E., Torné, M., and IAM Group, 1995. The Iberian Atlantic Margins (IAM) Project. EOS Trans. AGU (In press).

Badal, J., Corchete, V., Payo, G., Serón, F.J., Canas, J.A., and Pujades, L., 1991. Deep structure of the Iberian Peninsula determined by Raleigh wave velocity inversion. Geophys. J. Int., 107, 1-18.

Banda, E., 1988. Crustal parameters in the Iberian Peninsula. Phys. Earth Planet. Inter., 51, 222-225.

Banda, E., Suriñach, E., Aparicio, A., Sierra, J., and Ruiz de la Parte, E., 1981. Crust and upper mantle structure of the central Iberian Meseta (Spain). Geophys. J. Royal Astron. Soc., 67, 779-789.

Boillot, G., Winterer, E.L., Meyer, A.W., and Shipboard Scientific Party, 1987. Initial Reports, ODP 103, US Goverment Printing Office, Washington, DC.

Camelo, S.M.L., 1987. Analysis of bottom-hole temperature and preliminary estimation of heat flow in portuguese sedimentary basins. Revista Brasileira de Geofísica, 5, 139-142.

Christensen, N.I., and Shaw, G.H., 1970. Elasticity of mafic rocks from the Mid-Atlantic Ridge, Geophys. J. Royal Astron. Soc., 20, 271-284.

Duque, M.R. and Mendes-Victor, L.A., 1993. Heat flow and deep temperature in South Portugal. Studia geoph. et geod., 37, 279-292.

Duque, M.R. and Mendes-Victor, L.A., 1988. Heat flow and thermal gradients in Portugal. Proc. of the Vth Workshop "European Geotraverse (EGT) Project", Estoril, Portugal.

ETOPO5, 1986. Relief map of the Earth´s surface. EOS Trans, AGU, 67, 121.

Fenwick, D.K.B., Keen, M.J., and Lambert, A., 1968. Geophysical studies of the continental margins northeast of Newfoundland. Can. J. Earth Sci., 5, 483-500.

Fernàndez, M., Almeida, C. and Cabal, J. 1995. Heat flow adn heat production in Western Iberia. Proc. of the World Geothermal Congress, Florence, Italy (in press).

Haxby, W.F., 1988. Gravity field of the World's Oceans. National Geophysical Data Center, NOAA, Boulder, CO

Haxby, W.F., and Turcotte, D.L., 1978. On isostatic geoid anomalies. J. Geophys. Res., 83, 5473-5478.
ILIHA DSS Group, 1993. A deep seismic sounding investigation of lithospheric heterogeneity and anisotropy beneath the Iberian Peninsula. Tectonophysics, 221, 35-52.

Lachenbruch, A.H., Morgan, P., 1990. Continental extension, magmatism, elevation: formal relations and rules of thumb. Tectonophysics, 174, 39-62.

LASE Study Group, 1986. The structure of the U.S. east coast passive margin from Large Aperture Seismic Experiment (LASE). Mar. Pet. Geol., 3, 234-242.

Louden, K.E., Sibuet, J-C., and Foucher, J-P., 1991. Variations in heat flow across the Goban Spur and Galicia Bank continental margins. J. Geophys. Res., 96, 16131-16150.

Mauffret, A., Mougenot, D., Miles, P.R., and Malod, J.A., 1989. Cenozoic deformation and Mesozoic abandoned spreading centre in the Tagus Abyssal Plain (West of Portugal). Results of a multichannel seismic survey. Can. J. Earth Sci., 26, 1101-1123.

McKenzie, D., 1978. Some remarks on the development of sedimentary basins. Earth Planet. Sci. Let., 40, 25-32.

Mendes Victor, L.A., Hirn, A., and Veinante, J.L., 1980. A seismic section across the Tagus Valley, Portugal: possible evolution of the crust. Ann. Geophys., 36, 469-776.

Mueller, St., Prodehl, C., Mendes, A.S., and Sousa Moreira, V., 1973. Crustal structure in the southern part of the Iberian Peninsula. Tectonophysics, 20, 307-318.

Nafe, S.E., and Drake, C.L., 1961. Physical properties of marine sediments. In: M.N. Hill (Ed.), The Sea, 3. Interscience, New York N.Y., 794-819.

Parsons, B., and Sclater, J.G., 1977. An analysis of the variation of ocean floor bathymetry and heat flow with age. J. Geophys. Res., 82, 803-827.

Pierce, C., and Barton, P., 1991. Crustal structure of the Madeira-Tore Rise, eastern North Atlantic - results of a DOBS wide-angle and normal incidence seismic experiment in the Josephine Seamount region. Geophys. J. Int., 106, 357-378.

Pinheiro, L.M., Whitmarsh, R.B., and Miles, P.R., 1992. The ocean-continent boundary off the western continental margin of Iberia, II, Crutal structure in the Tagus Abyssal Plain. Geophys. J. Int., 109, 106-124.

Pollack, H.N., Suzanne, J., Hurter, and Johnson, J.R., 1991. The new global heat flow compilation. Department of Geological Sciences. The University of Michigan, Ann Arbor, Michigan (USA).

Purdy, G.M., 1975. The eastern end of the Azores-Gibraltar plate boundary. Geophys. J. Royal Astron. Soc., 43, 973-1000.

Rabinowitz, P.D., 1982. Gravity measurements bordering passive continental margins. In: Dynamics of Passive Margins, R. A. Scrutton (Editor), Geodynamics Series, 6, 91-115.

Sawyer, D.S., Whitmarsh, R.B., Kalus, A., and Leg 149 Shipboard Scientific Party, 1994. ODP Leg 149: Drilling a continent ocean transition at the Iberia Abyssal Plain. Workshop on Rifted Ocean-Continent Boundaries. Mallorca, 1994. Spain, 21-22.

Sibuet, J.C., Ryan, W.B.F. and Shipboard Scientific Party, 1979. Initial Reports DSDP, vol 47, US Goverment Printing Office, Washington, DC.

Sousa Moreira, V.S., Prodehl, C., Mueller, St., and Mendes A.S., 1983. Crustal structure of Western Portugal. In: E. Bisztricsany and G. Y. Szeidovitz (Eds.). Developments in Solid Earth Geophysics, 15. Proceedings of the 17th Gen. Assoc. Eur. Seismol. Comm., Budapest, 1980. Elsevier, Amsterdam, 529-532.

Stein, C.A., and Stein, S., 1992. A model for the global variation in oceanic depth and heat flow with lithospheric age. Nature, 359, 123-129.

Talwani, M., Worzel, J.L., and Landisman, L., 1959. Rapid computations for two-dimensional bodies with application to the Mendocino submarine fracture zone. J. Geophys. Res., 64, 49-59.

Watts, A.B. and Marr, C. Gravity anomalies and the thermal and mechanical structure of rifted continental margins. (This volumen)

Watts, A.B., 1978. An analysis of isostasy in the world's oceans, 1, Hawaiian-Emperor seamount chain, J. Geophys. Res., 83, 5989-6004.

White, R.S., Westbrook, G.K., Fowler, J.R., Spence, G.D., Barton, P.J., Joppen, M., Morgan, J., Bowen, B.W., Prestcott, C., and Bott, M.P.H., 1987. Hatton Bank (northwest U.K.) continental margin structure. Geophys. J. Royal Astron. Soc., 89, 265-272.

Whitmarsh, R.B., Pinheiro, L.M., Miles P.R. and Sibuet J.-C., 1993. Thin crust at the western Iberia ocean-continent transition and ophiolites. Tectonics, 12, 1230-1239.

Whitmarsh, R.B., Miles, P.R., and Mauffret, A., 1990. The ocean-continent boundary off the western continental margin of Iberia-I. Crustal structure at 40° 30' N. Geophys. J. Int., 103, 509-531.

Zeyen, H., and Fernàndez, M., 1994. Integrated lithospheric modeling combining thermal, gravity and local isostasy analysis: Application to the NE Spanish geotransect. J. Geophys. Res., 99, 18089-18102.

CONTINENTAL-OCEANIC CRUSTAL TRANSITION IN THE BRANSFIELD TROUGH AND THE SOUTH SCOTIA RIDGE (ANTARCTICA) ; PRELIMINARY RESULTS

RAMON VEGAS[1]
JUAN ACOSTA[2]
ELAZAR UCHUPI[3]

1) Universidad Complutense. Departamento de Geodinámica. 28040 Madrid
2) Instituto Español de Oceanografía. C/ C. de Maria 8. 28002 Madrid
3) Woods Hole Oceanographic Institution.Woods Hole, MA 02543, USA

ABSTRACT. A total of 1.500 km of multichannel seismic reflection profiles recorded aboard B/O Hesperides using a seven gun array and a 96 channel streamer during the austral summer of 1991-1992 were used to study four structural provinces on the Antarctic and Scotia plates. These provinces which differ in structural style and crustal character include:

I. Bransfield Trough, a half-graben in the back of the South Shetland trench/island arc with the master fault on the Antarctic Peninsula side and a chain of volcanic structures along its axis. It was formed either by a combination of continental rifting and back-arc sea-floor spreading and the volcanic chain along its axis represents a poorly organized spreading axis or by continental extension due either to a 110 ° bend in the Scotia-Antarctic plate boundary in the vicinity of Clarence Island or to complex plate interactions in the northern Antarctic Peninsula.

II. South Scotia Ridge, a chain of highs and depressions along the boundary between the Scotia and Antarctic plates. It consists of continental crustal fragments transported eastward from their original position along the South America-Antarctic isthmus during the opening of Drake Passage about 29 Ma ago.

III.The Scotia Sea. This structural province, north of the South Scotia Ridge has an oceanic crust. The surface of the crust is quite rough and in places it displays sub-crustal reflections. On two of the three profiles crossing of the South Scotia Ridge boundary there is evidence of compression due to local convergence along the Scotia-Antarctic plate boundary.

IV. Powell Basin. This basin located within the Antarctic plate was formed by back-arc spreading shortly before the collision of a spreading axis with a trench east of Jane Bank about 20 Ma. Except for a few local swells the basement surface lacks the relief typical of oceanic basement which is found in the Scotia Sea. This smooth surface could represent the top of a sequence of volcanics and sediments resting on oceanic crust.

Introduction

The development of the present geometry of the Shetland, Scotia and Antarctic plates and associated basins (Bransfield, Powell, Jane) and ridges (North and South Scotia Ridges; Fig. 1) is related firstly to Oligocene back-arc spreading associated with an east-facing arc trench at Jane Bank (King and Barker, 1988), secondly to the collision of a spreading axis with the trench east of

E. Banda et al. (eds.), Rifted Ocean-Continent Boundaries, 265–289.

266

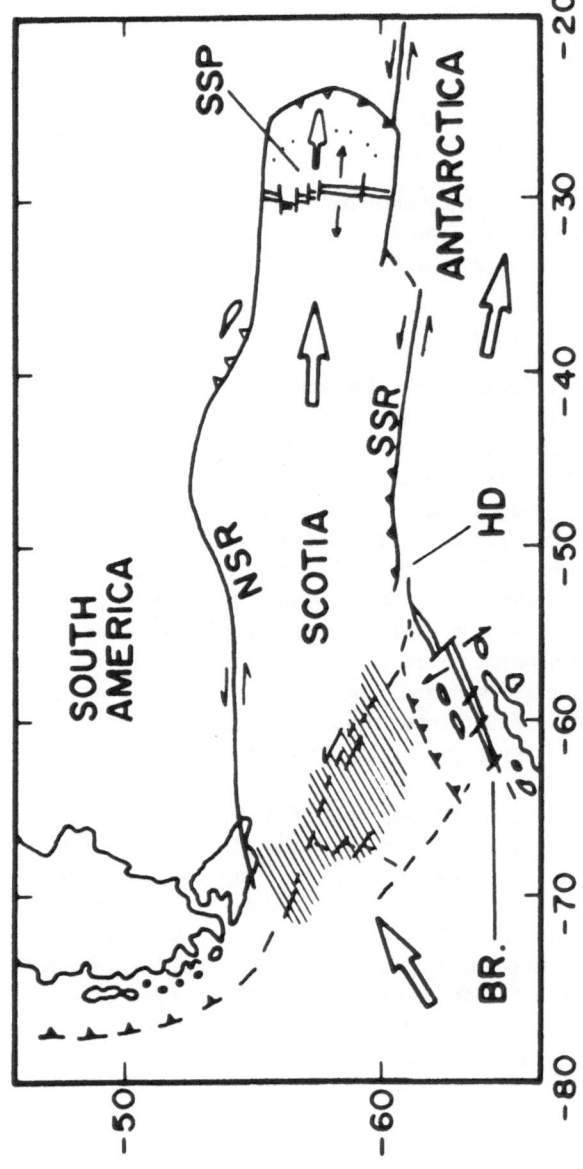

Figure 1 Structural setting of Scotia Plate and adjacent areas. Modified from Pelayo and Wiens (1989). SSP: South Sandwich Plate. NSR: North Scotia Ridge. SSR: South Scotia Ridge. BR: Bransfield Trough. HD: Heperides Deep. Blank arrows: relative motion of Antartica and Scotia Plates respect to South America. Hachured: diffuse plate boundary.

Bank in the early Miocene (Barker et al., 1984) and lastly to the eastward transport, during the opening of Drake Passage 28 to 6 Ma ago, of metamorphosed sedimentary and igneous rocks emplaced at the Pacific margin of Gondwanaland during the early Mesozoic (Dalziel and Elliot, 1973; Herron and Tucholke, 1976; Barker and Burrell, 1977; Barker and Dalziel, 1983; Dalziel, 1989).

In an attempt to clarify the geologic evolution of present and past tectonic processes along the Scotia-Antarctic plate boundary, the Instituto Español de Oceanografia has collected a fairly extensive network of seismic reflection profiles in the region (Acosta et al., 1989a, b, and c). Prior to 1992 (Antartida 86-11 and Exantarte 88-89 and 90-91 cruises) the seismic reflection profiles were obtained using a sparker having a capacity of 4,500 and 8,000 Joules discharged via 99 electrodes at 3 to 6 second intervals. Signals were received via a 24-element hydrophone single-channel array, amplified, filtered between 70 and 700 Hz and recorded on a EPC 3200 dry paper recorder. During the Scotia-92 cruise aboard B/O Hesperides a total of 1,500 km of multi-channel, magnetic and multibeam profiles were recorded in Bransfied Trough , the western end of the South Scotia Ridge, the Scotia Sea north of the ridge and Powell Basin south of the ridge (Fig. 2). During the cruise the bathymetry of a pull-apart basin between two parallel segments of the South Scotia Ridge also was mapped with the aid of the SIMRAD multi-beam echosounding system. The multi-channel profiles described in this paper were obtained using seven Bolt air-guns with a total volume of 936 in^3 at a pressure of 2,000 pounds in^2. The 50 m shot intervals were controlled by an integrated G.P.S. navigation system. The data were recorded on a Texas Instruments DFS-V digital recorder at a record length of 6 to 9 seconds with a sampling rate of 2 ms with a time delay in the deep areas. The streamer used consisted of 96 channel Teledyne type with a group length of 12.5 m and total length of active sections of 1200 m. Data were processed via Promax seismic software . The general processing sequence included demultiplex, geometry and statics corrections; gain recovery, resampled to 4 ms, adjacent trace summation; 96 trace/12 fold with a CDP spacing of 6.25;CDP gather;interactive velocity analyses;raw stack;frequency filtering and decovolution.

Bransfield Trough

The Bransfield Trough between the South Shetland Islands block and the Antarctic Peninsula is asymmetric in cross-section with its northwest slope which is much steeper than its southeast one.The floor of the trough is divided into three narrow depressions, (occidental basin, at the southwest end of the trough , King George Basin in the center and oriental Bransfield Basin at the northeast end). The northwest-trending highs, with the one between King George and North Bransfield basins rising above sea level to form Bridgeman Island (Fig. 1; Acosta et al., 1992). The floor of North Bransfield Basin becomes quite complex as it nears the southwest side of Clarence Island (Lawver and Villinger, 1989). Elephant Island west of the North Bransfield Basin is located at the site where the Shackleton Fracture zone, the present boundary of the Antarctic and Scotia plates, intersects the South Scotia Ridge. Near the intersection the fracture zone is characterized by left-lateral transpression (Dalziel et al., 1989; Klepeis et al., 1989). At the intersection itself , the Scotia-Antarctic plate boundary forms an angle of 110° with the South Scotia Ridge.Two linear troughs are flanked by ridges paralling the northwest trend of the Shackleton Fracture Zone and disrupt the ridge. One of the troughs is cut by multiple faults and the floor of the other one is crenulated and dissected by a series of small ridges (Klepeis et al.,

268

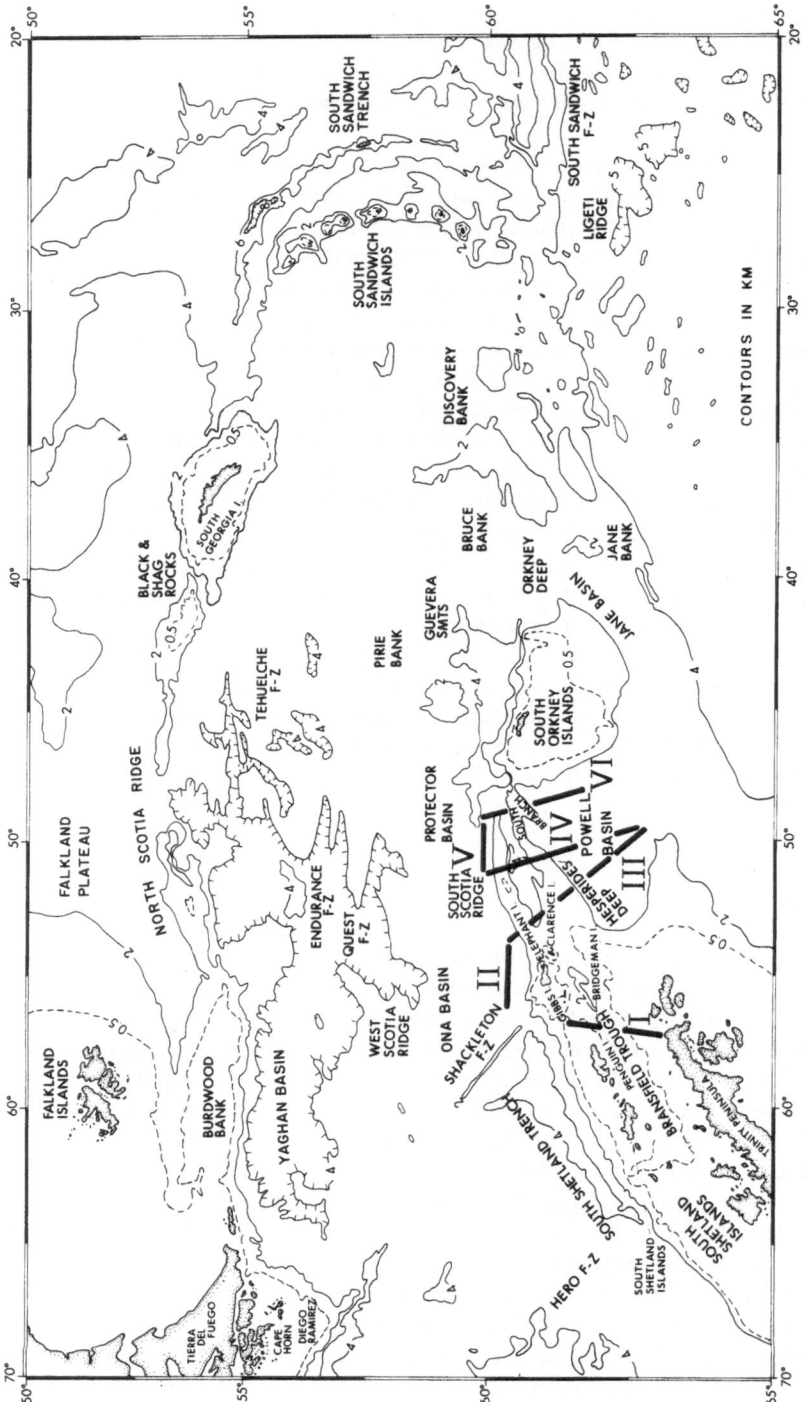

Figure 2 Chart locations of multi-channel seismic reflection profiles recorded during Scotia-92 cruise Bathymetry modified from GEBCO, 1981. I: profile 2, II: profile 3, III: profile 4, IV: profile 5, V: profile 6 and VI: profile 7

1989;1990). The bend in the plate boundary apparently has created a compressional regime north of Elephant Island and an extensional one south of Clarence Island (Lawver and Villinger, 1989).

The flat floor of Bransfield Trough is disrupted by a line of submarine peaks extending from Bridgeman Island to Deception Island and three Holocene volcanic islands (Bridgeman, Penguin and Deception islands;González-Ferrán and Katsui, 1970; Barker et al., 1975). The association of these submarine peaks with known volcanic features and the presence of a prominent magnetic anomaly, which Roach (1978) interpreted as being due to the emplacement of narrow strips of magnetized material along the flank of Bransfield Trough, led Ashcroft (1972), Barker (1976), Acosta et al. (1989a) and Jeffers et al. (1991) to propose that the features were of volcanic origin and represented an incipient spreading axis. Multi-beam data recorded with a SIMRAD system during the Gebra-93 cruise aboard the B/O Hesperides (Canals et al.,1992,1994) suggested that the continuity of the neovolcanic ridge is disrupted by small northwest-southeast trending transform faults and broad divides separating the basins in the trough..

Seismic refraction measurements indicate that the crust along the flanks of the Bransfield Trough is normal continental crust with a thickness of 30 to 35 km around the South Shetland Islands and 38 to 45 km along the coast of the Antarctic Peninsula (Ashcroft, 1972, Guterch et al., 1991a). The crust in Bransfield Trough itself is anomalous with a layer of 7.0-7.2 km/sec along the axis of the trough forming a massive diapir with its crest at a depth of 10 km, and a 7.6 km-7.7 km/sec at depths of 20 at the northern end and 25 km at its southern end of the trough; this layer may represent lower continental crust. More recent refraction measurements suggest that the Moho along the trough's axis is at a depth of >25 km (Guterch et al., 1991b). The crust above the 7 km/sec layer in the axis of the trough consists of a 1-2 km thick 6.0 to 6.8 km/sec layer at its base and a 1-2 km thick 4.0 km/sec layer at the top.(Ascroft, 1972 ; Guterch et al., 1991a). The 1-2 km thick sediments above the 4.0 km/sec crustal layer have velocities of 1.8 to 3.3 km/sec. Recent studies by Barker and Austin (1994) indicate that intracrustal diapiric igneous structures inferred by Gambôa and Maldonado (1990) and Acosta et al. (1992) along the southwest side of the trough trend at 5° to the axial nevolcanic zone, and are very young.

The seismic reflection profile (Fig.3 and 4) recorded during our investigation, which crossed the trough along the divide between the King George Basin and the North Bransfield Basin, displays a structure compatible with previous results as reported from the region. The Antarctic margin is disrupted by normal faulting while the syn-sedimentary sequence is separated from the more recent drift sediments by means of a prominent unconformity which proves the uplift and erosion that occurred during the onset of volcanism in the trough's axial zone (Fig. 3). The divide between the King George and North Bransfield basins along whose crest is Bridgeman Island, a Holocene volcano, is a massive volcanic terrain extending the length of the divide. Southeast of this massive volcanic pile are lava flows and the dikes/ plutons which Barker and Austin (1994) described. Farther southwestward, in the King George and Central Bransfield basins, the volcanic axial high is more linear and narrow, twin-peaked locally and throughout much of its length has topographic expression. Its continuity appears to be disrupted by small transform faults and massive volcanic outpourings along the basin divides.

All investigators agree that Bransfield Trough is a zone of extension forming behind the South Shetland Islands Arc (Barker et al., 1991), but what exactly caused this extension is yet to be resolved. Some have proposed that the opening of Bransfield Trough began about 4 Ma ,as a response to the cessation of Antartic-Phoenix spreading in western Drake Passage.As a result of the continued sinking of the subducted slab a roll-back in the hinge of subduction at the South

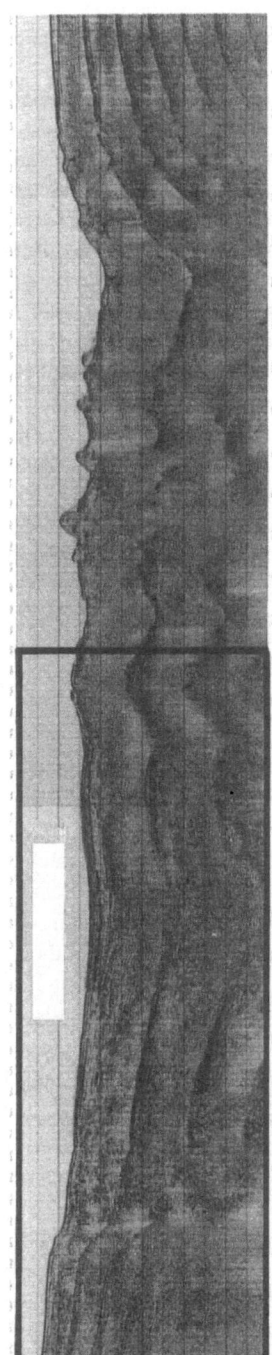

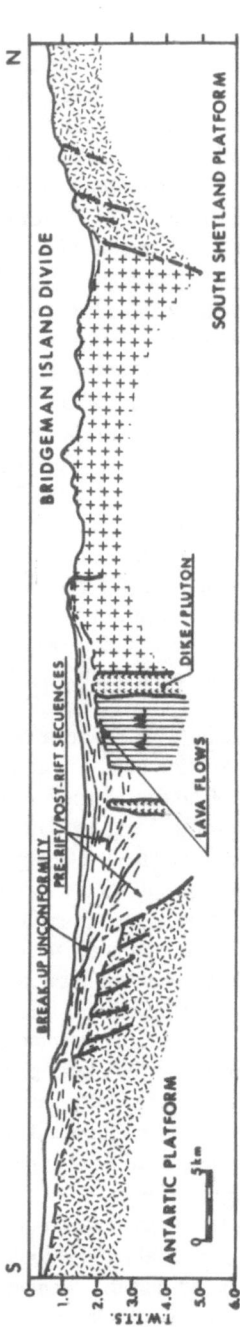

Figure 3 Line interpretation of multi-channel seismic reflection profile 2 recorded at the northeast end of Bransfield Trough. Crosses: neovolcanic ridge. stippled: continental crust. A.M.: acoustic masking. See fig.2 for location of profile.

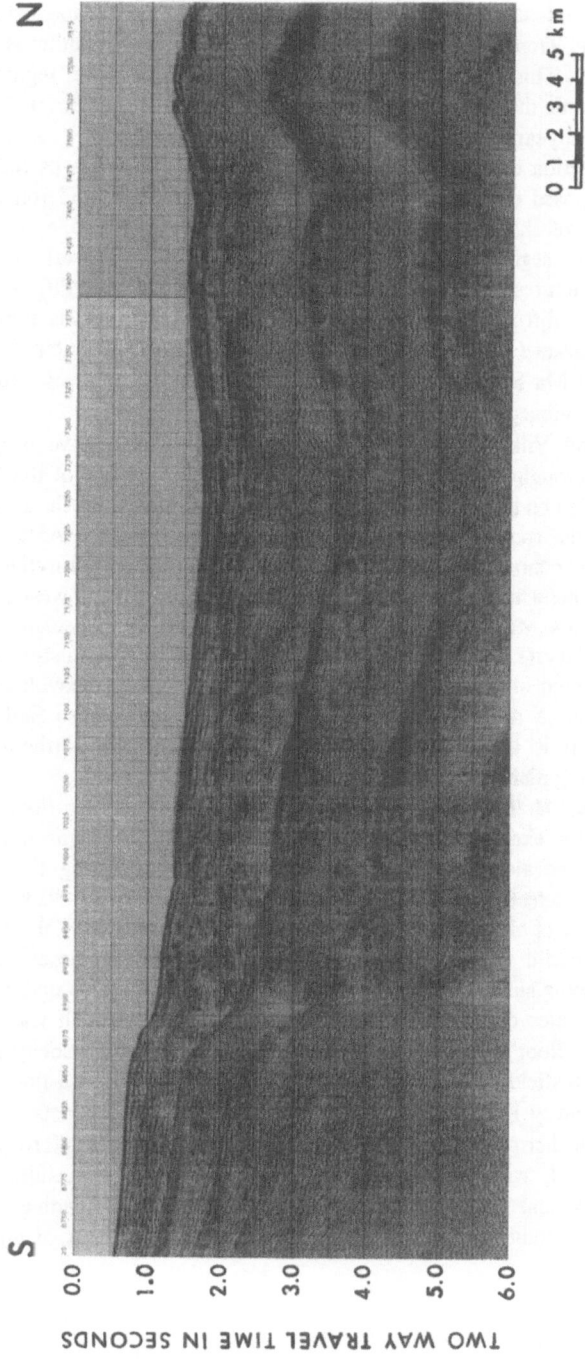

Figure 4 Section of profile 2 corresponding to box in figure 3

Shetland Trench , causing sea-floor spreading to began 1-2Ma ago (Roach, 1978; Barker and Dalziel 1983; Gamboâ and Maldonado, 1990; González-Ferrán, 1991). Subduction of the Phoenix plate along the South Shetland Trench supposedly began about 50 Ma at the southwest end of the Antartic plate (Barker, 1982). This subduction migrated northeastward causing segments of the mid-ocean ridge to collide with the Antarctic Peninsula (Herron and Tucholke, 1976; Barker, 1976, 1982) and ended 4.0 Ma years ago. Barker and Daziel (1983) and Barker et al. (1991) who noted that the extensional section of the Bransfield Trough is the same length as the surviving South Shetland Trench, proposed that the trough was formed by slab-pull, and rollback of the trench hinge. Fragmentation of the continental crust also is believed to be a result of this subduction. Apparently, as the segmented oceanic crust was subducted, the overlying crust was broken by transverse megafractures with each continental segment thus having different episodes of magmatism, erosion and uplift, postorogenic sedimentation and metalliferous mineralization (Hawkes, 1981).However, Barker (1982) stated that this segmentation of the Antarctic Peninsula was effective for only 10-30 Ma before the collision, because the fracture zones either did not extend to the trench or were oblique to the subduction direction.

More recently Lawver and Villinger (1989) and Lawver et al. (1990) have proposed that extension in the Bransfield Trough is not related to the subduction southwest of the Shackleton Fracture Zone. This subduction ended about 4 Ma ago when extension between the Antarctic and Phoenix plates ceased and the trench-fault-fault triple junction, between the Antarctic plates, disappeared along with the Phoenix Plate. Instead, they proposed that extension in the Bransfield Trough is related to a 110° bend in the Scotia-Antarctic plate boundary southwest of Clarence Island. This bend was created 4 Ma ago when the Shackleton Fracture Zone became part of the Antarctic-Scotia plate boundary (Lawver and Villinger, 1989). Barker and Austin (1994) also stated that the fan faulting noted in the Bransfield Trough indicated that the neovolcanic zone in the trough's axial zone is not a nascent spreading axis. But instead,a diffuse faulting in the trough.They proposed that it is a result of complex interactions in the northern Antarctic Peninsula, which are still taking place today.

Others have suggested that Bransfield Trough is the result of back-arc sea-floor spreading. Pelayo and Wiens (1989), for example, stated that study of focal mechanisms indicated that subduction is still taking place along the South Shetland Trench. Supporting the concept of present-day subduction is the detection of a recent earthquake at a depth of 50-80 km below the South Shetland Islands (Ibañez et al., 1993). This event could have resulted from the sinking of a subducted,cold oceanic slab and it does document present- day subduction processes. Because of the young age of the plates being subducted and the slow rate of subduction, this underthrusting is largely aseismic. Other advocates combine the concept that Bransfield Trough was formed by continental extension and sea-floor spreading and interpreted the crust with velocity of about 7 km/sec in the center of Bransfield Trough as oceanic crust emplaced by sea-floor spreading processes (Ashcroft,1972 ; Davey 1972 ; Roach, 1978; Jeffers et al., 1991 ; Acosta et al.,1992) . Acosta et al (1992; Fig. 4) further proposed that continental rifting in Bransfield Trough took the form of a simple shear model, with the master fault on the Antarctic Peninsula. They also suggested that the magnitude and depth of faulting along the back-arc spreading axis in the southern Bransfield Trough is indicative of diffuse extension rather than that of an organized spreading axis.

South Scotia Ridge

The South Scotia Ridge is aligned east-west parallel to the Scotia / Antarctic plate boundary where motion is left-lateral with a component of extension (Fig. 1 ; Pelayo and Wiens, 1989). East of the Shackleton Fracture Zone the ridge consists of two sub-parallel branches. They are separated by a V-shaped trough 5-10 Km wide, having over 5500 m in depth which we have called Hesperides Deep. The 20 km wide and 500 to 1000 m deep north branch of the South Scotia Ridge is convex in plan view. It extends 320 km eastward from Elephant Island and is separated from the South Orkney microcontinent by a north-south trending segment of Hesperides Deep (GEBCO, 1981; King and Barker, 1988). Its easternmost tip is characterized by spurs diverging to the northeast and southeast. The north branch's northern flank descends steeply into the Scotia Sea at a depth of 3000 m and its southern one drops more gradually to Hesperides Deep at a depth of 5500 m. The northward convex south branch is 20 to 60 km wide, 500 to 1000 m deep, extends 400 km eastward from Graham Land and is connected to the South Orkney Islands .The northern flank of this ridge descends to the central part of Hesperides Deep and the southern one to the ?500 m deep Powell Basin.

Seismic refraction and magnetic measurements on the South Scotia Ridge by Watters (1972) showed that the north branch is non-magnetic and its basement has a velocity of 5.1 to 5.6 km/sec. Above the basement are sediments with velocities 4.0 to 1.9 km/sec At about 120 km northeast of Elephant Island the basement appears to be downfaulted to the east with its lower part being filled with 0 to 0.5 km of 1.9 to 2.4 km/sec material and 1 to 2 km of material with a velocity of 3.2 km/sec.The basement west of this structural low is mantled with deposits having velocities of 4.4 to 3.6 km and 1.9 to 2.4 km/sec. The higher velocity layer has a thickness of 0 to 1 km and displays its maximum thickness in the vicinity of Elephant Island. The upper layer of lower velocity which extends the length of the branch is about 200 m thick. The South Scotia Ridge's Basement velocities through much of the length of the magnetic south branch are uniform and range from 4.9 to 5.1 km/sec. The eastern end of the south branch, however, is underlain with material with a velocity of 6.4 km/sec with a steep westerly dip (Watters, 1972). Above the basement is a 3.1 to 3.5 km /sec layer decreasing in thickness westward from 2.2 km to nothing southeast of Clarence Island. Above this is a layer with a velocity of 1.75 to 2.91 km/sec which does not exceed 1 km in thickness. Watters (1972) correlated the basement of the north branch with the parametamorphic rocks exposed on Elephant and South Orkney Islands. He also suggested that the lower velocity basement of the south branch may be correlated with the slightly metamorphosed Trinity Peninsula Series, which are overlain by gently folded Cretaceous sediments and intrusions of the Andean Intrusive Suite of the Antarctic Peninsula. He also stated that the widespread truncation of magnetic anomalies observed in the edges of the South Scotia Ridge's southern branch suggests that rifting of the ridge and formation of Hesperides Deep is a recent event, and that such processs may still be taking place today.

The multi-channel seismic reflection profiles recorded during Scotia-92 cruise clearly displayed the complexity of this segment of the South Scotia Ridge (Figs.5,6,9 and 11). As illustrated by profiles 5 and 7, the northern flank of the South Scotia Ridge is under compression. This compresion can be ascribed to the sligh convergence of the Antartic and South America plates trough the curved Scotia-Antartic boundary (Forsyth 1975). The intensification of deformation due to compression, increases eastward --compare profile 4 (Fig.6) where there is no evidence of compression with profile 7 (Fig.11)with its well-developed accretionary wedge-- where the plate boundary changes its trend from northeast to east and the two ridges converge. This zone of

274

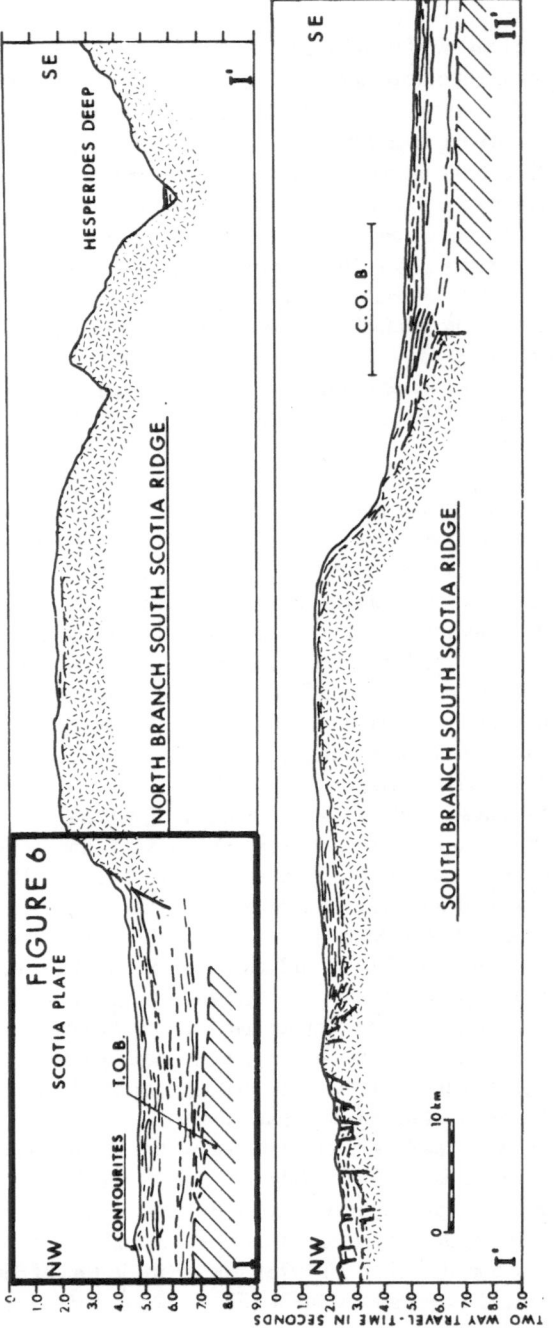

Fig.5 Line interpretation of profile 4 extending from the Scotia Sea to Powell Basin on the Antarctic Plate. Note that the sediment prism at the base of the South Scotia Ridge is not deformed, that the basement on the north branch is exposed, but that the basement on soth branch has an appreciable sediment cover. T.O.B: top of oceanic basement. C. O. B: continental / oceanic basement boundary.

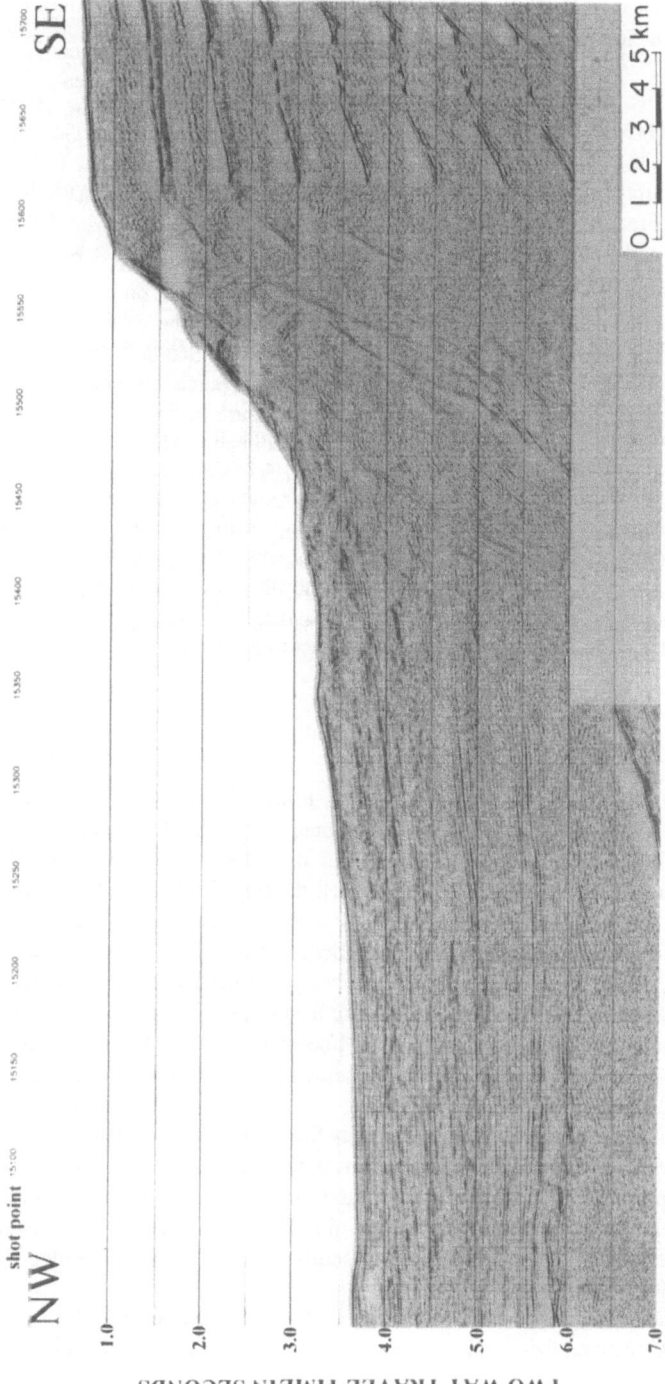

Figure 6 Section of multichannel seismic line 4 that correspond to the box marked in figure 5

accretion along the north side of the South Scotia Ridge appears to terminate at a north-south trending segment of Hesperides Deep at 47° 30'E. Beyond this low is a north-south trending spur which connects to the South Orkney Microcontinent; east of the spur is the northwest trending South Orkney Trough which connects with northeast trending Orkney Deep near 41°30'W (King and Barker, 1988). Two poor-resolution, single channel reflection profiles from the northeastern tip of the South Orkney Microcontinent at about 42°E do not display any evidence of compression (King and Barker, 1988). Possibly another zone of compression exists farther west at the point where the Scotia/Antarctic plate boundary changes trend as it becomes aligned with the northwest - oriented Shackleton Fracture Zone. Thus, deformation along the South Scotia Ridge is not as pervasive as Ludwig and Rabinowitz (1982) demonstrated for the North Scotia Ridge.

On profile 5 (Figs. 7 and 8), the accretionary wedge appears to rest on undisturbed flat-lying strata, but this will have to be verified by farther processing of the data. Sediments atop the accretionary wedge, near the crest of the north branch of the South Scotia Ridge , which arte along profile 7, appear to have partially collapsed therefore forming a perched basin which subsequently was filled with detritus derived nearby. Along profile 5 the north branch has the appearance of a tilted block with the less inclined side being covered with sediments, that are deformed at the foot of the block. Along profile 4 the north branch basement is fairly flat, being cut by an erosion surface less than 500 m deep (ORCA Group, 1994). The V- shaped Hesperides Deep between the north and south branches of the South Scotia Ridge along profiles 7 and 5 is devoid of recent sediment fill; only along profile 4 is there any evidence of sediment accumulation. Atop the south branch on profiles 4 and 5 is a well developed faulted and tilted sedimentary basin which narrows to the east and is flanked on the southeast side by a basement high. Along profile 4 the basement high is bordered on its southeast side by a broad northwest tilted sedimentary low.

Scotia Sea

The Scotia Sea north of the South Scotia Ridge was formed by sea-floor spreading during the opening of the Drake Passage and the eastward tectonic transport of the continental fragments making up the North and South Scotia Ridges. It has developed generally independently of the adjacent South America and Antarctic plates. The well developed magnetic anomalies associated with this oceanic basin belong to four spreading axes: a western one in Drake Passage which was formed 28 to 6 Ma by the separation of the South Scotia Ridge from the tip of South America (Barker and Burrell, 1977), a central segment shallower than the Drake Passage segment which has a thicker sediment cover and resembles a back-arc basin of uncertain age, a southeast segment formed by back-arc spreading during early to mid-Miocene (12-20 Ma), and an eastern segment formed by back-arc spreading behind the South Sandwich Islands arc during the last 5-7 Ma (Barker and Dalziel, 1983).

Seismic refraction measurements across the Scotian Sea near 42°W (Allen, 1966) showed the following crustal section : 1.0 to 2.2 km thick sediment sequence with assumed velocities ranging from 2 km at the top to 2.9-3.6 at the base and 3.3 to 6.0 km thick crustal layer with a velocity of 5.7 to 6.2 km/sec. North of the South Scotia Ridge this layer is underlain by a unit of unknown thickness with a velocity of 7.4 to 7.6 km/sec, and south of the North Scotia Ridge by the Moho with a velocity of 8.0 km/sec.

The nature of the oceanic crust in the Scotia Sea is shown by profiles 3 and 6 (Figs.10,11 and 12). The basement surface on these and the other profiles recorded in the zone, displays a broadly undulating topography characteristic of the top of Layer 2 (oceanic basement). An intra-crustal discontinuous reflector noted in places along the east-west trending Profile 3 at depth of 1.2 to 1.6 seconds below the surface of basement may represent the contact between the 5.7 to 6.2 km/sec (oceanic basement, Layer 2, and upper part of oceanic crust, Layer 3A) and 7.4-7.6km/sec (oceanic crust, layer 3B ; see Emery and Uchupi, 1984, p. 194-204 and references therein for a discussion on the seismic structure of the oceanic crust) layers noted in the seismic refraction measurements in the Scotia Sea. Along profile 6, trending parallel to strike of the South Scotia Ridge in the vicinity of Protector Basin, a low in the Scotia Sea (Fig.2), shows a fault that bound a broad swell along which the crust has been uplifted about 1 second (two-way travel time). This positive basement feature may correspond to a high displayed in the bathymetry and tectonic maps of the Scotia Arc (GEBCO, 1981 ; British Antarctic Survey, 1985). Toward the south, on lines 5 and 7 (Figs 5,6,9 and 11), the oceanic crust plunges along an east-trending hinge zone as it approaches the South Scotia Ridge and creates a structural low with about 3 seconds of relief. On profiles 5 and 7, this low is filled with sediment, but farther east the low along the base of the South Scotia Ridge has a topographic expression whose position is occupied by South Orkney Trough (GEBCO 1981; King and Barker 1988). Layer 1 (sediments) above oceanic basement along line 3, is as much as 2 seconds thick, and is composed of three sequences separated by well defined unconformities. The upper unit is transparent and internal-reflector-free and grades eastward to a unit with continuous reflectors whose geometry is affected by bottom current activity. The middle unit has continuous amplitude reflectors that pass westward into an acoustic chaotic facies. The basal unit which may have been affected by igneous activity fills lows in the oceanic basement and shows evidence of deformation concordant with that of the basement. On profile 7, the stratified unit north of the north branch of the ridge merges southward with the accretionary prism at the base of the South Scotia Ridge and is disrupted by a fault zone. Additional processing is needed to determine the nature of the contact between these two units. Along profile 5, the somewhat more acoustically transparent sediments atop a broad swell north of the hinge zone. It parallels the South Scotia Ridge plunge beneath the accretionary prism at the base of the ridge. The low created as a result of this plunge is filled with a flat-lying well-stratified sequence (fig.8).On the north-south trending line, profile 4 (fig.5 and 6), the well-stratified sediments at the base of the South Scotia Ridge dip gently northward away from the ridge.

Powell Basin; Antarctic Plate

Powell Basin south of the South Scotia Ridge on the Antarctic plate was formed by back-arc extension about 35 Ma (King and Barker, 1988). This extension was completed prior to the collision of a spreading axis with a trench east of the Jane Bank-Discovery Bank island arc (Fig. 2) about 20 Ma ago. Basement in Powell Basin on both Profiles 4 and 5 ramps onto the the southern flank of the South Scotia Ridge (figs.5 and 7). Along profile 4 the basement surface is subdued and characterized by discontinuous reflectors that pass southeastward to a basement whose surface is dominated by low amplitude hyperbolic echoes. Basement at the base of South Scotia Ridge on profile 5 displays high amplitude hyperbolic reflectors that may represent oceanic crust. It appears to plunge southward beneath a smooth-surfaced basement characterized by discontinuous reflectors. The acoustic nature of this smooth-surfaced basement suggests that it was not formed by sea-floor spreading processes. This basement may be the top of a basin fill sequence of some high reflective material; probably volcanics and igneous sills and flows intercalated with

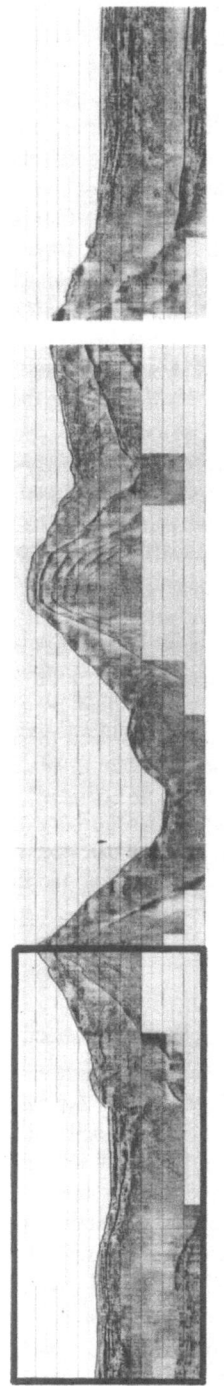

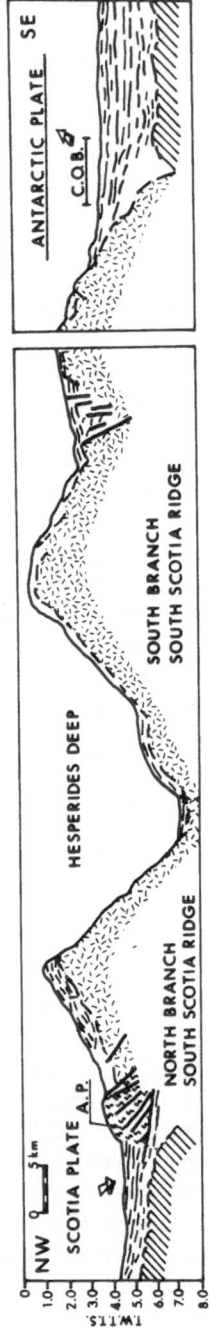

Figure 7 Line interpretation of profile 5 . See fig. 2 for location of profile. A. P.: Accretionary prism at the foot of the north branch of the South Scotia Ridge, the asymmetrical shape of the ridge with its gentler northwest side supporting a thick sediment cover. Powell Basin has a smooth basement surface with internal reflectors which may correspond to volcanics intercalated with sediments.

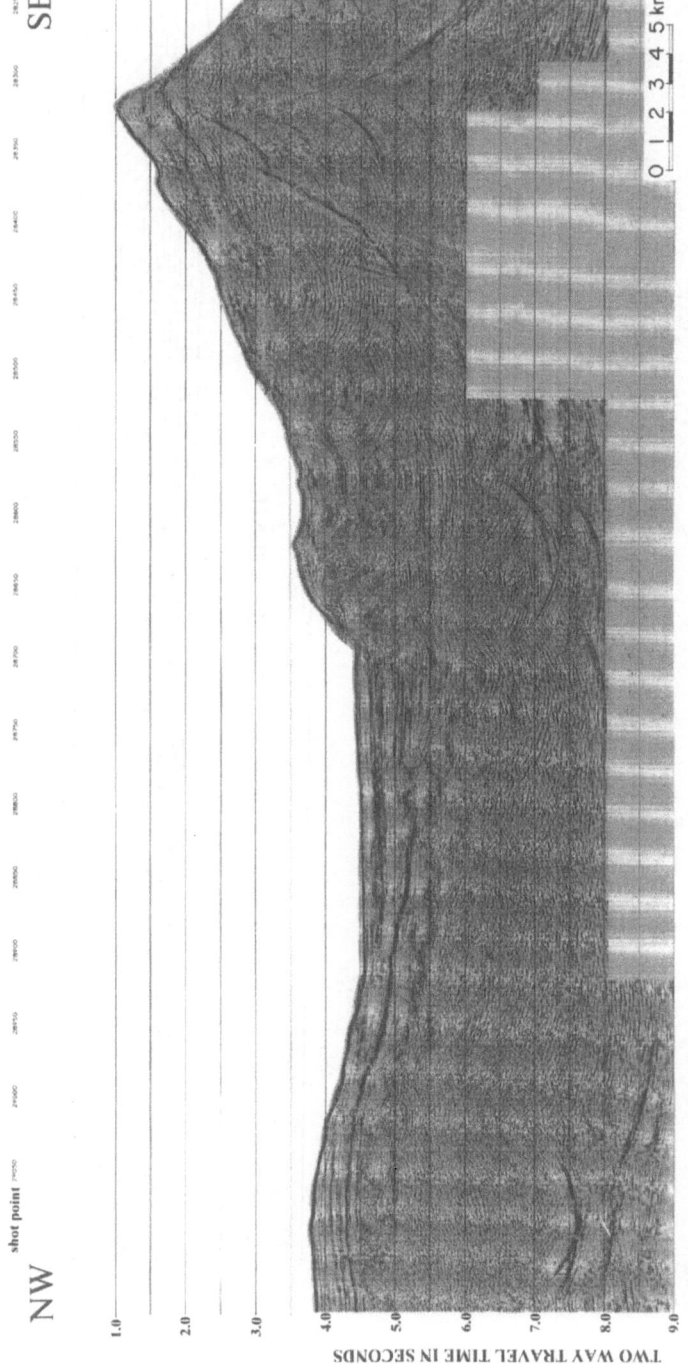

Figure 8 Migrated part of profile 5 (box in figure 7). Note the presence of reflectors below accretionary prism at north branch of South Scotia Ridge

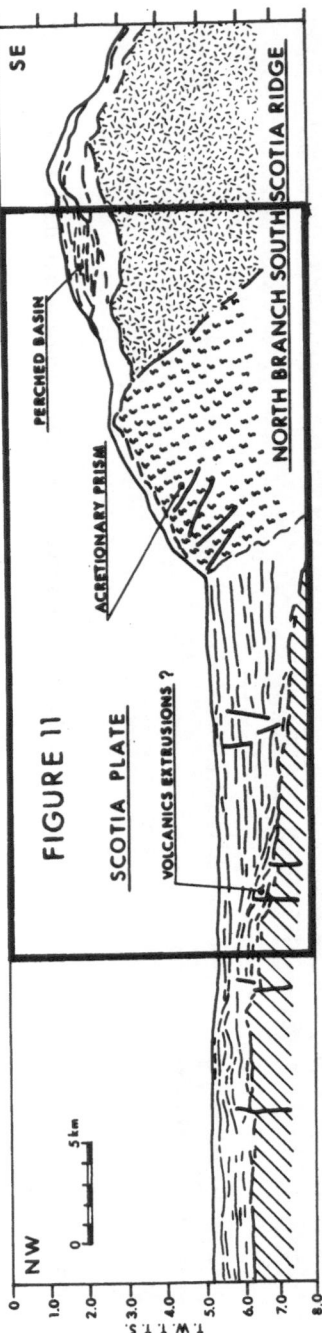

Figure 9 Line interpretation of profile 7. For location see fig.2 . Note well developed accretionay prism fronting the north branch of South Scotia Ridge. The extensive development of this deformed sequence reflects a local compressional component along the Scotia-Antarctic plate boundary. This compression is believed to be due to lateral motion along a slightly curved plate boundary.

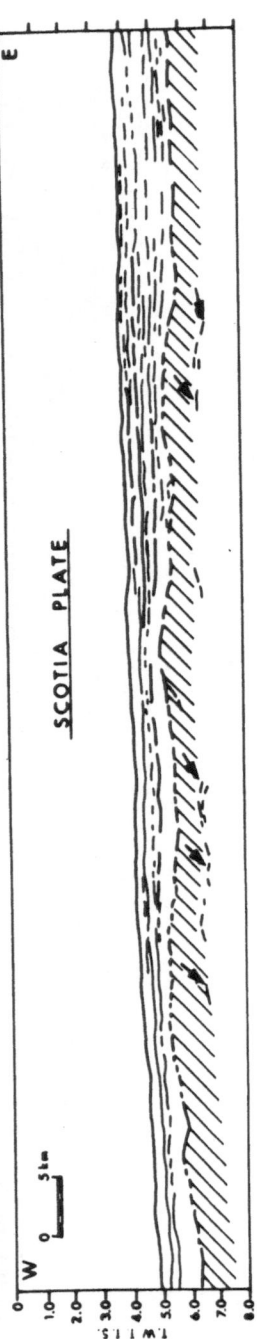

Figure 10 Line interpretation of profile 3. For location see fig. 2. Hatched: oceanic crust. Arrows: intra-crustal reflections.

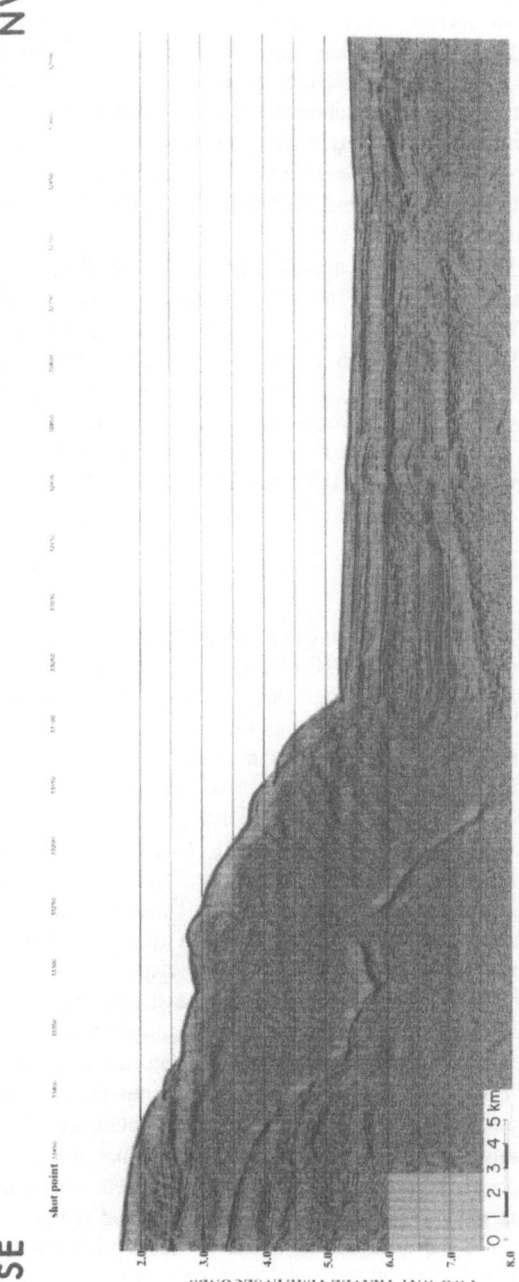

Figure 11 Section of profile 7 marked by a box in fig 9

sediments. These strata probably rest on the oceanic basement created by back-arc sea-floor spreading during the formation of Powell Basin. This buried oceanic basement may be represented by the sub-crustal high amplitude hyperbolic echoes which in places are nearly 1.5 seconds beneath the smooth basement surface. Broad basement highs rising above the surface of the smooth basement suggests that the oceanic basin may be subdivided into several subsidiary lows. Assuming a velocity of 4 km/sec for the unit interpreted as a volcanic/sediment, intercalations would imply that in places this unit is as much as 3 km thick and in others less than 1 km. Such excessive thickness would indicate, if our preliminary interpretation is correct, that early in its history this 35 Ma old back-arc basin experienced extensive effusion of volcanics from within and a massive input of sediments from its margins, the South Scotia Ridge, the South Orkney Island microcontinent and the Antarctic Peninsula.

The structure of Powell Basin resembles acoustically the crust of the eastern Caribbean. The Venezuela Basin in the eastern Caribbean also has a smooth basement (Horizon B) which sampling by the Deep Sea Drilling Project showed existence of basaltic sills and flows intercalated with Coniacian limestones. Sediments beneath the sills have velocities that suggest that they are a mixture of lithified sediments and volcanics, and basement beneath this sequence has velocities ranging from 5.30 to 6.2 km/sec which are characteristic of oceanic crust. The smooth basement in the Venezuela Basin also is flanked on the southeast by a rough surfaced basement that morphologically resembles oceanic basement. (Uchupi and Emery, 1984).

Conclusion

Around the periphery of the Scotia Plate, which has developed essentially independent of the South America and Antarctic plates during the last 40 Ma, are tectonic elements that were transported eastward from their original positions along the Pacific margin of Gondwanaland. What these original positions were and what routes the elements took to arrive at their present positions are problems yet to be resolved. These allochthonous tectonic elements and intervening younger deformed/undeformed sediments form the Scotia arc surrounding the Scotia Sea on the north and south, i.e. the North and South Scotia Ridges. The north-south trending eastern border of the Scotia plate; the South Sandwich spreading axis , is much younger having been formed during the last 7 Ma by subduction of the South America plate beneath the Sandwich plate. As has been demonstrated by Barker and Dalziel (1983) the North and South Scotia Ridges display a wide spectrum of present and past tectonic regimes, some of which are identifiable in the multi-channel seismic reflection profiles recorded during the present investigation.

South of the South Shetland Islands is the extensional northeast trending Bransfield Trough. This asymmetric trough is over 2000 m deep, is the site of Holocene volcanism, has a prominent axial magnetic anomaly associated with a line of volcanic peaks, and has a belt of diapiric structures that trend obliquely to the central neovolcanic zone on its southwest side. Some believed that Bransfield Trough began to form about 4 Ma in response to the cooling and sinking of a subducted crust, and that 1-2 Ma ago this continental rifting changed to sea-floor spreading to form the axial neovolcanic zone. Others have interpreted the low and its nascent spreading axis as due to subduction that is still taking place today. A third scenario proposed is that the trough is due to diffuse continental extension and that peaks in its center do not represent a spreading axis. This continental plate extension is believed, by some, to be in response to present-day complex plate interactions in the northern Antarctic Peninsula and by others it is believed to be related to a

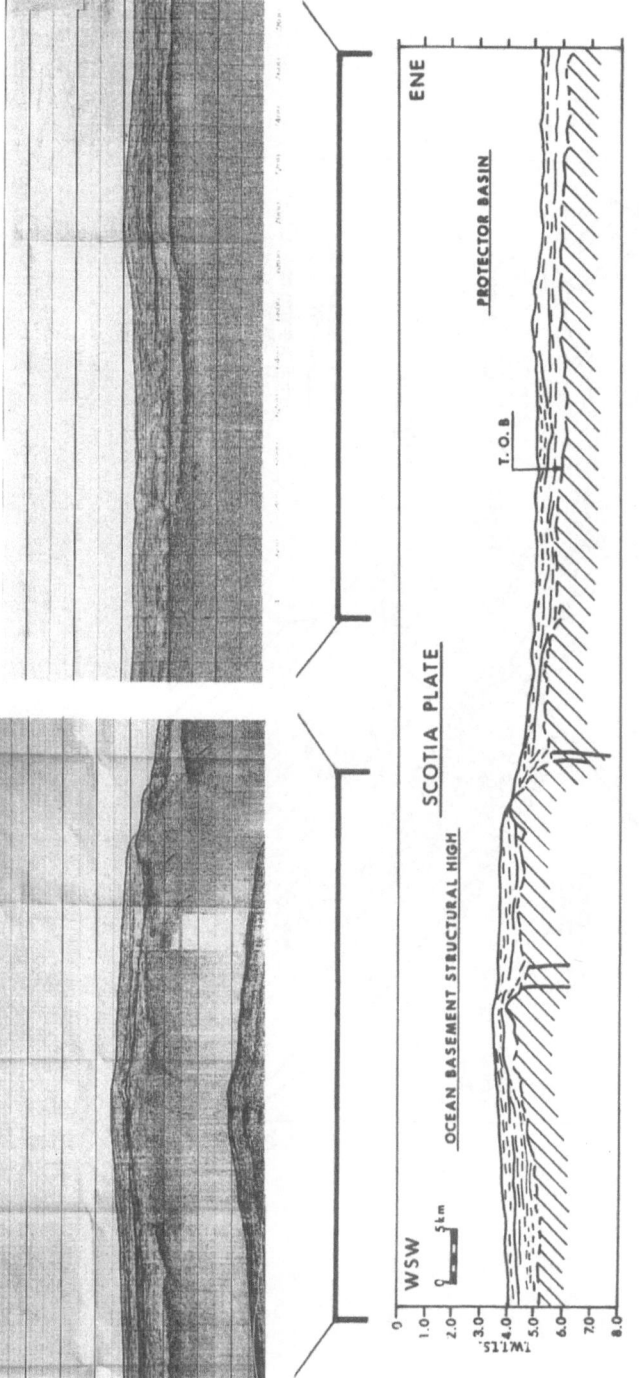

Figure 12 Top: photographs of two sections of profile 6. Their locations along profile are shown in the line interpretation in the lower panel. Bottom: line interpretation of profile 6. See fig.2 for location of profile. Point of interest on this profile is the fault bound basement swell southwest of Protector Basin.

284

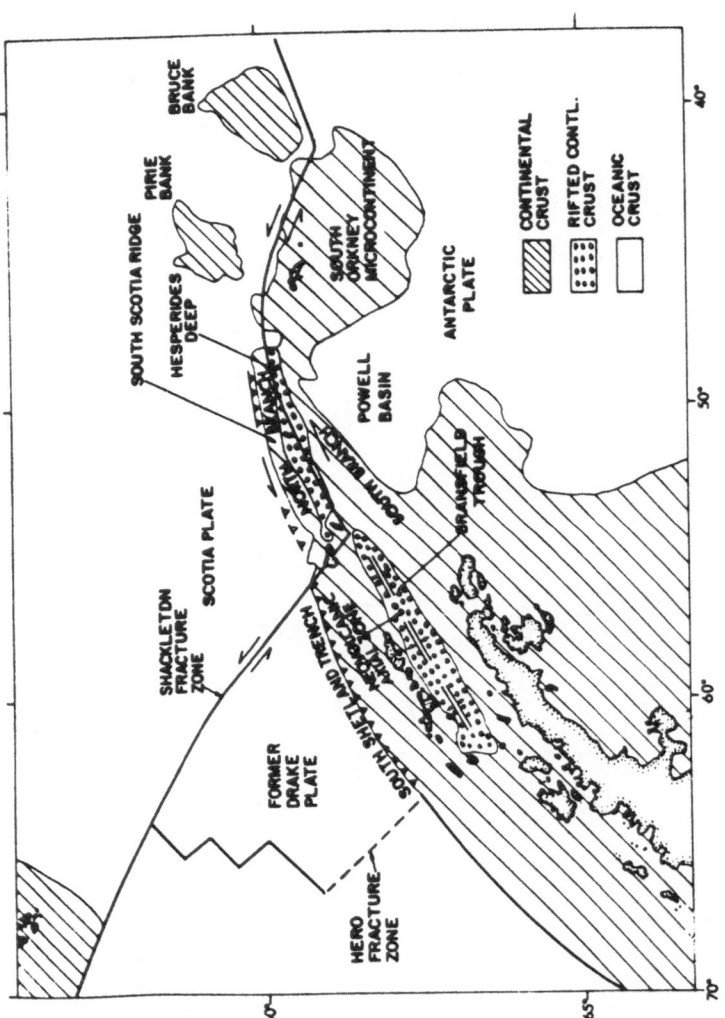

Figure 13 Tectonic map of the segment of the Scotia-Antarctic plate boundary investigated during present survey. Compiled from GEBCO(1981), Lawver and Villinger (1989) and the results from the present investigation. Hatchures: continental crust; dotted pattern: attenuated continental crust; blank areas: oceanic crust. A possible nascent spreading axis may exist along the center of the rifted Bransfield Trough.

change in trend of the Antarctic-Scotia plate boundary as it bends to follow the northwest orientation of the Shackleton Fracture Zone.

The crustal structure of the Scotia Sea is characteristic of normal oceanic regions that display prominent basement swells and may document that four distinct spreading systems have formed the basin. The segment of the South Scotia Ridge investigated by us between Elephant Island and the South Orkney Island Microcontinent is dominated and divided in two by the V-shaped Hesperides Deep which lacks any appreciable sediment fill (Fig.13). The nonmagnetic north branch was interpreted by King and Barker (1988) as a forearc structure, and the magnetic south branch as an arc high. Sediments at the base of the North Branch are deformed as a result of compression whose intensity increases eastward. This morphology of extension (Hesperides Deep) and local compression (accretionary prism at base of the South Scotia Ridge) is the result of the left-lateral strike-slip motion with a secondary extensional component . Powell Basin south of the South Scotia Ridge has been interpreted as an oceanic structure formed by back-arc spreading. Our profiles indicate that this oceanic basin has a different structure from the Scotia Sea north of the ridge. Its oceanic crust is overlain by a thick sequence of volcanics and sediments which may be capped by volcanic flows and/or sills to form a smooth acoustic basement closely resembling the Venezuela Basin in the eastern Caribbean Sea.

Acknowledgements

We wish to thank our collegues from Instituto Español de Oceanografia and the University of Barcelona for their assistance during Scotia-92 cruise. We also wish to thank Javier Escartin of the Woods Hole Oceanographic Institution-Massachusetts Institute of Technology Joint Program in Marine Geology and Geophysics, who also participated in the cruise, for making available to us his reprint collection on the geology and geophysics of the Scotia Sea region. We also are indebted to Colleen D. Hurter, Information Systems Associate II at Woods Hole Oceanographic Institution who made a literature search of the region using the computer facilities at WHOI. The skill of the captain and crew of the B/O Hesperides made possible the success of the sea-going operations in the hostile environment of the South Scotia Ridge for which we extend our gratitude. The cruise was organized by the Instituto Español de Oceanografia and the University of Barcelona who also made available funds they received from the Spanish Plan Nacional Antarctico (P.N.A.) and the Comision Interministerial de Ciencia y Tecnología (C.I.C.Y.T.) to pay expenses incurred during the cruise.

286

References

Acosta, J., Canals, M., Herranz, P. and Sanz, J.L.: 1989a, 'Investigación geológica y geofísica en el Arco de Scotia y Penínsulas Antártica. Resultados de la Campaña 'ANTARTIDA 8611". *Publ. Espc. Inst. Esp. Oceanografía* (2), 9-82.

Acosta, J., Canals, M., Harranz, P., and Sanz, J.L.: 1989b 'Perfiles sísmicos en la Shetlands del sur y estrecho de Bransfield. Estructura y dinámica reciente'.*Actas Plan Nac. Antartico CICYT*, 281-296.

Acosta. J. Canals, M., Diez, J.L., Fernandez Lopez, J.M., Herranz, P., Ortiz, R. and Sastre, J.C.: 1989c, 'Anomalias magnéticas al sur del mar de Bransfield'. *Actas Plan Nac. Antártico CICYT*, 254-257.

Acosta, J., Herranz, P., Sanz, J.L. and Uchupi, E.: 1992, 'Antarctic continental margin: Bransfield Trough, An Incipient Ocean Basin'. In: C.W. Poag and P.C. de Graciansky (Eds.). *Geologic Evolution of Atlantic Continental Rises*; New York, Van Nostrand Reinhold, 49-61.

Ashcroft, W.A.:1972,'Crustal structure of the South Shetland Islands and Bransfield Strait *'British Antarctic Survey scientific Report 66*, 1-43.

Allen, A.: 1966, 'Seismic refraction investigations in the Scotia Sea"' *British Antarctic Survey Scientific Reports*, **55**, 1-44.

Austin, J.A. and Uchupi, E.: 1982. 'Continental - oceanic crustal transitionm off Southwest Africa'. *Bull. Am. Ass. Petrol. Geol.*, v. **66**, 1328-1347.

Barker, D.H.N. and Austin, J.A., Jr.: 1994, 'Crustal diapirism in Bransfield Strait, west Antarctica - evidence for distributed extension in marginal basin formation'. *Geology*, **22**, 657-660.

Barker, P.F.: 1976. 'The tectonic framework in the Cenozoic volcanism in the Scotia Sea region, a review'. O. González-Ferrán (Ed.), *Symposium on Andean and Antarctic Volcanology Problems*. IAVCEI Special Series, Santiago, Chile, 330-346.

Barker, P.F.: 1982, 'The Cenozoic subduction history of the Pacific margin of the Antarctic Peninsula: ridge crest-trench intercations'. *Jour. Geol. Soc. London*, v. **139**, 787-802.

Barker, P.F. and Burrell, J.: 1977, 'The opening of Drake Passage'. *Marine Geology.*, v. **25**, 15-34.

Barker, P.F. and Dalziel, I.W.D.: 1983, 'Progress in geodynamics in the Scotia arc region'. In: S.J.R. (Ed.), *Geodynamics of the Eastern Pacific Region, Caribbean and Scotia Arcs*. American Geophysical Union Geodynamics Series, **9**, 137-170.

Barker, P.F., Barber, P.L. and King, E.C.: 1984, 'An early Miocene ridge crest-trench collision on the South Scotia Ridge near 36° W'. *Tectonophysics*, v. **102**, 315-332.

Barker, P.F., Dalziel, I.W.D. and Storey, B.C.: 1991, 'Tectonic development of the Scotia Arc region'. In: R.J. Tingey (Ed.), *Antarctic Geology*, Oxford University Press, Oxford, 215-248.

Baker, P.E., McReath, I., Harvey, M.R. and Roobol, M.J.: 1975, 'The geology of the South Shetland Islands: V. Volcanic evolution of Deception Island'. *British Antarctic Survey Scientifuc Reports* **78**, 81 pp.

British Antarctic Survey: 1985, 'Tectonic Map of the Scotia Arc'. Scale 1;300,000. BAS (misc) 3, Cambridge.

Canals,M, Acosta,J, Gracia,E, Escartin,J, and ORCA Group.: 1992, Caracterización Geológica de la región de enlace entre la cuenca de Bransfield y la dorsal sur de Scotia (Antártida)'. *Acta Geológica Hispánica* V.27 n° 3-4 ,89-110

Canals,M,Acosta,J,Baraza,J,Bart,P,Calafat,T,Casamor,J.L.,DeBatist,M,Ercilla,G,Farrán,M,Fran cés,G,Gracia, E,Ramos-guerrero,E,Sanz,J.L.,Sorribas,J,and Tassone,A. : 1994, 'La cuenca central de Bransfield (NW de la Península Antártica): Primeros resultados de la campaña GEBRA-93' *Geogaceta* ,16 (in press)

Dalziel, I.W.D. and Elliot, D.H.: 1973, 'The Scotia Arc and Antarctic margin'. In: A.E.M. Nairn and F.G. Stehli (Eds.), *The Ocean Basins and Margins, Volume 1, The South Atlantic*. Plenunm Press, New York, 171-246.

Dalziel, I.W.D , Birkenmajer, K., Mpodozis, C., Ramos, V., and Thomson, M.R.A. (Leaders): 1989, 'Tectonics of Scotia Arc, Antarctica: Punta Arenas, Chile to Ushuaia, Argentina'. In: P.M. Hanshaw (Ed.), *Field Trips for the 28th International Geological Congress* . American Geophysical Union, Washington DC. **Field Trip Guide Book T180**, 206 p.

Davey, F.J., 1972. 'Marine gravity measurements in Bransfield Strait and adjacent areas'. In: R.J. Adie (Ed.), *Antarctic Geology and Geophysics*. Universitetsforlaget, Oslo, 39-46.

Emery, K.O. and Uchupi, E.,: 1984, 'The Geology of the Atlantic Ocean'. Springer-Verlag, New York, 1050 p.

Forsyth, D.W.: 1975, 'Fault plate solutions and tectonics of the South Atlantic and Scotia Sea'. *Jour. Geophys. Research*, v. **80**, 1429-1443.

Gamboâ., L.A.P. and Maldonado, P.R.: 1990, 'Geophysical investigations in the Bransfield Strait and the Bellingshausen Sea,. Antarctica'. In: B. St. John (Ed.), *Antarctica as an Exploration Frontier: Hydrocarbon Potential, Geology and Hazards*. Am. Assoc. Petroleum Geologists Studies in Geology, **31**, 127-141.

González-Ferrán, O.: 1991, 'The Bransfield rift and its active volcanism'. In: M.R.A. Thomson, J.A. Crame and J.W. Thomson (Eds.), *Geological Evolution of Antarctica*. Cambridge University Press, New York, 505-509.

González-Ferrán, O. anf Katsui, Y., 1970. 'Estudio integral del volcanismo Cenozoico superior de las Islands Shetland del Sur, Antarctica'. *Instituto Antarctico Chile Series Cientificas*, **1(2)**, 123-174.

Guterch, A., Grad, M., Jasnik, T. and Perchuc, E.: 1991a, 'Tectonophysical models of the crust between the Antartic Peninsula and the South Shetland Islands'. In: M.R.A. Thomson, J.A. Crame and J.W. Thomson (Eds.), *Geological Evolution of Antartica*. Cambridge University Press, New York, 499-504.

Guterch, A., Shimamura, H. and Polish-Japan-Argentina Research Group, 1991b, 'An OBS-land refraction seismological experiment in the Brtansfield Trough, West Antarctica'. 1990/1991 *Abstracts, Sixth, International Symposium on Antarctic Earth Sciences*. Tokyo, Japan, 201-202.

Hawkes, D.: 1981, 'Tectonic segmentation of the northern Antartic Peninsula'. *Geology*, **v. 9**, 220-224.

Herron, E.M., and Tucholke, B.E.: 1976: 'Sea-floor magnetic patterns in the southwestern Pacific'. In: C.D. Hollister, C. Craddock et al., *Initial Reports of the Deep Sea Drilling Project* , **v. 35**, 263-287.

Ibañez, J.M., Ortiz, R., Blanco, R., Del Rey, R. and Morales, J. 1993, 'Terremotos profundos registrados en las islas Decepción y Livingston. Posible relaccion con la zona de subducción de las Shetlands del Sur'. *Vº Simposio de Estudios Antárticos*. Univ. Barcelona.

Jeffers. J.D., Anderson, J.B., and Lawver, L.A., : 1991, ' Evolution of the Bransfield Basin, Antarctic Peninsula'. In: M.R.A. Thomson, J.A. Crame and J.W. Thomson (Eds.), *Geological Evolution of Antarctica* . Cambridge University Press, Cambridge, 481-485.

Keller, R.A. and Fisk, M.R.: 1992, 'Quaternary marginal basin volcanism in the Bransfield Strait as a modern analogue of the southern Chilean ophiolites'. In: L.M.

King, E.C. and Barker, P.F.: 1988: 'The margins of the South Orkney microcontinent'.*Jour. Geol. Soc. London*, **v. 145**, 317-331.

Klepeis, K.A., Lawver, L.A.,Sandwell, D and Small, C.: 1989 'The morphology and tectonic structure of Shackleton Fracture Zone'.*Antarctic Journal*, **24**, 126-128,

Klepeis, K.A., Lawver, L.A, Zellers, S., Miller, J. and Nelson, G.: 1990, 'Bathymetry of the Shackleton Fracture Zone, Elephant Island and Clarence Island regions, Antarctica'. *Antarctic Journal*, **25**, 71-73.

GEBCO, 1981, General Bathymetric Chart of the Oceans. Chart 5.16. Scale: 1:10,000,000 at the equator. Canadian Hydrographic Service, Ottawa.

Lawver, L.A. and Villinger, A.: 1989, 'North Bransfield Basin: R/V Polar Duke cruise Pd VI-88 *Antarctic Journal*, **24**, 117-119.

Lawver, L.A., Klepeis, K., Dalziel, I.W.D. and Zellers, S., 1990, 'Intersection of the Shackleton Fracture Zone and the Antarctic, a complex continental margin' *EOS*, **71**, 1592.

Ludwig, W.L. and Rabinowitz, P.D., : 1982, 'The collision complex of the North Scotia Ridge'. *Jour. Geophys. Research*, **v. 87**, 3731-3740.

O.R.C.A. Group: 1992, 'Resultados preliminares de la campaña de geologia SCOTIA-92. Estudio geológico y geofisico de la sur del arco de Scotia'. In: J. Lopez-Martinez (Ed.), *Geologia de la Antártida Occidental*. III Congreso Geológico de España, 203-212.

O.R.C.A. Group: 1994. 'Bathymetry of the Hespérides Deep. Scotia Sea - South Scotia Ridge; Antartica'. Scale 1:200,000. Grup de Recerca en Geociéncias Marines, Universitat de Barcelona.

Pelayo, A.M. and Wiens, D.A.: 1989, 'Seismotectonics and relative plate motions in the Scotia Sea Region'. *Jour. Geophys. Research*, **v. 94**, 7293-7320.

Roach, P.J.: 1978, 'The nature of back-arc extension in Bransfield Strait'. *Royal Astronomical Society, Geophysical Journal*, **53**, 165.

Spray, J.: 1991,' Structure of the oceanic crust as deduced from ophiolites'. In: P.A. Floyd (Ed.), *Oceanic Basalts*, Van Nostrand Reinhold, New York, 49-62.

Watters, D..G.: 1972, 'Geophysical investigation of a section of the South Scotia Ridge'. In: R.J. Adie (Ed.), *Antartic Geology and Geophysics*, Universttetsforlaget, Oslo, 33-38.

MAPPING THE MOHO IN THE IBERIAN MEDITERRANEAN MARGIN BY MULTICOVERAGE PROCESSING AND MERGING OF WIDE-ANGLE AND NEAR-VERTICAL REFLECTION DATA

N. VIDAL, J. GALLART, J.J. DAÑOBEITIA AND J. DIAZ

Institute of Earth Sciences, Consejo Superior de Investigaciones Científicas, Martí i Franquès, s/n, 08028 Barcelona, Spain

ABSTRACT. Normal incidence seismic reflection data have been collected recently in the Iberian Mediterranean margin within the Spanish ESCI programme. The experiments include in-line recordings onshore of marine profiles, and provide coincident small and large-offset multicoverage of complex areas such as the onshore/offshore transition, where classic stacked images of the vertical reflectivity lack resolution and hamper a thorough comprehension of the structure and its lateral evolution. We obtained coherent and enhanced crustal images by developing a wide-angle multichannel processing analogous to the near-vertical conventional one and merging the final stacked and migrated sections.

All the wide-angle sections generated show clear images of the Moho reflectivity. In the Valencia trough, a combined transect documents the lateral evolution of the deep crustal reflectivity across strike of structures. A steady thinning of the crust is revealed in the Iberian margin, from a depth of 32 km beneath the Catalan ranges up to 19 km depth at 60 km seawards. A similar thinning rate is observed on the Balearic flank. In the central part of the trough the Moho is imaged at constant depths around 16-17 km. The lower crustal reflectivity as well as the velocity-depth and gravity results indicate that the thinning is accomodated mainly by the lower crust. Underplating features reported in continental passive Atlantic margins are not supported here by the low velocities of 6.4-6.5 km/s found in the lower crust. In the central part of the Betics-Alboran domain the wide-angle stacked section suggest, in agreement with previous gravity interpretations, that the thin Alboran Sea crust extends beyond the shoreline, up to 10-15 km inland where a Moho jump down of about 3 s TWT marks the southernmost limit of the internal Betics thick crust. The strong stretching rates inferred from the present crustal images in the Iberian Mediterranean margin are to be related with the interaction of extensional processes within a convergent regime between the African and Iberian plates, or even with shear tectonics in the Betics-Alboran domain.

1. Introduction

Normal incidence reflection marine seismic profiles collected with large air-gun sources are currently recorded at wide-angle offsets as well, either in land stations or OBS at sea (see examples in Meissner et al., 1992) with the aim to complement the near-vertical stacked images of the crustal reflectivity with the velocity-depth results of wide-angle forward modelling.

Consequently, some areas are sampled both at small and large offsets. For these areas, a general

E. Banda et al. (eds.), Rifted Ocean-Continent Boundaries, 291–308.

consistency is commonly reported in the derived seismic structure, particularly for prominent discontinuities such as the Moho (see example of the Alps in Valasek et al., 1991). After a more detailed inspection, significant mismatches may arise, however, in the structural signatures resulting from normal incidence and large aperture data, making it difficult to compare and discuss thoroughly the results. A number of possible sources of discrepancy are mentioned below.

Firstly, the results may be affected by uncertainties inherent to the independent approaches: normal incidence sections largely depend on the velocities used for stacking and migration, usually poorly constrained, and the large-offset forward modelling has to deal often with unreversed profiles.

Secondly, the forward modelling is based on the lateral correlation of seismic phases, and this, together with the low-frequency content of the signals, tends to enhance the lateral continuity of the reflective structures. In the normal incidence sections, the deep structure appears more heterogeneous, and the reflectors display shorter wavelengths. In the case of rifted margins the apparent lateral vanishing of deep reflectivity which is often observed may be partly due to the acquisition parameters and the processing sequence, that brings together data from different shot-gathers, which might be affected by local changes beneath the shotpoints perturbing the vertical penetration of energy.

Thirdly, in areas such as continental margins characterized by strong lateral variations of crustal structure, particularly in the upper sedimentary layers, the normal incidence data may completely lack resolution, whereas the large-aperture geometries undershoot the perturbed area and provide much better images at depth, especially if multifold can be achieved.

The aim of this research is to describe a comparable, unified multicoverage analysis of normal incidence and large-aperture data altogether in order to provide a straightforward comparison of the corresponding results, and to illustrate the ability of this technique to enhance the structural image of the Moho.

2. Data and geological settings

The data base of this study is the marine normal incidence reflection profiles from air-gun sources of the Spanish ESCI programme (Gallart et al. 1994; García-Dueñas et al., 1994). They were performed on the Mediterranean margin of the Iberian Peninsula, across the Valencia trough and the Alboran Sea, together with piggy-back recordings on land of 6 portable stations in line with the marine profiles (Fig. 1). A second profile across the Valencia trough from explosive sources and recorded on land by 50 in-line stations in Catalunya and 20 in Mallorca is also considered.

The Valencia trough and the Alboran Sea are two small basins in the western Mediterranean, caused by Neogene extension developed in a region of overall convergence between the African and European plates. They are located at the western end of the Alpine orogenic belt, and linked respectively to linear and arcuate ranges of the Pyrenees and the Betics-Rif systems. Both basins are characterized by a rather small size, stretched continental crustal features and presumably high-average strain rates during rifting (Watts and Torné, 1992; Watts et al., 1993).

A number of geological and geophysical results on the crustal structure of the two basins have been established after multidisciplinary research during the last decade (Banda and Santanach, 1992; Watts et al., 1993), but key parameters for geodynamical modelling of the continent/ocean transition, such as the onshore/offshore variations in the reflectivity of the lower crust and geometry of the Moho are still not controlled in detail. Constraints on this are investigated by the new normal incidence and large-aperture reflection data presented here, first examined independently and afterwards attempting an unified multicoverage analysis.

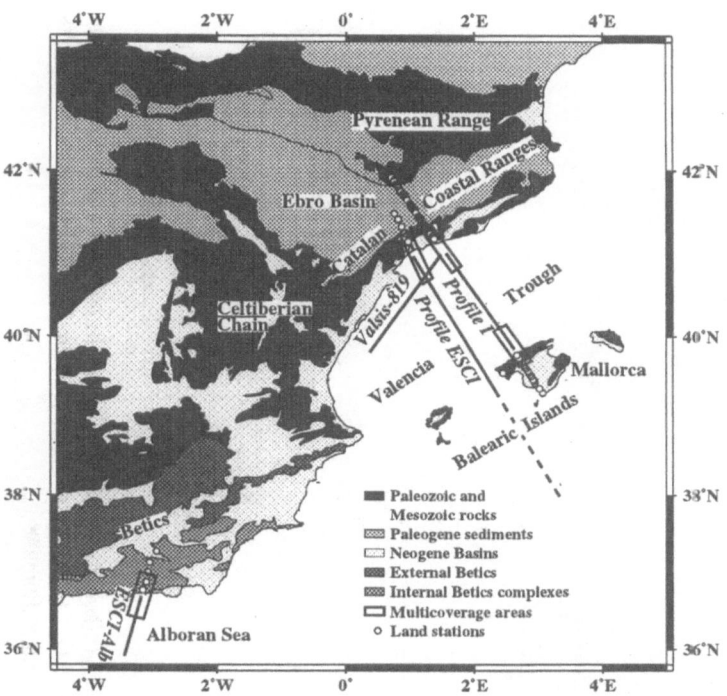

<u>Figure 1</u>. Geological scheme of the Iberian Mediterranean margin showing location of land stations and wide-angle multicoverage areas.

3. Independent analysis of near-vertical and wide-angle data sets in the NE Iberian margin

3.1. NORMAL INCIDENCE STACKED SECTIONS

The continent/ocean transition in NE Spain is investigated by an ESCI seismic reflection profile made up of two segments. A 50 km-long profile on land (Gallart et al., 1994) connects the southern end of the ECORS-Pyrenees profile (Choukroune et al., 1989) with the Mediterranean coast. The continuation at sea forms a 400 km-long profile across the Valencia trough and its flanks, up to the South Balearic basin. In this study we considered 220 km of the marine line, from the Catalan coast to the Balearic promontory (Fig. 1).

A composite section of the two profiles is shown in Figure 2. Differences in frequency content of the signals are related to the type of seismic sources: explosives on land and air-guns at sea. In Fig. 2 the trace interval is 30 m on the land segment and 25 m in the marine part. The processing sequences applied to the land and marine normal incidence reflection data are reported in Table I. They were geared towards making the on-land and at-sea parts of the section directly comparable.

The northwesternmost land segment displays a strong north-dipping lower crustal reflectivity,

294

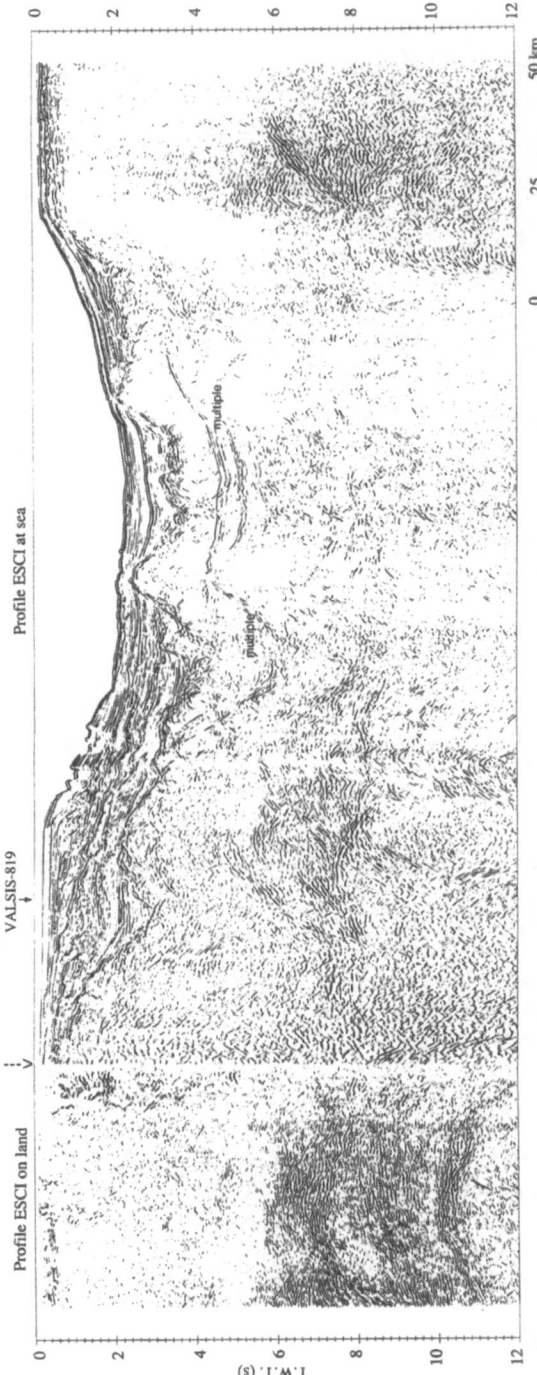

Figure 2. Composite section of the normal incidence ESCI profiles on the Iberian mainland (50 km on the left side) and at sea, across the Valencia trough up to the Balearic promontory (220 km on the right side). A lateral-coherency filtering was applied on the unmigrated section. For other processing parameters, see Table I. The arrow marks the shoreline.

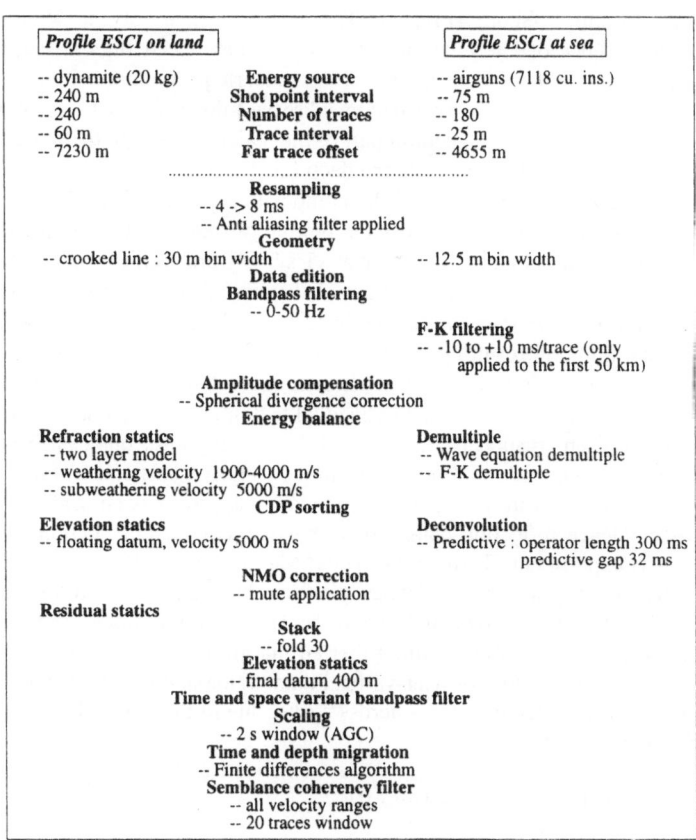

Table I. Processing sequence for near-vertical ESCI seismic profiles on land and at sea. In the middle part are listed common steps for both profiles. Specific parameters are within each column.

between 5.5 and 11 s TWT, in agreement with the image at its northward continuation within the ECORS line (Suriñach et al., 1993). This reflectivity becomes subhorizontal beneath the Catalan Ranges, where the Moho corresponds to the deepest and very strong reflector of the section, as controlled by wide-angle modelling (see § 3.2). The upper crust is basically transparent below the sediments. Towards the shoreline, the crustal reflectivity is less coherent, which may be related to the higher attenuation, scattering and lower frequency content of the energy propagating from shots on the low-velocity Neogene sediments of the Reus basin.

The 220 km-long marine segment from the Catalan coast to the Balearic promontory is displayed on the right side of Fig. 2. The sedimentary sequence is resolved along the profile, and extensional features in the Catalan flank of the Valencia trough can be inferred from a close inspection of the upper crustal reflectivity. This study focuses, however, on the seismic signature of the deep crust, the reflectivity of which is, in general, weaker on the marine profile than on the land segment. A 3 s-thick reflective band between 5 and 8 s TWT is observed between 30 and 60 km offshore, where the reflector at 8 s TWT marks the base of the crust. A similar reflective image was found in the ENE-WSW VALSIS profile 819 (Torné et al., 1992) which crosses the

ESCI sea line at 35 km offshore (Fig. 1). The Balearic flank of the Valencia trough is characterized by an absence of laterally coherent deep reflections. A high reflectivity beneath 5 s TWT is concentrated on the northwestern side of the Balearic promontory. Contrary to the image on the Catalan side, the diffuse reflective pattern observed down to 10-11 s TWT perturbs the identification of the Moho here. In the central part of the Valencia trough, the presence of remnant water multiple energy perturbs the image of the deep crustal structure as compared to the flanks.

A fundamental, unresolved question from the composite section of Fig. 2 concerns the lateral evolution of the lower crust and Moho across the Iberian onshore/offshore transition. Inland, a 5 s-thick lower crust and a crust-mantle boundary at 11 s TWT are well expressed. Seawards, after 20 km without clear deep reflections, the bottom of the crustal reflectivity is observed around 8 s TWT and, where present, the lower reflective band has a thickness of only 3 s.

Instead of a crustal thinning resulting from extensional processes, the normal incidence image of Fig. 2 rather suggests an imbrication and shortening of two crusts of different nature and thickness, that could be related to shear and collisional tectonics, in a manner similar to the image of the Iberian and European crusts beneath the North Pyrenean Fault Zone (Suriñach et al., 1993). However, the poor resolution at depth found beneath the Catalan coast and within 20 km seawards should be carefully considered in lithospheric geodynamic approaches (subsidence analysis, etc.) based on normal incidence reflection data. Other near-vertical seismic sections available in the area, from the Valsis experiment (Torné et al., 1992; Maillard et al., 1992) have similar features: clear deep reflective images along the Catalan platform but lack of resolution across strike of structures. Probably, the short-wavelength lateral changes of the upper sedimentary seismic properties in the strongly blocked and faulted Iberian margin (Roca and Guimerà, 1992) perturb significantly the stacking procedure of near-vertical data. Wide-angle reflection data appear more likely to constrain the deep crustal features across strike, due to the very different sampling paths and energy/frequency content.

3.2. VELOCITY-DEPTH FORWARD MODELLING

The air-gun shots of the marine ESCI profile were recorded at far offsets on land at 6 portable stations along the ESCI land profile, and at one station in Mallorca. Record-sections (receiver gathers) were built up and interpreted for each station. Forward modelling was carried out, based on well defined Pg and Pn refracted phases, as well as PiP and PmP reflections correlated in the sections (Gallart et al., 1994).

In order to check the lateral consistency of the velocity-depth results, an almost parallel, 30 km apart, refraction profile (labelled Profile I in Fig. 1) was also interpreted. This profile, made up from explosive sources spaced 1.5 km, was recorded by 20 stations in Mallorca and 50 on the Catalan mainland, spaced every 2 km. Two OBS along the line also provided data (Dañobeitia et al., 1992).

Although the requirement of fitting the data (arrival times and amplitudes) for all the land stations severely restricts the acceptable models, the velocities within the sedimentary sequence and the upper-lower crustal transition in the Valencia trough are not fully constrained, because the profiles are unreversed. In all the cases, significant crustal thinning beneath the trough was inferred. The Moho shallows seawards, from about 32 km beneath the mainland, either continuously up to 13-14 km depth in the Valencia trough axis, or keeping rather constant depths around 16-17 km in the trough (Figure 3). The thinning affects mainly the lower crust, which is reduced to 5-6 km thickness in the trough, or may almost disappear beneath the axis. Rather low-velocities of 6.4-6.5 km/s are found in the lower crust, and best-fitting velocities in the uppermost mantle range between 7.8 and 8.0 km/s.

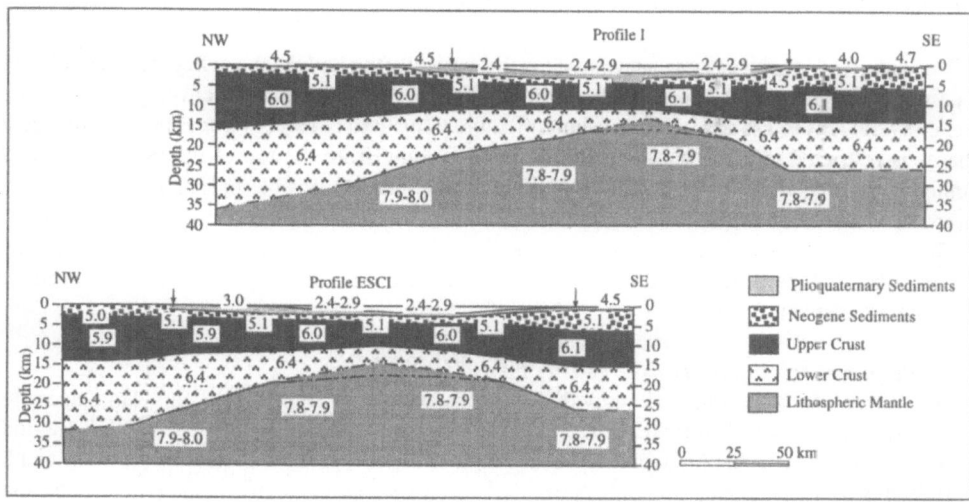

Figure 3. Forward modelling results for profiles I and ESCI across the Valencia trough and its flanks which fit best the recorded data on the land stations of Fig. 1. Near the axis of the trough, the thickness of the lower crust and the Moho are not well constrained: the broken lines indicate possible variations of Moho location, and a velocity gradient up to 0.1 s⁻¹ in the lower crust is compatible with the data. Inset numbers are velocities in km/s. The arrows mark the shoreline.

4. Towards a unified analysis of seismic reflection data

In the previous section, near-vertical and wide-angle data of the Valencia trough case-history were studied independently. Results of both interpretations appear to be complementary, but their combination into a single, detailed seismic section of the crustal structure is disturbed by the uncertainties inherent in each method (velocity analysis on the normal incidence section, phase identification or initial models on the forward analysis, etc.). Readily consistent results are to be expected when more comparable analysis of the reflection data sets can be performed. In this respect, multicoverage at large-offsets is achieved in some areas from the wide-angle recordings onshore at several in-line stations. Therefore, a multichannel wide-angle analysis can be undertaken in some onshore/offshore areas of the eastern Iberian Peninsula. A processing sequence as similar as possible to the conventional one for normal incidence reflection data has been developed, so that the final stacked and migrated sections can be compared and merged into a single crustal transect image.

4.1. MULTICOVERAGE WIDE-ANGLE PROCESSING

The wide-angle multicoverage analysis concerns basically the PmP Moho-reflected phase, clearly seen out to ranges of 100 km in all the profiles. For the PmP phase, multicoverage is achieved along several tens of km around the onshore/offshore transitions. Multicoverage for other reflected phases is restricted to a PiP phase in profile ESCI coming from the top of the lower crust, and limited to an area close to the Catalan shoreline, as this phase is visible only along 20-30 km in the receiver gathers and the station spacing is 10 km.

The main steps in the processing sequence performed with the ITA-Landmark software package are described below. If not stated otherwise, results refer to the marine profile ESCI recorded on the Catalan mainland.

4.1.1. Preprocessing in the receiver-gathers and CMP sorting

The classic record-sections of individual stations were first transformed into large-aperture receiver-gathers in SEG-Y format. A careful introduction of the geometry and bin-width definition is essential, because of the large and different range of distances. After several trials, a bin-width of 1000 m was adopted, as a trade-off between the shot spacing of 75 m and the receiver spacing of 10 km. This bin-width avoids destructive interference during stacking, given the apparent velocities and frequency content of the signals recorded (Flueh and Dickmann, 1992).

Editing of traces, muting, energy equalization and 3-20 Hz band-pass filtering were also applied to the receiver-gathers. Predictive deconvolution using long operator lengths of 1 to 2 s and predictive gaps of 0.1 s, attenuated the ringing of the signal and sharpened up the arrivals. In most of the gathers, refracted energy (Pg and Pn phases) was clearly apparent and had to be eliminated before Common Mid Point (CMP) sorting. Attempts to remove the refractions through f-k filtering were unsuccessful, as a significant distortion was introduced in the reflected energy, probably due to similar wavenumbers and frequencies on both types of energy at large offsets. Therefore, since the refracted phases were well identified and shifted from the reflected energy, we applied a direct mute of refractions in the receiver-gathers.

After this preprocessing, CMP gathers are built up. According to the bin-width considered, an average folding of 27 was obtained, with traces coming from 3-4 different stations on land.

4.1.2. Stacking: static and dynamic corrections.

In order to build up a wide-angle stacked section, static and dynamic corrections had to be applied to the CMP gathers. The static topographic corrections to account for station elevations and shot-water columns were considered, but residual statics algorithms did not improve the stacked section as they were not well adapted for time shifts up to several tenths of a second.

Dynamic corrections, i.e. Normal Move-Out (NMO) were most essential for wide-angle data involving very different, large offsets. An example of the NMO correction over a CMP is shown in Fig. 4. We performed the dynamic correction considering an one-layer model for the whole crust. If more complex velocity models are considered, the full NMO correction must be applied for large-aperture data, instead of two-terms approximations valid for small offset/depth ratios. The NMO is very dependent here on the average velocity considered, which can be constrained up to 0.1 km/s, as illustrated in Fig. 4. After the NMO correction, far-offset traces may be highly affected by stretching in the first seconds, and mute must be applied correspondingly.

The stacked section obtained after these processing stages is shown in Fig. 5.

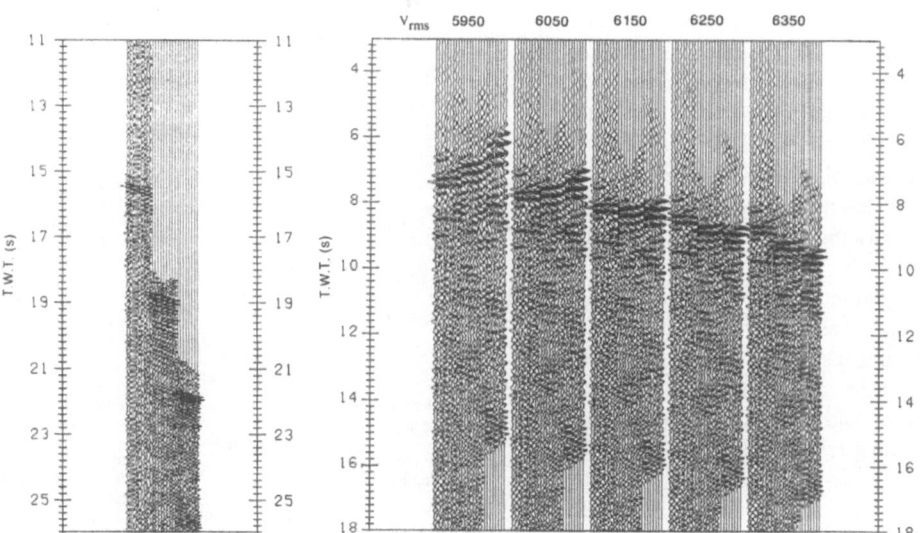

Figure 4. Example of a wide-angle CMP gather for profile ESCI, near the Catalan shoreline, made up of 27 traces coming from 3 different stations (bin width 1000 m). Left side: original CMP. Right side: effect of NMO corrections considering 5 different velocities. A 6.15 km/s velocity appropriately places the Moho reflections around 8 s TWT (panel in the middle). Small velocity variations of 0.1 km/s result in clear NMO mismatches.

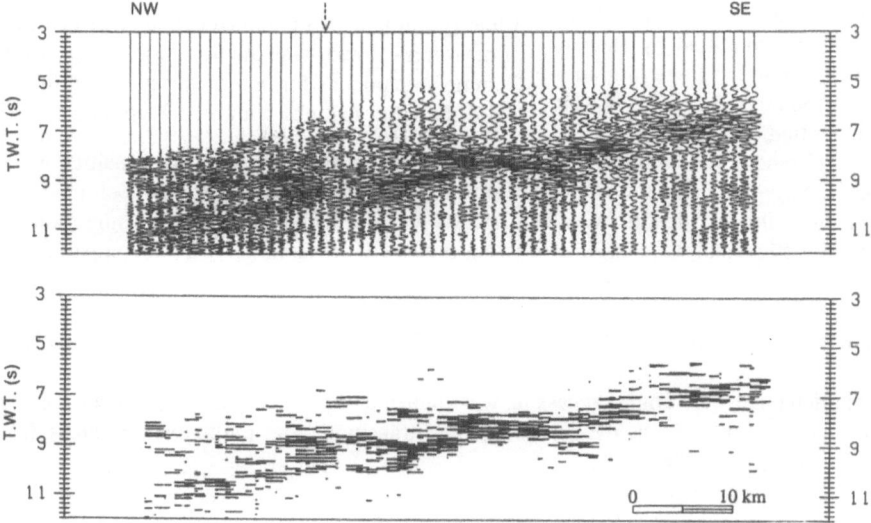

Figure 5. Wide-angle stacked section for profile ESCI in the Catalan margin. A continued reflectivity from the bottom of the crust is clearly imaged across the onshore/offshore transition (the arrow marks the shoreline). Semblance-coherency filtering (lower panel) enhances the signature of the Moho and the crustal thinning seawards. See Table II for processing details.

4.1.3. Post-stack processing. Migration.

In the normal incidence stacked sections, a semblance coherency filter considering all velocity ranges is usually applied to enhance the crustal reflectivity on the final section. This filter was also used satisfactorily in our wide-angle stacks, despite the large spacing of 1 km between the CMPs.

Migration of the wide-angle stacked sections was undertaken, to obtain depth images of the Moho comparable with those of the normal incidence sections and with the velocity-depth modelling results. Although migration of a wide-angle section may be affected by spatial aliasing or noise smearing, different time and depth-migration algorithms were applied satisfactorily to our high-energy low-frequency wide-angle data set. Best results were obtained for algorithms based on time-shift methods, using a depth-varying laterally-constant velocity in agreement with the forward modelling results. The reflectivity pattern was not modified by changes of up to 10% in the average migration velocity, although the reflectors were placed at shifted depths.

Near vertical processing		*Wide angle processing*
-- 4 ms	**Sampling rate**	-- 8 - 16 ms
Resampling		
-- 30 m bin width	**Geometry**	-- 1000 m bin width
	Data edition	
-- 0-50 Hz	**Bandpass filtering**	-- 3 - 20 Hz
	F-K filtering	-- Not useful for refracted phases
Amplitude compensation		**Mute of refracted phases**
-- Spherical divergence correction		-- Pg and Pn
	Energy balance	
Refraction statics		
Demultiple	**CDP sorting**	
-- 90 m - 7200 m	-- Distances	-- 10 km - 140 km
	Elevation statics	
	-- floating datum plane	
	Predictive deconvolution	
-- operator length 300 ms predictive gap 32 ms		-- operator length 1 s predictive gap 0.1 s
	NMO correction	
-- maximum correction (30 km depth) 0.07 s		-- maximum correction (30 km depth) 15 s
		-- High resolution in velocities
	-- mute application	
Residual statics	**Stack**	
-- fold 30		-- fold 10 - 30
	Elevation statics	
	-- final datum 400 m	
Time and space variant bandpass filter		
	Scaling	
	Time and depth migration	
-- Finite differences algorithm		-- Time shift algorithm
	Semblance coherency filter	
-- 20 traces	-- all velocity ranges	-- 3 traces

Table II. Summary of parameters and steps characteristic of near-vertical and wide-angle multichannel processing. Common techniques for both data sets are in the middle, and differential features within each column.

Finite-differences migration algorithms which allowed laterally varying velocities were also considered. Migrated images with some distortions and artifacts on rather monocromatic signals were obtained, although results were consistent with those from the time-shift migrations.

The near-vertical and wide-angle processing sequences are summarized and compared in Table II.

4.2. WIDE-ANGLE STACKED SECTIONS IN THE IBERIAN MEDITERRANEAN MARGIN

The multichannel wide-angle processing was applied to three data sets sampling the Catalan and Balearic margins, and one at the Betics-Alboran transition. The multicoverage areas for each profile are indicated in Fig. 1.

4.2.1. Wide-angle sections at the Valencia trough flanks

The stacked section from marine profile ESCI piggy back recorded along the land segment is displayed in Fig. 5. A 30 fold coverage was achieved for most of the section. Despite the spacing of 1 km between CMPs, strong laterally coherent energy coming from the bottom of the crust is observed along the profile. The Moho shallows steadily from 11 s TWT at 20 km inland up to 6 s TWT about 40 km seawards.

Consistency on the wide-angle stacked image of the Moho across the onshore-offshore transition at the Catalan margin of the Valencia trough is supported by results from an analogous analysis performed on the parallel profile I, 30 km to the NE (Gallart et al., 1994). Profile I, shot with explosives, had a more homogeneous spacing between shots (1.5 km) and stations (2 km) than profile ESCI, but a lower coverage ranging from 5 to 20 fold was achieved from bin-widths of 1000 m. Folding fluctuations were originated by the existence of noisy traces which were eliminated. CMPs may consist of rather heterogeneous traces, and energy equalization is essential in the processing. The stacked section is shown in Fig. 6, and the Moho-reflections display a pattern similar to profile ESCI: they are located at 11 s TWT inland, and shallow across the shoreline to about 6 s TWT at about 30 km offshore. At larger distances seawards, the reflectivity becomes diffuse, probably due to the low-fold and highly stretched NMO-corrected traces in that area.

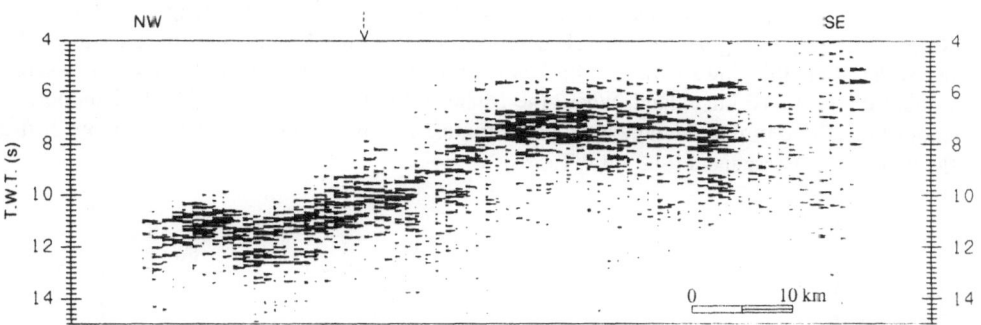

Figure 6. Wide-angle stacked section for profile I in the Catalan margin, after semblance-coherency filtering. The images of Moho reflectivity and crustal thinning are comparable to those of profile ESCI (Fig. 5). In the last 15 km seawards, resolution decreases in the CMPs due to low-fold and high stretching after NMO correction. The arrow marks the shoreline.

On the Balearic margin of the Valencia trough, wide-angle multichannel analysis is feasible for Profile I recorded in 20 stations on the island of Mallorca. However, some of the stations have not been included in the processing, as they were affected by local noisy conditions (wind, distortions within uppermost sediments, etc.). Using the best defined PmP data, a fold between 5 and 15 is achieved along the

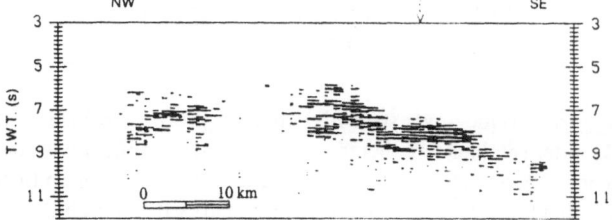

Figure 7. Wide-angle stacked section for profile I in the Balearic margin, after semblance-coherency filtering. Although the coverage is low (fold between 5 and 10), a continued crustal thinning is inferred towards the axis of the trough. The arrow marks the Mallorca shoreline.

onshore-offshore transition, in a section covering 15 km landwards and 35 km seawards (Fig. 7).

Deep reflectivity is reasonably imaged for most of the section, displaying a continued crustal thinning towards the Valencia trough. Moho-reflections are less sharply identified than in the Catalan margin, but range from 9-10 s beneath Mallorca to around 6 s at the western end of the section.

4.2.2. A wide-angle stacked section at the Betics-Alboran transition

As part of the Spanish ESCI programme, some marine and terrestrial deep reflection profiles have been collected recently in the Betics-Alboran domain (García-Dueñas et al., 1994). A crustal transect mapping the evolution from the Betics range to the central Alboran basin is envisaged by a NNE-SSW profile made up of two segments on land and at sea, each about 100 km-long (Fig. 1). The marine profile was simultaneously recorded along the land segment by 5 portable stations, each at a distance of 10 km. The analysis of these wide-angle data should highlight the onshore/offshore transition, which is likely to be poorly constrained by the normal incidence sections (since the first sections provided by the contractor were perturbed by artifacts and had no resolution in areas near the coast; further processing is still on going).

The multichannel processing sequence described in § 4.1. was applied to the wide-angle reflections present in the 5 receiver-gathers. A final stacked section 60 km-long is shown in Fig. 8. In the 35 km-long segment at sea, clear reflections from the Moho were imaged around 6 s TWT with a slight thinning seawards. This band of reflections can be followed continuously across the shoreline up to 10 km inland, where the Moho is located at about 8 s TWT. This horizon cannot be identified further inland, on the northernmost 15 km of the section, where a more diffuse reflectivity appears around 11-12 s TWT.

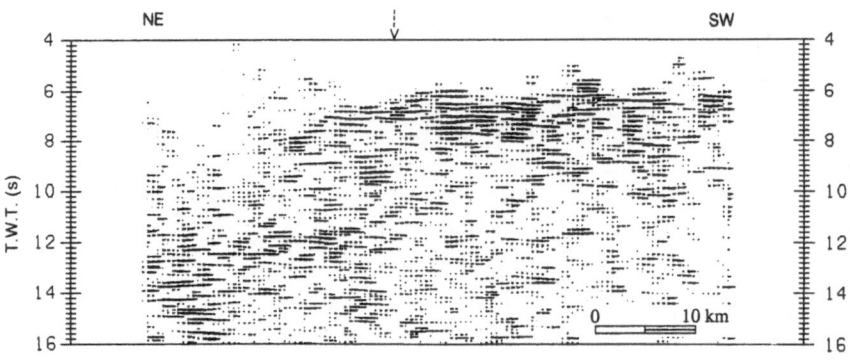

Figure 8. Wide-angle stacked section for profile ESCI across the Betics-Alboran sea transition (Fig. 1), after semblance-coherency filtering. Moho reflections around 6 s are visible in the marine part, with a moderate shallowing seawards. They continue up to 10-15 km inland, where a jump down of about 3 s is observed. The arrow marks the shoreline.

Therefore, the wide-angle image of the crustal-mantle transition in the Betics-Alboran transect significantly differs from the one obtained with the same technique in the Valencia trough. Instead of a laterally continuous clear thinning of the crust from land to sea, a rather sharp step in the Moho of about 3 s is suggested around 10-15 km inland, followed by a moderate thinning seawards.

4.3. A NORMAL INCIDENCE AND LARGE-APERTURE MERGED STACKED SECTION ACROSS THE VALENCIA TROUGH

The final step of our processing sequence is to produce unified stacks combining the normal incidence and large-aperture sections. Superimposing both data sets is presented for the Valencia trough, and will be carried out for the Betics-Alboran domain as soon as final normal incidence sections be available.

Differences in sampling rates, trace spacing or energy content between the near-vertical and wide-angle sections had to be compensated to obtain a final, homogeneous section. Therefore, sampling rates of 8 ms were used throughout, and a decimation applied to the normal incidence data, to get a common CMP spacing of 1 km. Strong energy differences, especially related to the overcritical wide-angle reflections were compensated through energy balancing laterally and at depth.

Regarding migration procedures, best images for the normal incidence sections, especially on the upper sedimentary sequence, were obtained using finite-differences time and depth migration algorithms that allowed us to consider lateral and dip changes in velocities. On the other hand, the same migration algorithms tested on the wide-angle sections produce some distortions and artifacts on the Moho-reflectivity pattern. The deep reflections are better migrated using time-shift algorithms with laterally constant velocities, even though an overall consistency in time and depth results is found after both types of migration. Therefore, we present the final time and depth migrated merged sections combining the results from the best migration of each type of data.

The final merged stacked and migrated sections across the Valencia trough and its flanks are shown in Fig. 9 The combined crustal image is based on the profile ESCI land and marine normal incidence reflection sections (upper panel of Fig. 9), together with the wide-angle sections in the Catalan margin along profile ESCI and in the Balearic margin along profile I. In this latter case, we assume a lateral continuity of structures along strike between profiles I and ESCI.

The combined section shows a remarkable consistency and continuity between normal incidence and large aperture reflectivity. The areas in Catalan and Balearic margins where near-vertical data lack resolution at depth are covered by a strong wide-angle reflectivity. The deep crustal evolution across strike at the NE Iberia stretched margin is thus mapped in detail for the first time. A continued thinning of the crust is revealed across the flanks of the Valencia trough and the onshore/offshore transition.

In the Catalan mainland the Moho is located at 32 km depth, and shallows steadily up to 19 km depth at about 40 km seawards. On the Balearic side, a crustal thinning of around 10 km occurs over a horizontal distance of about 40 km. In the Valencia trough, weaker reflections at depth are observed along profile ESCI, indicating a Moho at fairly constant depths around 16-17 km.

5. Conclusions and discussion

Coincident near-vertical and wide-angle reflection data sets are now available in the NE Iberian margin. A classic, independent analysis of the former in terms of stacked crustal sections, and of the latter through velocity-depth forward modelling was performed as a first step. A thorough comparison of the results is perturbed by the lack of resolution in complex areas, such as the onshore/offshore transitions. Therefore, a more homogeneous, readily comparable analysis of normal incidence and large-aperture data has been attempted. However, some degree of uncertainty or discrepancy of the corresponding results should be considered inherent in the two surveying techniques as they involve different sampling (horizontal versus vertical) of the physical parameters of the crust.

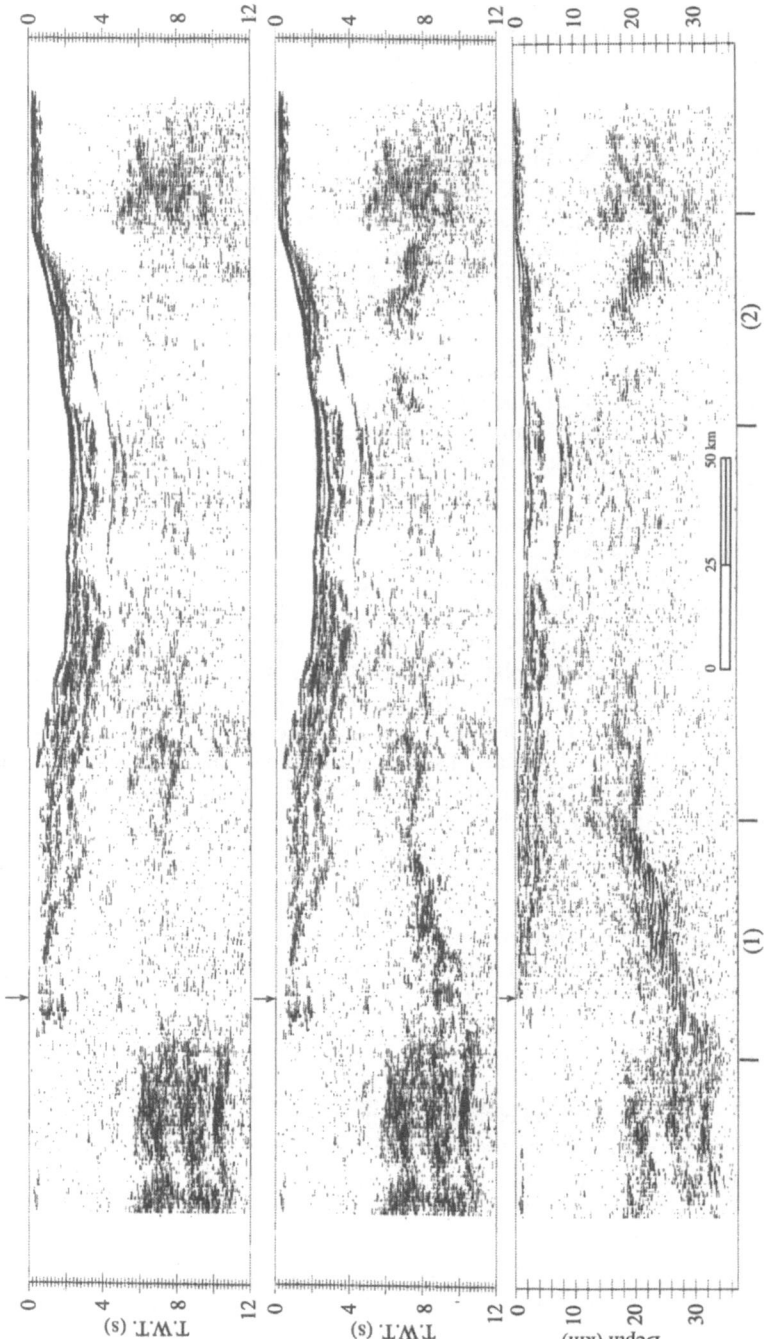

Figure 9. Near-vertical and wide-angle merged stacked section across the Valencia trough and its flanks. The upper panel shows the near-vertical reflection ESCI section alone, decimated to 1 trace per km to be compatible with the wide-angle trace interval. In the middle panel, the wide-angle stacks from profile ESCI in the Catalan margin and profile I in the Balearic margin have been included. The lateral evolution of the Moho and the continuous crustal thinning seawards are revealed in this composite section. The lower pannel corresponds to the depth-migrated section, taking into account velocities in agreement with forward modelling results (Fig. 3).

A wide-angle multichannel processing sequence was developed, and applied to different data sets in the Mediterranean margin of the Iberian Peninsula. Basic steps of this processing include energy equalization, band-pass and deconvolution filtering, muting of refracted energy. 1000 m bin-widths, static and dynamic corrections, time/depth migration and semblance-coherency filtering. In particular, NMO corrections are most essential, and may constrain stacking velocities of up to 0.1 km/s.

In all the 4 examples considered, the wide-angle stacked sections display clear images of the Moho-reflectivity within the multifold area. The crustal depths obtained in the migrated sections of profiles I and ESCI in the Valencia trough are in agreement with the velocity-depth results of the forward modelling (compare Fig. 3 and Figs. 5,6,7,9), and support the consistency of the two independent wide-angle interpretation approaches.

A direct inspection of the coincident normal incidence and large-aperture stacked sections across the Valencia trough (Figures 2 and 5) indicates that the latter maps strong reflections from the bottom of the crust in areas where the former lacks resolution in the middle and lower crust. Therefore, an enhanced crustal transect along the Iberian and Balearic margins of the Valencia trough was produced by merging the near-vertical and wide-angle stacked and migrated sections. This results in the first transect which documents with continuity the lateral evolution of the deep crustal reflectivity across strike of structures. An important, continued thinning of the crust is revealed in the Iberian margin, ranging from depths of 32 km beneath the Catalan Coastal ranges to depths of 18 km at about 60 km seawards. On the Balearic side, a similar thinning rate is observed, although a precise identification of the Moho beneath Mallorca is not easy due to the diffuse ending of the deep reflectivity.

In the central part of the Valencia trough, the Moho reflections found around 8 s TWT in our section (16-17 km depth) fully agree with recent results by Collier et al. (1994) from a two-ship common offset profile of the Valsis experiment sampling a very close area to the NE, and by Torres et al. (1993) from unmigrated CMP line-drawing of the same line. Results from classic interpretations of available large-offset data, either refraction profiles crossing the Valencia trough at 3 different azimuths (Dañobeitia et al., 1992) or expanding spread profiles centered in several specific zones (Torné et al., 1992; Pascal et al., 1992), suggest shallower Moho depths around 13-14 km in the axis of the trough. Our best fitting velocity-depth models for profiles I and ESCI, 30 km apart, show differences of 2-3 km in crustal depths along the trough. This apparent discrepancy may be due to poor large-offset constraints in the central part of the basin, or may correspond to short-wavelength variations in crustal thickness along strike of structures. Similar lateral changes in Moho depths were detected in areas of the continental platform (Torné et al., 1992; Maillard et al., 1992) and around the shoreline (Gallart et al., 1990). Comparison of available seismic results in the Gulf of Lions and Valencia trough areas (Mauffret et al., 1992) indicates a general tendency of crustal thickening from NE to SW, towards the Betics domain.

Features of the lower crust such as its thickness or reflective signature are not imaged with continuity along our crustal transect. Wide-angle multicoverage information is restricted to a 15-20 km-large zone around the Reus basin (Fig. 5) and the shoreline, where it prolongs the 5 s-thick strongly reflective pattern observed on the Catalan mainland. Along the marine normal incidence section, lower crustal reflectivity is not observed within the continental shelf. Instead, a 2s-thick band of reflections is detected on the continental slope, between 30-60 km seawards. In the central part of the Valencia trough and its Balearic flank, removing of strong sea-water multiples developed there may significantly distort lower crustal energy arrivals. Reflectivity below the Balearic promontory is concentrated in a 5 s-thick, 20 km-large band at its NW side.

In the velocity-depth models described in § 3.2, the crustal thinning towards the Valencia trough is accomodated mainly by the lower crust. Although poorly constrained in the axis of the trough, a 5-6 km-thick lower crustal layer is considered to be present along the Valencia trough. Rather low velocities of 6.4-6.5 km/s, well controlled on the flanks of the trough, are derived for the lower crust. This seems to discard the occurrence of underplating structures which are characteristic of some continental passive Atlantic margins (Holbrook et al., 1992).

Forward modelling of gravity data provides additional constraints. Models developed along an Iberian-Balearic transect (Watts and Torné, 1992) or on the continental flank of the trough (Gallart et al., 1994) are in agreement with our seismic section, and support a minimum 50% thinning of the lower crust seawards. Features of the basement (Bartrina et al., 1992) and all the geophysical constraints now available for the Valencia trough favour the idea of a continental, highly stretched crust.

Data from another Western Mediterranean basin, the Alboran Sea and its transition to the Betics have been considered in our study. Crustal thicknesses reported up to now suggest structural differences, and lateral variations in Moho depths between the Betic domains. Values down to 38 km depth are found on a profile along the internal Betics (Banda et al., 1993), parallel to the shoreline and 30 km inland. Furthermore, a Moho around 25 km depth was obtained from a profile along the coast (Barranco et al. 1990). Very shallow values around 15 km depth were interpreted in its westward continuation in an area which could be affected by local ultramafic structures. In the Alboran basin, velocity-depth results are poorly constrained by refraction data; Moho depths of 15-18 km were reported in experiments in the 70ies (Hatzfeld, 1976).

A NNE-SSW crustal transect from the Betics to the Alboran basin was designed in the Spanish ESCI programme. First interpretations of the land segment (García-Dueñas et al., 1994) show clear reflectivity in the normal incidence section down to 11s TWT. However, the southernmost 20 km of the profile deserve further processing due to distortions and artifacts, and the northernmost part of the marine section also lacks resolution at its present processing stage. Therefore, constraints on the onshore/offshore transition can only be provided at present by the piggy-back recordings on land. The wide-angle stacked section obtained in our multichannel processing is significantly different from those in the Catalan margin. A moderate crustal thinning is observed between 10-15 km inland and 35 km seawards, Moho reflections lying between 8 and 6 s TWT. This horizon vanishes further inland, where a more diffuse deep reflectivity is found around 11-12 s TWT.

The wide-angle stacked image suggests that in the central part of the Betics-Alboran domain, the thin Alboran Sea crust goes beyond the shoreline, up to 10-15 km inland, where a jump down of about 3 s in the Moho marks the southernmost limit of the internal Betics-thick crust. This seismic image is in full agreement with the gravity interpretation of the strong Bouguer anomaly gradient of about 10 mGal/km in that area (Torné and Banda, 1992). A steplike transition along 10-20 km is needed in the density models, in contrast to the more gradual transition along 100 km which fits the pattern of the gravity anomalies in the Eastern Alboran basin. Subsidence analyses (Docherty and Banda, 1992; Watts et al, 1994) also indicate major subsidence rates towards the Western Alboran basin.

The seismic crustal images presented for the Catalan-Balearic and Betics-Alboran domains differ significantly from non-volcanic Atlantic margins, such as the Iberian Atlantic (Whitmarsh et al., 1993) or the Eastern Canadian margin (Bassi et al., 1993), in that stretching of the continental crust by a factor of 2 is reached over very short lateral distances. This suggests that in addition to extensional rifting processes, the interaction with the surrounding convergent regime and the compressional periods could play an important role in increasing the differences in crustal thicknesses. In the Betics-Alboran domain, the steplike Moho transition would probably result from the additional interaction of shear tectonics. Lateral westward movements that juxtapose two crustal blocks of different thicknesses should be taken into consideration, prior to the Alpine-to-recent compressional and extensional tectonics between the African and Iberian plates.

Acknowledgements

The ESCI seismic programme is supported by the Spanish "Plan Nacional de I+D". This work has been partially financed by CICYT projects n° GEO89-0858-E; GEO90-0733; GEO90-0617. Additional funding comes from DGICYT project n° PB86-0619. N. Vidal benefited from a CICYT PhD-grant.

References

-Banda, E. and P. Santanach, 1992. The Valencia trough (Western Mediterranean): an overview, *Tectonophysics, 208,* 183-202.

-Banda, E., Gallart, J., García-Dueñas, V., Dañobeitia, J.J. and J. Makris, 1993. Lateral variation of the crust in the Iberian Peninsula. New evidence from the Betic cordillera. *Tectonophysics, 221,* 53-66.

-Barranco, L.M., Ansorge, J. and E. Banda, 1990. Seismic refraction constraints on the geometry of the Ronda peridotitic massif (Betic cordillera, Spain), *Tectonophysics, 184.* 379-392.

-Bartrina, M.J., Cabrera, L., Jurado, M.J., Guimerà, J. and E. Roca, 1992. Evolution of the central Catalan margin on the Valencia trough (Western Mediterranean), *Tectonophysics, 203,* 219-248.

-Bassi, G., Keen, C.E. and P. Potter, 1993. Contrasting styles of rifting: models and examples from the eastern canadian margin, *Tectonics, 12,* 639-655.

-Choukroune, P. and ECORS Team, 1989. The ECORS Pyrenean deep seismic profile: reflection data and the overall structure of an orogenic belt, *Tectonics, 8,* 23-29.

-Collier, J.S., Buhl, P., Torné, M. and A.B. Watts, 1994. Moho and lower crustal reflectivity beneath a young rift basin: results from a two-ship, wide-aperture seismic reflection experiment in the Valencia trough (Western Mediterranean), *Geophys. J. Int., 118,* 1. 159-180.

-Dañobeitia, J.J., Arguedas, M., Gallart, J., Banda, E. and J. Makris, 1992. Deep crustal configuration of the Valencia trough and its Iberian and Balearic borders from extensive refraction and wide-angle reflection profiling, *Tectonophysics, 203,* 37-55.

-Docherty, J.I.C. and E. Banda, 1992. A note on the subsidence history of the northern margin of the Alboran sea basin, *GeoMar. Lett., 12,* 82-87.

-Flueh, E.R. and T. Dickmann, 1992. Technical aspects of wide-angle data collection and processing, In: Meissner R., Snyder, D., Balling, N. and E. Staroste (Eds.), The BABEL Project, C.E.C: report, 123-130.

-Gallart, J., Rojas, H., Díaz, J. and J.J. Dañobeitia, 1990. Features of deep crustal structure and the onshore/offshore transition at the Iberian flank of the Valencia trough (Western Mediterranean), *J. Geodyn., 12,* 233-252.

-Gallart, J., Vidal, N., Dañobeitia, J.J., and the ESCI-Valencia Trough Working Group, 1994. Lateral variations in the deep crustal structure at the Iberian margin of the Valencia trough imaged from seismic reflection methods, *Tectonophysics, 232,* 59-75.

-Garcia-Dueñas, V., Banda, E., Torné, M., Córdoba, D. and ESCI-Béticas Working Group, 1994. A deep seismic reflection survey across the Betic chain (southern Spain): first results, *Tectonophysics, 232,* 77-89.

-Hatzfeld, D., 1976. Etude sismologique et gravimétrique de la structure profonde de la mer d'Alboran: mise en évidence d'un manteau anormal, *C.R. Acad. Sci. Paris, 283,* 1021-1024.

-Holbrook, S.W., Reiter, E.C., Purdy, G.M. and M.N. Tocsöz, 1992. Image of the Moho across the continent-ocean transition, U.S. east coast, *Geology, 20,* 203-206.

-Mauffret, A., Maillard, A., Pascal, G., Torné, M., Buhl, P. and B. Pinet, 1992. Long listening multichannel seismic profiles in the Valencia trough (VALSIS 2) and the Gulf of Lions (ECORS):

a comparison, *Tectonophysics, 203*, 285-304.

-Meissner, R., Snyder, D., Balling, N. and E. Staroste (Editors), 1992. The BABEL Project, Commission of European Communities, 155 pp.

-Pascal, G., Torné, M., Buhl, P. Watts, A.B. and A. Mauffret, 1992. Crustal and velocity structure of the Valencia trough (Western Mediterranean), Part II. Detailed interpretation of five Expanded Spread Profiles, *Tectonophysics, 203,* 21-35.

-Roca, E. and J. Guimerà, 1992. The Neogene structure of the eastern Iberian margin: structural constraints on the crustal evolution of the Valencia trough (Western Mediterranean), *Tectonophysics, 203*, 203-218.

-Suriñach, E., Marthelot, J.M., Gallart, J., Daignières, M. and A. Hirn, 1993. Seismic images and evolution of the Iberian crust in the Pyrenees, *Tectonophysics, 221*, 67-80.

-Torné, M. and E. Banda, 1992. Crustal thinning from the Betic cordillera to the Alboran sea, *GeoMar. Lett., 12,* 76-81.

-Torné, M., Pascal, G., Buhl, P., Watts, A.B. and A. Mauffret, 1992. Crustal and velocity structure of the Valencia trough (western Mediterranean), Part I. A combined refraction/wide angle reflection and near vertical reflection study, *Tectonophysics, 203,* 1-20.

-Torres, J., Bois, C. and J. Burrus, 1993. Initiation and evolution of the Valencia trough (western Mediterranean): constraints from deep seismic profiling and subsidence analysis, *Tectonophysics, 228,* 57-80.

-Valasek, P., Mueller, St., Frei, W. and K. Holliger, 1991. Results of NFP 20 seismic reflection profiling along the Alpine section of the European Geotraverse (EGT), *Geophys. J. Int, 105,* 85-102.

-Watts, A.B. and M. Torné, 1992. Subsidence history, crustal structure, and thermal evolution of the Valencia trough: a young extensional basin in the western Mediterranean, *J. Geophys. Res., 97,* B13, 20021-20041.

-Watts, A.B., Platt, J.P. and P. Buhl, 1993. Tectonic evolution of the Alboran Sea basin, *Basin Research, 5,* 153-177.

- Whitmarsh, R.B., Pinheiro, L.M., Miles, P.R., Recq, M. and J.C. Sibuet, 1993. Thin crust at the western Iberia ocean-continent transition and ophiolites, *Tectonics, 12,* 1230-1239.

STRUCTURE OF THE MARMARA SEA BASIN IN THE NORTH ANATOLIAN FAULT ZONE

Mustafa ERGÜN, Erdeniz ÖZEL and Coşkun SARI
Dokuz Eylül University
Faculty of Engineering
Department of Geophysics
35100 Bornova-İzmir / TURKEY

ABSTRACT.   The Marmara Sea basin should be considered to form part of the North Anatolian Fault (NAF) and the Aegean crustal regime in which a listric-faulted upper crustal section overlies a lower crust that was thinned in a ductile manner, the overall crustal thickness being about 25-30 km. The local Moho upbulge of about 5 km is consistent with Bouguer gravity anomalies as shown by modelling. As the North Aegean area, the Sea of Marmara is undergoing a combination of right-lateral strike-slip and north-south extension with the formation of pull-apart basins. As a result of the collision of the Arabian and Anatolian land masses during the Middle Miocene, the westward escape of the Anatolian block gave rise to E-W compression in western Turkey, the relief of which was produced by N-S extension. In northern Turkey and towards the North Aegean Sea: the NAF splits into several fault strands defining a broad tectonic zone with associated high swarmlike seismic activity. The Marmara Sea basin is the extension of the Thrace basin in the north and northwest. During the Middle Eocene, the subsidence of basement occurred, creating the Thrace basin. Therefore it could be assumed that the extensional basins of the Sea of Marmara have existed since the Eocene According to interpretation of geological, geomorphological and geophysical data, the Sea of Marmara can be divided into five different blocks which are controlled by two sets of fault systems: (i) almost E-W trending normal faults (the Northern and Southern Boundary faults); and (ii) NE-SW oriented subvertical strike-slips. The blocks are undergoing relative vertical motions and rotations. The Marmara Sea basin accommodates strike-slip and extensional movements. The east-west trending normal fault systems of the Sea of Marmara is a diffuse zone of crustal thinning associated with an estimated 30 percent of north-south extension since the Tortonian.

1. Introduction

The Sea of Marmara is an inland sea lying between the Thrace basin and Anatolia in northwestern Turkey with an area of about 11,350 km². It is connected with the straits of

E. Banda et al. (eds.), Rifted Ocean-Continent Boundaries, 309–326.
© 1995 *Kluwer Academic Publishers. Printed in the Netherlands.*

310

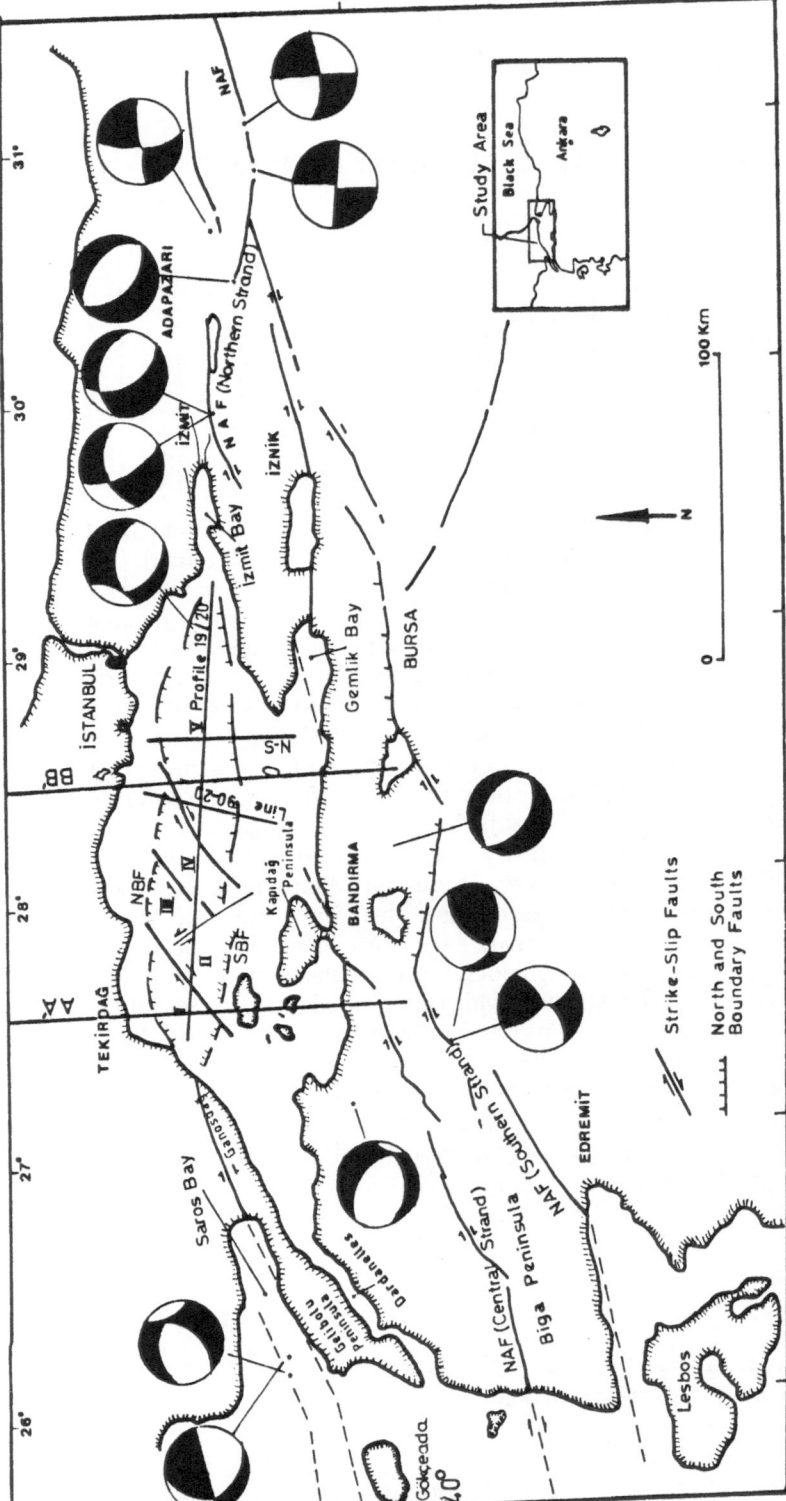

Fig. 1: Fault plane solutions and the main tectonic framework of the Sea of Marmara and surroundings (modified from Barka, 1992). Basinal blocks: I, III and V; Uplifted blocks: II and IV in the Sea of Marmara within the limits of Northern and Southern boundary faults (NBF and SBF). Locations of Interpreted N-S geological section, High resolution Seismic Profile (19/20), the Seismic Line 90-20, and Gravity Profiles of AA' and BB' are shown here.

Bosphorus and Dardanelles to the Black Sea in the north and to the Aegean Sea in the south. It has a very large continental shelf area with many islands. The southern shelf area is much larger than the northern one. There are three deep basins running in an E-W direction. These depressions have the following depths: 1,112 m in the west; 1,220 m in the middle; and 1,238 m in the east.

The Sea of Marmara occupies a key position lying over the direct westward continuation of the NAF zone. This zone, about 1500 km long, is seismically active, right-lateral transform whose relative motion between the Anatolian and Black Sea blocks is taken up (Fig. 1). To the west, through the Sea of Marmara, the NAF splits into several fault strands defining a broad tectonic zone with associated high seismicity (Crampin and Evans, 1986; Barka and Kadinsky-Cade, 1988; and Barka, 1992). This diffuse zone of shallow seismicity with a clustering of microseismic activity along major fault strands is similar to the pattern of seismicity in the north Aegean Sea (Mc Kenzie, 1978; Lyberis and Deschamps, 1982; Eyidoğan, 1988; Taymaz et al., 1991; and Barka, 1992). The east-west line of the NAF to the east of the Sea of Marmara passes westwards into a graben defining İzmit Bay and a zone of deep water in the eastern and central part of the Sea of Marmara. Water depths which exceed 1.000 m in places are comparable with depths encountered along the North Aegean trough.

The motion of Anatolia with respect to the Black Sea is achieved by the right lateral motion along the NAF. This is evident from the fault plane mechanism of the large earthquakes in this region (Taymaz et al., 1991; Barka, 1992; Rotstein, 1985). Thus, the NAF in western Turkey and its extension in the north Aegean Sea seem to play approximately the same role as in Anatolia; a relatively uninterrupted transform motion between the Aegean block and Eurasia is taking place in this area. Therefore, effective motion of the Aegean block in western Turkey is achieved through lithospheric extension. The motion of the Aegean block with respect to Eurasia in the north Aegean Sea seems to be achieved primarily along strike-slip faults.

The seismicity of western Turkey typically displays swarm-type activity with marked clustering in time and space of low magnitude earthquakes (Üçer et al., 1985). The major swarm lineaments in western Turkey outline a triangular area bounded by the direct continuation of the NAF across the Sea of Marmara and by the İzmir-Adapazarı swarm lineament (Üçer et al., 1985). On the basis of the energy release maps (Crampin and Evans, 1986; and Eyidoğan, 1988), it is seen that the Marmara Block has different seismic characteristics from the rest of western Turkey and the energy released in the Marmara province is significantly higher than the energy released in southwestern Turkey.

Crampin and Evans (1986), using fault mechanism solutions, suggest that the Marmara block has been rotated and sheared to accommodate the right-lateral motion of the NAF and the extensional tectonics of southwestern Turkey. This extension may result in graben formation and probably listric faulting. The focal depths of earthquakes in western Turkey, in general, are up to 50 km although the major earthquakes are not deeper than 10-15 km. Eyidoğan (1988) suggests that the right-lateral shear in the Marmara province is associated with the E-W trending strike-slip faults. The N-S extension of 7.1 mm/yr and E-W contraction of 10.0 mm/yr were calculated with the thinning of seismogenic layer, equivalent to 0.13 mm/yr (Eyidoğan, 1988).

Heat flow values higher than normal are confined mainly around the Aegean Sea and western Turkey including the Sea of Marmara and the Saros basin (Jongsma, 1974). The results of these computations clearly show that the temperature distribution below the Aegean Sea and western Turkey is higher by a factor of 2 to 3 than the temperatures computed from the deep Ionian Basin and the Black Sea Basin. The Aegean Sea is characterized by high heat flow values and is influenced by a strong vertical uplift which has led to a general expansion of the area (Le Pichon et. al., 1984).

The Saros basin in the NE corner of the Aegean Sea is a half-graben bounded by a large normal fault system along its northern margin (Le Pichon et. al., 1984). The fault that crosses the Gelibolu peninsula, is a continuation of the ENE trend of the Saros trough. The faulting associated with the Saros trough is a semicontinuous feature from about 25° E to the western basin of the Sea of Marmara at about 27.5° E. In the western part of the trough along Ganosdağ, bounding the NW side of the deep offshore basin in the western Sea of Marmara, the faults dip SE, and north of the Gelibolu peninsula they dip NW. Such a change in the polarity of tilting and faulting along the strike of a basin is a common feature of extensional graben.

In the north of Saros Bay, there is another ENE-WSW trending graben, parallel to the Saros trough, but now filled with Plio-Pleistocene sediments (Biju-Duval et. al., 1978; Lyberis 1984). It could continue eastwards into a Neogene basin on land in eastern Thrace (Burke and Uğurtaş, 1974; Mercier et. al., 1991). The faults bounding these basins are thought to be active (Mercier et. al., 1991). The lack of bathymetric expression offshore is probably related to the great volume of sediment that is washed into this area by the rivers. It might be that the Saros trough was activated after the extension ceased along the northeastern branch.

2. Geology and Stratigraphy of the Sea of Marmara and the Surrounding Areas

The structures of the Sea of Marmara have morpho-tectonic characteristics comparable to those of the North Aegean trough. The Ganosdağ area of active faults joins the North Aegean trough to the Sea of Marmara. Although there are still some controversies (Alvarez et al., 1984; Brooks and Kiriakıdis, 1986; Le Pichon et al., 1984; Lyberis, 1984; McKenzie, 1978; and Dewey and Şengör, 1979), the general framework of this region (The North Aegean trough through the Saros Bay to the Marmara Sea basin) is relatively young and superimposed on older structures of an orogenic belt of alpidic origin.

The Marmara Sea basin is the extension of the Thrace basin (Fig. 2) which is a Tertiary basin where the subsidence of the platform took place during the Middle Eocene (Ketin, 1983). Neogene lateritic depositions are present at the borders of this basin (İlhan, 1976). The Marmara islands on the southern shelf of the Sea of Marmara, are made up of crystalline rocks of marbles and granites. The Istranca massif at the north and northeast of the Thrace basin is made up of metamorphic and crystalline rocks. Istanbul and the Anatolian side peninsula regions are characterized by the Paleozoic rocks with a progressive transgression towards the end of the Silurian from clastics to limestones. The Devonian of the region is made up of nodular limestones, lydites and marls. These facies change to a marine facies in the Lower Carboniferous (Doust and Arıkan, 1974).

The geology of Saros basin can be explained in terms of the geology of the Thrace basin in the north and the geology of the Gelibolu peninsula in the south (Fig. 2). The sequence of recifal limestones of the Middle Eocene is present at the basement of the ridge along a NW-SE direction in the north of Saros Bay. A similar structure is also present further east (Ganosdağ) changing its direction to WSW-ENE. The Gelibolu peninsula is constrained in the north by the Ganosdağ strike-slip and its southwestern extension in the Saros Bay. This peninsula is represented by a monoclinic fold striking in the SW-NE direction. The northwestern side is overthrusted over the southeastern side with a reverse fault. The sedimentary sequence of the Gelibolu peninsula which starts with the Upper Cretaceous-Lower Eocene sequence forms the basement for the overlying Middle Eocene rocks.

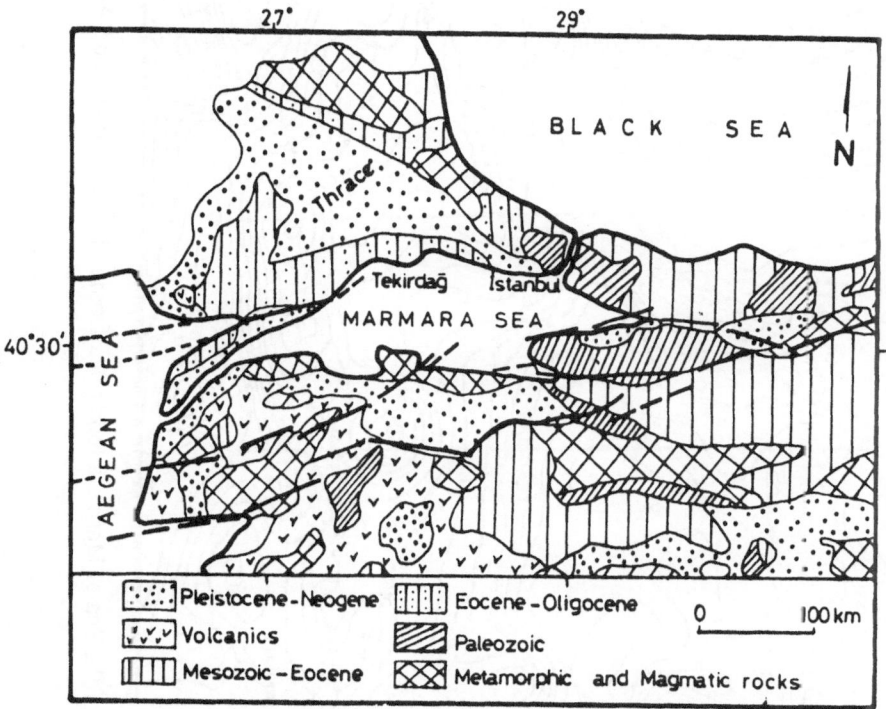

Fig. 2: Simplified geological map of the Sea of Marmara and surrounding areas (modified from the geological map of Turkey).

3. Gravity and Magnetic Data of the Sea of Marmara

The Turkish mainland and the central parts of western Turkey show negative Bouguer gravity anomalies in the range of 60 to 70 mGals (Ekşioğlu, 1991). The Bouguer gravity values decrease from west to east. In the middle of the western Black Sea basin maximum values in range of 140-150 mGals are reached. The zero contour of the Bouguer gravity anomaly follows the trend of NAF in the eastern parts of the Sea of Marmara. This trend bends in a NE-SW direction just south of the İznik Bay. Therefore

314

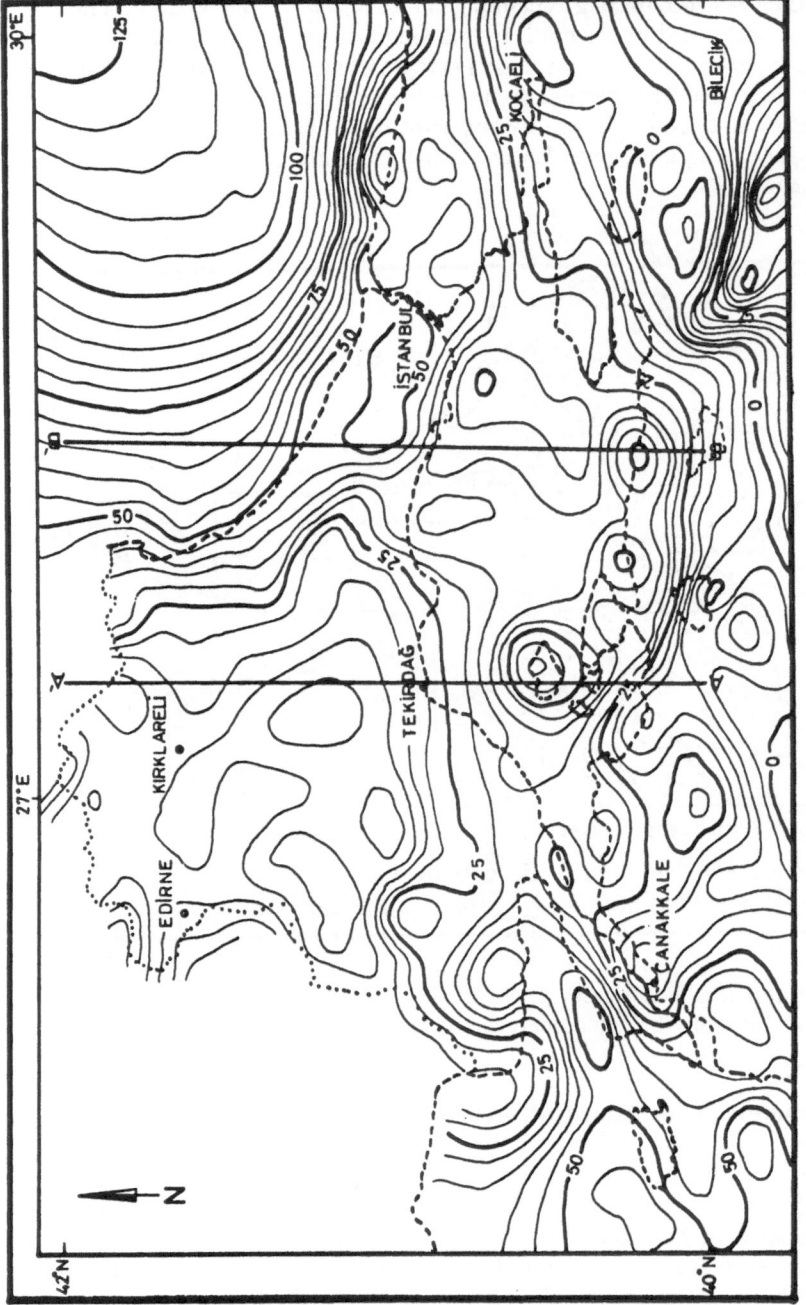

Fig. 3: Bouguer gravity anomaly map of the Sea of Marmara (compiled from MTA and TPAO, and Adatepe, 1988). Contour interval: 5 Mgal.

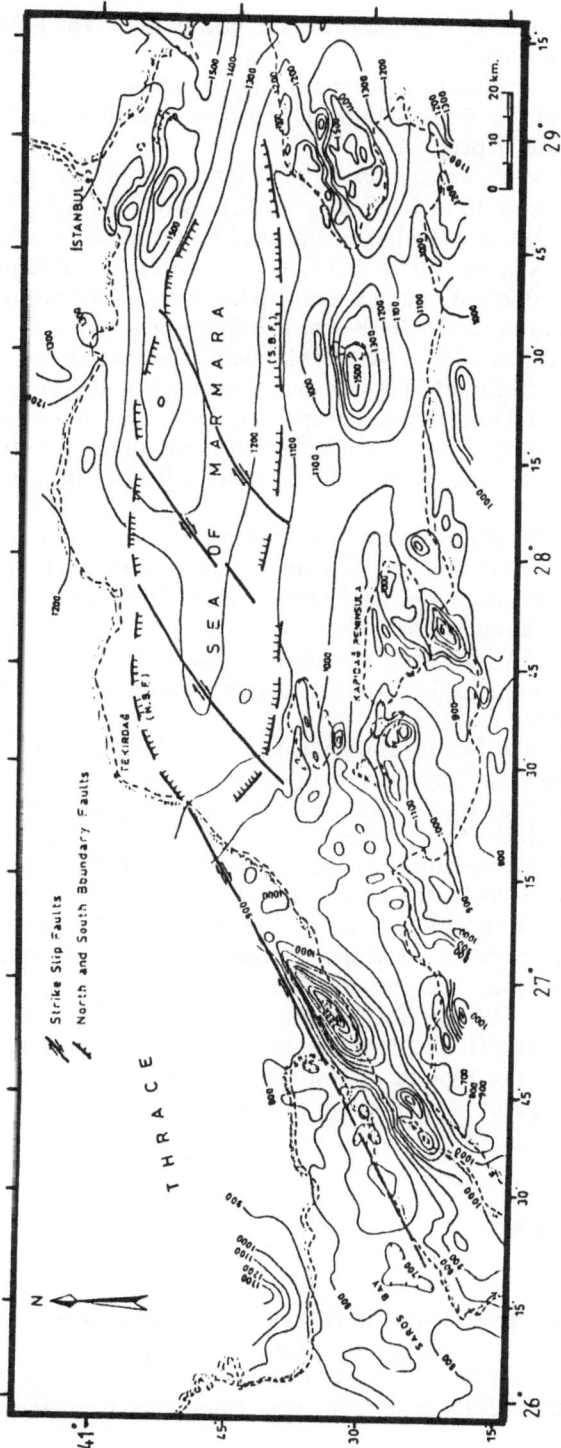

Fig. 4: Aeromagnetic map of the Sea of Marmara (modified from Kale, 1985). Contour interval: 100 nT.

the general trend of the Bouguer gravity anomalies are mainly E-W. The E-W trends of the Rhodope massif seem to dominate in the Thrace basin. This trend also continues in the north Aegean trough. It is marked by a free-air gravity low (Makris, 1977; Brooks and Kiriakidis, 1986) owing to the thick sedimentary fill of the steep morphological depression which is up to 1500 m deep.

However, the Bouguer gravity anomaly map (Fig. 3) for the Sea of Marmara which is compiled from the data available from the Geological Survey of Turkey (MTA) and the General Command of Mapping, shows a large number of small anomalies which disturb the regional gravity trend (Adatepe, 1988; and Ekşioğlu, 1991). The Bouguer gravity values increase from about 20-25 mGal at the southern boundary of the Sea of Marmara to 50-55 mGal in the north towards the Black Sea. The Bouguer gravity values decrease towards the Thrace basin in the NW of Sea of Marmara. The main trend of positive Bouguer gravity anomaly in the Sea of Marmara starts from the Kapıdağ Peninsula in the south and runs in a SSW to NNE direction towards the north. This anomaly coincides with the uplifted block (Block IV). The western and eastern sides of this positive anomaly are the negative anomaly zones which are coincident with the basinal areas in the Sea of Marmara.

Magnetic anomalies are generally associated with magmatic rocks. The major magnetic anomalies trend E-W on the north side of the Marmara Sea basin (Fig. 4). These have a long wavelength owing to the great burial depth of the sources. It has been calculated that the bodies causing magnetic anomalies have depths of about 3-3.5 km (Ergün, 1990; and Kale, 1985) for the main E-W trending northern magnetic anomalies. These E-W trending magnetic anomalies are displaced to the south by branches of strike-slip faults which are the northern strand of the NAF.

The southern platform of the Sea of Marmara is covered by the short wavelength magnetic anomalies mainly caused by the granites and volcanics of the shelf and the Eocene-Miocene volcanics of the Biga Peninsula further south (Ergün, 1977). These are mainly andesitic volcanics which are thus inferred on the basis of magnetic anomalies to be associated with observed active faulting in the region due to the distension tectonics. These young volcanics seem to be the cause of these anomalies. However, whilst considering this question, it should be pointed out that the ophiolites which are outcropping on the Armutlu Peninsula, could also be contribute to these anomalies.

No regular magnetic anomalies are present in the Saros Bay. The main elongated magnetic anomalies run in a SW-NE direction on the southern side of the Ganosdağ strike-slip which constrain the Gelibolu peninsula in the north. The cause of these magnetic anomalies are the outcropping ophiolithes especially on the west coast of Marmara Sea. The volcanic rocks in the north of the Saros Bay are also associated with some magnetic anomalies.

3. Interpretation of Gravity Data

The Marmara Sea basin can be evaluated in terms of a localized manifestation of the brittle upper crustal response to regional lithospheric stretching. Therefore it does not require to be underlain by a very thin crust and an anomalous upper mantle. The crustal thickness, ranging between 22-25 to 40 km, was calculated from the studies of

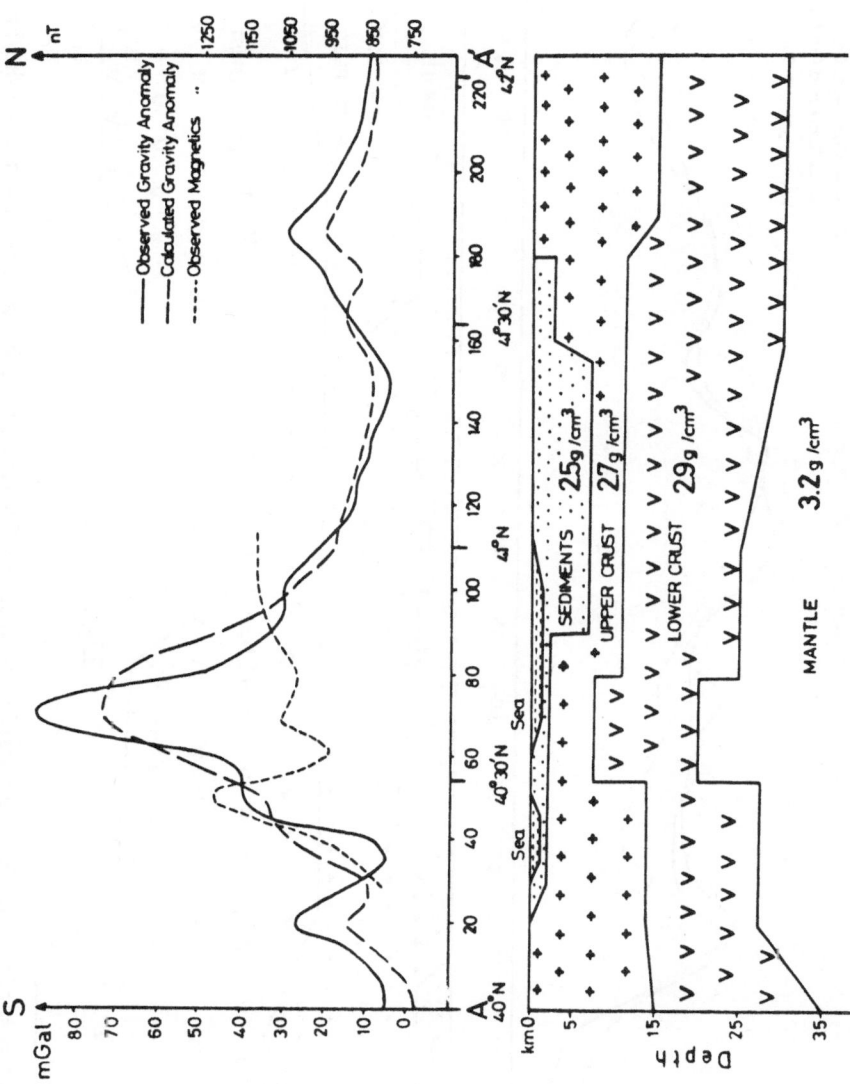

Fig. 5: Intrepretation of the AA' Bouguer gravity anomaly profile (from the Biga peninsula in the south through the Sea of Marmara to Thrace in the north) through 27° 30' E longitude (Aeromagnetic data covers only the Sea of Marmara).

318

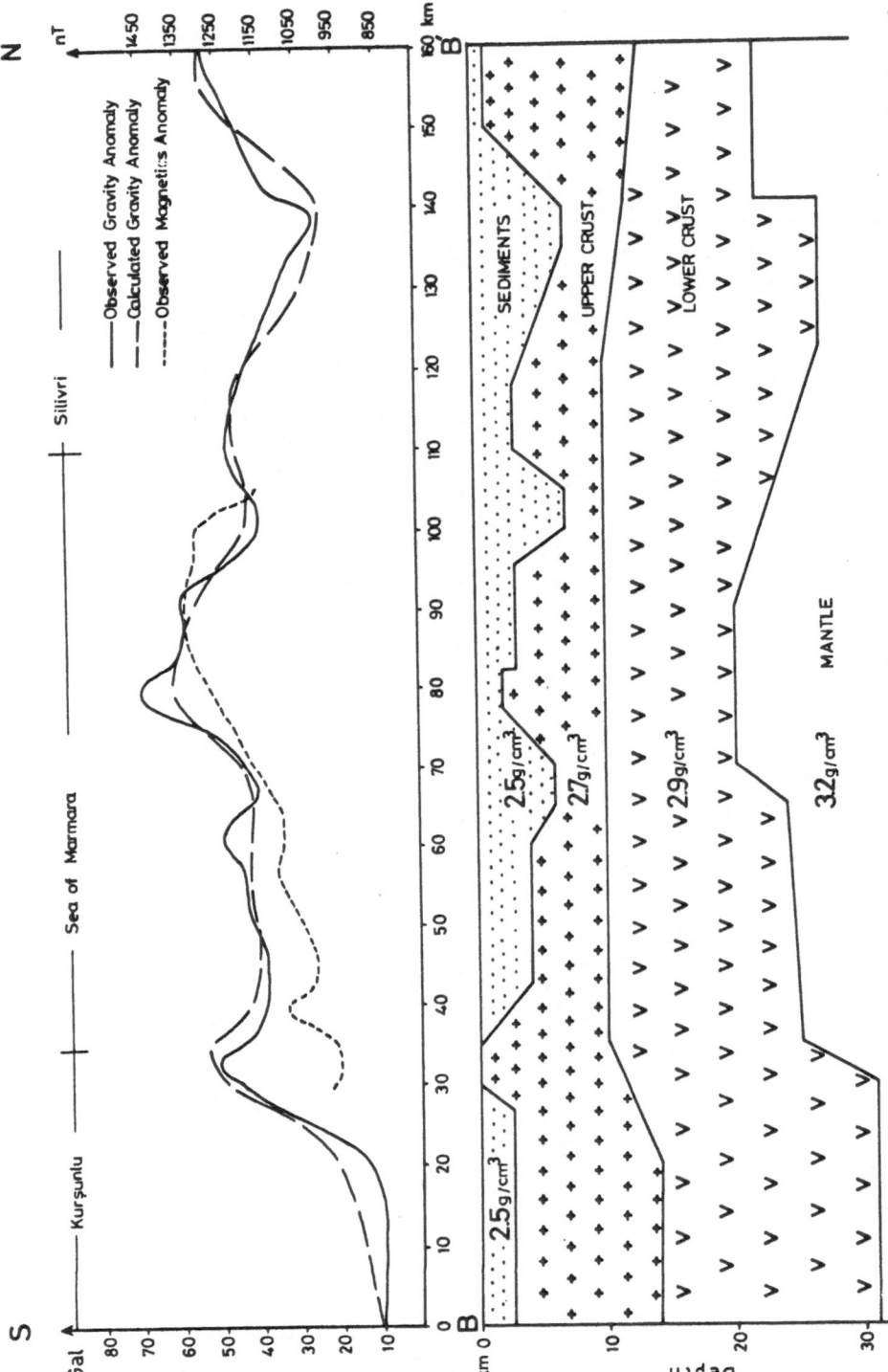

Fig. 6: Intrepretation of the BB' Bouguer gravity anomaly profile (from the Biga peninsula in the south through the Sea of Marmara to Thrace in the north) through 28° 15' E longitude (Aeromagnetic data covers only the Sea of Marmara).

seismology and magnetotellurics (Kalafat et al., 1987). The crust is thinner under the Sea of Marmara. The interpretation here involves some crustal thinning but the amplitude of the local Moho upbulge is only about 5 km (Ergün, 1990) under the Sea of Marmara. The northern side of the Sea of Marmara was modelled as an uplifted area whereas the Thrace basin in the NW and the southern shelf areas of the Sea of Marmara were modelled as the basinal areas (Ergün, 1990; and Ekşioğlu, 1991; and Özel, 1992).

The N-S profiles (Figs. 5 and 6) run from the middle of the Biga Peninsula in the south to the middle of the Thrace basin crossing the Sea of Marmara at the longitudes of 27° 30' E (Profile BB') and 28° 15' E (Profile AA'). The Profile AA' (Fig. 5) shows the general trend of Bouguer gravity anomalies increase towards the Sea of Marmara. The relative positive Bouguer gravity anomaly over the Marmara Sea basin coincides with the ridge between the eastern deep basin and the middle one. The densities used are: 3.2, 2.9, 2.7 and 2.5 g/cm^3 for the mantle, lower crust, upper crust and sediments respectively. The interpretation here involves some crustal thinning being the amplitude of the local Moho upbulge only of about 5 km (Ergün, 1990). These modelling interpretations should be considered carefully because the structures are not exactly two dimensional here. However, the resolution of this subject awaits the implementation of some wide angle reflection/refraction seismic experiment as well as heat flow measurements.

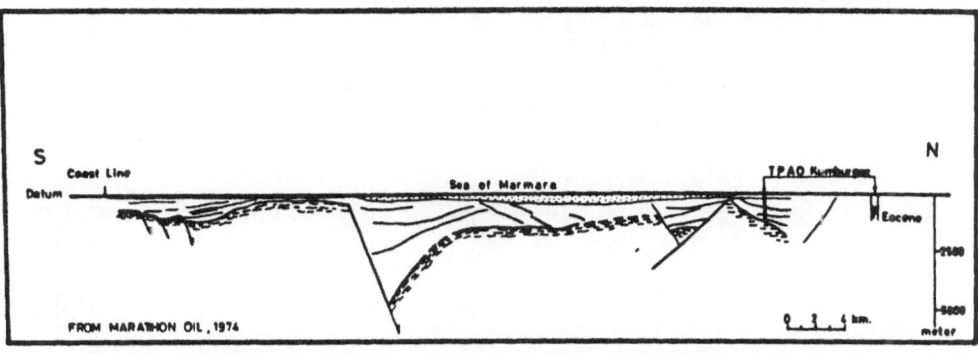

Fig. 7: The N-S section showing main structural features in the Sea of Marmara with the interpreted growth faults and roll-over continuing Eocene deposits (modified from Barka and Gülen, 1988).

4. Marine Seismic Data of the Sea of Marmara

The Sea of Marmara has been surveyed with seismic reflection by some oil companies (Turkish Petroleum Company, Marathon Oil, and Turkish Geological Survey "M.T.A."). The simplified N-S cross-sections of interpreted profile is given in Fig. (7) crossing the main structural lineaments in the Sea of Marmara (Barka and Gülen, 1988). The sediments are 5 km thick and some of them here been interpreted as Eocene (Barka and Gülen, 1988). During the Middle Eocene, the subsidence of the basement of the platform occurred creating the Thrace basin where the Sea of Marmara is the southeastern

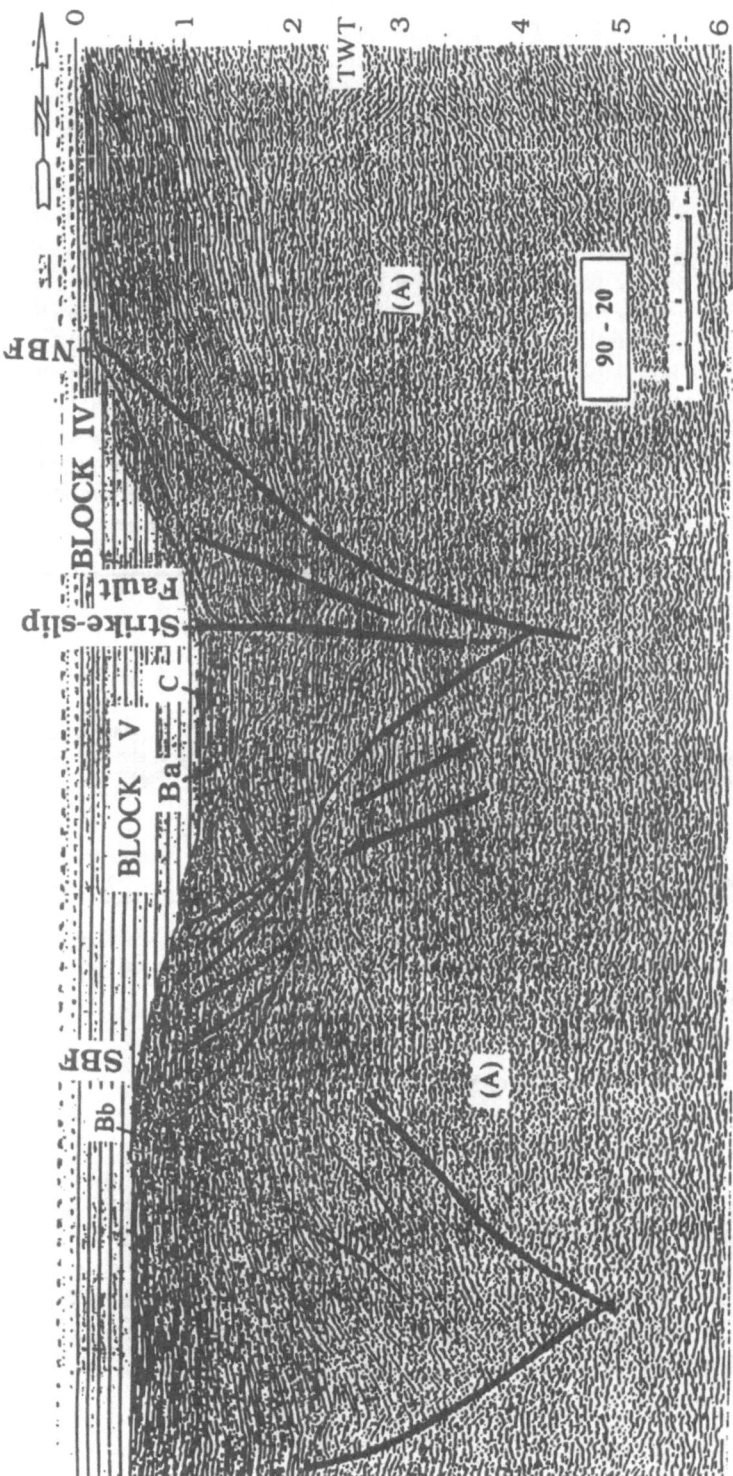

Fig. 8: Interpreted multichannel seismic profile (Line 90-20) in the Sea of Marmara (Location is given in Fig. 1). SBF: Southern Boundary Fault; NBF: Northern Boundary Fault; Block V: Basinal area; Block IV: Uplifted area; A: Pre transform facies; B: Syn-transform facies (a: Upper and b: Lower); C: Recent facies.

extension of this basin. Therefore, it could be assumed that these extensional basins of the Sea of Marmara have existed since Eocene times. The interpretation of these structures were made by Özel (1992) as the "negative flower structures" created by the northern strand of the NAF which uses the older fault planes of the Thrace tensional basin. Structural style depends on the tectonic setting, especially the location with respect to plate boundaries and the type of boundary.

The Sea of Marmara is influenced by two fundamental fault systems: (i) E-W trending normal faults (the Northern and Southern Boundary Faults), together with a number of secondary, normal subparallel faults; (ii) NE-SW oriented subvertical strike-slips which are largely dextral (Barka, 1991). These two fault systems divide the Sea of Marmara into roughly five different rhomboidal pull-apart structures (Barka and Kadinsky-Cade, 1988; and Özel, 1992) (Fig. 1). Three basinal blocks correspond to three deep basins. The basins are separated by two high blocks which about 600 m higher than the surrounding area and are marked by a relatively high Bouguer gravity anomaly (Fig. 3). All the blocks are subsiding but the subsidence rates are higher for the basins.

Three different seismic facies have been recognized by Özel (1992). These are: (A) The facies are the sedimentary facies deposited before the activation of the NAF (pre-Upper Miocene) occurring in the shelf areas outside the northern and southern boundary faults. These are defined as the pre-transform facies. (B) Syn-transform facies which were deposited concurrently with the evolution of NAF, are made up of two different facies units (a and b) in the basins and horst areas respectively within the boundary faults. The uppermost part of this facies (B-a) is characterized by well-stratified, almost parallel and well-defined reflections in the subsiding basins reaching thicknesses of about 1 km in places. The second unit of the syn-transform facies occurs over the uplifted blocks (B-b) consisting of reasonably well-stratified lower but completely distorted upper parts. (C) The upper-most facies are the most recent thin sedimentary cover over the first unit of syn-transform facies (B-a) in the deeper basinal areas. These facies are very thin and not seen in the seismic sections.

The interpretation of the seismic line (Line 90-20) can be explained in terms of a strike-slip system with an inverse flower structure (Fig. 8) using the normal fault assemblages detached from the Thrace region in the north (Barka and Gülen, 1988; and Özel, 1992). The Line 90-20 (SW-NE) traverses the Blocks IV and V which are the dominant uplifted block and the eastern basin in the Sea of Marmara respectively. It is mainly made up of the lower syn-transform facies (B-b). This profile cuts both the Northern and Southern Boundary Faults as well as the strike-slip fault separating Blocks IV and V.

The seismic section 19/20 (Fig. 9) crosses the deeper part of Sea of Marmara from east to west. All five blocks are shown in this section. The easternmost basin (Block V) is covered by the first unit of the syn-transform facies (B-a). Block IV is the main uplifted part of the Sea of Marmara. It is made up of the second unit of the syn-transform facies (B-b). This is a very dominant uplifted block having a positive SW-NE trending Bouguer gravity anomaly (Fig. 3). This structure is cut by a strike-slip fault inferred from the changes in the reflection patterns. The first unit of the syn-transform facies is well stratified in Blocks I and III which are the small basinal areas at the western end of the Sea of Marmara. Block II is located in between these two basinal areas, made up of the second unit of the syn-transform facies (B-b). There are normal and reverse fault structures in this small uplifted block (Block II).

322

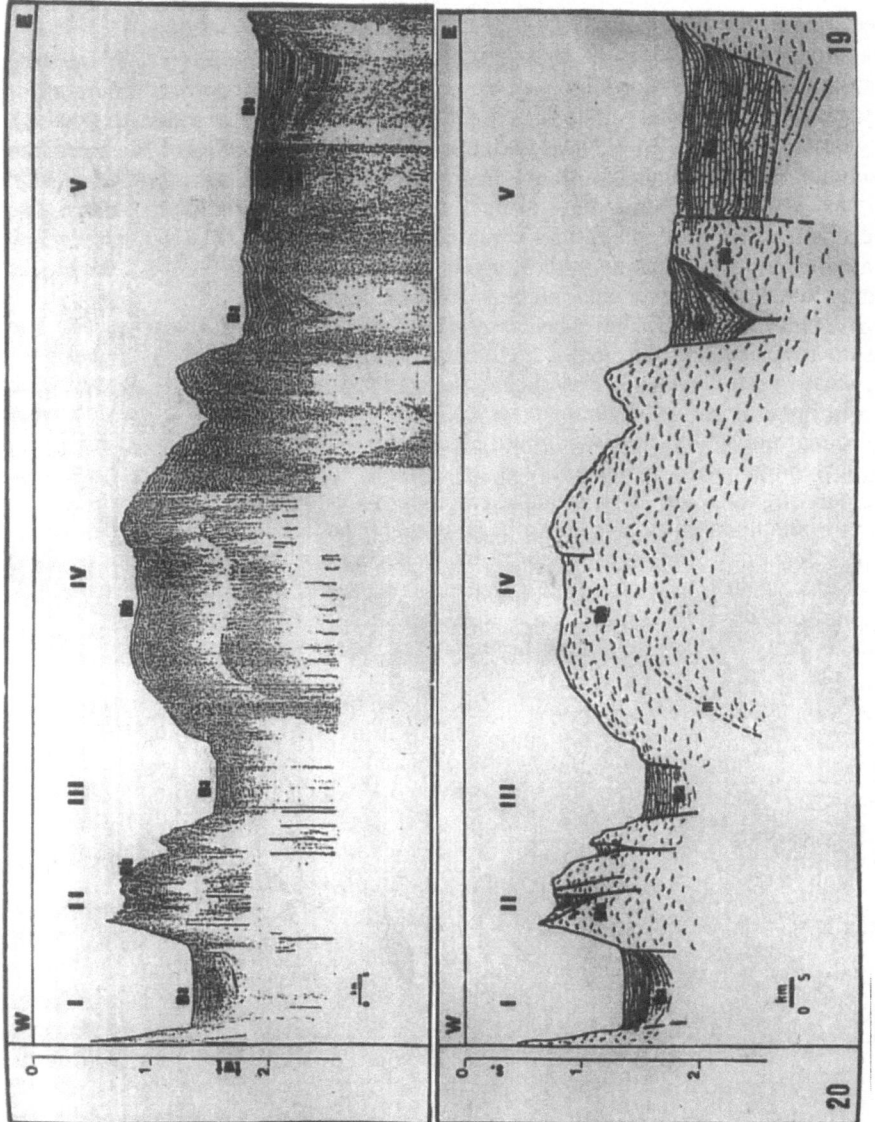

Fig. 9: Interpreted Profile 19/20 of the airgun high resolution seismic data of Piri Reis (from Özel, 1992) (Location is given in Fig. 1). Blocks I, III and V: Basinal areas; Blocks II and IV: Uplifted areas; A: Pre-transform facies; B: Syn-transform facies (a: Upper and b: Lower); C: Recent facies.

Strike-slip fault assemblages may be associated with the main part of the NAF. The predominant motion is strike-slip although a small component of throw may also be evident. These faults have most probably developed contemporaneously with the deposition of the Thrace basin since the Middle Eocene as evidenced by thickening of the sediments on the downthrown side of the faults. Rotation of the downthrown blocks in the growth faulting produces rollover structures.

5. Discussion

The relationship of the areas of the northern Aegean and northwestern Turkey with the adjacent Eurasia-Black Sea plate is best described by Jackson and McKenzie's (1988) model of a deforming zone comprising rotating blocks with non-vertical fault margins. "This is the simplest model of faulting ... which can accommodate strike-slip and extensional movement without requiring movement of material along the strike of the zone". This provides a specific mechanism for "damping" the right-lateral displacement along the NAF and translating the westward motion of the Turkish block into a north-south extension in the Aegean area (Kissel et al., 1989).

Geologically the Marmara Sea basin is the extension of the Thrace basin which is a Tertiary basin where the subsidence of basement took place during the Middle Eocene. The neotectonic structural framework in the Marmara Sea basin is primarily controlled by the north strand of NAF which became active in the Upper Miocene. In fact the NAF uses the earlier structural system of the tensional Thrace basin within the Sea of Marmara. The NAF must have occurred along these areas of weakness in the Sea of Marmara and should have been active before the Middle Eocene, with the additional effects of counter-clockwise rotations of blocks in western Turkey and the Aegean Sea (Rotstein, 1985; Le Pichon and Angelier, 1980; Şengör, 1987; and Kissel et al., 1989).

Among various crustal models postulated to characterize extensional regimes e.g., the Aegean area we can include: (1) individual grabens locally isostatically compensated by an underlying zone of an intensely thinned crust (Le Pichon et. al., 1984); (2) low-angle normal faults extending down through the entire lithospere so that extension is due to discrete shear between large, coherent sheets; and (3) graben flattened off into a basal decollement in the mid crust with extension in the lower crust being accommodated by penetrative ductile stretching on a regional scale. The high heat flow values for this region indicate crustal thinning in this region.

6. Conclusions

The neotectonic structural framework in the Marmara Sea basin is primarily controlled by the northern strand of NAF which became active in the Upper Miocene. Because the interpretation seismic sections often involves some ambiguity, knowledge of the structural style appropriate to the region can help in making the interpretation consistent with all available data, not merely with the seismic data alone. The Marmara Sea basin is related to an uplift of the Moho. This basin accommodates both strike-slip and extensional movements. The east-west trending normal fault systems of the Sea of Marmara is a diffusive zone of crustal thinning. However, it is obvious that the NAF affects the region at depth within the upper mantle with the formation of "negative flower structures".

According to the interpretation of the geological, geomorphological and geophysical data, the Sea of Marmara can be divided into five different blocks, which are controlled by two sets of fault systems: (i) E-W trending normal faults (the Northern and Southern Boundary faults); and (ii) NE-SW oriented sub vertical strike-slips. The blocks are undergoing vertical motions and rotations relative to one another. The whole area is subsiding although the subsidence rate is higher in the basinal areas than in the blocks. Three deep basins of the Sea of Marmara are related to the basinal blocks which are covered by the horizontally well-layered upper syn-tranform sedimentary sequences. These are separated by two uplifted blocks which are up of lower syn-transform facies that suffered compressional stresses. The strike-slip faults bring together the different sedimentary sequences horizontally. The middle uplifted block has a positive Bouguer gravity anomaly. The northern Boundary Fault zone has a magnetic anomaly of deep origin and dissected by the strike-slip faults trending NE-SW. The magnetic anomalies are also present on the south side of the Southern Boundary Fault and they are the short wavelength anomalies of shallower sources most possibly related to the intrusive volcanic rocks in this region.

6. References

Adatepe, M.F., 1988, Marmara Denizi jeofizik verilerinin değerlendirilmesi, Ph.D. Thesis, İstanbul University, Turkey.

Alvarez, F., Virieux, J. and Le Pichon, X., 1984, Thermal consequences of lithosphere extension over continental margins: the initial stretching phase, Geophys. J.R. Astr. Soc., London, 143, 23-7.

Barka, A.A., 1992, The North Anatolian fault zone, Annales Tectonicae, Special Issue to Volume VI, 164-195.

Barka, A.A. and Gülen, L., 1988, New constraints on age and total offset of the North Anatolian fault zone: Implications for tectonics of the eastern Mediterranean region, METU Journal of Pure and Applied Sciences (Ankana-Turkey), 21, 39-63.

Barka, A.A. and Kadinsky-Cade, K., 1988, Strike-slip fault geometry in Turkey and its influence on earthquake activity, Tectonics, 7, 663-684.

Biju-Duval, B., Letouzey, J. and Montadert, L., 1978, Variety of margins and deep basins in the Mediterranean, Am. Assoc., Petrol. Geol., Memoir 29, 293-317.

Brooks, M. and Kiriakidis, L., 1986, Subsidence of the North Aegean trough: an alternative view, Journ. Geol. Soc., London, 143, 23-27.

Burke, W.F. and Uğurtaş, G., 1974, Seismic interpretation of Thrace basin, Proc. Second Petroleum Congress of Turkey, p. 229-249, TPAO Internal report, Ankara.

Crampin, S. and Evans, R., 1986, Neotectonics of the Marmara Sea regions of Turkey, Journ. Geol. Soc., London, 143, 343-348.

Dewey, J. F., and Şengör, A.M.C., 1979, Aegean and surrounding regions: complex multiplate and continuum tectonics in convergent zone, Geol, Soc. Am. Bull. Part I, 90, 84-92.

Doust, H. and Arıkan, Y., 1974, The geology the Thrace basin: in Okay H. and Dileköz E. (eds.). "The Association of Turkish Petroleum Geologists Bulletin", Second Petroleum Congress of Turkey, p. 119-136.

Ekşioğlu, G., 1991, Marmara Denizi yapısının jeofizik verilerle incelenmesi, M.Sc. Thesis, D.E. University, İzmir-Turkey.

Ergün, M., 1977, Magnetic studies in Cyprus and the Biga peninsula, Turkey, Ph. D. Thesis, University of Leicester, England, pp. 224.

Ergün, M., 1990, Geophysical framework of the Sea of Marmara, Rapp. Comm. Int. Mer. Medit., 32, pp 137.

Eyidoğan, H., 1988, Rates of crustal deformation in western Turkey as deduced from major earthquakes, Tectonophysics, 148, 83-92.

İlhan, E., 1976, Türkiye Jeolojisi, O.D.T.Ü. Yayın No: 51, Ankara, Turkey.

Jackson, J.A. and Mc. Kenzie, D.P., 1988, The relationship between plate motions and seismic moment tensors and the rate of active deformation in the Mediterranean and Middle East, Geophys. J., 93, 45-73.

Jongsma, D., 1974, Heat flow in the Aegean Sea, Geophys. J. R. Astron. Soc., 37, 337-346.

Kalafat, D., Gürbüz, C., and Üçer, S.B., 1987, Batı Türkiye'de kabuk ve üst manto yapısının araştırılması, DAB, 59, 43-64.

Kale, B., 1985, Manyetik anomalilerinin ters çözüm yöntemiyle analizi ve Marmara Denizi verilerine uygulanması, M. Sc. Thais, Dokuz Eylül University, İzmir-Turkey.

Kissel, C., Laj, C., Poisson, A., Savaçın, Y, Simeakis, K. and Mercies, J. L., 1989., Paleomagnetic evidence for rotational deformations in the Aegean domain, Tectonics, 5, 783-795.

Ketin, İ, 1983, Türkiye Jeolojisine genel bir bakış, İ.T.Ü. Kütüphanesi, Sayı No: 1259, İstanbul, Türkiye.

Le Pichon, X. and Angelier, J., 1979, The Hellenic arc and trench system: a key to the neotectonic evolution of the eastern Mediterranean area, Tectonophysics, 60, 1-42.

Le Pichon, X., Lyberis, N. and Alvarez, F, 1984, Subsidence history of the North Aegean trough, in Dixon, J.E. and Robertson, A.H.F. (eds). "The geological evolution of the eastern Mediterranean", Spec. Publ. Geol. Soc., London, 17, 727-246.

Lyberis, N., 1984, Tectonic evolution of the North Aegean trough, in Dixon, J.E. and Robertson. A.H.F. (eds.). "Geological evolution of the eastern Mediterranean", Spec. Publ. Geol. Soc. London, 17, 709-725.

Lyberis, N. and Deschamps, A., 1982, Sismotectonique du fosse Nord Egeen; relations avec la faille Nort-Anatolienne, C. R. Acad. Sci. Paris, 295, Ser. 2, 625-628.

Makris, J., 1977, Geophysical investigations of Hellenides, Hamburger Geophysikalische Einzelschriften, Nr. 33: 128p.

McKenzie, D., 1978, Active tectonics of Alpine-Himalayan belt: the Aegean and surrounding regions, Geophys. J.R. Astr. Soc., London, 55, p. 217-254.

Mercier, J-L, Vargeley, P., Simeakis, T., Kissel, C. and Laj, C., 1991, The continuation of the North Anatolian dextral strike-slip fault into the oblique fault zone of the North Aegean Trough (W. Turkey and N. Greece): timing, tectonic regimes, fault kinematics and rotations, Tectonics, 10, 254-271.

Özel, E., 1992, Marmara Denizi'nin neotektonik yapısının jeofizik yöntemlerle incelenmesi, Ph. D. Thesis, Dokuz Eylül University, İzmir-Turkey.

Rotstein, Y., 1985, Tectonics of the Aegean block: Rotation, side arc collision and crustal extension, Tectonophysics, 117, 117-137.

Şengör, A. M.C. , 1987, Cross-faults and differential stretching of hanging walls in regions of low-angle normal faulting: examples from western Turkey, Geological Society Special Publications (London), No: 28, 575-589.

Şengör, A. M. C., Görür, N. and Şaroglu, F, 1985, Strike-slip faulting and related basin formation in zones of tectonic escape: Turkey as a case study, in Biddle, K. T. and Christie-Blick, N. (eds) "Strike-slip faulting and Basin Formation", Soc. econ. Paleontol.and Mineral., Spec. Pub, 37, 227-64.

Taymaz, T., Jackson, J. and McKenzie, D., 1991, Active tectonics of north and central Aegean Sea, Geophys. J. Int., 106, 443-490.

Üçer, S.B., Crampin, S., Evans, J.R., Miller, A. and Kafadar, N., 1985, The Marnet radiolinked seismeter network spanning the Marmara Sea and the seismicity of western Turkey, Geophys. J.R. Astr. Soc.,London, 83, 17-30.

THE COTE D'IVOIRE - GHANA TRANSFORM MARGIN: AN EXAMPLE OF AN OCEAN-CONTINENT TRANSFORM BOUNDARY

J. MASCLE
GEMCO (URA CNRS 718)
Laboratoire de Géodynamique Sous-Marine
B.P. 48 - 06230 Villefranche-sur-Mer
France

Christophe BASILE
LGA (URA 69)
Institut Dolomieu - Université J. Fourier
38031 Grenoble
France

B. PONTOISE
ORSTOM (UR 1F)
Laboratoire de Géodynamique Sous-Marine
B.P. 48 - 06230 Villefranche-sur-Mer
France

F. SAGE
GEMCO (URA CNRS 718) et ORSTOM (UR 1F)
Laboratoire de Géodynamique Sous-Marine
B.P. 48 - 06230 Villefranche-sur-Mer
France

ABSTRACT. Transform continental margins have been subject of recent attention for two reasons. Firstly, transform faults represent the third category of major plate boundaries and are less understood than the two other major plate boundaries, (divergent and convergent). Secondly, the precise study of transform margins can help in better constraining the structure and evolution of the ocean-continent transform boundary, particularly the history of the deformations, vertical movements and their effects on the sedimentary records.

We briefly present and discuss a set of multichannel seismic reflexion and wide-angle seismic refraction data that were recently gathered over the Cote d'Ivoire - Ghana transform margin.These data substantiate that this continental margin includes two segments. A divergent segment is characterized by a gradually stretching continental crust and accompanying rifting structures; a transform segment is chiefly underlined by a prominent bathymetric marginal ridge located along the Ocean Continent Boundary. Between the 20 km-thick marginal ridge and an abnormally thin oceanic crust, the Ocean Continent transition is remarkably sharp, in the order of a few km. Such a sharp transition is also well supported by gravity modelling.

327

E. Banda et al. (eds.), Rifted Ocean-Continent Boundaries, 327–339.

1. İntroduction

The transform continental margin as a specific type of continent-ocean boundary has gradually come to be accepted since the 1970s. Geophysical-geological data from several transform margins clearly demonstrate that these continental borderlands are drastically different from divergent margins in terms of crustal structure, deformation, subsidence and sedimentation history.

Over the past 20 years, several marine geophysical cruises (mainly gravity, refraction, and reflection seismic profiling) have been devoted to few transform margins, namely a) Agulhas Margin and its conjugate the Falkland Margin (Ewing et al., 1971 ; Emery et al., 1975 ; Scrutton, 1976, 1979, 1982 ; Lorenzo et al., 1989), b) southern Newfoundland Margin (Hayworth and Keen, 1979 ; Todd et al., 1989 ; Keen et al., 1990), c) southern margin of the Exmouth plateau (Lorenzo et al., 1991), d) Spitzbergen Margin (Lowell, 1972 ; Eldholm and Talwani, 1982 ; Eldholm et al., 1987), and e) equatorial African transform margins (Fail et al., 1970 ; Arens et al., 1971 ; Delteil et al., 1974 ; Emery et al., 1975 ; Mascle, 1976 ; Blarez, 1986 ; Blarez et al., 1987 ; Mascle and Blarez, 1987 ; Mascle et al., 1988 ; Basile, 1990, Basile et al., 1992 ; Mascle et al, 1993 ; Basile et al., 1993) and their conjugates, the northern Brazilian Margin (Ponte and Asmus, 1978 ; Zalan et al., 1985 ; Costa et al., 1990).

From these results, features typical for transform margins can be recognized. They are illustrated here for the Côte d'Ivoire-Ghana (CIG) Transform Margin (Fig. 1A, 1B) which results from a rather simple motion between two plates (Africa and South America).

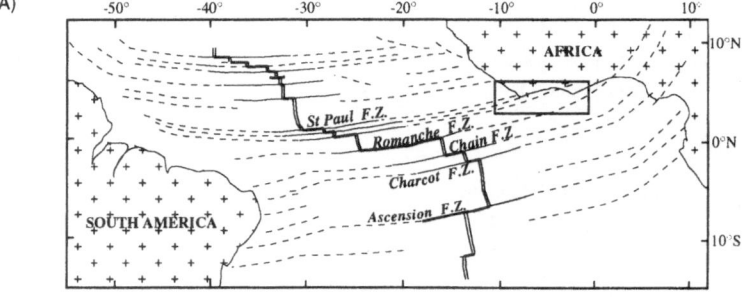

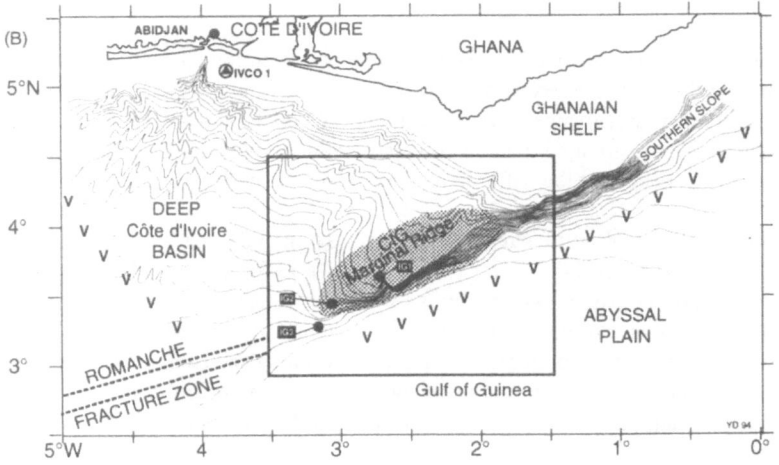

Figure 1 Geodynamic, geological and bathymetric framework of the CIG Transform Margin.

1 A : Fracture zones in the Equatorial Atlantic and associated continental margins.

1 B : Simplified bathymetry and main morpho-structural domains of the CIG Transform Margin.

The dots show location of ODP Leg 159 holes, the triangle the location of hole IVCO2

These features are **a)** lateral structural continuity between a major oceanic fracture zone and a continental transform margin (the Romanche Fracture Zone and CIG Margin). **b)** a very steep and narrow continental slope (20-30 km) between a continental shelf and an adjacent oceanic abyssal plain, indicating a very sharp crustal transition between thick or partially thinned continental crust and oceanic lithosphere, **c)** a morphologically well expressed marginal ridge, bounding the transform margin, along an adjacent extensional basin (Côte d'Ivoire Basin).

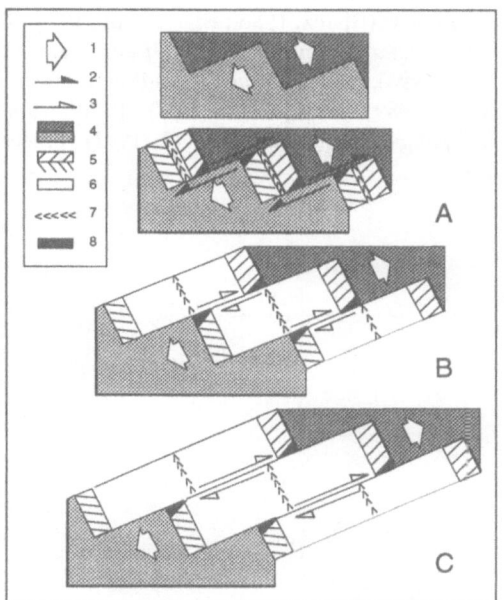

Figure 2 - Evolution sketch of a transform margin. Main stages of a rift transform margin.
(1) divergence
(2) and (3) transform motion respectively between continental and oceanic crust
(4) continental crust
(5) thinned continental crust
(6) oceanic crust
(7) ridge axis
(8) marginal ridge.

All these features result from a translational stress acting between two continents along active transform faults. During rifting and opening, such evolution can be schematized into three main stages (Fig. 2) (Mascle and Blarez, 1987).

Stage 1 - An intra-continental active transform fault (stage A on Fig. 2). This results in a contact between two thick continental plates of different thickenesses. In this case, the transform boundary of the extensional basin is subjected to shear stresses accompanied by potential uplift to create a bordering marginal ridge.

Stage 2 - Continent-ocean active transform fault contact (stage B on Fig. 2). In this setting, the proximity of the hot oceanic lithosphere may induce important vertical readjustment of the nearby continental margin border.

Stage 3 - Inactive continent-ocean transform fault (stage C on Fig. 2). This is tectonically a passive transform margin, undergoing relatively uniform thermal subsidence.

Such evolution implies successive crustal/lithospheric contacts which could lead to strong thermal and pressure contrasts.

These should have been recorded during the margin creation and subsequent sedimentary evolution (Sage, 1994).

2. Outline of the regional geology of the Côte d'Ivoire Ghana transform margin

The CIG margin results from a major transform motion between two plates. This motion is still active today along the Romanche Fracture Zone, which offsets the Mid-Atlantic Ridge by 945 km (Fig 1A) (Fail et al., 1970). The area of recent investigation is the Côte d'Ivoire Marginal Ridge, which corresponds to a sharp transition between a laterally thinned continental crust and an adjacent oceanic crust. The setting of the present-day marginal ridge includes a fossil ridge that connects laterally with the extinct Romanche Fracture Zone (Fig. 1A and 1B).

The seismic stratigraphy and tectonics of the area are mainly based on investigations (Fig. 3) by the 1983 *Equamarge I* cruise (Blarez, 1986 ; Blarez et al., 1987 ; Mascle and Blarez, 1987 ; Mascle et al., 1988), the 1990 *Equasis* cruise, the *Equaref* cruise and *C. Darwin* cruise 55 (1990 and 1991, respectively) (Sage, 1994), the 1988 *Equamarge* II cruise (Mascle, Auroux et al., 1987 ; Basile et al, 1989 ; Popoff et al., 1989 ; Pontoise et al, 1990 ; Basile, 1990 ; Basile et al., 1992 ; Basile et al., 1993), the 1992 *Equanaute* cruise (Mascle et al., 1993) ; processing of MCS and refraction data are still is in progress.

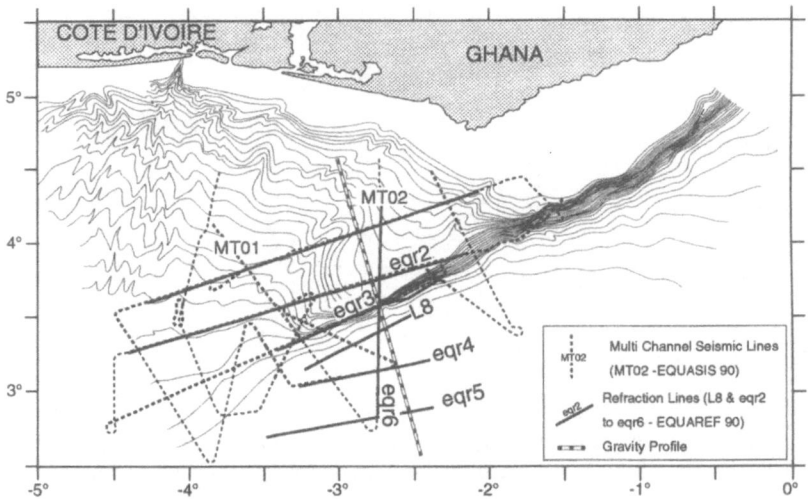

Figure 3 - Recent geophysical data collected on the CIG Transform Margin shown on a simplified bathymetric map of the margin. Dotted lines: multichannel seismic reflection lines; thick black lines show location of the wide angle OBS sections. The Equanaute dives were performed along the Southern slope of the CIG Marginal Ridge in two areas surveyed in swath bathymetry. Black and white thick dotted line indicates location of the gravity section used for modelling.

SEISMIC STRATIGRAPHY

The seismic stratigraphy was defined on the basis of angular relationships between several sedimentary units, especially along the northern slope of the CIG marginal ridge. Six main units, A to F (Fig. 4) (Basile et al, 1993) are distinguished.

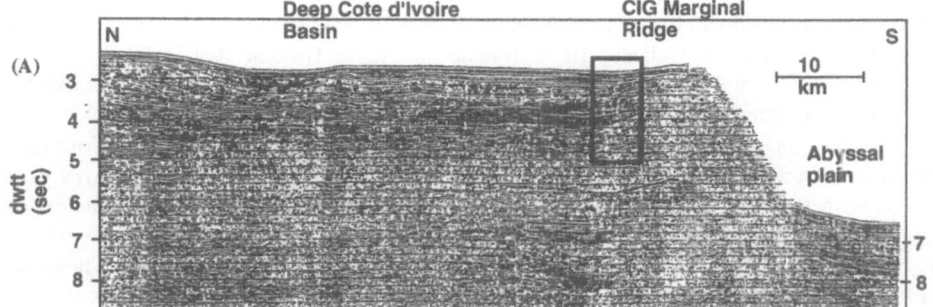

Figure 4 - Multichannel line MT02.
(**A:**) A typical MCS (24 channels) profile across the deep Ivorian Basin and the Marginal Ridge (to the South) line MT02, Location on Figure 3. Vertical scale in second two way travel time.
(**B**): Enlargement of a section of the same line (processed on 96 channels) showing the main litho-acoustic facies as discussed in text.

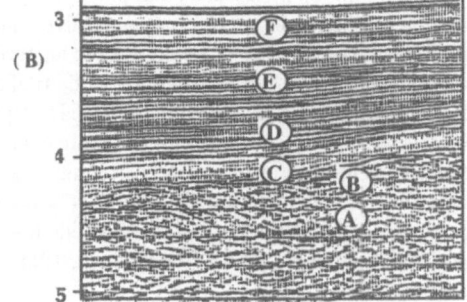

- The relationship between basal Unit A and the underlying acoustic basement are still not clear. This unit is deformed in both the divergent Ivorian Basin and the transform marginal ridge. It is divided into subunits in A0, A1 and A2. A0 seems to be ante-rift in the whole area. A1 is syn-rift in the divergent Ivorian Basin. A2 is post-rift in the Ivorian Basin, but appears to be deformed within the transform margin domain.
- Unit B clearly corresponds to the post-rift sediments, and is not deformed within the transform margin. It unconformably overlies both the A sequence of the divergent Ivorian Basin and the A2 (syn transform) sequence of the transform margin.
- Units C to F lie conformably on the previous units, within and along the eastern side of the transform margin. However, they lie unconformably on the B and A2 sequences, which constitute most of the northern marginal ridge slope. All units lie almost horizontally in the deep Ivorian Basin and may progressively pinch out against the marginal ridge, possibly due to coeval ridge uplift.
- Sedimentary units C and D were deposited both by aggradation within the Ivorian Basin and by progradation originating from the marginal ridge summit. The upper part of the C sequence onlaps the ridge top, whereas the lower D sequence is restricted to the deepest part of the Ivorian Basin. Such progressive restriction of the sedimentation area also characterizes E and F sequences.

LITHOLOGY - AGES

So far only few samples have been recovered along the transform margin itself. These consist of detrital sediments (sandstones or siltstones). However, during the *Equanaute* dives, 14 geological cross sections were made along the southern slope of the marginal ridge, and 165 samples were collected (Mascle et al., 1994). Most of these are also terrigenous and probably belong to the same thick sandy-clayey formations (mainly sedimentary units A1 and A2). These rocks include fine-grained to coarse-grained sandstones, greenish, lenticular to wavy-bedded, siltstones and black shales, which show numerous syndiagenetic microfaults and slumps (Mascle et al., 1993).

Locally, near the foot of the continental slope (probably sedimentary unit A1), orthoquartzites and indurated shales, characterized by slaty cleavage, were sampled.

In our opinion, the *Equanaute* dives chiefly documented the A2 unit, made up of massive sandstones and siltsones interbedded with pelites. Here, in situ observations reveal silty-clayey strata interbedded with large-scale trough cross-bedded sandstones. Both the sedimentary facies and structures favour a very shallow marine environment of a deltaic type.

Only two samples (one core and one dredge) provided reliable age informations on the sedimentary units. Both indicate a middle to uppermost Albian age (Klingebiel, 1976; Grosdidier, pers. comm., 1989).

Tentative dating on microfauna and palynology on a few shales sampled during deep dives was also attempted. Microfauna from *Equanaute* dive 1 indicates a probable Barremien to Cenomanian age (Moullade pers. comm., 1994) ; pollens give an early Cretaceous age (Dejax, pers. comm., 1994). Preliminary results from the fission track of detrital apatites give the following ages of cooling for crossing the isotherm 60°C : 68 M.y. (dive EN1, depth 3479 m), 52 Ma (dive EN4, depth 2405 m), and 44 Ma (dive EN9, depth 3905 m) (Bouillin et al., 1994).

In these latitudes, the lack of clear oceanic magnetic anomalies does not facilitate an accurate chronology of oceanic opening. The proposed kinematic models are thus chiefly based on fracture zone geometry and tentatively fitted to magnetic reversal chronology derived from the South Atlantic (Klitgord and Schouten, 1986 ; Cande et al., 1988; Scotese et al., 1988; Shaw and Cande, 1990; Nürnberg and Müller, 1991).

All reconstructions show that the rifting of the equatorial Atlantic occurred during the early Cretaceous, possibly in Neocomian-Barremian times according to Doyle et al. (1982) from the Keta Basin, Popoff (1990) and Brunet et al. (1991) from the Benue trough. The various reconstructions suggest that kinematic parameters were modified in Santonian times. This reorganization was tentatively correlated with the final disruption of the African and Brazilian cratons within the equatorial area (Mascle and Blarez, 1987).

TECTONICS AS DERIVED FROM MULTICHANNEL SEISMIC DATA

To the north of the marginal ridge (Fig. 1B), the deep Ivorian Basin is an extensional margin, thinned by dip-slip faults trending north-south to northeast-southwest. These faults bound tilted blocks and half-grabens, which are infilled by thick syntectonic sediments (unit A1). To the south, the fossil marginal ridge which is 130 km long and 25 km wide, towers over the Ivorian basin by 1300 m and over the adjacent oceanic crust by more than 4000 m. The regional structure of the transform margin itself is linked to the westward thinning of the adjacent divergent basin. Eastwards the transform border is first expressed by a narrow shelf. Between 2°10'W and 2°45'W, the transform zone is chiefly expressed by the marginal ridge. West of 2°45'W, the ridge top dips progressively westwards similar to the bordering divergent Ivorian Basin (Fig. 1B).

Migrated multichannel seismic lines have greatly improved our knowledge the marginal ridge (Figs. 4 and 5). East of 2°10'W, transform motion has generated flower structures in the Ivorian Basin - transform margin transition. To the north, these strike-slip become transpressional and exhibit reverse slip. The associated folds are arranged en échelon, indicating on right-lateral movement. Southwards, a wide (> 10 km) and deep (1 km) channel may represent an infilled intracontinental transform valley, such as the Jordan valley in the Dead Sea transform. Its southern limit probably lies now near the top of the present continental slope where the deformations observed during the *Equanaute* scientific dives appear to be more important (folding, vertical fracturing, fracture cleavage) than southward, cut of the transform valley (strata tilted S-E).

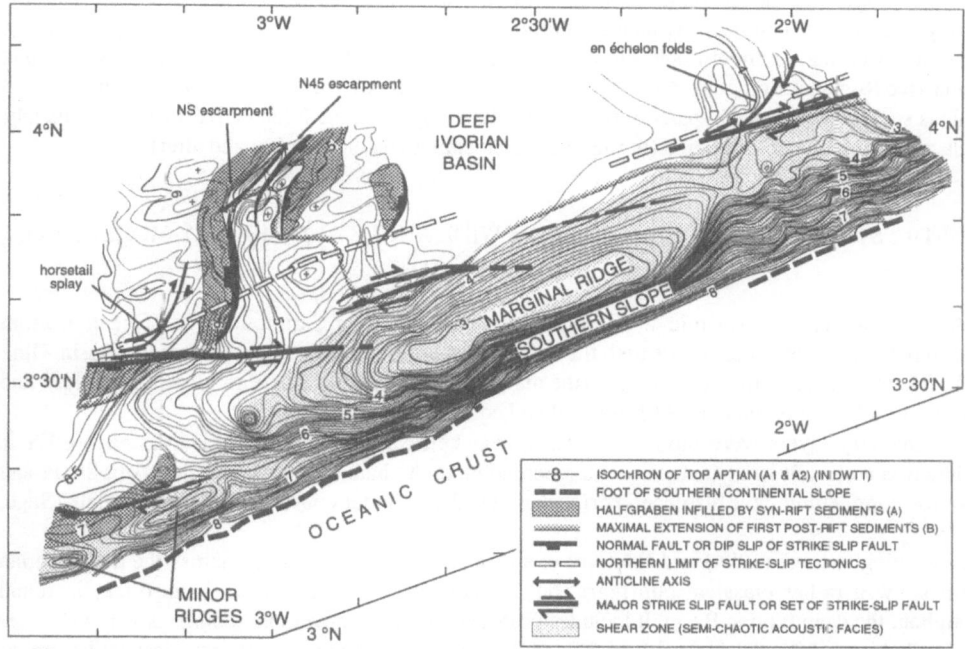

Figure 5 - Simplified structural sketch of the CIG Marginal Ridge and surroundings. Depth is shown in two-way travel time to the top of the synrift unit A.

The strike-slip fault activity was recorded during the sedimentation of A1, A2 and B units. The reverse strike-slip faults were active from the deposition of A1 sequences (uplifted during the sedimentation), up to the deposition of A2 (cut by faults).

Between 2°10'W and 2°45'W (MCS line MT2), along a N-W cross section of the marginal ridge (Fig. 4), three tectonic domains can be distinguished including the western prolongation of the transform valley observed eastwards (see above). There, the transform valley is located at the boundary between the deep Ivorian Basin and the marginal ridge itself. The feature eroded the deepest unit (A0) and is partly infilled by the faulted A1, and mainly the unfaulted A2 units. The northern marginal ridge slope consists of an important thickening of the A2 unit to the south. This thickening at the top resulted in the edification of a sedimentary swell with sedimentary lenses prograding to the north, and distal fans along its northern slope.

Equanaute dive data lend support to the hypothesis that deformation increases upslope towards the ridge top (fracturing and folding) where strike-slip activity was probably concentrated.

The seismic lines (Fig. 4) also show that the B unit is restricted to the deep Ivorian Basin and northern ridge slope. Thus, sedimentation took place simultaneously by both aggradation in the Ivorian basin (sediments originating from African coasts) and progradation (probably sediments derived from the ridge top erosion).

West of 2°45'W (Fig. 5), to the west of the marginal ridge, the southern border of the transform margin includes several minor acoustic ridges arranged en echelon. These tower over the oceanic crust to the south and over a thick (about 2 km) syn-transform basin to the north. These ridges are believed to here originated in connection with transform motion ; different hypotheses can be made concerning their nature and deformation history. Are these ridges continental or oceanic, basement or sediments ? Did these ridges appear during the rifting of the Ivorian Basin, during intracontinental transform faulting or during the continent-ocean transform faulting ?

These lower margin minor ridges are very similar to "en echelon" ridges mapped to the west, along the inactive Romanche Fracture Zone (Honnorez et al., 1993). Are they similar in nature and origin ? If this is so, this indicates that parts of the continental margin basement (and cover) were tectonically displaced during transform motion (during intracontinental transform contact and after).

CRUSTAL STRUCTURE AS DEDUCED FROM WIDE ANGLE AND GRAVITY MODELLING

A dense network of seismic wide angle data, recorded using digital OBSs (Ocean Bottom Seismometers) was obtained across both the divergent margin segment (the deep Ivorian Basin - line eqr2 : fig. 3), its transform margin border (the marginal ridge and the base of its slope - lines eqr3 and L8) and the adjacent oceanic crust (lines eqr4 and eqr5).

Sedimentary layers were modellised using near vertical reflection recorded for each OBS in conjunction with well identified reflectors recorded on multichannel seismic profiles. Amplitude and wave form modelling have also been performed in order to better constraint the final models (Sage, 1994).

According to Sage (1994) and Pontoise et al. (in prep.), crustal sections across the deep Ivorian Basin show a rather classical configuration. A sedimentary cover, 3 to 4 km-thick, is found throughout the deep Ivorian Basin. It includes post-tectonic and syn-tectonic sediments with velocities ranging between 1.7 - 2.1 $km.s^{-1}$ and 2.8 - 4.0 $km.s^{-1}$ respectively, overlying a crustal basement characterized by a sharp increase in seismic velocity, as attested by recorded strong reflexions from the interface. The continental basement can itself be divided into two units; an upper continental crust and a lower continental crust, having velocities of 5.2 to 5.5 $km.s^{-1}$ and 6 to 6.9 $km.s^{-1}$ respectively. The upper crustal layer is characterized by a relatively important vertical velocity gradient of 0.25 s^{-1} whereas the lower section exhibits a rather low vertical gradient of 0.06 s^{-1}. Continental crust progressively thins westwards from about 18 km to 7-8 km in 200 km. Such a crustal structure is in good agreement with a westward directed stretching expressed, at the level of the upper crust and sedimentary cover, by block tilting and syn-rift grabens. A comparable crustal structure (fig. 6) characterizes the transform marginal ridge itself, where crustal thickness also varies from 18-20 km eastwards to 7-8 km westwards. Both, on crustal and surficial structural grounds, the transform marginal ridge can be interpreted as a tectonized and uplifted fragment of the proximal divergent Ivorian Basin.

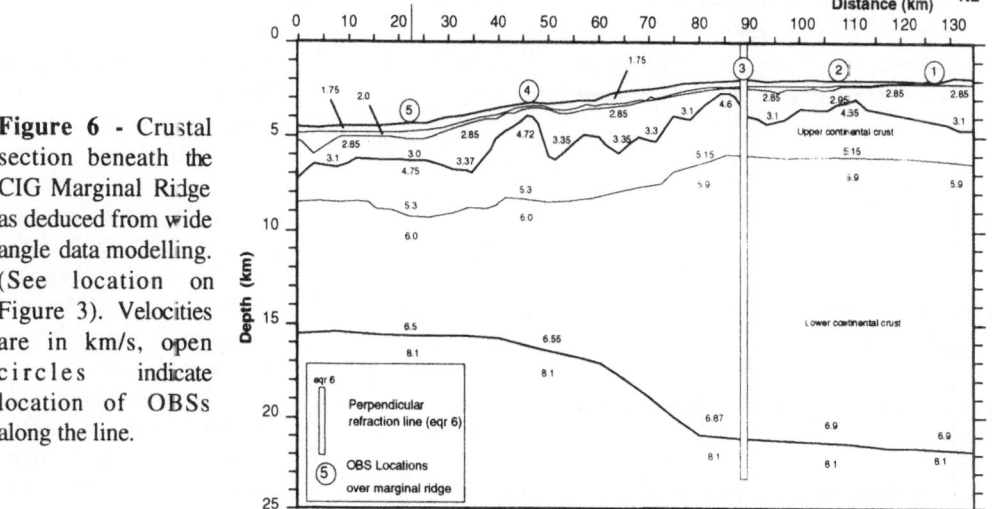

Figure 6 - Crustal section beneath the CIG Marginal Ridge as deduced from wide angle data modelling. (See location on Figure 3). Velocities are in km/s, open circles indicate location of OBSs along the line.

Seismic and gravity modelling were attempted along a line perpendicular to the marginal ridge (eqr6 - fig. 7). Seismic velocities were estimated using both forward and inverse arrival-times modelling together with seismic velocities constraints from existing orthogonal sections (eqr 2 to 5 and L8 lines). Density values of the different layers were deduced from velocity-depth functions using Barton's law (1986). For the upper continental crust, rock density measurements on samples dredged along the southern marginal ridge slope are in good agreement with deduced density. The observed gravity data are characterized by a wide anomaly (90 mGal peak to peak) located over the marginal ridge. Over the deep Ivorian Basin (between 60 to 83 km) the shape of the anomaly is controlled by the geometry of the sedimentary and crustal structure. In another hand, the 90 mGal peak to peak anomaly is constraint by the Moho geometry and best fitted by a very sharp ocean-continent transition located just south of refraction line L8. Further south, the crustal structure is found to be characteristic of oceanic type, with oceanic layers II and III, although the crustal thickness is uncommonly thin (2 times thinner than expected for a so-called "Normal Atlantic Crust"). According to Sage (1994) and Pontoise et al. (in prep.), the width of the transition between the continental crust rooted marginal ridge and the nearby southern oceanic crust can be estimated in the order of 5 km. This value is deduced from the computed RMS between observed and calculated gravity anomalies for different ocean-continent transition geometry.The model clearly shows that both the slope and the location of the ocean-continent transition range between 70 and 80 deg. in a narrow zone less than 5 km-wide. Preliminary lithostatic pressure estimates, through the transition, indicate a maximum horizontal lithostatic gradient, from the continent to the ocean, in the order of 60 MPa at 9 km depth. This value appears below the limit of brittle rheology for a continental domain (Ranalli and Murphy, 1986).

At depth, for example at 20 km, the horizontal lithostatic gradient is in the order of 10 MPa.

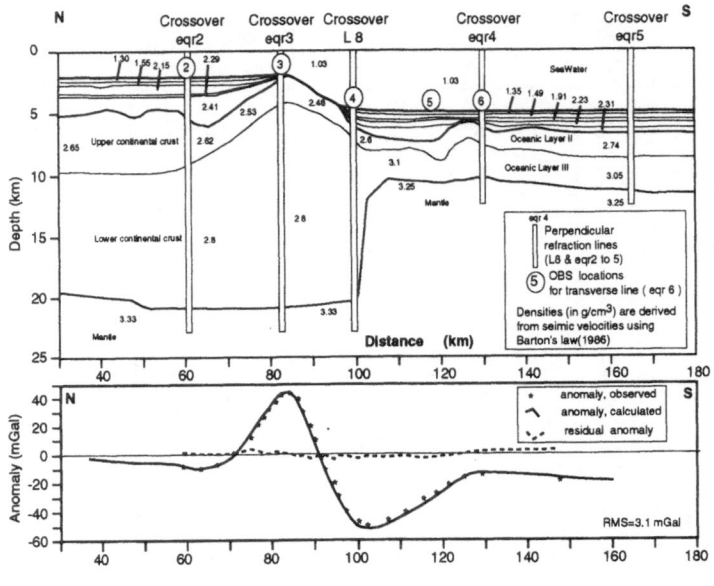

Figure 7

- 7 A : Crustal model across the Côte d'Ivoire - Ghana Transform Margin (see location on Fig. 3). The crustal model is constrained by both measured gravity anomalies and wide angle data. eqr2, eqr3, L8, eqr4, eqr5 indicate location of orthogonal wide angle lines used to constrain the model.

- 7 B : Observed and calculated gravity anomalies along a line perpendicular to the transform margin .

3. Conclusions

Transform margins clearly represent a specific ocean/continent boundary between a laterally thinned continental crust and an adjacent oceanic crust. The transition between both crustal domains is expressed at three levels :

- on superficial/morphological grounds, the transition between the continental margin and the oceanic crust is underlined by a prominent marginal ridge. This ridge, which towers over both the nearby divergent basin and the oceanic abyssal plain, is laterally prolonged by a steep continental slope (towards the continent) and connects with a fossil major fracture zone (the Romanche Fracture Zone in the case of Côte-d'Ivoire - Ghana).

- on geological/structural grounds, the transform marginal ridge corresponds to a highly deformed wedge of clastics where strike slip, dip slip faults and "en echelon" folding were active during the transform motion along the margin.

At the base of the lowermost continental slope, the transition between the fossil oceanic fracture zone and the marginal ridge consists of a series of "en echelon" arranged minor structural ridges. These may represent tectonically displaced continental margin crustal blocks and their sedimentary cover.

- finally, on crustal grounds, the transition between the transform margin and the oceanic lithosphere appears to be particularly sharp. In the case of the Côte-d'Ivoire - Ghana transform, this transition occurs in less than five km, juxtaposing a twenty km-thick continental crust against a five km-thick oceanic crust.

This lead us to question **(1)** the role and significance of thermal exchanges between two lithospheres at different temperatures as already inferred by Mascle and Blarez (1987) and tentatively modelled by Todd and Keen (1989), **(2)** the nature of the mechanisms able to compensate, through time, the constraints generated between two adjacent, oceanic and continental, domains in sharp transition.

We acknowledge the captains, crews and technical staff of the R.V.Jean Charcot and R.V. le Nadir (IFREMER - GENAVIR) on board of which most of the data discussed in this paper have been collected between 1988 and 1992.
Thanks to CNRS - INSU (Geosciences Marines) and ORSTOM for providing financial support for data processing.

Contribution number 664 of the GEMCO - URA CRNS - UMPC 718.
Contribution number 1F201 of the ORSTOM - UR 1F

REFERENCES

Arens, G., Delteil, J.R., Valery, P., Damotte, B., Montadert, C. and Patriat, P., 1971. The continental margin of the Côte-d'Ivoire and Ghana. *Inst. Geol. Sci. London, Rep.* 70, 61-78.

Barker, P.F., Dalziel, I.W.D. et al., 1976. *Init. Repts. DSDP*, 36 : Washington (U.S. Govt. Printing Office).

Barton, P.J., 1986. The relationship between seismic velocity and density in the continental crust - a useful constraint ? *Geophys. J. R. Astr. Soc.*, 87, 195-208

Basile, C., 1990. Analyse structurale et modélisation analogique d'une marge transformante : l'exemple de la marge profonde de Côte-d'Ivoire-Ghana. *Mém. Doc. CAESS Rennes*, 39, 220 pp.

Basile, C., Brun, J.P. and Mascle, J., 1992. Structure et formation de la marge transformante de Côte-d'Ivoire-Ghana : apports de la sismique réflexion et de la modélisation analogique. *Bull. Soc. Géol. France*, 163, 207-216.

Basile, C., Mascle, J., Auroux, C., Bouillin, J.P., Mascle, G., Gonzalves de Souza, K. & le groupe Equamarge, 1989. Une marge transformante type : la marge continentale de Côte-d'Ivoire-Ghana : résultats préliminaires de la campagne Equamarge II. *C. R. Acad. Sci. Paris*, t. 308, Série II, 997-1004.

Basile, C., Mascle, J., Popoff, M., Bouillin, J.P. and Mascle, G., 1993. The Côte-d'Ivoire-Ghana transform margin : a marginal ridge structure deduced from seismic data. *Tectonophysics*, in press.

Blarez, E., 1986. La marge continentale de Côte-d'Ivoire - Ghana : structure et évolution d'une marge continentale transformante. Ph. D. Thesis, Univ. Pierre et Marie Curie, 188 pp.

Blarez, E., Mascle, J., Affaton, P., Robert, C., Herbin, J.P. and Mascle, G., 1987. Géologie de la pente continentale ivoiro-ghanéenne : résultats de dragages de la campagne Equamarge. *Bull. Soc. Géol. France*, 5, 877-885.

Bouillin, J.P., Poupeau, Riou, L., Sabil, N., Basile, C., Mascle, J., Mascle, G., et the Equanaute Scientific Party, 1994. La marge transformante de Côte-d'Ivoire-Ghana : premières données thermo-chronologiques (campagne Equanaute, 1992). *C.R. Acad. Sci. Paris*, t. 318, 1365-1370.

Brunet, M., Dejax, J.,Brillanceau, A., Congleton, J., Downs, W., Duperon-Landoueneix, M., Eisenmann, V., Flanagan, K., Flyn, L., Heintz, E., Hell, J., Jacobs, L., Jehenne, Y., Ndjeng, E., Mouchelin, G. and Pilbeam, D.,1991. Mise en évidence d'une sédimentation précoce d'âge barrémien dans le fossé de la Bénoué en Afrique Occidentale (bassin du Mayo Oulo Léré, Cameroun), en relation avec l'ouverture de l'Atlantique sud. *C. R. Acad. Sc. Paris*, 306, 1125-1130.

Cande, S. C., LaBrecque, J. L. and Haxby, W. F. 1988. A high resolution seafloor spreading history of the South Atlantic. *Jour. Geophysi. Res.* 93 13479-13492

Costa, I.G., Beltrami, C.V. and Alves, L.E.M., 1990. A evoluçao tectono-sedimentar e o habitat do oleo da bacia do Ceara. B. *Geoci. PETROBRAS*, 4, 65-74.

Delteil, J.R., Valery, P., Montadert, C., Fondeur, C., Patriat, P. and Mascle, J., 1974. Continental margin in the northern part of the gulf of Guinea. In C.A. Burk and C.L. Drake (Editors), *Geology of continental margins. Springer, New York* : 297-311.

Doyle, J.A., Jardiné, J. and Doerenkamp, A., 1982. *Afropollis*, a new genus of early angiosperm pollen, with notes on the Cretaceous palynostratigraphy and paleoenvironments of northern Gondwana. *Bull. Centres Rech. Explor.-Prod. Elf Aquitaine*, 1, 39-117.

Eldholm, O., Faleide, J.I. and Myhre, A.M., 1987. Continent-ocean transition at the western Barents Sea/Svalbard continental margin. *Geology*, 15, 1118-1122.

Eldholm, O. and Talwani, M., 1982. The passive margins of northern Europe and East-Greenland. In R.A. Scrutton (Editor), Dynamic of passive margins. *Geodynamics Series*, 6, 30-44.

Emery, K.O., Uchupi, E., Phillips, J., Bowin, C. and Mascle, J., 1975. Continental margin off Western Africa: Angola to Sierra Leone. *Am. Assoc. Petrol. Geol. Bull.*, 59, 2209-2265.

Ewing, J.I., Ludwig, W.J., Ewing, M. and Eittreim, S.L., 1971. Structure of the Scotia Sea and Falklands plateau. *J. Geoph. Res.*, 76, 7118-7137.

Fail, J.P., Montadert, L., Delteil, J.R., Valery, P., Patriat, P. and Schlich, R., 1970. Prolongation des zones de fractures de l'océan Atlantique dans le golfe de Guinée. *Earth and Planet. Sci. Letters*, 7, 413-419.

Honnorez, J., Villeneuve, M. and Mascle, J., 1993. Old continent-derived metasedimentary rocks in the Equatorial Atlantic: an acoustic basement outcrop along the fossil trace of the Romanche transform fault at 6°30'W. *Marine Geology*, in press.

Keen, C.E., Kay, W.A. and Roest, W.R., 1990. Crustal anatomy of a transform continental margin. *Tectonophysics*, 173, 527-544.

Klingebiel, A., 1976. Sédiments et milieux sédimentaires dans le golfe de Bénin. *Bull. Centre Rech. Pau-SNEAP*, 1, 129-148.

Klitgord, K.D. and Schouten, H., 1986. Plate kinematics of the Central Atlantic. In: P.R. Vogt and E. Tucholke (Editors), The western Atlantic regions, *Geol. Soc. Am.*, 351-377.

Lorenzo, J.M. and Mutter, J.C., 1989. Seismic stratigraphy and tectonic evolution of the Falklands/Malvinas plateau. *Rev. Bras. Geoc.*, 18, 191-200.

Lorenzo, J.M., Mutter, J.C., Larzon R.L., and the Northwestern Australia Study Group, 1991. Development of the continent-ocean transform boundary of the southern Exmouth Plateau. *Geology*, 19, 843-846.

Lorenzo, J.M. and Vera, E.E., 1992. Thermal uplift and erosion across the continent - ocean transform boundary of the southern Exmouth Plateau. *Earth and Planetary Sci. Lett.*, 108, 79-92.

Lowell, J.D., 1972. Spitsbergen Tertiary orogenic belt and the Spitsbergen fracture zone. *Geol. Soc. Am. Bull.*, 83, 3091-3102.

MacDonald, K.C., Castillo, D.A., Miller, S.P., Fox, P.J., Kastens, K.A. and Bonatti, E, 1986. *J. Geophys. Res.*, Vol. 91, 3334-3354.

Mascle, J., 1976. Le golfe de Guinée : un exemple d'évolution de marge atlantique en cisaillement. *Mém. Soc. Géol. France*, nouvelle série, 128, 104 pp.

Mascle, J., Auroux, C. and the shipboard scientific team, 1989. Les marges continentales transformantes ouest-africaines (Guinée, Côte-d'Ivoire-Ghana) et la zone de fracture de la Romanche : Campagne Equamarge II (Février-Mars 1988). *Campagnes Océanographiques Françaises. Publications IFREMER* 150 pp.

Mascle, J. and Blarez, E., 1987. Evidence for transform margin evolution from the Côte- d'Ivoire - Ghana continental margin. *Nature*, 326, 378-381.

Mascle, J., Blarez, E. and Marinho, M., 1988. The shallow structures of the Guinea and Côte-d'Ivoire-Ghana transform margins : their bearing on the equatorial Atlantic Mesozoic evolution. *Tectonophysics*, 155, 193-209.

Mascle, J., Guiraud, M., Basile, C., Benkhelil, J., Bouillin, J.P., Cousin, M. and Mascle, G., 1993. The Côte-d'Ivoire-Ghana transform margin : preliminary results from the Equanaute cruise (June 1992). *C. R. Acad. Sci. Paris*. t. 316, p. 1255-1261.

Mascle, J. and the Scientific Party, 1994. *Les marges continentales transformantes ouest-africaines - Côte-d'Ivoire, Ghana, Guinée*, IFREMER, Série Repères Océan. 5. 135 pp.

Nürnberg, D. and Müller, R.D.1991.The tectonic evolution of the South Atlantic from Late Jurassic to present. *Tectonophysics*.191 - 27-53

Pontoise, B., Bonvalot, S., Mascle, J. and Basile, C.. 1990. Structure crustale de la marge transformante de Côte-d'Ivoire-Ghana déduite des observations de gravimétrie en mer. *C. R. Acad. Sc. Paris*, 310: 527-534.

Popoff, M.. 1990. Deformations intracontinentales gondwanienne : rifting Mésozoïque en Afrique. Thesis Univ. Marseille III, 358 pp.

Popoff, M., Raillard, S., Mascle, J., Auroux, C., Basile, C., and Equamarge group. 1989. Analyse d'un segment de marge transformante du Ghana : résultats de la campagne Equamarge II (Mars 1988). *C. R. Acad. Sc. Paris*, 309, 481-487.

Ranalli, G. and Murphy, B.C., 1986. Rheological stratification of the lithosphere. *Tectonophysics*, 132,281-295

Sage, F., 1994. Structure Crustale d'une Marge Transformante et du Domaine Océanique Adjacent : exemple de la marge de Côte-d'Ivoire-Ghana. *Ph D Thesis*, Université Paris VI, 339 p.

Scotese, C.R., Gahagan, L.M. and Larson, R.L., 1988. Plate tectonic reconstructions of the Cretaceous and Cenozoic ocean basins. *Tectonophysics*, 155: 27-48.

Scrutton, R.A., 1976. Crustal structure at the continental margin of South Africa. Geophys. *J. R. Astron. Soc.*, 44, 601-623.

Scrutton,, R.A., 1979. On sheared passive continental margins. *Tectonophysics*, 59, 293-305.

Scrutton, R.A., 1982. Crustal structure and development of sheared passive continental margins. In : R.A. Scrutton (Editor), Dynamics of passive margins. *Geodynamics Series*, 6. 133-140.

Shaw, P.R. and Cande, S.C.1990.High-resolution inversion for South Atlantic plate kinematics using joint altimeter and magnetic anomaly data. *Jour. Geophysi. Res.* 90, 2625-2644

Todd, B.J., Reid, I.and Keen, C.E., 1988. Crustal structure across the Southwest Newfoundland transform margin. *Can. J. Earth Sci.*, 25, 744-759.

Todd, B.J. and Keen, C.E. 1989. Temperature effects and their geological consequences at transform margins. *Can. J. Earth Sci.*, 26, 2591-2603.

Zalan, P.V., Nelson, E.P., Warme, J.E. and Davis, T.L., 1985. The Piaui basin: rifting and wrenching in an Equatorial Atlantic transform basin. In : K.T. Biddle and N. Christie-Blick (Editors), *Strike-slip deformation, basin formation and sedimentation. N. Soc. Econ. Paleont. Miner., Spec. Publ.*, 37. 143-158.

Backarc Evolution and Ophiolite Formation

Hajimu KINOSHITA
Earthquake Research Institute
University of Tokyo
Yayoi, Bunkyo, Tokyo, 113 Japan

ABSTRACT. Data from the Japan Sea, between Japan and the Siberian Continent, and from a part of the Ophiolite belt along the southern rim of Japan are presented to discuss a possible relationship between the opening of the Japan Sea and the formation of the Ophiolite belt. It is speculated that the Ophiolite belt was formed by slicing of the lithosphere and uplift of serpentinized mantle peridotite in Miocene times. Slicing of the lithosphere was induced by the opening of the Shikoku Basin (a part of the Philippine Sea Plate) shortly after the opening of the Japan Sea.

1. Introduction

The northwestern Pacific is characterized by the presence of a number of backarc basins along the Eurasian continental margin. One of the most intensively studied basins is the Japan Sea (Figure 1) between Japan and Eurasia. The opening of the Japan Sea and the occurrences of a conspicuous Ophiolite belt in Central Japan is believed to be connected in terms of their geologic and tectonic environments (e.g., Arai and Okada, 1991).

The formation scheme of the Japan Sea was first discussed by Terada (1934) using bathymetry. This author suggested that the basin was formed behind the drifting Japanese landmass off the Eurasian continent leaving banks and highs as remnant continental fragments. Since then, the Japan Sea and its surrounding area have been the subject of extensive geoscientific studies (e.g.,Tamaki, Pisciotto, Allan et al., 1990).

The magnetic anomaly patterns of this area were studied systematically by Isezaki (1986). Seama and Isezaki (1990) and Nakasa and Kinoshita (1994) reported weak but clear lineation patterns and apparent fracture zones accompanied by the magnetic lineation discontinuity in the northeastern corner of the Japan Basin.

Gravity data collection from the Japan Sea and its adjacent area was made from surface gravity data by Tomoda (1973). Yoshii (1979) made a compilation of gravity data along a line across entire Japan Sea region and suggested upper mantle heterogeneity. Teleseismic and surface wave study of this area revealed Moho depth (e.g., Evans et al., 1978) and a seismic anisotropy (Okada et al., 1978). Tamaki (1986) discussed a possible age for the formation of the Japan Sea by bathymetric depth. Kono and Furuse (1989) collected gravity data around the Japanese Islands including part of the Japan Sea. Nakasa and Seno (1994) and Hirata (1994) discussed the gravitational stability of the lithosphere of the Japan Sea.

Heat flow data were accumulated extensively (e.g.,Yoshii, 1972). These data were used to consider the thermal regime of the upper mantle. Tamaki (1986) estimated the age of the Japan Sea using heat flow data (e.g., Kuramoto, 1989). It is noted that the Japan Sea can be divided into three regions based on average heat flow values and that the western part of the Japan Sea is the youngest. This fact has led to a speculation that the Japan Sea was formed passively under the influence of shearing forces between the Eurasian Plate and the Pacific Plate due to obliquity of their relative motion (Jolivet et al., 1991). Results of ODP (Ocean Drilling Program) Legs 127/128 (Tamaki, Pisciotto and Allan et al., 1990; Ingle, Suyehiro and Breymar et al., 1990) imply that the Japan Sea was formed as early as 20-25 Ma BP.

Seismic velocity structure of the crust of the Japan Sea is one of the key parameters for understanding the tectonic evolution. The seismic structures of the Japan Sea were studied by

E. Banda et al. (eds.), Rifted Ocean-Continent Boundaries, 341–354.

342

a number of research groups (e.g., Murauchi, 1966 and 1972; Ludwig et al., 1975; Hirata et al., 1987; Tokuyama et al., 1987; Katao, 1988; Chung et al., 1990; Karp et al., 1992; Hirata et al.,1992; Bikkenina et al., 1994; Hirata, 1994). The Japan Sea consists of a combination of typical oceanic crust mainly in the Japan Basin (northern part) and a number of blocks of continental crust under premature extension. The tuffaceous layer is considered to be distributed widely in the entire region of the Yamato Basin (southern part) implying an occurrence of powerful volcanic activities in the past.

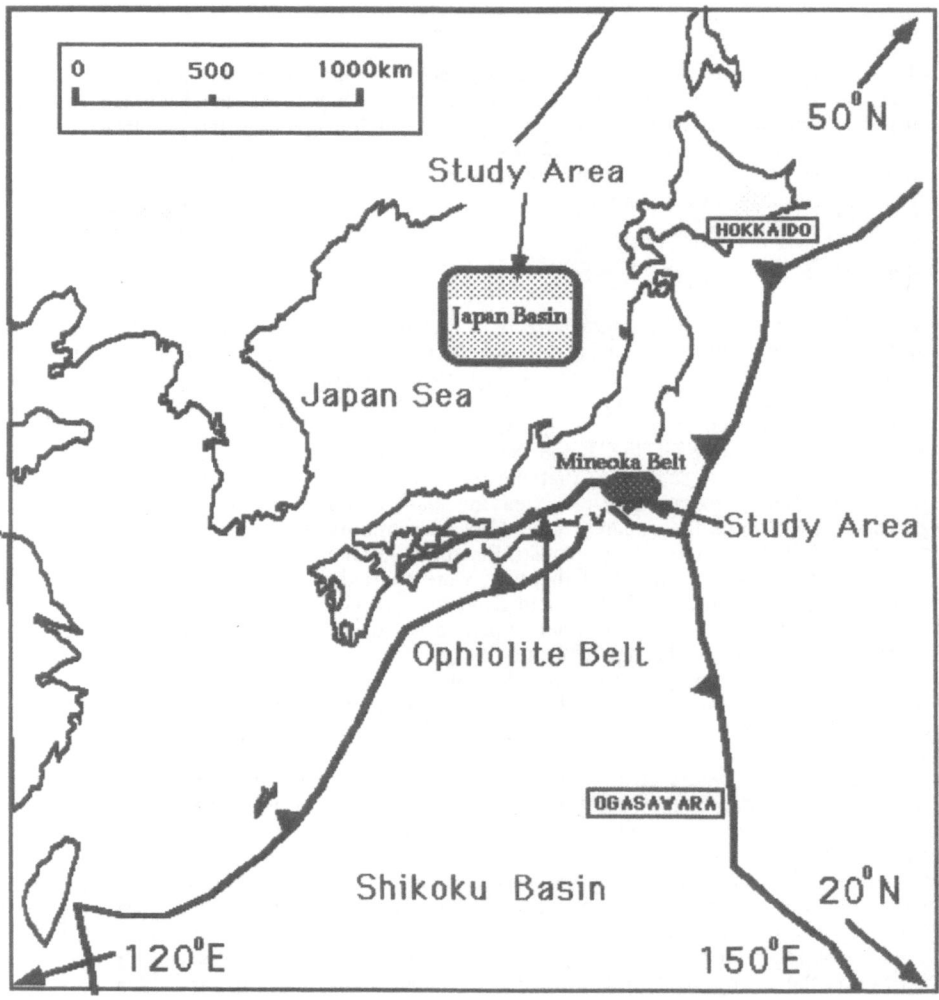

Figure 1.- An index map showing approximate positions of the Japan Sea, magnetic study region, and the Mineoka Ophiolite. Thick curve on the southern rim of the Central Japan is an outline of outcrops of the Ophiolite belt. Trench and trough systems are shown by solid curves with triangles.

Magnetic characteristics of the basement layers of the oceanic lithosphere is studied using ODP Hole 504B (e.g., Kinoshita et al., 1989). The mechanical properties change almost in a parallel fashion and there is a number of detachment faults at its deepest part (e.g., Alt, Kinoshita, Stokking et al., 1993). It is inferred that the basement magnetic layer could be scraped off by tectonic shearing forces such as accretion due to subduction (Kinoshita and Matsuda, 1989).

Magnetic structures of the Mineoka Ophiolite belt (Figure 1) in the southern part of Japan suggest a shallow burial of a thin and elongated magnetized body. The Ophiolite belt consists of serpentinized ultrabasic rocks and their brecciated cumulates. Its main body is restricted to a narrow band of 15 km. Geology (e.g., Arai and Okada, 1991) shows that the Mineoka body intruded along the strike slip fault zone between the Shikoku Basin, the northern part of the Philippine Sea plate, and the Japan landmass in conjunction with a large tectonic development, such as the opening of the Japan Sea followed by the opening of the Shikoku Basin. Mineoka Ophiolite body is buried within the upper part of the crust and causes a geomagnetic anomalies in the region (e.g., Kinoshita et al., 1994). Arai and Okada (1991) suggested that the Mineoka body was formed by accretion of serpentine diapirs in the forearc region. Similarly, Soh et al.(1991) proposed that the Mineoka belt was produced by tectonic fragmentation of the crust of the Izu-Bonin arc by arc-arc collision. These studies were supplemented by geophysical data on faults (Research Group for active faults of Japan, 1991), gravity (Kono and Furuse, 1989), and seismic structure (Asano et al., 1979).

Magnetic field surveys were carried out mostly using airborne proton magnetometers (e.g., Geological Survey of Japan, 1980; Jica, 1983; Nakai et al., 1987). It is suggested that the thickness and depth of the Mineoka Ophiolite belt is small (e.g., Kinoshita et al., 1994). It is likely that the wide low of the magnetic field is produced by the deeper part of the basement beneath the Mineoka hill.

2. The Japan Sea

2.1 GRAVITY AND HEAT FLOW

The entire region of the Japan Sea is isostatically compensated. Free air gravity anomaly in the Japan Basin traces exactly the bathymetric topography except around conspicuous seamounts. However, the Residual Gravity Anomaly implies that the upper mantle beneath this region consists of less dense materials (Yoshii, 1972). Heat flow value distribution shows insignificant variability within the entire basin area of the Japan Sea with exceptionally high heat flow values at around the boundary between the Japanese landmass and the Japan Basin where the shallow earthquakes and fault lines are observed. The gravitational and heat flow stability implies a thermal and dynamic equilibrium of the shallow part of the upper mantle (Nakasa and Seno, 1994; Hirata, 1994).

2.2 UPPER CRUST AND SEDIMENTARY LAYERS

Several measurements and maps on sediment thickness were produced (e.g., Tamaki, 1988). The basement contours were drawn by solving finite difference equations (Akima, 1970, 1974; Briggs, 1974) after applying empirical depth corrections to the sound velocity (Ludwig et al.,1975). The surface of the basement feature of the Japan Basin is fairly rugged covered by sediments as thick as 3 km and there is a number of isolated seamounts (Nakasa and Kinoshita, 1994).

Seismic structure of the transient region between Siberia and the Japan Basin was studied by Bikkenina et al. (1994). The author states; "It is suggested that the basin was formed in connection with stretching and destruction of the continental crust". The transition shows a thinning of the continental crust gradually into the Japan Basin structure.

Average thickness of the Japan Sea lithosphere was studied by surface waves (e.g., Evans et al., 1978). Marine seismological surveys were run by many works (e.g., Hirata et al., 1992; Hirata, 1994). The crust consists of three oceanic layers: sedimentary layer, layer 2 (3.3-5.9 km/s), and layer 3. The Moho surface is almost flat in the eastern part and slightly deeper in

344

the western part (Yoshii, 1972; Hirata et al., 1992). The surface of the sediment layer is almost flat in contrast to the rugged topography of the basement.

The Yamato Basin to the south of the Japan Basin has a semi-continental crustal structure and its crust is twice as thick (14 km) as that of the Japan Basin on average (e.g., Chung, 1992; Hirata, 1994). These data allow us to draw a two dimensional cross section of the lithosphere from the continent (Siberia) over the backarc basin (Japan Sea), and from the island arc (Japan) to the trench (Japan or Izu Ogasawara) systems (Figure 2).

2.3 AGE

The formation ages of the Japan Sea, estimated by different approaches, range from 10 to 30 Ma (e.g., Kaneoka et al., 1990; Tamaki et al., 1990). Most of these age estimates also refer to various data sources from land masses around this region. The ODP Leg127 reached basement volcanic layers which give the radiometric ages as 15-19 Ma (e.g., Kaneoka et al., 1990). It is debatable whether these layers reached by ODP is part of the basement massive flow units or not. The basaltic rock sample from the Bogorov Seamount (indicated by B in Figure 3) gave K-Ar age 18 Ma (Sahno and Vasiliev, 1974).

The age of the formation of the entire Japan Sea is not clearly defined yet. It should be noted however, that the formation of the Japan Basin occurred earlier than 20 Ma before the formation of the Yamato Basin. For instance, the last rifting episodes of the Yamato Basin occurred at around 15 Ma (e.g., Tamaki et al., 1990), significantly later than the formation of the Japan Basin. Indirect evidence of the opening of the Japan Sea is provided by the bending of the Japanese Island arc at about 19-20 Ma (e.g., Otofuji and Matsuda, 1983).

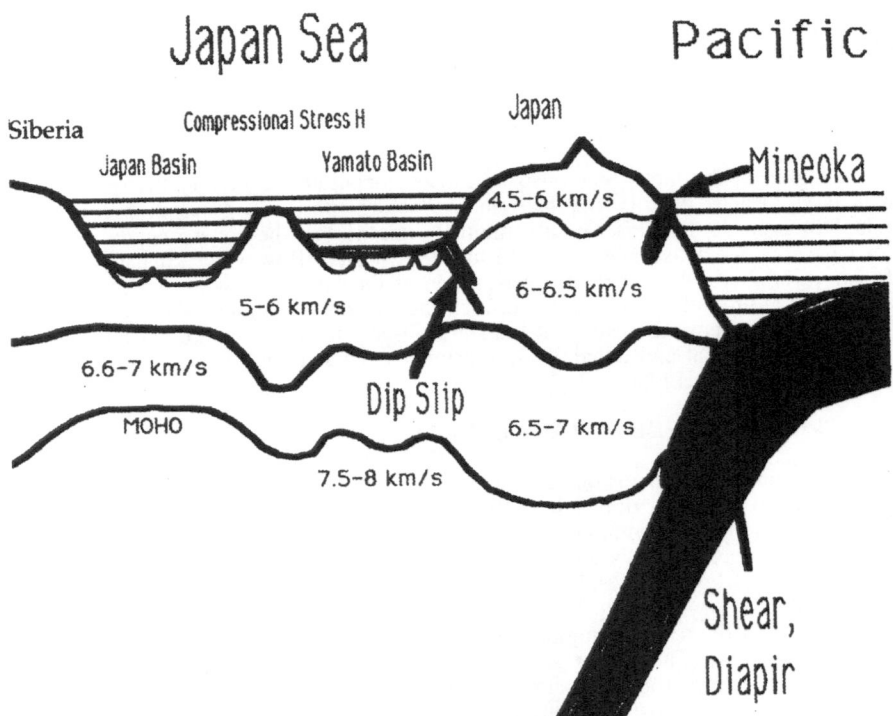

Figure 2.- A cross section of a continent-backarc basin-island arc-trench system based on the data from Yoshii (1972), Hirata et al. (1992), Shinohara and Suyehiro (1992) and Bikkenina et al. (1994). Representative seismic velocity values are given in km/s.

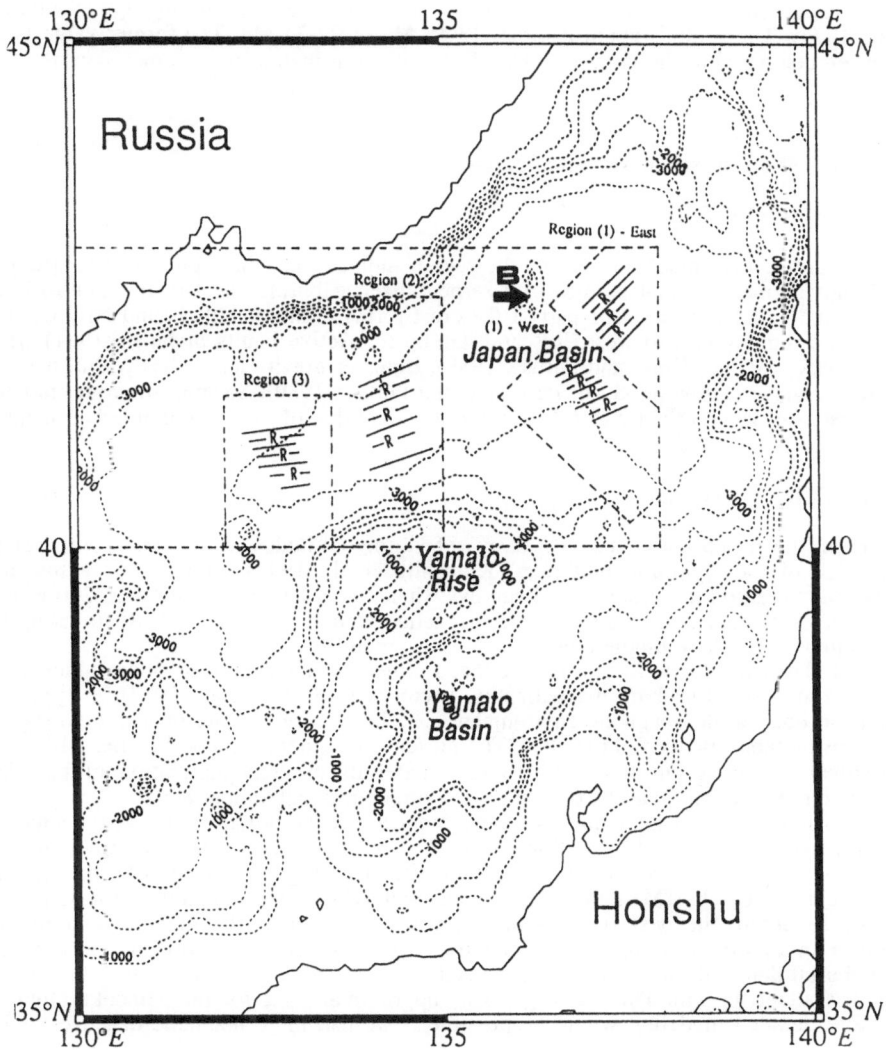

Figure 3.- An index map of magnetic the anomaly of the Japan Basin. Solid lines are trends of the magnetic lineation patterns, and Positive and Negative (R) magnetic anomalies in three rectangle regions (dotted line) correspond to the Study Area of Figure 1. The entire area is divided into three parts on the basis of three different directions of the magnetic lineations. A character B with a thick arrow shows the location of the Bogorov Seamount.

Clear magnetic lineation patterns are observed only in the Japan Basin (Figure 3). The geomagnetic data were used to estimate the age of the opening of the Japan Basin. The analysis was restricted to a comparatively small rectangle (e.g., Kinoshita et al., 1994)

The model simulation was performed to identify anomaly chron (Harland et al.,1989). There was no unique combination among parameters due to their trade off. The alignment of magnetic anomalies is shown in Figure 3; eastern part region 1, age 22-23 Ma, or 23-24 Ma; western part of region 1, age 16-18 Ma. A dredge haul rock sample from the Bogorov

Seamount located in the western part of region 1 gave a compatible absolute age (Sahno and Vasiliev, 1974). The central region is either 13-15 Ma or 22-24 Ma. The former value is too young when compared to the latter estimate. In the western region the basement age is 13-15 Ma.

3. Ophiolite belt on the southern rim of Japan

3.1 CRUSTAL FEATURES

Tectonic and petrographical studies of the region indicate that the Ophiolite is a group of isolated and fragmented small bodies of a serpentinized gabbroic sheet of the Cenozoic era (e.g., Arai and Okada, 1991). This part of the crust forms a line of hills bound on both sides by fault lines running east to west (Research Group for Active Faults in Japan, 1991) which are seismically inactive (Kono and Furuse, 1989). Bouguer gravity anomalies within the fault region reveal value as low as -30 mGal (Kono and Furuse, 1989). Seismic structure indicates that the basement layer velocity (5-6 km/s) is lower than that of the surrounding basement by about 8-10 % (Asano et al., 1979).

3.2 MAGNETIC ANOMALY

The ground level and aeromagnetic surveys of various altitudes were carried out over the eastern edge of the Ophiolite belt (e.g., Kinoshita et al., 1994). Figure 4 is a magnetic anomaly map based on the ground level surveys where wave lengths shorter than 200 m have been cut off. A characteristic feature of the magnetic anomaly field is a NW-SE trending low at high altitudes and sharp dipole type anomalies at ground level.

A low of the magnetic field can be produced either by a magnetized body which has a nearly horizontal south seeking magnetization vector (e.g., Nakai et al., 1987) or by a large nonmagnetic body in the magnetic environment. The latter case is more plausible taking into account seismic and gravity field studies. The presence of a nonmagnetic basement body can be explained by either burial of sedimentary materials or demagnetization of basement formations due to tectonic reasons (brecciation, hydrothermal alteration, etc.).

A 3-dimensional inverse modeling was carried out to fix the initial model. After removing long wavelength magnetic field variations, a block model was used for forward modeling. The modeling was achieved by assuming a structural situation (Figure 5) where the position and extent of central blocks (MB: magnetic block, and NMD: non-magnetic depression) are readjusted to obtain the best fit. A schematic cross section of the magnetic anomaly is reproduced in Figure 6 along an artificial track line. According to this model the most probable burial depth of the top face of the magnetized prism is between 0.5 to 1 km and the thickness is about 0.5 km. This value gives an approximate size for the Mineoka Ophiolite body. More detailed modeling would be possible if we had more magnetic data from wider regions.

4. Discussion

The speculation on the formation of the Japan Sea is mainly based upon geometrical considerations related to the number of apparent continental fragments detached from the coast of Siberia (left side of Figure 7). The splitting of the continental blocks was followed by ocean floor spreading due to external shear forces (right side of Figure 7) and then by the crustal stretching of the southern part of the Japan Sea.

A mantle return flow induced by subducting the Pacific Plate led to anatexis, dehydration, upwelling of partial melt, and to the formation of wet magma sources, or to a plume from dry partial melt from subducted slabs (e.g., Hasebe et al., 1970; Tatsumi, 1989). The wedge flow is probably governed by thermal energy rather than mantle drag. The mobility of the light elements is discussed in the light of hydrofracturing processes as to reduces the effective viscosity (e.g., Nakashima, 1993) which is a major factor of a high mobility of fluids within the mantle.

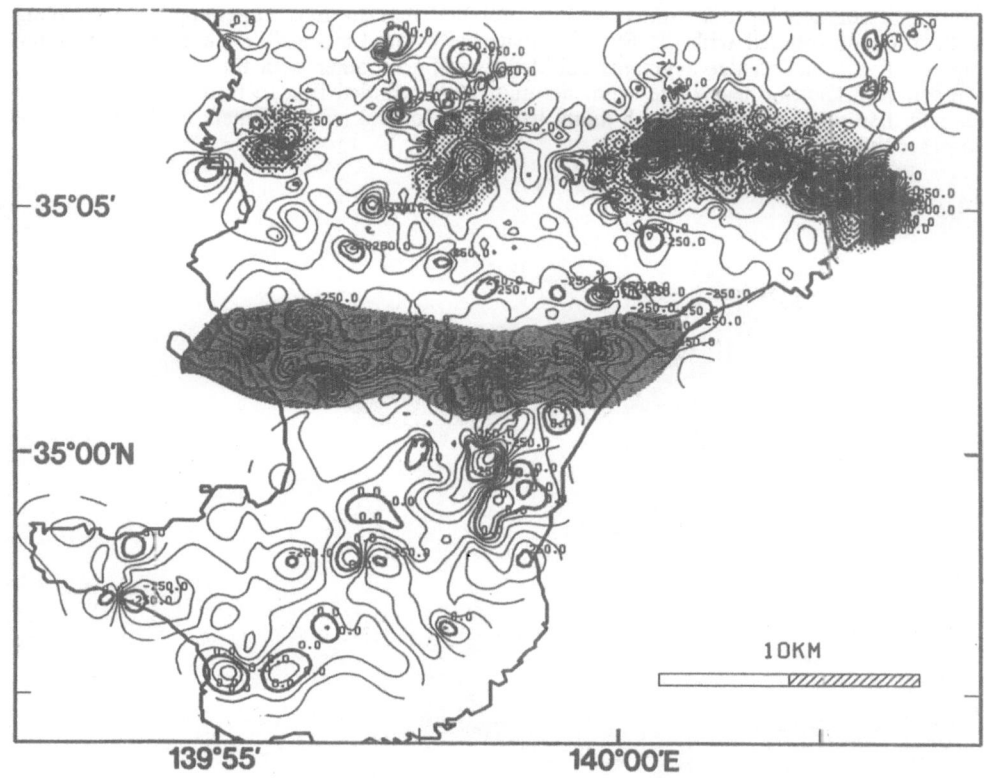

Figure 4.- A magnetic anomaly map of the Mineoka Ophiolite belt region measured at ground level. Contour interval is 40 nT. Lightly dotted areas indicate magnetic highs and dark mesh areas indicate magnetic lows.

A retreat of the subducting lithosphere due to the differential motion in the upper and lower part of the asthenosphere may be possible. The upper part of the mantle beneath the Japan Sea is unique in a way that the mantle material over the subducting Pacific Plate consists of lighter materials (Yoshii, 1979). This implies either that the mantle beneath the Japan Sea has comparatively high temperatures or that it consists of light materials. The high temperature scheme seems to be advocated by the distribution of fairly high heat flow values in the Japan Sea compared to the cooling plate model. Volcanic activity indicates that the variation of the geometry of the mantle return flow shifts back and forth. It is difficult to maintain a persistent return flow in a wide area for a long period of time. Hot spot upwelling is unlikely because the subduction and deep rooted hot spot must coexist.

Considering a persistent magma upwelling in an area comparable to the Japan Basin, we figure out one of the mechanisms of the backarc basin formation as induced by collision of a ridge system with a continental fringe. It is speculated here that collisions of a number of active ridges of the Kula Plate took place for the following reasons although the time of the consumption of the spreading center is little known. From the north near Hokkaido to the south near Ogasawara it stretches over 2000 km from 110 to 180 Ma (e.g., Nakanishi et al.,1989). It is estimated that the half spreading rate of the plate is 30-40 mm/year. If this rate was persistent from 0 to 110 Ma, the stretched 3000 km part would be consumed by subduction in the latest 30 Ma. A number of backarc basins along western Pacific were

348

formed between 60-20 Ma ago. Their occurrences are interconnected although it is hard to show their relationship quantitatively. It is assumed that an active opening center moving northwestward collided with the Eurasian Plate and caused a drastic change in the tectonic scheme. The oceanic ridge segments collided one after another with the continental fringe. The southern and northern ridge segments must have collided first to form the South China Sea and then the Bering Sea which might have been formed by entrapment of the Kula plate (Marlow and Cooper, 1983).

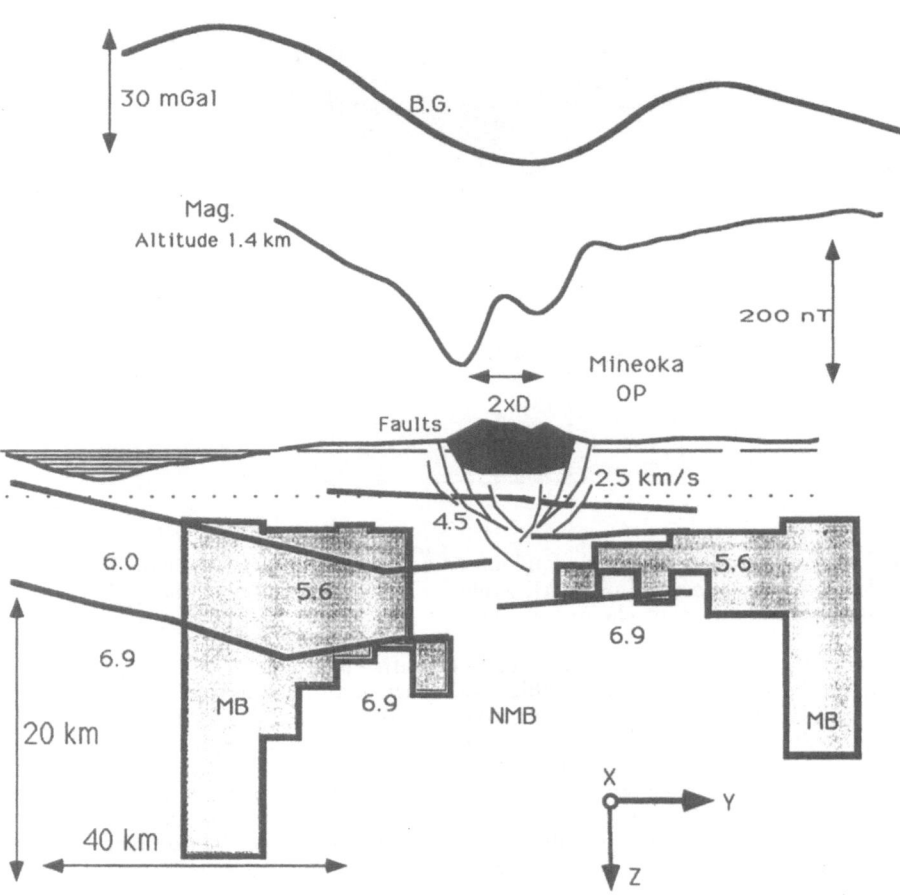

Figure 5.- A schematic two-dimensional cross section model of the tectonic and structural configuration of the Mineoka Ophiolite belt and its adjacent area. Seismic velocities (km/s) of the crust are from Asano et al. (1979), Bouguer Gravity (BG) curve from Kono and Furuse (1989), ultrabasic emplacement (a dark block near surface in the center) from Arai and Okada (1991), and fault planes (dipping lines from surface) from Research Group for Active Faults of Japan (1991). Dotted sections (MB) are magnetized prism (J_{NRM} = 1.5 A/M) parallel to the local geomagnetic field. Blank region (NMD) is nonmagnetic due to some tectonic or depositional environments.

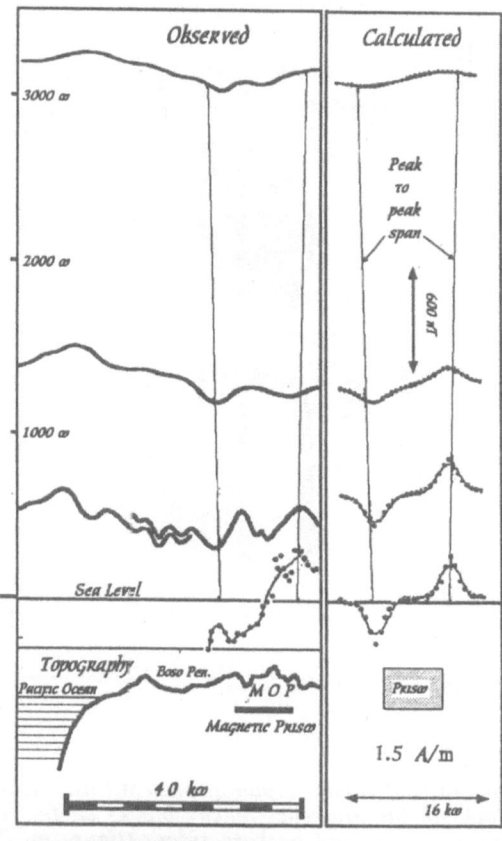

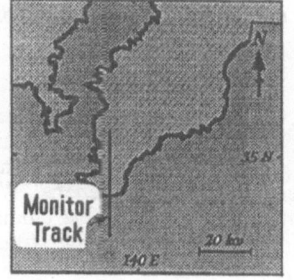

Disturbance
from a
small scale
buried body

Figure 6.- Summary of the magnetic anomaly along a Monitor Track Line (bottom inset) equal to a full extent of the horizontal axis of the top left figure. Topography, and composite magnetic field profiles at various altitudes (ordinates) along the monitor track line are shown. From top to bottom on the left; magnetic field cross section of 3,150 m ; profile at 1,350 m (both from Nakai et al., 1987); profile at 430 m altitude (Utashiro and Kono, 1972; Geological Survey of Japan, 1980); ground level magnetic field (Kinoshita et al., 1994). Altitude is given in m, horizontal distance in km. Right side shows altitude dependence of total force anomaly from a two-dimensional single block of a magnetized body (Jn = 1.5 A/m) extending perpendicular to the face of this page. Relative value of the magnetic total force of 600 nT is given by vertical thin arrows on top of the figure. Width of the buried body is 8 km. Thickness of the body is about 500 m. Magnetic anomaly fields at various altitudes are shown. A pair of thin funnel shaped vertical lines in the left figure is a copy of those lines connecting between peaks of the anomaly on the right figure.

A cross section of the upper mantle along a line across islandarc-basin-continent is drawn in Figure 2 (Yoshii, 1972; Hirata et al., 1992; Shinohara and Suyehiro, 1992; and Bikkenina et al., 1994) to discuss the formation of backarc basin and intrusion of Ophiolite in the forearc region. A two-dimensional feature of the evolution of this system can be proposed with the following assumptions: (1) Subduction of the oceanic lithosphere by slab pull. (2) Liquid squeezed out of the subducting slab makes lower solidus of asthenospheric materials. (3) Differentiation of magma in the isolated reservoir. A simulation study on stretching, break-up of continental lithosphere followed by ocean floor spreading was performed by Takeshita and Yamaji (1990) to give more realistic basin formation history.

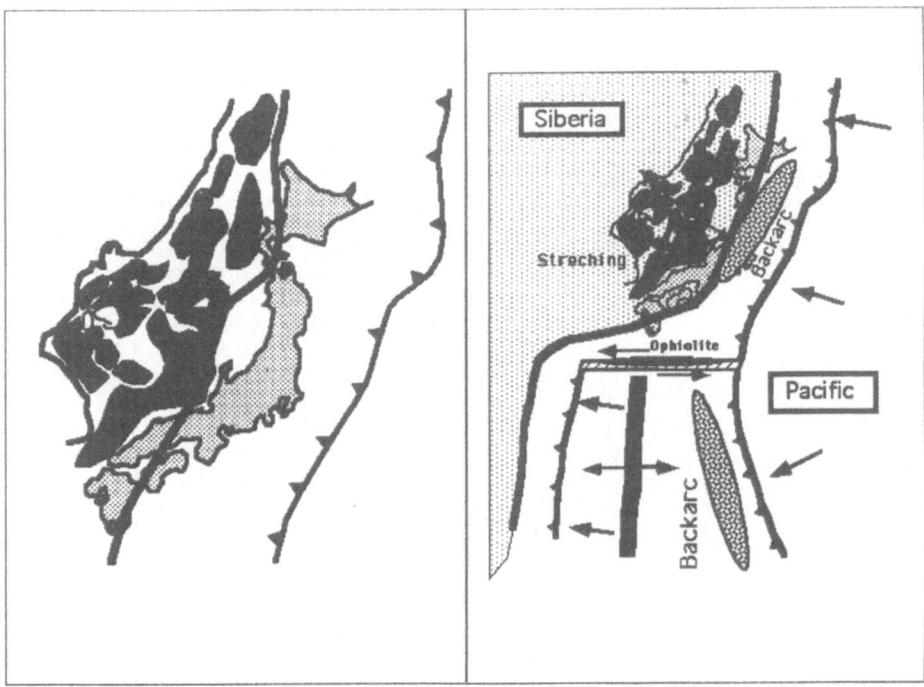

Figure 7 Distribution of continental fragments around Japan. Japan at present (gray dotted part on the left figure) and fragmented blocks in the Miocene (patchy blocks on the left figure), with number of tectonic lines; trenches, faults and backarc ridges (figure on the right), activated by tectonic development in accordance with the Japan Sea and the Shikoku basin openings (modified from Arai and Okada, 1991).

The big events occurred during the Japan Sea development. The age can be determined by the magnetic anomaly of the Japan Basin. The formation age of the Japan Basin reexamined by model experiments gives 13-24 Ma for the opening of the Japan Basin. Radiometric ages, estimates from heat flow and bathymetric depths range between 18-26 Ma in general. While the Japan Sea was formed, a megashear motion on the southern rim of the Japanese landmass took place (e.g., Arai and Okada, 1991) and diapir occurred along this shear zone (right of Figure 7).

Basement layers of the oceanic lithosphere is covered by hydrothermally altered basaltic and ultrabasic rocks. The magnetism of the basement rocks decays at depth 1 km from their surface (e.g., Kinoshita et al., 1989). The mechanical properties change in a similar fashion as observed by changes in sound velocity of basement formations (e.g., Aali, Kinoshita and Stokking et al., 1993). The basement magnetic layer can be easily scraped off by accretion and subsequent underplating (Kinoshita and Matsuda, 1979). The accreted part of the ultrabasic materials may play a major role in diapirism of Ophiolite bodies.

The amplitude of the magnetic anomalies observed at ground level decreases rapidly at higher altitudes implying that the entire size of the Mineoka Ophiolite body is small and confined to a narrow band. A simplified two-dimensional crustal model of ultrabasic bodies around the Mineoka belt indicates that the dimension of the intrusive mass is no thicker than 0.5 km and its horizontal extent is about 8 -10 km. Magnetic intensity of this body is significantly stronger than that of the surrounding formations, which explains the broad band of the geomagnetic low extending further to the NW over the Boso Peninsula at higher altitudes. The present two dimensional model is sufficiently simplified to show only an approximate dimension of the buried body. The root of the diapir had to be much deeper when it intruded. The rest of the diapir sank back deep into the upper mantle leaving only a topmost part of the intruded body. The three-dimensional forward modeling can only be obtained with detailed information of the entire Mineoka area. The Mineoka Ophiolite body is probably floating in non-magnetic environments. This implication is interesting to understand the tectonic development of these ultramafic bodies.

5. Summary

The trends of magnetic anomaly lineations in the eastern, central, and western areas of the Japan Basin are slightly different in their orientation of alignments. The age, spreading rate, thickness of magnetic basement, and magnetic intensity values seem to differ from the eastern to the western areas of the basin. The easternmost part (region 1; Figure 3) indicates that the the age of study region is around 22-24 Ma whereas to the west the age is 16 to 18 Ma . The central region has two possible ages; 13-15 or 22-24 Ma. The westernmost region was obviously formed at around 13-15 Ma, the youngest age of the whole basin. All these values cover a similar range comparable to other age estimates based on radiometric dating, bathymetric features and heat flow values. Although the magnetic modeling does not show very clear features of the Japan Basin formation and evolution, it seems likely that the opening of the Basin in the eastern and central parts occurred almost at the same time. These opening centers propagated westward, with varying migration speeds during 24 and 13 Ma. Diapir in the forearc region may have occurred in conjunction with the Japan Sea opening. Ground magnetic surveys delineate the dimension of the intrusive bodies. It is concluded that the size of the diapir located on land is comparatively small 0,5 km x 8 km in its cross section.

Acknowledgments

The author would like to express thanks to Prof. T W. C. Hilde, Prof. S. Uyeda, Texas A & M University, and Dr. E. Kikawa, Geological Survey of Japan, for their critical arguments and suggestions. The authors is indebted to Prof. K. Suyehiro, Ocean Research Institute, Prof. N. Hirata, Earthquake Research Institute for providing data and scientific comments. This program was partly supported by the Grant-In-Aid for International Scientific Research Program, Ministry of Education, Science and Culture.

References

Akima, H., 1970, A new method of interpolation and smooth curve fitting based on local procedures, J. Acm (Association for computing machinery), 17, 589-602.

Akima, H., 1974, A method of bivariate interpolation and smooth surface fitting based on local procedures, Comm. Acm., 17, 18-20.

Alt, J., H. Kinoshita, L. Stokking and Shipboard Scientific Party, 1993, Costa Rica Rift. Proc. ODP (Ocean Drilling Program), Init. Repts.148 , 27-122.

Arai, S. and H. Okada, 1991, Petrology of serpentine sandstone as a key to tectonic development of serpentine belts, Tectonophysics, 195, 65-81.

Asano, S., Y. Ichinose, I. Hasegawa, S. Iizuka, and H. Suzuki, 1979, Crustal structure in the southern Kanto district derived from explosion seismic data, J. Seismol. Soc. Japan, 2-32, 41-55.

Bikkenina, S.K., E.G. Zhiltzov, K.F. Sergeev, and V.N. Solivjev. New Seismic structural model of earth's crust in the northern part of of the Japan Sea. J. Phys. Earth. (submitted).

Briggs, I. C., 1974, Machine contouring using minimum curvature, Geophysics, 39, 1, 39-48.

Chung, T.W., N. Hirata and R. Sato, 1990, Two-dimensional P-and S-wave velocity structure of the Yamato Basin, the southeastern Japan Sea, from refraction data collected by an ocean bottom seismographic array, J.Phys. Earth, 38: 99-147.

Chung, T.W., 1992, A quantitative study of seismic anisotropy in the Yamato Basin, the southeastern Japan Sea, from refraction data collected by an ocean bottom seismographic array, Geophys. J. Int., 109, 620-638.

Evans, J.R., K. Suyehiro and I.S. Sacks, 1978, Mantle structure beneath the Japan Sea - A re-examination, Geophys. Res. Lettr., 5, 478-490.

Geological Survey of Japan, 1980, 1:200,000 Aeromagnetic charts.

Harland, W. B., R. L. Armstrong, A. V. Cox, L. Craig, A. G. Smiths, and D. G. Smith, 1989, A geologic time scale, Cambridge, 263pp., Cambridge Univ. Press.

Hirata, N., H. Kinoshita, K. Suyehiro, M. Suyematsu, N. Matsuda,T. Oichi, H. Katao, S. Koresawa, and S. Nagumo 1987, Report on DELP 1985 cruise in the Japan Sea, Part II. Seismic refraction experiment conducted in the Yamato Basin, Southeast Japan Sea, Bull. Earthq. Res. Inst.,Univ. Tokyo, 62, 347-365.

Hirata, N., B. Y. Karp, T. Yamaguchi, T. Kanazawa, K. Suyehiro, J. Kasahara, H. Shiobara, M. Shinohara and H. Kinoshita, 1992, Oceanic crust in the Japan Basin of the Japan Sea by the 1990 Japan-USSR expedition, Geophys. Res. Lett., 19, 2027-2030.

Hirata, N., Seismic crustal structure of the marginal basins in the western Pacific: New results by OBS observation. (This volume).

Ingle, J. C., K. Suyehiro, M.T. Von Breymann and Shipboard Scientific Party., 1990, Proc. ODP (Ocean Drilling Program), Init. Repts., Scientific Results, Vol. 127/128, Pt. 2, 849-859.

Isezaki, N., 1986, A magnetic anomaly map of the Japan Sea, J. G. Geomag. Geoelectr., 38: 403-410.

Jica (Japan International Cooperation Agency), 1983, 1:200,000 Aeromagnetic charts over Uraga Strait.

Jolivet, L., P. Huchon, J.P.Brun, X. Le Pichon, N.Chamot-Rooke and J.C. Thomas, 1991, Arc deformation and marginal basin opening: Japan Sea as a case study, J.Geophys. Res., 96, 4367- 4384.

Kaneoka, I., Y. Takigami, N. Takaoka, S. Yamashita and K. Tamaki, 1990, $^{40}Ar-^{39}Ar$ Analysis of Volcanic rocks recovered from the Japan Sea floor: constraints on the age of formation of the Japan Sea, in Proceedings of the Ocean Drilling Program, Scientific Results, Vol.127/128, Pt. 2, 819-836.

Karp, B.Ya, N. Hirata, H. Kinoshita, K. Suyehiro, V.V. Zdorovenin and V.N. Karnaukh, 1992, USSR-Japan seismic experiment in the Japan Sea, preliminary results, Pacific Geology, No.5, 138-147.

Katao, H., 1988, Seismic structure and formation of the Yamato Basin, Bull. Earthq. Res. Inst., 63: 51-86.

Kinoshita, H. and N. Matsuda, 1989 Seismic structure and geomagnetic anomaly in the Nankai Trough related to subduction of the Philippine Sea Plate, J. Geomag. Geoelectr., 41, 161-173.

Kinoshita, H., T. Furuda and J. Pariso, 1989, Downhole magnetic field measurements and paleomagnetism, Hole 504B, Costa Rica Ridge, Init. Repts. ODP Leg 111, 147-156.

Kinoshita, H., R. Morijiri, Y. Nakasa and T. Fujiwara, Backarc basins and ophiolite formations, J. Geography (submitted).

Kono, Y. and N. Furuse, 1989, Gravity anomaly map in and around the Japanese islands, Univ. Tokyo Press (with sheet maps).

Kuramoto, S., 1989, Age and mode of the Japan Sea Opening, a review. Month. Earth, 11:503-512.

Ludwig, W. J., S. Murauchi and R.E. Houts, 1975, Sediments and structure of the Japan Sea, Geol. Soc. Am. Bull., 87: 651- 664.

Murauchi, S., 1966, Explosion seismology, in Second Progress Report on the Upper Mantle Project of Japan (1965-1966), Natl. Committee for UMP, Sci. Council of Japan, pp.11-13.

Murauchi, S., 1972, Crustal Structure in the Sea of Japan from Seismic Exploration, Kagaku, 42: 367-375

Nakai, J., M. Komazawa and Y. Okubo, 1987, Distribution of gravity anomalies and aeromagnetic anomalies in the Kanto district, Chigaku-zasshi, 96-4, 185-200.

Nakanishi, M., M. Tamaki and K. Kobayashi, 1989, Mesozoic magnetic anomaly lineations and seafloor spreading history of the northwestern Pacific, J. Geophys. Res., 94, 15437-15462.

Nakasa, Y. and T. Seno, 1994 , Compensation mechanism of the Yamato Basin, the Japan Sea, J. Phys. Earth, 42, 187-195.

Nakasa, Y. and H. Kinoshita, 1994, A supplement to magnetic anomaly of the Japan Basin, J. Geomag. Geoelectr., 46-6, 481-500.

Nakashima, Y., 1993, Static stability and propagation of a fluid-fluid edge crack in rock: Implication for fluid transport in magmatism and metamorphism, J. Phys. Earth, 41, 189-202.

Okada, H., T. Moriya, T. Matsuda, T. Hasegawa, S. Asano, N. Kasahara, A. Ikami, H. Aoki, Y. Sasaki, N. Harukawa and K. Matsumura, 1978, Velocity anisotropy in the Sea of Japan as revealed by big explosions, J. Phys. Earth, 26, 491-502.

Otofuji, Y. and T. Matsuda, 1983, Paleomagnetic evidence for the clockwise rotation of southwest Japan, Earth. Planet. Sci. Lett., 62: 349-359.

Research Group For Aactive Faults of Japan, Active faults in Japan, Rev. edit., 1991.

Sahno, B. G. and B. I. Vasiliev, 1974, Basaltoids of the Japan Sea bottom. Ed. by N. P. Basilkovsky and B. Ya. Karp, Far East Sci., 52-55.

Seama, N. and N. Isezaki, 1990, Sea-floor magnetization in the eastern part of the Japan Basin and its tectonic implications, Tectonophysics, 181, 285-297.

Shinohara, M. and K. Suyehiro, 1992, Seismic structure and dynamics of the upper mantle of island arc - trench systems, Earth monthly, 14-6, 341-347.

Soh, W., K. T. Pickering, A. Taira and H. Tokuyama, 1991, Basin evolution in the arc-arc Izu collision zone, Mio-Pliocene Miura group, central Japan, Jour. Geol. Soc. London, 148, 317-330.

Takeshita, T. and A. Yamaji, 1990, Acceleration of continental rifting due to a thermomechanical instability, Tectonophysics, 181, 307-320.

Tamaki, K.,1986, Age estimation of the Japan Sea on the basis of stratigraphy, basement depth, and heat flow data, J. G. G., 38: 427-446.

Tamaki, K., 1988, Geological structure of the Japan Sea, Bull. Geol. Surv. Japan, 39-5, pp.269-365.

Tamaki, K., K.A. Pisciotto, J. Allan and Shipboard Scientific Party, 1990, Proc. ODP (Ocean Drilling Program), Init. Repts.127 , 844pp.

Terada, T., 1934, On bathymetrical features of the Japan Sea, Bull. Earthq. Res.Inst., 12: 650-656.

Tokuyama, H., M. Suyematsu, K. Tamaki, E. Nishiyama, S. Kuramoto, K. Suyehiro, H. Kinoshita, H. and A. Taira, 1987, Report on DELP 1985 cruises in the Japan Sea Part 3:Seismic reflection studies in the Yamato Basin and Yamato Rise Area, Bull. Earthq. Res. Inst., 62: 367-390.

Tomoda, Y., 1973, Maps of Free air and Bouguer gravity anomalies in and around Japan, Univ. Tokyo Press. Uyeda, S. and Vacquier, V., 1968, Geothermal and geomagnetic data in and around the Island Arc of Japan, Geophysical Monogr., 12: 349-366.

Utashiro, S. and T. Kondo, 1972, Aeromagnetic surveys over the southern part of Kanto District in Japan, Rep. of Hydrographic researches, 18, 1-12.

Yoshii, T., 1972, Terrestrial Heat Flow and Features of the Upper Mantle Beneath the Pacific and the Sea of Japan, J. Phys. Earth, 20: 271-285.

Yoshii, T., 1979, Compilation of geophysical data around the Japanese Islands-1, Bull. Earthq. Res. Inst., 54: 75-117.

SEISMIC CRUSTAL STRUCTURE OF THE JAPAN SEA: NEW RESULTS BY OCEAN BOTTOM SEISMOGRAPHIC OBSERVATIONS

NAOSHI HIRATA
Earthquake Research Institute, the University of Tokyo
Tokyo 113, Japan

EIJI KURASHIMO
Department of Earth Sciences, Chiba University
Chiba 263, Japan

ABSTRACT. A series of seismic refraction and reflection experiments was conducted in the Japan Sea located in the back-arc of the Japanese Island-arc-trench system (western Pacific). In September 1992, we carried out a survey in the southwestern part of the Japan Sea including the northern Tsushima Basin, the Kita-oki Bank, and the western Yamato Basin. The area contains a bathymetric high which is supposed to be a stretched continental fragment in the Japan Sea. Seventeen ocean bottom seismometers (OBSs) were deployed on a 110-km-long line in a north-south direction and a 210-km-long line in an east-west direction. We used explosives and airguns as seismic sources.

The crustal structure deduced in the Tsushima Basin is neither typically oceanic nor continental: the crustal thickness is about 13 km including a 1.5-km-thick sediment. The structure is similar to that found in the Yamato Basin. Along the east-west line crossing the bank area, the crustal structure varies horizontally. Beneath the Kita-oki Bank the crust is of a continental type and has a thickness of 23 km, whereas in the Tsushima and Yamato Basins the crust is 13 km thick.

We interpreted the Tsushima Basin structure as a transition from continental to oceanic crust. We compiled seismic crustal structures obtained in the Japan Sea and the Okinawa Trough to test a simple stretching model of formation of extensional basins. We found a linear relationship between crustal thicknesses and backstripped depth of the basement, indicating that the crust is in isostatic equilibrium. Subsidence due to a relative thinning of the crust is found to be in between what is expected from an initial subsidence and a total subsidence as deduced from McKenzie's model. The result is consistent with an age of about 20 Ma or more for the Japan Sea.

1. Introduction

A series of seismic refraction and reflection experiments was conducted in the Japan Sea and the Okinawa Trough to improve our understanding of the structure and evolution of back-arc basins, both of which are located in the back-arc of trenches in the western Pacific (Figure 1). The Japan Sea is supposed to be formed by a back-arc spreading due to subduction of the Pacific Plate beneath the Eurasian Plate [e.g., Tamaki, 1985; Jolivet *et al.*, 1989; Tamaki, Pisciotto, Allan *et al.*, 1990; Ingle, Suyehiro, von Breymnn *et al.*, 1990].

E. Banda et al. (eds.), Rifted Ocean-Continent Boundaries, 355–369.

356

Various types of crust were found in the Japan Sea [Ludwig *et al.*, 1975]. Only a part of the Japan Sea has a typical oceanic crust even though the area has experienced sea floor spreading. The northern part of the Japan Sea has an oceanic crust with a total thickness of about 8 km including the sedimentary layer [Hirata *et al.*, 1992]. In the central part of the Japan Sea the water depth is about 100 m. In the areas where bathymetry is less than 3000 m the crust exhibits a transitional characteristic from continental to ocean [Hirata *et al.*, 1987, 1989]. Since these areas have a thick sedimentary column and a thick crust, it is not easy to determine the entire crustal structure [Karp *et al.*, 1992]. Recent developments of ocean bottom seismometers (OBS) have revealed the details of the deep crustal structure.

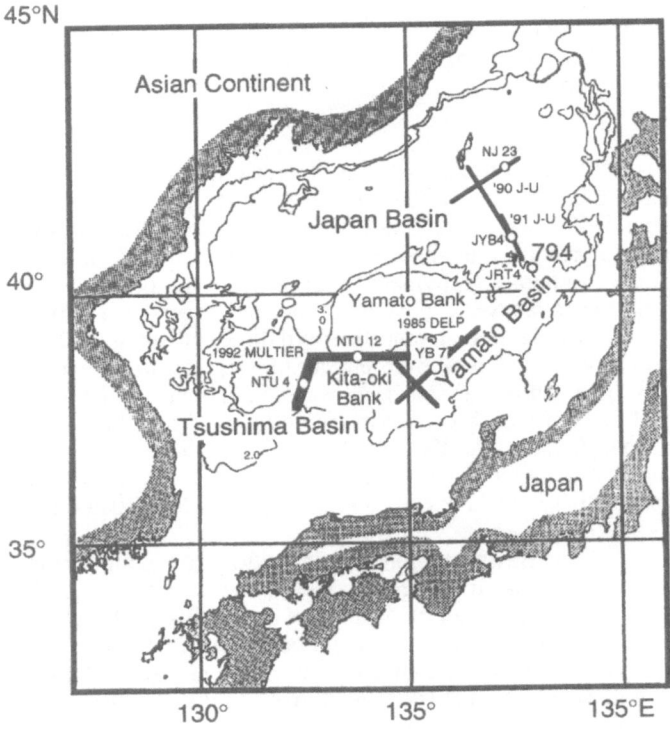

Fgiure 1. Location of recent seismic experiments in the Japan Sea.
Isobath is in km.

We summarize results of recent seismic source experiments using ocean bottom seismometers (OBSs). We will show here that the crust in the Japan Sea has a simple relationship in its thickness and depth to the basement indicating isostatic compensation in the area. We will discuss the applicability of a simple stretching model for the formation of an extensional sedimentary basin.

2. New seismic results from the 1992 experiment

We describe only the outline of the 1992 experiment and the data to discuss a seismic crustal structure in the southwestern Japan Sea. Detailed explanation of the survey and analysis of the data appear elsewhere [Kurashimo *et al.*, in preparation].

2.1 EXPERIMENT

In September 1992, we carried out a survey in the southwestern part of the Japan Sea including the northern Tsushima Basin, the Kita-oki Bank, and the western Yamato Basin (Figure 2) [Hirata *et al.*, 1993]. The area contains a bathymetric high which is supposed to be a stretched continental fragment in the Japan Sea. Seventeen ocean bottom seismographs (OBSs) were deployed on a 110-km-long line in a north-south direction (Line 1) and a 210-km-long line in an east-west direction (Line 2). We used a total of 3 t of explosives (4 x 200 kg, 110 x 20 kg) and airguns as seismic sources.

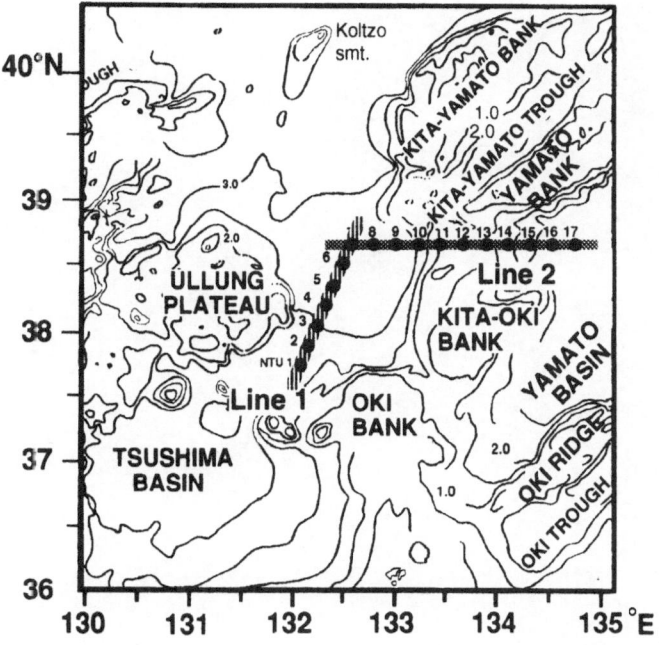

Figure 2. Location of the 1992 seismic experiment in the southwestern Japan Sea. Solid circles show the positions of the ocean bottom seismographs. Shaded lines are refraction/reflection lines.

2.2 DATA

We obtained three sets of seismic data from the experiment: single channel reflection data, airgun generated refraction data recorded by OBSs, and explosive source signals recorded by OBSs. We shot two 9-liter airguns every 40 s spaced at 100 m. We recorded airgun signals by the OBSs up to an offset distance of more than 40 km (Figure 3).

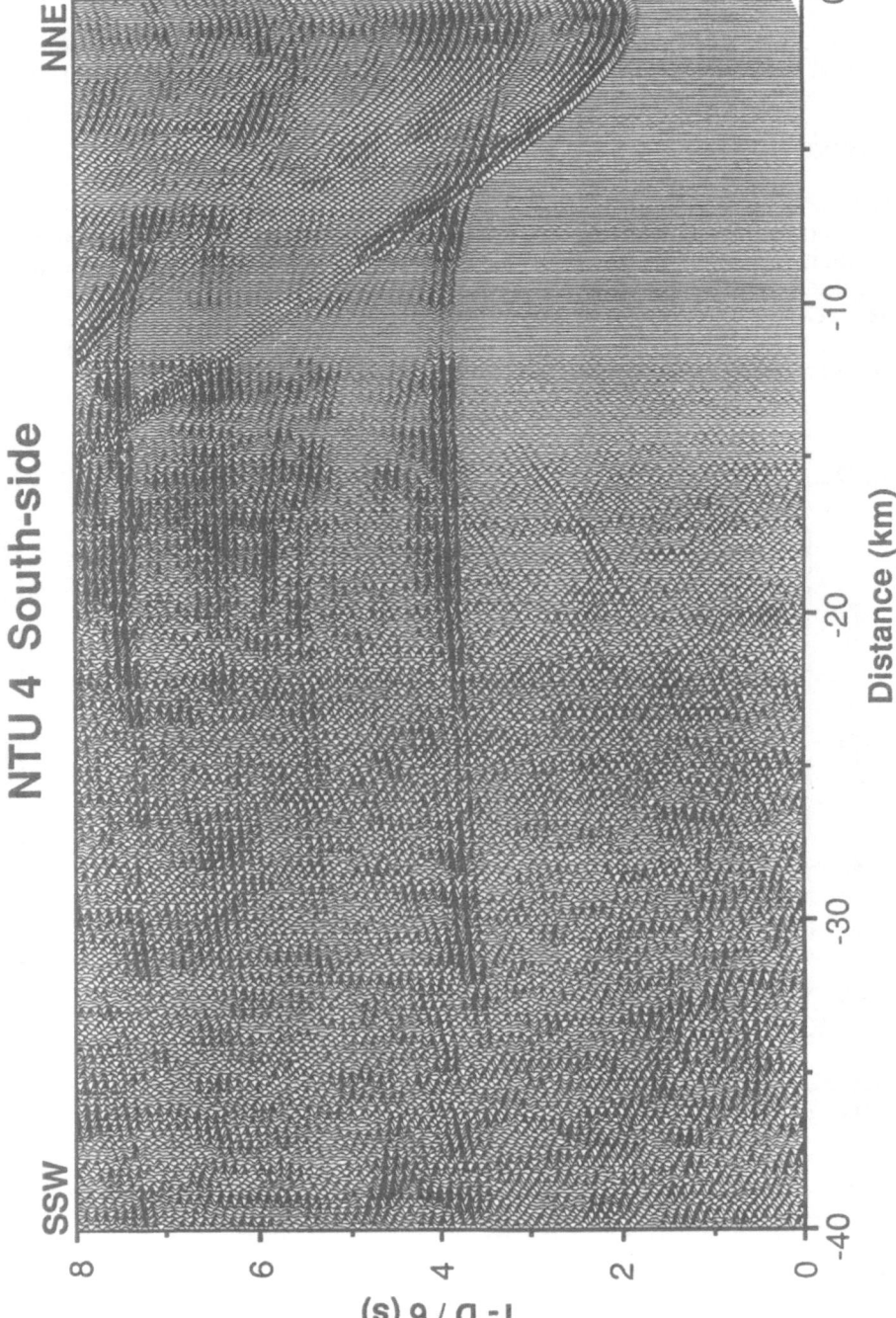

Figure 3. Example of seismograms for airgun shooting recorded by OBS NTU4 on Line 1. Vertical component of band-pass (5 -25 Hz) filtered signals is shown. Amplitudes are normalized by maximum amplitude of each trace.

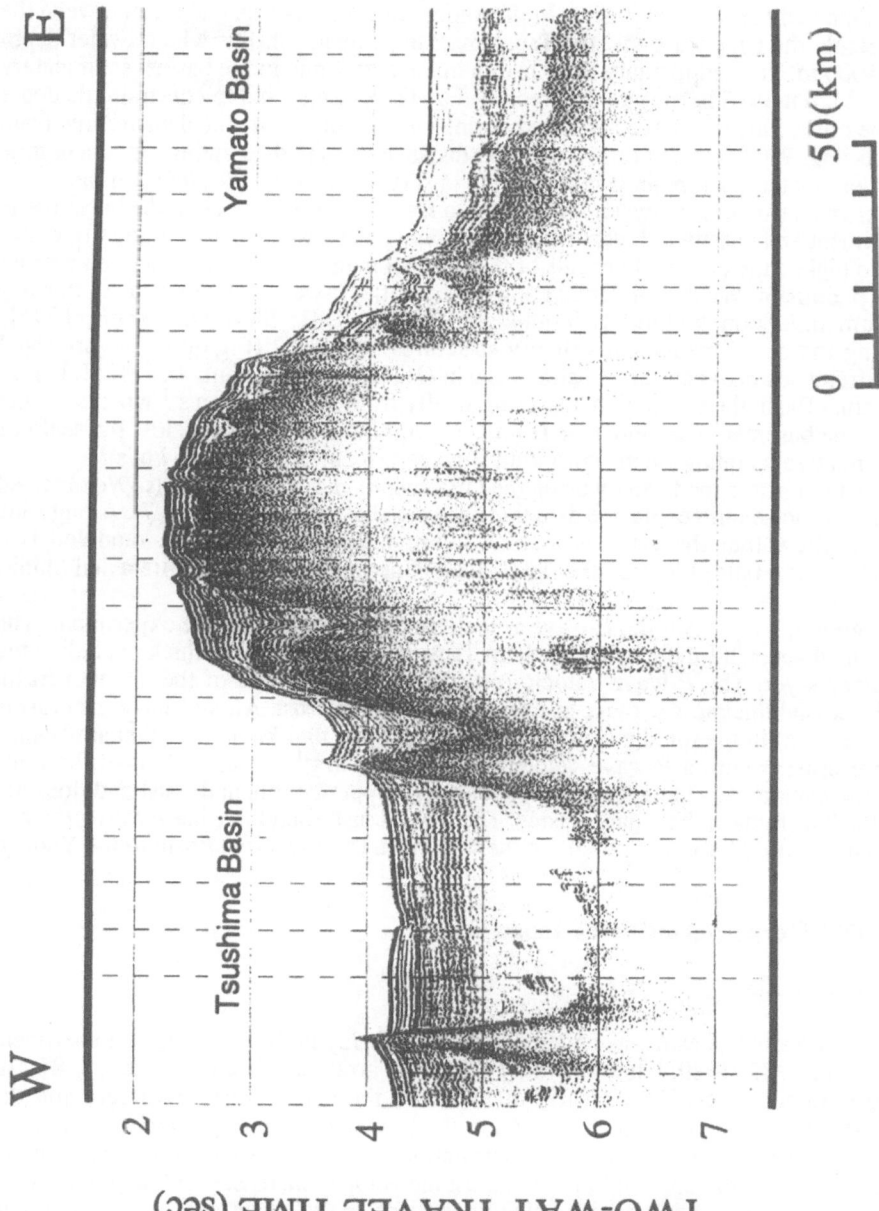

Figure 4. Single channel reflection profile along Line 2 cutting across the Tsushima Basin and the Yamato Basin.

The reflection data clearly documented the structure of the sediment and the basement layers. Figure 4 shows variations of the sedimentary structure along Line 2, including both the Tsushima and the Yamato Basins. In the middle of Line 2 is a saddle area between the Yamato Bank, the Kita-oki Bank, the Tsushima, and Yamato Basins, where water depth is about 1000 m. The sedimentary layer is thin in the area while in the basins sedimentary layers are 2 km thick. The airgun data recorded by OBSs were used to constrain the upper part of the crust. Since we have a dense shooting spacing of 100 m, the data are free from spatial aliasing. We first transformed the time-distance domain data into the intercept time (τ) - ray parameter (p) domain (Figure 5). We observed a marked difference in the shallow crustal structures between the Tsushima Basin and the bank area: the basin has a thick sediment and basement layers where the P-wave velocity increases gradually with depth. We picked the principal arrivals of the refraction and wide angle reflection arrivals in the τ - p domain. We then inverted them to obtain the P-wave velocities of depth using the τ -sum inversion method [Diebold and Stoffa, 1981; Shinohara et al., 1994]. Comparing the one-dimensional velocity structures on Line 2 (Figure 6) we observed maijor differences between the Tsushima Basin (NTU8) and the bank area (NTU11). In the Tsushima Basin the velocity increases gradually from 1.5 to 6.5 km s^{-1} whereas in the bank area the basement has velocities from 1.5 to 4.5 km s^{-1}. At 1 km below the seafloor there is a relatively homogeneous layer with P-wave velocities from 5 to 6 km s^{-1}.

The entire crust is modeled by using the OBS recorded explosive signals. We obtained high signal-to-noise ratio explosive data throughout the experiment (Figure 7). Along Line 2 the data show that the crust is a thick continental type crust. We modeled two dimensional wave fields by using the ray method [Cerveny et al., 1977; Hirata and Shinjo, 1986].

In Figure 8, we present the P-wave velocity model obtained by the experiment. The crust obtained beneath Line 1 in the northern Tsushima Basin is 13 km thick, including the sedimentary layer. The P-wave velocity profile is similar to that of the Yamato Basin where the crustal thickness is twice as thick as the oceanic crust. Along Line 2 the crust is thickest (23 km) in the middle of the line between the Kita-oki and the Yamato Banks. The upper crust, where a P-wave velocity is about 6 km s^{-1}, occupies half of the crust beneath the bank area. The layer corresponds to the upper continental crust and does not exist in the Tsushima and Yamato Basins. The lower crust found in Line 2 with a P-wave velocity of 6.7 km s^{-1} exists throughout the line from the Tsushima Basin to the Yamato Basin.

3. Recent OBS experiments in the Japan Sea

3.1 YAMATO BASIN

The southern Yamato Basin was covered by an OBS array in the 1985 DELP experiment [Hirata et al., 1987, 1989, Chung et al., 1990]. The array consisted of twenty OBSs on two perpendicular lines. They clearly documented the entire crustal structure and the uppermost mantle. The crust is 11 to 14 km thick, which is twice as thick as a normal oceanic crust. By contrast, the velocity distribution in the crust is similar to that found in the oceanic crust: in the upper 4 km of the crust the velocity increases with depth from 3.8 km s^{-1} to 5.4 km s^{-1} showing a steep velocity gradient. In the lower crust, at a depth range of 4.5 to 13 km below sea floor, the velocity increases from 6.6 to 7.2 km s^{-1}.

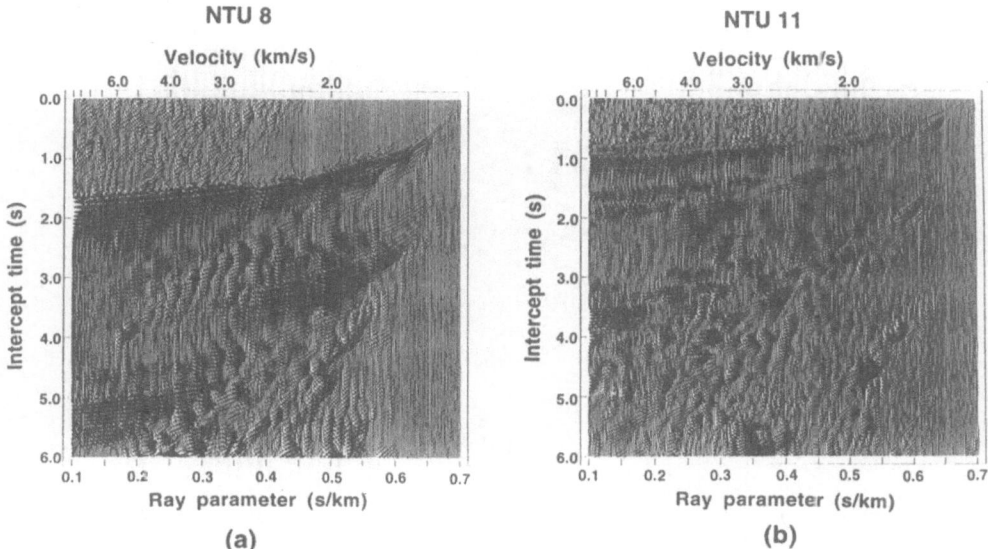

Figure 5. τ-p transform of seismograms for Line 2. We calculated semblance of slant stacked seismogram with different pairs of intercept time (τ) and ray parameter (p). We used vertical component signal of a geophone within an offset range of 15 km. Semblance window is 0.1 s long. (a)NTU8, (b) NTU11.

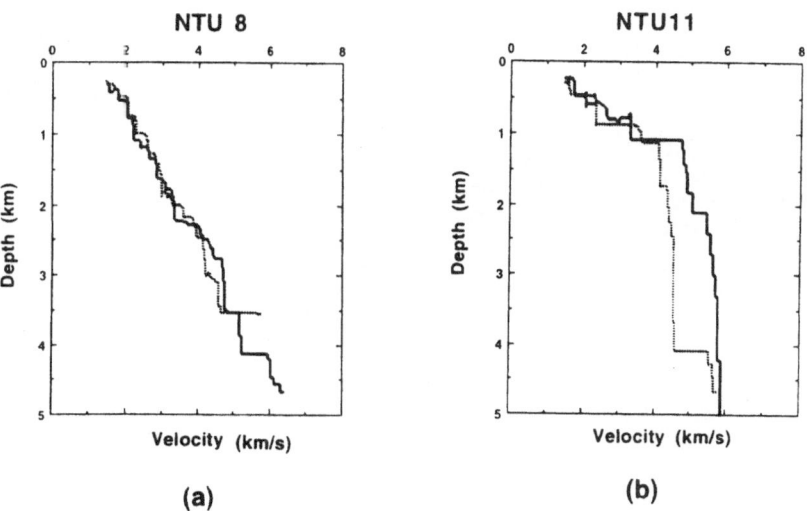

Figure 6. One-dimensional velocity-depth functions for Line 2. Principal arrivals in the τ-p domain are inverted into velocity-depth profiles by the τ-sum inversion method. Thick lines are determined by data shot from the east of the OBS, and dashed lines are those from the west.

362

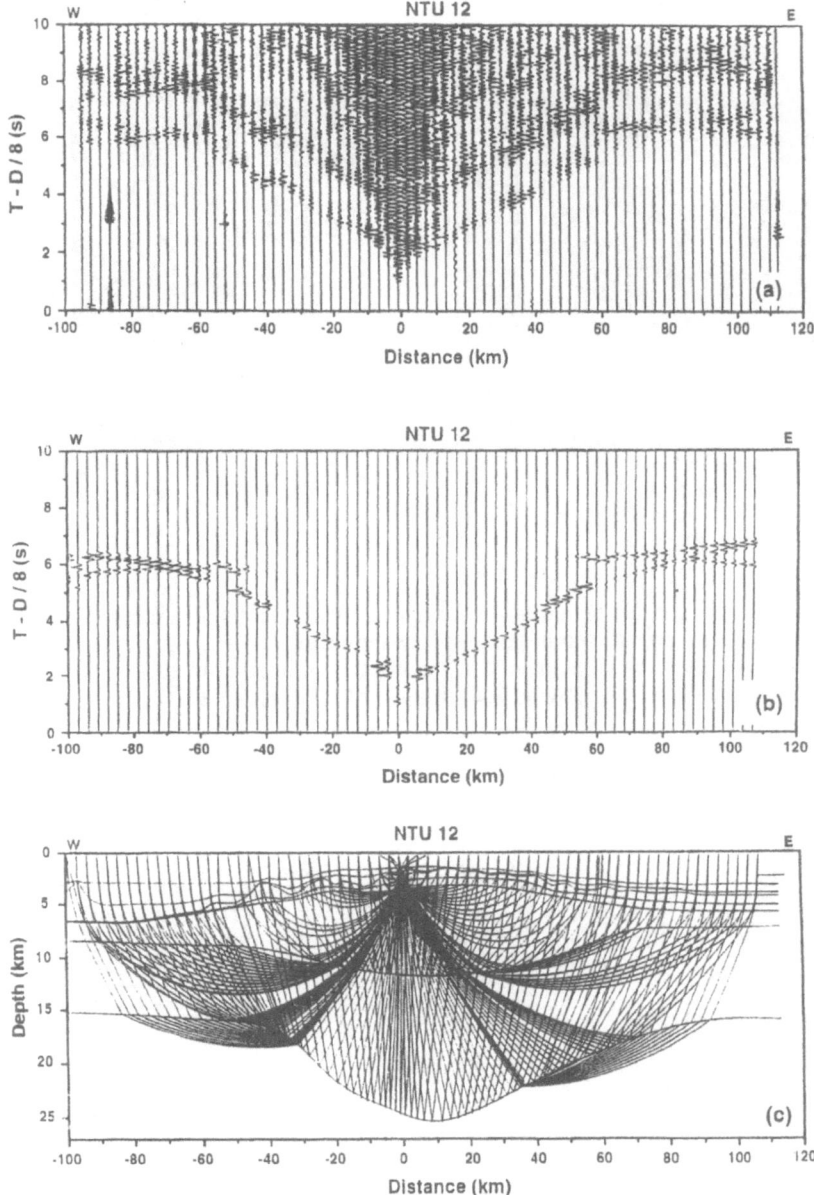

Figure 7. Example of OBS recorded explosive signals and forward modeled wave forms.
(a) Vertical component of the geophone signal. (b) Ray theoretical synthetic seismograms.
(c) Ray paths used for calculating the synthetic seismogram. We used the model shown in
Figure 8. Distances are from the OBS NTU 12 which is located in the middle of Line 2.

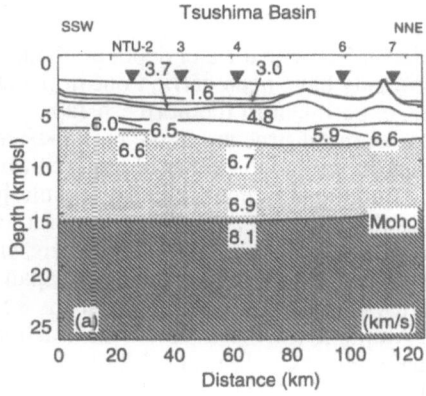

Figure 8. Two-dimensional velocity structure in the Tsushima Basin. (a) Line 1. (b) Line 2. Numerals are *P*-wave velocities in km s⁻¹.

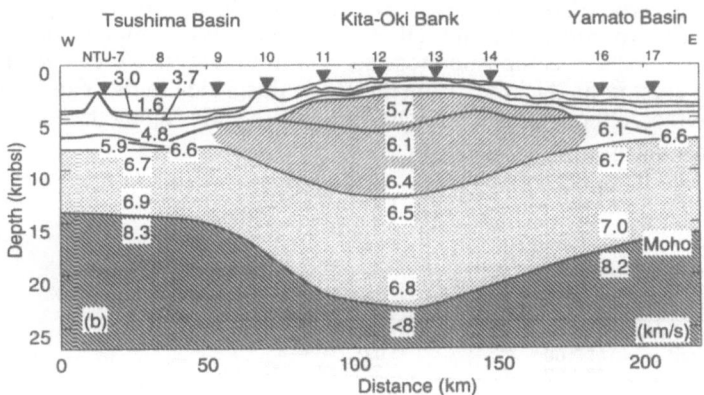

At the bottom of the crust is a 1-km-transition layer from the crust to the mantle with an increasing velocity from 7.2 to 7.4 km s⁻¹. Because the crust has no distinct layer with a *P*-wave velocity around 6 km s⁻¹, we interpret that its upper part is not of a continental type. Hirata *et al.* (1989) interpreted that the thick crust was composed of a large amount of igneous material injected into the oceanic type crust after the backarc spreading. If this is the case, relatively high velocity lower crust may suffer from igneous activity.

The northern Yamato Basin was covered by the real-time seismic experiment during Ocean Drilling Program (ODP) Cruise Leg 128 in 1989 [Shinohara *et al*, 1992]. They used a broad band downhole seismometer at Hole 794D and nine ocean bottom seismometers to record airgun signals. They found that the crust is 14 km thick and is divided into the upper and lower crust. The upper crust has a large velocity gradient with depth, whereas the lower crust is relatively homogeneous. The velocity structure is in general similar to that found in the southern Yamato Basin.

3.2 JAPAN BASIN

The Japan Basin is the largest basin in the Japan Sea with a water depth of about 3500 m. During the last decades a number of experiments have been carried out in the study area. The most recent is the 1990 Japan-USSR experiment in the northern Japan Basin [Hirata *et al.*, 1992]. A controversy surrounded the crustal structure of the basin before the 1990 experiment. Russian scientists claimed that the crustal thickness of the Japan Basin was about 16 km suggesting that the seismic structure was not typically oceanic [e.g., Kovylin *et al.*, 1966; Karp *et al.*, 1992] whereas the Japanese and USA group indicated that there was 8.5-km thick crust in the Japan Basin [Murauchi, 1972; Ludwig *et al.*, 1975]. The 1990 experiment, which was conducted by a group from both Japan and Russia, clarified the existence of the oceanic crust in the Japan Basin: the crust is 8.5 km thick including 2 km of the sedimentary layer. Although the magnetic anomaly pattern in the Japan Sea is chaotic and amplitude of the anomaly is low [Isezaki, 1986], the seismic study suggests an opening and spreading of the sea floor. The complex anomaly pattern was interpreted as a result of many pseudo-faults associated with spreading ridge propagation [Tamaki and Kobayashi, 1988].

The area between the Japan Basin and the northern part of the Yamato Basin was also investigated by an OBS array [Lu *et al.*, 1992]. The array was located in the area between the 1990 Japan-USSR experiment and the ODP Hole 794D. We compiled the crustal structure in the area to make a composite cross section in Figure 9. We can see lateral heterogeneity of the crust in its thickness from the Japan Basin side to the Yamato Basin side. We also notice that although the total crustal thickness varies along the line, the *P*-wave velocity in the crust dose not vary much.

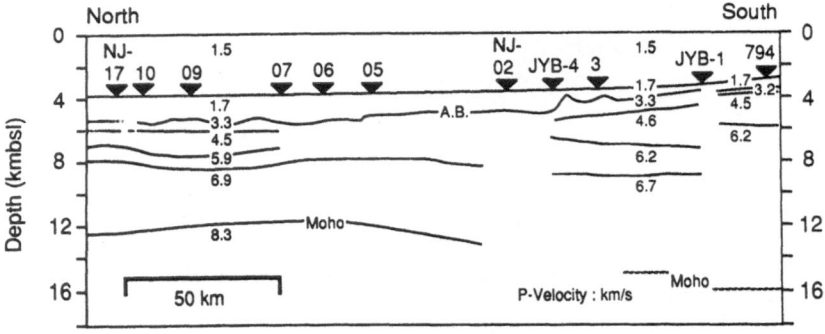

Figure 9. Composite crustal section from the northern Japan Basin to the ODP Hole 794D in the northern Yamato Basin. NJ denotes OBS in the 1990 experiment [Hirata *et al.*, 1992], JYB stands for the 1991 experiment [Lu *et al.*, 1992], and 794 for the Hole 794D [Shinohara *et al.*, 1992].

4. DISCUSSION

4.1 ISOSTATIC EQUILIBRIUM

We compile crustal structures obtained in the Japan Sea in Figure 10. The Japan Sea generally has a thick sedimentary layer. The sea has a high rate of sedimentation probably because it is surrounded by lands that supply a large amount of land material. Hence sediment loading on the crust is important to discuss the subsidence of the crust. We corrected the depth of the basement by assuming an isostatic equilibrium in the sediment and upper crust.

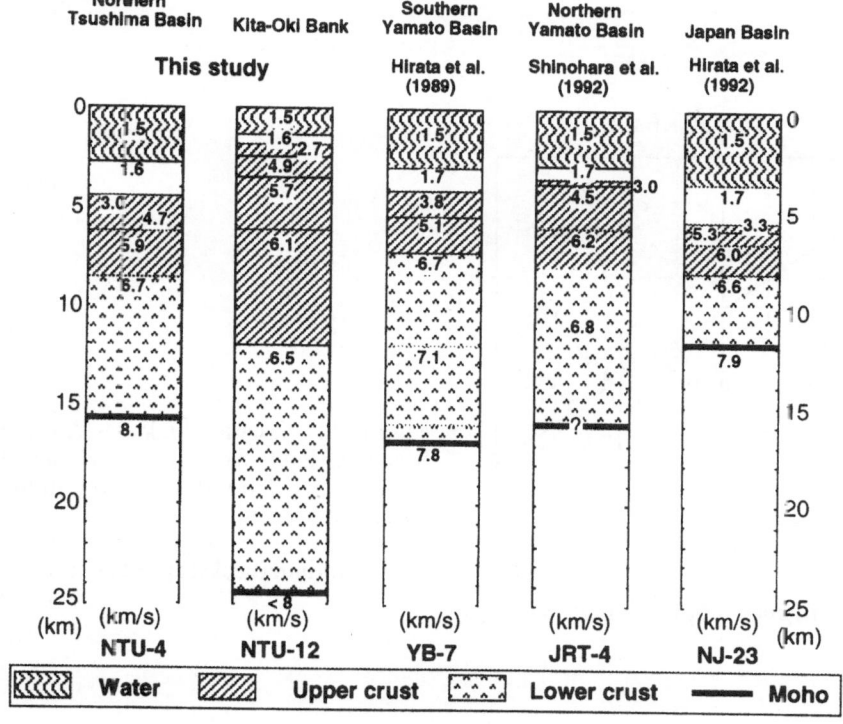

Figure 10. Comparison of the crustal structures in the Japan Sea.

We divided the crust into a sedimentary layer and an igneous part of the crust according mainly to the P-wave velocity determined by the refraction method using OBSs. Acoustic basements seen in reflection profiles are sometimes not the interface between the sediment and the crystalline crust. For example, Leg 127 of the Ocean Drilling Program in the Japan Sea found that the bottom part of the sedimentary layer consisted of igneous materials interbedded with sediments. Low frequency reflectors seen below the acoustic basement were found to be sills in the uppermost part of the crust. The layer usually has velocities of 2.5 to 3.5 km s^{-1}. We therefore classify the layers below the sediment-sill complex as the crystalline crust.

We calculated a backstripped depth of basement: we assumed the density of the mantle as 3.3 x 10^3 kg m^{-3}, the density of water as 1.0 x 10^3 kg m^{-3}, and the density of the

366

sediments as 2.5×10^3 kg m^{-3}. We show the relation between the crustal thickness (H) and the backstripped depth of the basement (S) in Figure 11. Here we can sea a clear linear relationship in the data. A least squares fitted line to the data is

$$H = 30 - 5.0\,S, \tag{1}$$

where H is in km and S is in km below sea floor. Equation (1) predicts that the crustal thickness at zero water depth is 30 km. This is consistent with the observation that the crustal thickness of Honshu in Japan is about 30 km [e.g., Yoshii, 1977]. We now predict that the deepest part of the Moho, which corresponds to the Yamato Bank, is 30 km below sea level although the depth has not yet been directly determined by the seismic method.

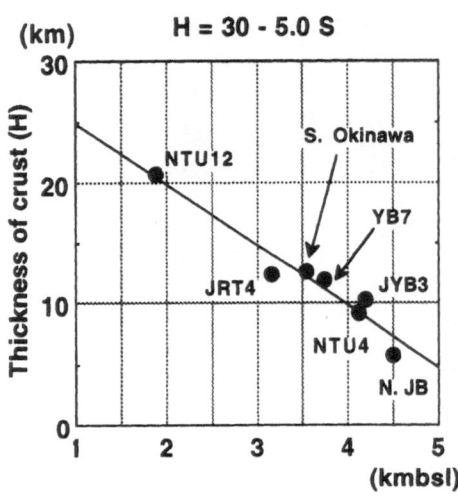

Figure 11. Crustal thickness(H) and basement depth (S) in the Japan Sea and the Okinawa Trough. The depth to the basement is corrected by the unloading of sediments assuming isostatic equilibrium (backstripped depth).

4.2 EXTENSIONAL CRUST?

We now discuss whether the 14-km crust found in the Tsushima Basin and the Yamato Basin is thick oceanic crust or thinned continental crust. If we assume that the initial thickness of the crust was 30 km, we can correlate the backstripped depth of basement, S, and the relative thinning of the crust, $1-1/\beta$, where β is a relative stretching factor. In Figure 12 we plotted data from the Japan Sea and the Okinawa Trough crusts. The best regression line is

$$S = -0.04 + 6.1\,(1 - 1/\beta). \tag{2}$$

Equation (2) indicates that S is larger than the initial subsidence due to the stretching of the crust and isostatic compensation. We show in Figure 12 lines for initial subsidence and total subsidence at infinite time accounting for thermal subsidence [Le Pichon and Sibuet, 1981]. Clearly, S is smaller than expected by the thermally compensated lithosphere at infinite time [McKenzie, 1978]. So, we can say if the simple stretching model is applicable to the Japan Sea crust formation, then Equation (2) is consistent with a minimum age of 20 Ma for the Japan Sea [Kaneoka, 1990].

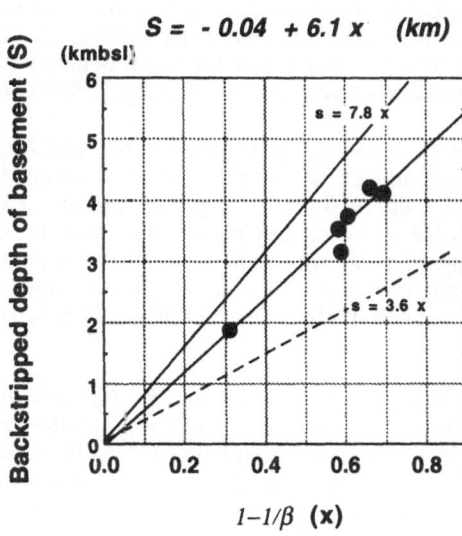

Figure 12. Subsidence (S) and relative thinning of the crust $(1 - 1/\beta)$, where β is a stretching factor. The initial thickness of the crust (H_0) is assumed to be 30 km $(1/\beta = H/H_0)$. Solid circles are observed data from the Japan Sea and the Okinawa Trough. The dashed line is the initial subsidence due to sudden stretching and the thin line is the total subsidence with thermal subsidence at infinite time [Le Pichon and Sibuet 1981].

It should be pointed out, however, that Equation (2) is largely dependent on the data at Line 2 of the 1992 Tsushima Basin experiment. Since the crustal thickness in the bank area is well constrained by the seismic data, the area is revealed to have continental characteristics in seismic velocity. The relationship shown in Equation (2) indicates that crust of the Japan Sea is in isostatic equilibrium and that thermal subsidence is taking place. Namely, the basins were thermally developed as if the simple stretching model took place. However, the relationship does not mean that the Tsushima and Yamato Basins are formed only by the stretching of the continental crust. In fact, the seismic velocities of the basins do not show continental type layer of a P-wave velocity of around 6 km s^{-1}. Consequently, the other process than the simple stretching should take place. The point of the relationship in Equation (2) is that the water depth is controlled by density contrast between the crust and the Mantle that are cooling after formation of the crust. So, we speculate that at the beginning of the Japan Sea formation a part of the sea was opened due to the backarc spreading and an other part was thinned and underwent igneous intrusions. This process is essential to explain the oceanic type velocity in the Tsushima and Yamato Basins.

5. Conclusions

Recent OBS measurements indicate that there are two types of crust: an oceanic type (Japan Basin) and a continental type (Yamato Bank). The 1992 seismic experiment in the area between the Yamato Bank and the Kita-oki Bank demonstrated a 23-km thick crust with *P*-wave velocities similar to those found in the Asian continent on the Japan Sea. The structure suggests that the Yamato and the Kita-oki Banks were formed by extensional tectonics during the opening of the Japan Sea. By contrast, although the Tsushima and the Yamato Basins have a crust that is twice as thick as that of a normal ocean basin, the velocity distribution beneath the basins is similar to that of an oceanic crust.

A linear relationship between water depth and Moho depth is well established in the Japan Sea. The relationship predicts that the crustal thickness at zero water depth is 30 km. After correcting the sedimentary loading, the backstripped basement depth is also well correlated with the crustal thickness. The observation indicates that the Japan Sea is in isostatic equilibrium.

If we assume that the present day crust in the Japan Sea is thinned due to stretching and the original thickness is 30 km, then the subsidence, S, and the relative thinning, $1-1/\beta$, have the expression: $S = 6.1(1-1/\beta)$, where β is a stretching factor. The coefficient factor of the expression is between what is expected by initial subsidence due to stretching and the total subsidence including thermal subsidence. Since the seismic velocities in the Yamato and Tsushima Basins suggest igneous activity, the basins undergo both crustal thinning and intrusion of magmatic material during the stretching of the Japan Sea crust.

Acknowledgments. The present study is a part of the *Multisphere Interaction, Evolution, and Rhythm of the Earth Program* (MULTIER). The cruise was conducted under the agreement of joint research in the Ulleung Basin between Korea Ocean Research and Development Institute (KORDI) and Chiba University. We thank B. C. Suk and N. Isezaki for their effort on the program. We thank the Shipboard Scientific member of the R/V *Wakashio-maru* and R/V *Eador*. We thank M. Shinohara, T. Sato, J. Kasahara, S. Abe and N. Isezaki for their help in OBS operations. We thank students of Chiba University onboard for their help in underway geophysical measurements and OBS operations. S. Miura contributed the initial stage of the data analysis, which we appreciate. We thank Captains, officers and crew of the R/V *Wakashio-maru*, and R/V *Eador*. This research is partly supported by the Grant-in-Aid for Scientific Research numbers 04402011 (Research A), from the Ministry of Education, Science and Culture of Japan.

References

Cerveny, V., I. A. Molotkov, and I. Psencik , *Ray Method in Seismology*, Univerzita Karlova, Prague, 215pp, 1977.
Chung, T.W., N. Hirata, and R. Sato, Two-dimensional *P*- and *S*- wave velocity structure of the Yamato Basin, the southeastern Japan Sea, from refraction data collected by an ocean bottom seismographic array, *J. Phys. Earth*, *38*, 99 - 147, 1990.
Diebold, J. B., and P. L. Stoffa, The traveltime equation, tau-p mapping, and inversion of common midpoint data, *Geophysics*, *46*, 238 - 254, 1981.
Hirata, N. and N. Shinjo, SEISOBS - modified version of SEIS83 for ocean bottom seismograms (in Japanese), *Zisin* (*J. Seismo. Soc. Jpn.*), *39*, 317 - 321, 1986.
Hirata, N., H. Kinoshita, K. Suyehiro, M. Suyemasu, N. Matsuda, T. Ouchi, H. Katao, S. Koresawa, and S. Nagumo, Report on DELP 1985 cruise in the Japan Sea Part II: Seismic refraction experiment conducted in the Yamato Basin, southeast Japan Sea, *Bull. Earthq. Res. Inst., Univ. Tokyo*, *62*, 347 - 365, 1987.

Hirata, N., H. Tokuyama, and T.W. Chung, An anomalously thick layering of the crust of the Yamato Basin, southeastern Sea of Japan: the final stage of back-arc spreading, *Tectonophysics, 165*, 303 - 314,1989.

Hirata, N., B. Ya. Karp, T. Yamaguchi, T. Kanazawa, K. Suyehiro, J. Kasahara, H. Shiobara, M. Shinohara, and H. Kinoshita, Oceanic crust in the Japan basin of the Japan sea by the 1990 Japan - USSR expedition, *Geophys. Res. Lett., 19*, 2027 - 2030, 1992.

Hirata, N., S. Miura, E. Kurashimo, N. Takahashi, T. Yamaguchi, S. Abe, and M. Shinohara, Seismic crustal structure of the northern Tsushima Basin, *Prog. Abstr. seismol. Soc. Jpn., 1* . 173, 1993.

Ingle, J. C. , K. Suyehiro, M.T. von Breymann *et al.*, *Proc. ODP, Init. Repts., 128*, Ocean Drilling Program, College Station, TX, 1990.

Isezaki, N., A magnetic anomaly map of the Japan Sea, *J.Geomagn.Geoelectr., 38*, 403 - 410, 1986.

Jolivet, L., P. Huchon, and C. Rangin, Tectonic setting of western Pacific marginal basins, *Tectonophysics, 160*, 23 - 48, 1989.

Kaneoka, I., Radiometric age and Sr isotope characteristics of volcanic rocks from the Japan Sea floor, *Geochem. J., 24*, 7 - 19, 1990.

Karp, Y. B., N. Hirata, and H. Katao, Crustal structure of the Japan Sea, in *Geology and Geophysics of the Japan Sea (Japan - USSR Monograph series*, vol. 1), edited by Isezaki, N., K. Tamaki, and Y. B. Karp, in press, Terrapub, Tokyo, 1992.

Kovylin, V. M., B. Y. Karp, and R.B. Shayakhmetov, The structure of the crust and sedimentary layer of the Japan Sea according to seismic data, *Dokl. Akad. Nauk SSSR, 168*, 1048 - 1051, 1966.

Le Pichon, X. and J. C. Sibuet, Passive margins: a model of formation, *J. Geophs. Res., 86*, 3708 - 3720, 1981.

Lu, R., N. Hirata, T. Yamaguchi, T. Kanazawa, H. Tokuyama, H. Kinoshita, Seismic crustal structure in the boudary zone between the Japan Basin and the Yamato Basin by the 1991 Japan-USSR experiment (in Japanese), *Prog. Abstr. seismol. Soc. Jpn., 2*, 22, 1992.

Ludwig, W. J., S. Murauchi, and R.E. Houtz, Sediments and structure of the Japan Sea, *Geol. Soc. Am. Bull., 86*, 651 - 664,1975.

McKenzie, D., Some remarks on the development of sedimentary basins, *Earth Planet. Sci. Lett., 40*, 25 -32, 1978.

Murauchi, S., Crustal structure of the Japan Sea (in Japanese), *Kagaku. 42*, 367 - 375, 1972.

Shinohara, M., N. Hirata K., H. Nambu, K. Suyehiro, T. Kanazawa, and H. Kinoshita, Detailed crustal structure of northern Yamato Basin, *In* Tamaki, K., K. Suyehiro, J. Allan, M. McWilliams, *et al.*, *Proc. ODP, Sci. Repts., 127/128*, Pt.2, College Station, TX (Ocean Drilling Program), 1075 - 1106, 1992.

Shinohara, M., N. Hirata, and N. Takahashi, High resolution velocity analysis of ocean bottom seismometer data by the τ–p method, *Marin. Geophs. Res., 16*, 185 - 199, 1994.

Tamaki, K., Tow modes of back arc spreading, *Geology, 13*, 475 - 478, 1985.

Tamaki, K., and K. Kobayashi, Geomagnetic anomaly lineation in the Japan Sea (in Japanese), *Mar.Sci.Mon., 20*, 705 - 710, 1988.

Tamaki,K., K.Pisciotto, J. Allan, *et al.*, *Proc. ODP, Init. repts., 127*, College Station, TX (Ocean Drilling Program), 1990.

Yoshii, T., Crust and upper-mantle structure beneath northeastern Japan (in Japanese), *Kagaku, 47*, 170 - 176, 1977.

KEY WORD/SUBJECT INDEX

THEMATIC INDEX